Units of Volume

Unit	Abbreviation	Equivalent
Liter	L*	1
Milliliter	mL	10^{-3} L (1 mL = 1 cm^3 = 1 cc)
Microliter	μL	10^{-6} L

Volume conversions

1 tsp = 5 mL	1 mL = 0.034 fl oz
1 tbsp = 15 mL	1 L = 2.11 pt
1 fl oz = 30 mL	1 L = 1.057 qt
1 cup = 0.24 L	1 L = 0.265 gal
1 pt = 0.474 L	1 L = 33.78 fl oz
1 qt = 0.946 L	1 cubic meter (m^3) = 61000 cubic inches
1 gal = 3.77 L	1 cubic meter (m^3) = 35.3 cubic feet
	1 cubic meter (m^3) = 0.00973 acre-inches

*Note: Many people use an upper case "L" for the symbol for liter to avoid confusion with the number one (1). Similarly milliliter is written mL.

Units of Weight

Unit	Abbreviation	Equivalent
Tonne (metric ton)	t	10^3 kg
Kilogram	kg	10^3 g
Gram	g	1
Milligram	mg	10^{-3} g
Microgram	μg	10^{-6} g
Nanogram	ng	10^{-9} g
Picogram	pg	10^{-12} g

Weight conversions

1 oz = 28.4 g	1 g = 0.0352 oz
1 lb = 454 g	1 kg = 2.205 lb
1 lb = 0.454 kg	
1 U.S. ton = 0.91 tonne	1 tonne = 1.102 U.S. tons

Temperature conversions

$$°C = \frac{(°F - 32)}{9} \times 5$$

$$°F = \frac{(°C \times 9)}{5} + 32$$

Some equivalents

0°C = 32°F

37°C = 98.6°F

100°C = 212°F

ENVIRONMENTAL
Science

ENVIRONMENTAL

Science

A Study of Interrelationships

Tenth Edition

Eldon D. Enger

Delta College

Bradley F. Smith

Western Washington University

with contributions by

Anne Todd Bockarie

Philadelphia University

Higher Education

Boston   Burr Ridge, IL   Dubuque, IA   Madison, WI   New York   San Francisco   St. Louis
Bangkok   Bogotá   Caracas   Kuala Lumpur   Lisbon   London   Madrid   Mexico City
Milan   Montreal   New Delhi   Santiago   Seoul   Singapore   Sydney   Taipei   Toronto

The McGraw·Hill Companies

Higher Education

ENVIRONMENTAL SCIENCE: A STUDY OF INTERRELATIONSHIPS, TENTH EDITION

Published by McGraw-Hill, a business unit of The McGraw-Hill Companies, Inc., 1221 Avenue of the Americas, New York, NY 10020. Copyright @ 2006, 2004, 2002, 2000, 1998, 1995, 1992, 1989, 1986, 1983 by The McGraw-Hill Companies, Inc. All rights reserved. No part of this publication may be reproduced or distributed in any form or by any means, or stored in a database or retrieval system, without the prior written consent of The McGraw-Hill Companies, Inc., including, but not limited to, in any network or other electronic storage or transmission, or broadcast for distance learning.

Some ancillaries, including electronic and print components, may not be available to customers outside the United States.

This book is printed on recycled, acid-free paper containing 10% postconsumer waste.

1 2 3 4 5 6 7 8 9 0 DOW/DOW 0 9 8 7 6 5 4

ISBN 0–07–252829–X

Publisher: *Margaret J. Kemp*
Senior Developmental Editor: *Kathleen R. Loewenberg*
Outside Developmental Services: *Kennie Harris*
Marketing Manager: *Tami Petsche*
Senior Project Manager: *Mary E. Powers*
Lead Production Supervisor: *Sandy Ludovissy*
Senior Media Project Manager: *Jodi K. Banowetz*
Senior Media Technology Producer: *Jeffry Schmitt*
Senior Designer: *David W. Hash*
Cover/Interior Designer: *Jamie E. O'Neal*
Senior Photo Research Coordinator: *Lori Hancock*
Photo Research: *LouAnn K. Wilson*
Supplement Producer: *Brenda A. Ernzen*
Compositor: *Cenveo*
Typeface: *10/12 Times*
Printer: *R. R. Donnelley Willard, OH*

The credits section for this book begins on page C-1 and is considered an extension of the copyright page.

Library of Congress Cataloging-in-Publication Data

Enger, Eldon D.
  Environmental science : a study of interrelationships / Eldon D. Enger, Bradley F. Smith. — 10th ed.
    p.   cm.
  Includes bibliographical references and index.
  ISBN 0–07–252829–X (hard copy : alk. paper)
  1. Environmental sciences—Textbooks. I. Smith, Bradley F. II. Title.

GE105.E54   2006
363.7—dc22   2004019843
                                                             CIP

www.mhhe.com

Dedication

To Judy, my wife and friend,
for sharing life's adventures.

Eldon Enger

For Ian, from that first
perch out of Prentiss Bay
to helping preserve the
future of the world's
fisheries. I am proud of you.

Brad Smith

About the Cover

Front cover: Artist Terry Isaac is a native Northwesterner, living in the Willamette Valley of Oregon between the Cascade Mountain Range and the Pacific Ocean. Isaac paints the wildlife and landscapes of North America. Although he received a formal art education, graduating with honors, Isaac believes that his best training has come from being outdoors and from studying the work of his favorite wildlife artists.

Isaac particularly admires the art of 19th century painters Thomas Moran and Albert Bierstadt, who painted the dramatic light and landscapes of the West. Isaac, painting predominantly in acrylics, strives to capture not only the characteristics of his wildlife subjects but also the light that surrounds them. He takes a great deal of care to depict both his subjects and their habitats with accuracy.

Isaac has been selected to participate in several important art exhibitions, including the Leigh Yawkey Woodson Art Museum's prestigious "Birds in Art" exhibitions. He was also commissioned to produce 14 waterfowl drawings for the *Audubon Bird Handbook,* published in 1987, and is author of *Painting the Drama of Wildlife Step by Step,* published by North Light Books.

A collector favorite, Isaac has been chosen as winner of the People's Choice Award at galleries in North Carolina and Ohio. He was a Special Guest Artist at the Southeastern Wildlife Expo in 2000 and was named Artist of the Year in 1998 at both the Florida Wildlife Expo and Pacific Rim Wildlife Art Show. In 1999 he was named a Friend of the National Zoo as Artist of the Year. Hired as a visual consultant, Isaac helped to create the concept art for the main character in the Disney movie, *Dinosaur.*

Brief Contents

Contents

PART II
Ecological Principles and Their Application 60
Imitating Mother Nature in a Florida Lake 61

PART IV
Resource Management 244
Natural Community Conservation Planning 245

Preface

Ten Respected, Successful Editions

As we produce the tenth edition of this text, it is appropriate to celebrate its success and reflect on how the field has changed over the past two decades. Since *Environmental Science: A Study of Interrelationships* was first published in 1983, it has been translated into Chinese and Korean, and we estimate that nearly a million students have used the text. Over the 10 editions, we have seen the field of environmental science change significantly. Initially, most of the courses were designed to raise awareness about environmental issues. Today, the field of environmental science has become a major discipline at many colleges and universities. Students who enroll in an introductory environmental science course may move in several directions that were not present over 20 years ago. Fields such as environmental law and environmental engineering are new disciplines. In addition, other traditional disciplines such as landscape planning, urban planning, agriculture, and industrial engineering now pay significant attention to environmental concerns.

It has been an interesting intellectual exercise to keep up-to-date in this rapidly changing field. For example, in the early 1980s, air and water pollution by industry was a key issue in the developed countries. Today, however, most industries have made major improvements in controlling pollution and are no longer the main sources of air and water pollution. The actions of individual citizens now have become major sources of pollution. Automobile use is responsible for much of the air pollution that affects cities, and runoff from lawns, city streets, and farms is an important source of water pollution. Many new environmental problems have also arisen. Climate change, the AIDS pandemic, genetically modified crops, and concerns about the loss of biodiversity have become core issues. Over time, as these issues have assumed greater importance, they have been incorporated into the text.

Why "a Study of Interrelationships"?

Environmental science is an interdisciplinary field. Because environmental disharmonies occur as a result of the interaction between humans and the natural world, we must include both when seeking solutions to environmental problems. It is important to have a historical perspective, appreciate economic and political realities, recognize the role of different social experiences and ethical backgrounds, and integrate these with the science that describes the natural world and how we affect it. *Environmental Science: A Study of Interrelationships* incorporates all of these sources of information when discussing any environmental issue. This text is written for a one-semester, introductory course taken by students with a wide variety of career goals. The central theme of the book is interrelatedness. No text of this nature can cover all issues in depth. Although many facts are presented in charts, graphs, and figures to help illustrate the scope of environmental issues, the text's main focus is on identifying major issues and giving appropriate examples to illustrate the complex interactions that are characteristic of all environmental problems. The facts are provided for the person who wants them, but they do not obscure the general concepts and principles being described. The authors have endeavored to present a balanced view of issues, diligently avoiding personal biases and fashionable philosophies. It is not the purpose of this textbook to tell you what to think. Rather, our goal is to provide access to information and the conceptual framework needed to understand complex issues so that you can comprehend the nature of environmental problems and formulate your own views.

New to this Anniversary Edition!

New Author Brings Expertise and Unique Perspective

A special contributing author, Anne Todd Bockarie, updated chapters 2 and 16 for the tenth edition. She is an Assistant Professor of Biology at Philadelphia University and teaches Ecology, Biodiversity, Environmental Science, Introductory Biology, and Special Topics: Tropical Conservation. She helped develop the Environmental Science program at the university as well as a new major in Environmental and Conservation Biology. She has created new courses in ecological field methods, coral reef assessment in Jamaica, and wildlife management in Yellowstone National Park. She holds a Research Associate position at Yale University and has supervised graduate research on the ecological and social impacts of large-scale park restoration for the past six years. Dr. Bockarie's M.S. and Ph.D. degrees in reforestation and forestry extension are from the University of Florida. She has extensive international consulting and training experience in agriculture, forestry, and parks management in Africa and the Caribbean.

New Biodiversity Chapter

Chapter 12, "Biodiversity Issues," discusses how population growth and the ability of people to exploit resources led to current concerns about biodiversity. This important topic is examined at the genetic, species, and ecosystem levels, and the value of biodiversity is described from several points of view: ethics, direct economic values, and services provided by organisms in functioning ecosystems. A discussion of the threats to biodiversity looks at habitat loss, overexploitation, introduction of exotic species, and control of pest organisms. The chapter ends with an overview of efforts to protect biodiversity and includes a discussion of legal protection and steps being taken to ensure that sustainable management practices are used to preserve biodiversity.

In-Depth Coverage on Sustainability

Sustainability is a core concept in this text and is covered in several chapters. In chapter 2, sustainability is discussed in the context of ethical considerations. Chapter 3 expands on the idea by looking at the topic from an economic point of view. Sustainability is also discussed in the chapters on energy use, biodiversity, water management, agriculture, and at many other points in the text.

New Feature Examines Environmental Change

When we look at environmental issues in general, we find that the same basic issues exist today as in 1983. Although significant progress has been made toward solving some environmental problems, progress in others has been slow, and new threats or concerns have emerged. To help the reader evaluate progress and recognize how difficult some problems are, we have instituted a new feature with this edition. Each chapter begins with a "Past, Present, Future" feature that shows how specific topics discussed in the chapter have changed over more than 20 years.

New Art Adds Depth and Realism

There are over 40 new figures and tables in this tenth edition, many of which have been drawn in a more realistic, three-dimensional style. Numerous other pieces have been revised to reflect updated data or content.

"Risk and Cost" Chapter Completely Rewritten

Chapter 3, "Risk and Cost: Elements of Decision Making," has a new introduction to the topics of risk and economics, a new section on how risk is characterized, and a new table on the causes of accidental death. The chapter now describes renewable and nonrenewable resources, and includes a major section on assigning value to natural resources. There are also greatly expanded sections on deferred costs, external costs, and subsidies.

All New "Issues—Analysis" Case Studies

Every chapter in the tenth edition features a new "Issues—Analysis" reading, written to reflect current issues and/or new data pertaining to the chapter subject matter. In addition, there are 10 other new readings chosen to complement new text content.

New Foldout World Map

This unique, oversized foldout piece features full-color ecological regions and political world maps and allows students to refer to either map while reading, helping them to better understand and appreciate global environmental issues.

Special Introduction to Critical Thinking

In preparation for the critical thinking questions found at the end of every chapter, this important preface to Chapter 1 explains the value of correctly evaluating available information to make informed decisions. Six key characteristics of critical thinking are examined.

Guest Essays Provide Regional Examples

In this tenth edition, each part opens with a new guest author's essay highlighting an environmental issue close to their home. These articles offer expert insight into problems similar to those in students' own backyards.

- Part I opens with an essay entitled **"Can We Save Both Blue Crabs and Watermen?"** which describes the regulation of the blue crab fishery in Chesapeake Bay by looking at a specific problem and the ramifications of regulatory action. *Author: Jennifer Rhode, Georgia College and State University*

- The Part II opening composition, **"Imitating Mother Nature in a Florida Lake,"** examines the cleanup of a lake in central Florida and its effects on ecosystems and businesses downstream. This essay is an excellent transition to the chapters in this unit, which provide an understanding of the ecological principles that are basic to organism interactions and the flow of matter and energy in ecosystems. *Authors: Blase Maffia and Lisa Ganser, University of Miami*

- Part III focuses on energy, beginning with its opening article, **"Winds of Change,"** which follows the events that led an electric utility that relies on nuclear power to invest in wind power in Minnesota. *Author: John C. Cronn, St. Cloud State University*

- Part IV begins with the essay, **"Natural Community Conservation Planning,"** which describes a California policy designed to preserve biodiversity. *Author: Morgan Barrows, Saddleback College*

- The Part V essay, **"Lake Champlain: It Isn't Easy Stayin' Clean,"** introduces the topic of pollution by tracing the effects of human activities on the Lake Champlain ecosystem and provides a good introduction to the conflict between human impacts on the environment and efforts to reduce damage. *Author: Alan McIntosh, University of Vermont*

Significant Revision

As with previous editions, reviewers' suggestions have been incorporated into the text. Some of these suggestions required small changes in text or figures to improve clarity and accuracy. Others required major revisions in the content of certain chapters. Specific changes include:

- Chapter 1 has two new boxed readings: Environmental Close-Up: "Traditional Resource Use and Conflict Management in Keoladeo National Park, India—Science Versus Policy" and Global Perspective: "Biodiversity, Human Welfare, and Economic Development."

- Chapter 2 has a new table that lists the accomplishments of selected international treaties and a new Environmental Close-Up: "What's in Your Backyard?" that allows students to assess the level of pollution in their neighborhood. The sections dealing with the CERES principles, environmental justice issues and international trade in endangered species have all been expanded.

- Chapter 3 has been completely revised and includes major new sections dealing with renewable and nonrenewable resources, environmental costs and pollution, and subsidies. These changes are supported with new tables and figures.

- Chapter 5 features a new Environmental Close-Up entitled "Human Health and Exotic Species."

- Chapter 6 includes a new Environmental Close-Up on "Nonnative Invasive Aquatic Plants."

- Chapter 7 provides new material on lemming cycles in the Arctic and the concept of cultural carrying capacity.

- Chapter 8 has been updated with recent information on human population.

- Chapter 9 features a major new section on the politics and economics of energy use, which incorporates fuel efficiency standards and OPEC, and many new and revised figures.

- Chapter 10 now includes expanded coverage on renewable energy as well as energy efficiency and conservation.

- Chapter 11 has a new section explaining the politics of nuclear power and additional material on world demand for nuclear power.

- Chapter 12 has been completely rewritten to emphasize biodiversity issues. Its new Global Perspective, "Biodiversity 'Hot Spots,'" has an extensive table and a map detailing the characteristics and locations of these important ecosystems. The chapter also includes new material on the genetic, species, and ecosystem diversity aspects of biodiversity; information on the value of biodiversity; a major new section on threats to biodiversity that focuses on habitat loss, overexploitation, introduction of exotic species, and control of pest organisms; and a new section on legal protection of endangered species.

- Chapter 13 contains a new Environmental Close-Up entitled "Farmland Preservation in Pennsylvania." The chapter also includes additional coverage on urban sprawl and sustainable growth.

- Chapter 14 features two new Environmental Close-Ups: "Desertification and Global Security" and "The Conservation Security Program."

- Chapter 15 has a new example and table showing the development of pesticide resistance by insects and a new figure on the amount of pesticide used on agricultural land.

- Chapter 16 offers updated information on efforts to save the Aral Sea, the long-term impact of the *Exxon Valdez* oil spill, and water use throughout the world.

- Chapter 17 was updated throughout and modified to allow for a better flow of content. The chapter includes new information on the deaths in Europe due to the 2003 heat wave.

- Chapter 18 has a much-expanded section on mining waste and includes an Environmental Close-Up entitled "Mountain Top Removal."

- Chapter 19 has been updated throughout and includes a new pie graph that shows industries responsible for toxic releases.

- Chapter 20 has expanded coverage on the societal benefits of environmental compliance and a new section explaining the difference between *governance* and *government*. There are also two new tables and a new Global Perspective on "ISO Standards for Environmental Management Systems" in this chapter.

Acknowledgements

The creation of a textbook requires a dedicated team of professionals who provide guidance, criticism, and encouragement. It is also important to have open communication and dialog to deal with the many issues that arise during the development and production of a text. Therefore, we would like to thank Publisher Marge Kemp; Developmental Editors Kathy Loewenberg and Kennie Harris; Marketing Managers Lisa Gottschalk and Tami Petsche; Project Manager Mary Powers; Production Supervisor Sandy Ludovissy; Photo Research Coordinator Lori Hancock; and Designer David Hash for their suggestions and kindnesses.

We'd like to express our gratitude to the guest essay authors; their contributions add an interesting regional perspective to each of the five units: Jennifer Rhode, *Georgia College and State University;* Blase Maffia and Lisa Ganser, both of the *University of Miami;* John C. Cronn, *St. Cloud State University;* Morgan Barrows, *Saddleback College;* and Alan McIntosh, *University of Vermont.* Finally, we'd like to thank our many colleagues who have reviewed all or part of *Environmental Science: A Study of Interrelationships.* Their valuable input has continued to shape this text and help it meet the needs of instructors around the world. In particular, we'd like to thank: Saleem H. Ali, *University of Vermont;* Frank Bartell, *Community College of Philadelphia;* Donna Bivans, *Pitt Community College;* Daniel Capuano, *Hudson Valley Community College;* Richard Clements, *Chattanooga State Tech College;* John C. Cronn, *St. Cloud State University;* Peter Konovnitzine, *Chaffey College;* Julie Phillips, *De Anza College;* Lauren Preske, *University of Southern Indiana;* Jennifer

Rhode, *Georgia College and State University;* Daniel Sivek, *University of Wisconsin-Stevens Point;* Kristen Jensen Sullivan, *De Anza College;* Sara Topf, *Parks College;* Mike Toscano, *San Joaquin Delta College;* Arlene Westhoven, *Ferris State University;* and Jeff White; *Lake Land College.*

Eldon D. Enger
Bradley F. Smith

Supplements

Digital Content Manager (DCM) CD-ROM

This indispensable supplement is a set of digital visual resources that offers instructors the freedom to use any piece of line art, photo, or table from the textbook, along with a number of additional resources (over 100 animations and dozens of new photos not printed in the text), to customize classroom presentations.

Instructor's Testing and Resource CD-ROM

This cross-platform CD-ROM provides a computerized test bank utilizing software to quickly create customized exams. The user-friendly program allows instructors to sort questions by format; edit existing questions or add new ones; and scramble questions for multiple versions of the same test.

Transparencies

A set of 100 transparencies is available to users of the text. These acetates encompass key figures from the text, including new art from this edition.

Online Learning Center
(http://www.mhhe.com/environmentalscience)

This comprehensive website offers numerous resources for both students and instructors.

Student Resources—Everything you need in one place:
- Practice quizzing
- Additional readings
- Study tools
- Guide to electronic research
- Regional perspectives (case studies)
- Animations
- Career information
- How to contact your government officials
- Access Science (offers the advantage of an online, interactive encyclopedia)

Instructor Resources—In addition to all of the above, you'll receive:
- Answers to review questions
- Class activities
- Questions for your Classroom Performance System (CPS)
- PowerPoint lectures
- Interactive world maps
- PageOut (create your own course website)

Packaging opportunities

McGraw-Hill offers packaging opportunities that not only provide students with valuable course-related material, but also a substantial cost savings. Instructors, ask your McGraw-Hill sales representative for information on discounts and special ISBNs for ordering a package that contains one or more of the following:

Field and Laboratory Activities Manual by Enger and Smith

This lab manual provides hands-on exercises that are relevant, easy to understand, and applicable to students' lives. The experiments are designed to be concise, unique, inexpensive, and easily tailored to any course.

New! *Exploring Environmental Science with GIS* by Stewart et al.

This short book provides a set of exercises aimed at students and instructors who are new to GIS but are familiar with the Windows operating system. The exercises focus on improving analytical skills, understanding spatial relationships, and understanding the nature and structure of environmental data. Because the software used is distributed free of charge, this text is appropriate for courses and schools that are not yet ready to commit to the expense and time involved in acquiring other GIS packages.

Annual Editions: Environment by Allen

This small text is a compilation of current articles that explore the global environment, the world's population, energy, the biosphere, natural resources, and pollution.

Taking Sides: Clashing Views on Controversial Environmental Issues by Goldfarb

This represents the arguments of leading environmentalists, scientists, and policy makers. The issues reflect a variety of viewpoints and are staged as pro and con debates. Topics are organized around four core areas: general philosophical and political issues, the environment and technology, disposing of wastes, and the environment and the future.

Sources: Notable Selections in Environmental Studies, 2/e by Goldfarb

This volume brings together primary source selections of enduring intellectual value—classic articles, book excerpts, and research studies—that have shaped environmental studies and our contemporary understanding of it. The book includes carefully edited selections for the works of the most distinguished environmental observers, past and present. Selections are organized topically around the following major areas of study: energy, environmental degradation, population issues and the environment, human health and the environment, and environment and society.

Student Atlas of Environmental Issues by Allen

The *Student Atlas of Environmental Issues* is an invaluable pedagogical tool for exploring the human impact on the air, waters, biosphere, and land in every major world region. This informative resource provides a unique combination of maps and data helping students understand the dimensions of the world's environmental problems and the geographical basis of these problems.

Interactive World Issues CD-ROM

This CD explores environmental issues that affect various geographic regions. For example, you'll visit Oregon and investigate water rights of the Columbia River. You'll listen to Native Americans whose living depends on salmon fishing and then to the farmers who need water to irrigate their crops. Additional case studies discuss migration in Mexico, apartheid in South Africa, population issues in China, and farming in urban Chicago.

You Can Make a Difference: Be Environmentally Responsible, 2/e by Getis

This book is organized around the three parts of the biosphere: land, water, and air. Each section contains descriptions of the environmental problems associated with that part of the biosphere. Immediately following each problem or "challenge" are suggested ways that individuals can help solve or alleviate them. This book has been written to provide the reader with some easy and practical ways to protect the Earth and to help understand why the task is so important.

Custom publishing

Did you know that you can design your own text or laboratory manual using any McGraw-Hill text *and* your personal materials to create a custom product that correlates specifically to your syllabus and course goals? Talk to your McGraw-Hill sales representative about this creative option.

About the Authors

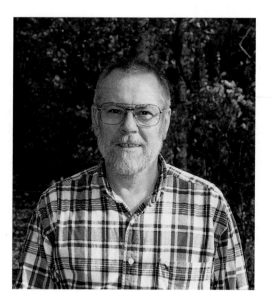

Eldon D. Enger is an emeritus professor of biology at Delta College, a community college near Saginaw, Michigan. He received his B.A. and M.S. degrees from the University of Michigan. Professor Enger has over 30 years of teaching experience, during which he has taught biology, zoology, environmental science, and several other courses. He has been very active in curriculum and course development. A major contribution to the curriculum was the development of an environmental technician curriculum and the courses that support it. He was also involved in the development of learning community courses in stream ecology, winter ecology, and plant identification. Each of these courses involved students in weekend-long experiences in the outdoors that paired environmental education with physical activity—stream ecology and canoeing, winter ecology and cross-country skiing, and plant identification with backpacking.

Professor Enger is an advocate for variety in teaching methodology. He feels that if students are provided with varied experiences, they are more likely to learn. In addition to the standard textbook assignments, lectures, and laboratory activities, his classes included writing assignments, student presentation of lecture material, debates by students on controversial issues, field experiences, individual student projects, and discussions of local examples and relevant current events. Textbooks are very valuable for presenting content, especially if they contain accurate, informative drawings and visual examples. Lectures are best used to help students see themes and make connections, and laboratory activities provide important hands-on activities.

Professor Enger received the Bergstein Award for Teaching Excellence and the Scholarly Achievement Award from Delta College and was selected as a Fulbright Exchange Teacher twice—to Australia and Scotland. He has participated as a volunteer in several Earthwatch Research Programs. These include: studying the behavior of a bird known as the long-tailed manakin in Costa Rica, participating in a study to reintroduce endangered marsupials from islands to mainland Australia, and efforts to protect the leatherback turtle in Costa Rica. He also served as a participant in a People to People program which allowed for an exchange of ideas between U.S. and South African environmental professionals. While traveling he has spent considerable time visiting coral reefs, ocean coasts, mangrove swamps, alpine tundra, prairies, tropical rainforests, cloud forests, deserts, temperate rainforests, coniferous forests, deciduous forests, and many other special ecosystems. This extensive experience provides the background to look at environmental issues from a broad perspective.

Professor Enger is married, has two grown sons, and enjoys a variety of outdoor pursuits such as cross-country skiing, hiking, hunting, fishing, camping and gardening. Other interests include reading a wide variety of periodicals, beekeeping, singing in a church choir, and preserving garden produce.

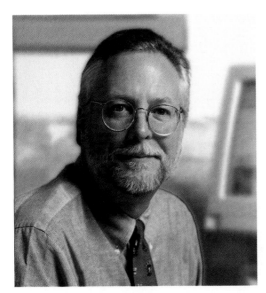

Bradley F. Smith is the Dean of Huxley College of Environmental Studies at Western Washington University in Bellingham, Washington. Prior to assuming the position as Dean in 1994, he served as the first Director of the Office of Environmental Education for the U.S. Environmental Protection Agency in Washington, D.C. from 1991 to 1994. Dean Smith also served as the Acting President of the National Environmental Education and Training Foundation in Washington, D.C. and as a Special Assistant to the EPA Administrator.

Before moving to Washington, D.C., Dean Smith was a professor of political science and environmental studies for fifteen years, and the executive director of an environmental education center and nature refuge for five years.

Dean Smith has considerable international experience. He was a Fulbright Exchange Teacher to England and worked as a research associate for Environment Canada in New Brunswick, Canada. He is a frequent speaker on environmental issues worldwide and serves on the International Scholars Program for the U.S. Information Agency. He also served as a U.S. representative on the Tri-Lateral Commission on environmental education with Canada and Mexico. In 1995, he was awarded a NATO Fellowship to study the environmental problems associated with the closure of former Soviet military bases in Eastern Europe. Dean Smith is an Adjunct Professor at Far Eastern State University in Vladivostok, Russia and is a member of the Russian Academy of Transport. He also serves as a commissioner for the International Union for the Conservation of Nature (IUCN) and is the President of the World Conservation Learning Network for the IUCN.

Nationally, Dean Smith serves as a member/advisor for many environmental organizations' boards of directors, advisory councils, and executive committees. He is the co-chair of the Washington State Sustainabily Council and President of the Council of Environmental Deans and Directors. He previously served President Clinton's Council for Sustainable Development (Education Task Force).

Dean Smith holds B.A. and M.A. degrees in Political Science and Public Administration and a Ph.D. from the School of Natural Resources and Environment at the University of Michigan.

Dean Smith lives with his wife Daria, daughter Morgan, son Ian, and English setter Skye, along Puget Sound, south of Bellingham. He is an avid outdoor enthusiast.

Guided Tour

The features of this anniversary edition are unique.

The organization and principle features of this book were planned with the students' holistic learning and comprehension in mind. The text discussion is supported by unique learning aids that offer assistance in studying, reinforce the understanding of key concepts, and promote critical thinking skills.

New! World Maps

Two oversized maps on a dual foldout page allow students to use both a political map, showing countries of the world, and an ecological regions map, indicating various biomes, while studying from any section of the textbook. The maps are accompanied by a guest essay that offers map reading guidelines.

Guest Author Essays

Guest authors have again contributed part-opening essays written about environmental issues pertaining to their geographical regions. This range of perspectives allows students to identify with concerns close to home while also realizing that other parts of North America experience different challenges unique to their area.

Conceptual Study Aids

A brief list of objectives and an outline open each chapter, giving the reader a broad overview of the interrelated forces involved in the material to be discussed. Students are encouraged to refer to these resources while reading and reviewing the chapter.

New! "Past, Present, Future" Charts

When we look at environmental concerns in general, we find that the same basic issues exist today as in 1983, when this book was first published. Although there has been significant progress in addressing some environmental problems, improvement in others has been slow, and new threats or worries have emerged. To help the reader evaluate progress in solving environmental problems and also to understand the complexity of these issues, a new feature entitled, *Past, Present, Future* appears at the beginning of each chapter. These charts show how specific topics discussed in the pages to come have changed over the past 20-plus years.

PAST PRESENT FUTURE

1983 U.S. proved oil reserves were about 35.1 thousand million barrels.

1983 World proved oil reserves were about 677 thousand million barrels.

1983 The price of a barrel of oil (in 2002 US$) was about $50 per barrel.

1983 World energy consumption was about 6714 million metric tons of oil equivalent.

1983 World per capita energy consumption was about 1.43 metric tons of oil equivalent.

2003 U.S. proved oil reserves were about 30 thousand million barrels (about a 15 percent decrease).

2003 World proved oil reserves were about 1,050 thousand million barrels (about a 55 percent increase).

2003 The price of a barrel of oil (in 2002 US$) was about $26 per barrel (about a 50 percent decrease).

2003 World energy consumption was about 9500 million metric tons of oil equivalent (about a 41 percent increase).

2003 World per capita energy consumption was about 1.50 metric tons of oil equivalent (about a 5 percent increase).

New Three-dimensional Art

There are over 40 new figures and tables in this tenth edition, many of which have been drawn in a more realistic, three-dimensional style. Numerous other pieces have been revised to reflect updated data or content.

Figure 14.10 Major Soil Types There are thousands of different soil types, but many of them can be classified into three broad categories. Soils formed in the grasslands have a deep A horizon. The shallow B horizon does not have sufficient nutrients to support root growth. In forest soils, the A horizon is thinner, and leaching transfers many nutrients to the B horizon. Thus, roots are found in both the A and B horizons. Desert soils have very thin A horizons.

Relevant Readings

Context is so very important in the world of environmental science. Boxed readings help students understand that variables surrounding an environmental issue can affect both the problem and the solution. Looking beyond our borders helps to gain a different perspective on nuclear power, or water pollution—just two of the topics discussed in the **"Global Perspectives." "Environmental Close-Ups"** are case study-like readings that offer the reader a closer examination of a topic described in the surrounding text. Every chapter in the tenth edition also features a <u>new</u> **"Issues—Analysis"** reading, written to reflect current issues and/or new data pertaining to the chapter subject matter.

Digital Content Manager (DCM) CD-ROM

This indispensable supplement is a multimedia collection of visual resources that allows instructors to utilize artwork from the text and other sources to create customized course tools.

ALL of the Art from the Book

Over 275 illustrations and 300 photos, already in PowerPoint format, are at your disposal to use in classroom presentations. These include all of the new, realistic drawings that were rendered for this tenth edition, representing some of the most important concepts in environmental science.

Protozoa, nematodes, earthworms, insects, algae, bacteria, and fungi are typical inhabitants of soil. (See figure 14.8.) The role of protozoa in the soil is not firmly established, but they seem to act as parasites and predators on other forms of soil organisms and, therefore, help to regulate the populations of those organisms.

Figure 14.7 Pore Spaces and Particle Size (a) Soil that is composed of particles of various sizes has spaces for both water and air. The particles have water bound to their surfaces (represented by the colored halo around each particle), but some of the spaces are so large that an air space is present. (b) Soil composed of uniformly small particles has less space for air. Since roots require both air and water, the soil in (a) would be better able to support crops than would the soil in (b).

Digital Content Manager

Animations

One hundred full-color animations illustrating many different concepts covered in the study of environmental science are available for your use in creating classroom lectures, testing materials, or online course communications.

PowerPoint Lecture Outline

Don't have time to create your own classroom presentation? The Digital Content Manager offers a customized presentation combining art from the text with instructor-written lecture notes, covering all 20 chapters.

Additional Photo Library

Over 300 full-color photographs *additional* to those already in the textbook are also included in this CD-ROM. Chosen for their specific contribution to topics in environmental science, these photos are searchable by content and will add interest and contextual support to your lectures.

Active Art

These special art pieces consist of key environmental science illustrations that are converted to a format that allows you to break the art down into core elements, and then group the various pieces and create customized images. This is especially helpful with difficult concepts as they can be explained to students step by step.

Table Library

In addition to hundreds of illustrations and photos, this valuable CD also includes every table in the textbook, available in ready-to-use digital files.

"See It in Motion" Videos

For those of you who want more coverage on oceans in your environmental science course, a special collection of live underwater videos is included in the Digital Content Manager. "See It in Motion" video clips display interesting habitats and behaviors for many animals in the ocean.

Global Base Maps

Frustrated that your students can't identify cities/states/countries beyond their immediate surroundings? In support of the authors' commitment to global environmental concern with the environment, we've developed 22 base maps for all world regions and major sub-regions. These maps are offered in four versions: black-and-white and full-color, both with labels and without labels. All of these choices allow you flexibility in planning class activities, quizzing opportunities, study tools, and Power-Point enhancements. Help your students understand that what happens in Africa or Europe affects all of us.

Critical Thinking

We live in an age of information. Computers, e-mail, the Internet, CD-ROMs, instant news, and fax machines bring us information more quickly than ever before. A simple search of the Internet will provide huge amounts of information. Some of the information has been subjected to scrutiny and is quite valid, some is well-informed opinion, some is naive misinformation, and some is even designed to mislead. How do we critically evaluate the information we get?

Critical thinking involves a set of skills that helps us to evaluate information, arguments, and opinions in a systematic and thoughtful way. Critical thinking also can help us better understand our own opinions as well as the points of view of others. It can help us evaluate the quality of evidence, recognize bias, characterize the assumptions behind arguments, identify the implications of decisions, and avoid jumping to conclusions.

Characteristics of Critical Thinking

Critical thinking involves skills that allow us to sort information in a meaningful way and discard invalid or useless information while recognizing that which is valuable. Some key components of critical thinking are:

Recognize the importance of context.

All information is based on certain assumptions. It is important to recognize what those assumptions are. Critical thinking involves looking closely at an argument or opinion by identifying the historical, social, political, economic, and scientific context in which the argument is being made. It is also important to understand the kinds of bias contained in the argument and the level of knowledge the presenter has.

Consider alternative views.

A critical thinker must be able to understand and evaluate different points of view. Often these points of view may be quite varied. It is important to keep an open mind and to look at all the information objectively and try to see the value in alternative points of view. Often people miss obvious solutions to problems because they focus on a certain avenue of thinking and unconsciously dismiss valid alternative solutions.

Expect and accept mistakes.

Good critical thinking is exploratory and speculative, tempered by honesty and a recognition that we may be wrong. It takes courage to develop an argument, engage in debate with others, and admit that your thinking contains errors or illogical components. By the same token, be willing to point out what you perceive to be shortcomings in the arguments of others. It is always best to do this with good grace and good humor.

Have clear goals.

When analyzing an argument or information, keep your goals clearly in mind. It is often easy to get sidetracked. A clear goal will allow you to quickly sort information into that which is pertinent and that which may be interesting but not germane to the particular issue you are exploring.

Evaluate the validity of evidence.

Information comes in many forms and has differing degrees of validity. When evaluating information, it is important to understand that not all the information from a source may be of equal quality. Often content about a topic is a mix of solid information interspersed with less certain speculations or assumptions. Apply a strong critical attitude to each separate piece of information. Often what appears to be a minor, insignificant error or misunderstanding can cause an entire argument to unravel.

Critical thinking requires practice.

As with most skills, you become better if you practice. At the end of each chapter in the text, there are a series of questions that allow you to practice critical thinking skills. Some of these questions are straightforward and simply ask you to recall information from the chapter. Others ask you to apply the information from the chapter to other similar contexts. Still others ask you to develop arguments that require you to superimpose the knowledge you have gained from the chapter on quite different social, economic, or political contexts from your own.

Practice, practice, practice.

PART
I

Environmental Science in a Social Context

A commercial fisherman hoists pot of crabs aboard his boat from the Chesapeake Bay.

Can We Save Both Blue Crabs and Watermen?

Jennifer M. Rhode

Georgia College and State University

Chesapeake Bay lies in the mid-Atlantic region of the East Coast of the United States, surrounded by Maryland and Virginia. In 2001, commercial watermen harvested over 270 million kilograms (600 million pounds) of Chesapeake Bay seafood; blue crabs made up about 10 percent of this harvest. At its peak, the Chesapeake Bay blue crab seafood fishery was worth more than $150 million per year. Since the mid-1990s, however, the bay's crab population has dropped precipitously and now hovers around its all-time recorded low.

In response to alarming declines in crab abundance, the Bi-State Blue Crab Advisory Committee, a group of scientists and managers from Maryland and Virginia, recommended a 15 percent reduction in baywide harvest between 2001 and 2003. Implementation has been slow, and the future of this species, the fishery, and the watermen who harvest crabs for a living remains in limbo.

Increased fishing pressure has contributed to the decline in blue crab populations. Chesapeake Bay oysters are too scarce to produce a commercially viable yield, so watermen who historically fished both oysters and crabs now focus their efforts exclusively on crabs. Technological improvements have made commercial crab harvesting more efficient, increasing the rate of crab capture. Crabbing has also become a more popular pastime among the burgeoning population of Chesapeake Bay's watershed.

Other sources of crab mortality have also reduced the number of Chesapeake Bay blue crabs. First, the numbers of predators have increased. Both Maryland and Virginia imposed a total moratorium on striped bass fishing from the mid-1980s until 1989, allowing fish populations to increase. Since crabs compose up to 50 percent of the striped bass diet, increasing numbers of these voracious predators might be expected to cause a concomitant decrease in blue crab numbers. Second, crabs have become more vulnerable to predators. Juvenile blue crabs rely on seagrass beds for camouflage to protect them from predation, but increased nutrient pollution of Chesapeake Bay caused mass die-offs of seagrasses in the 1960s and 1970s. Although these plant populations have begun to recover, seagrasses now cover less than 3 percent of the bay's bottom. Finally, nutrient pollution can result in dramatic decreases in oxygen levels, causing crabs to suffocate. Anecdotal evidence supports this, as the incidence of "crab jubilees" (mass exoduses of crabs from water onto land) seems to be increasing.

Declining crab populations might also be a result of decreasing crab birthrates. Some studies have shown that there are too few female crabs to sustain historic population numbers, and the number of eggs per female has decreased. Male sperm counts have also decreased. Either of these factors might negatively impact the number of new crabs produced each year.

So, the current decline in crabs could be attributed to a number of individual factors but is probably a combination of more than one of these. Who is to blame for this problem? How can crab populations be restored? Opinions vary widely.

The states that border Chesapeake Bay have criticized one another for contributing to the blue crab crisis. Maryland claims that Virginia has had a disproportionate impact on crab populations by allowing the capture of mature female and egg-carrying crabs, as well as the capture of smaller crabs that have yet to reproduce, and permitting year-round crabbing. Virginia, in turn, cites its establishment of spawning sanctuaries (started in 2000) and the fact that Maryland catches more crabs as evidence that Virginia is not negatively impacting crabs.

Watermen claim that the decline in crabs is cyclical, exacerbated by the harsh winter of 2002 and Hurricane Isabel in 2003. Restricting crabbing or trying to convert the industry to aquaculture will destroy the traditional way of life that they and their families have maintained for hundreds of years, they say. The watermen fault the recreational fishermen for catching too many crabs and blame state regulatory agencies for allowing striped bass populations to increase.

Recreational crabbers claim the right to use state waters, citing that they already have to buy crabbing licenses and obey strict size regulations and catch limits. They counter the claims of watermen with statistics: For example, the recreational catch in the 2001–2002 season was less than 13 percent of the commercial harvest.

Scientists acknowledge the watermen's observation that crab abundance is often cyclical. However, they say that these depressed population levels have lasted too long to be due to natural causes. They also present data showing that, in the past, populations of both striped bass and blue crabs were higher. Thus, the presence of these fish does not automatically depress crab numbers. Scientists think that the solution to declining crab populations is to create sanctuaries and habitat corridors, places where crabs can live and reproduce without threat of capture. Some environmentalists go even further, stating that use of this natural resource should cease completely, and be replaced with aquaculture.

What Do You Think?

1. Closing the striped bass fishery allowed populations of that animal to recover in less than a decade. Should the Bi-State Blue Crab Advisory Committee take similar action with the blue crab fishery?

2. Should Maryland and Virginia pay watermen *not* to harvest crabs until their populations have recovered?

3. If recreational crabbers adopted a catch-and-release program, would that help crab populations recover? Would such a program be popular?

4. What sorts of scientific evidence would help you decide whether crab populations have recovered?

1

Environmental Interrelationships

Chapter Outline

Objectives

After reading this chapter, you should be able to:

• Understand why environmental problems are complex and interrelated.

• Realize that environmental problems involve social, ethical, political, and economic issues, not just scientific issues.

• Understand that acceptable solutions to environmental problems are not often easy to achieve.

• Understand that all organisms have an impact on their surroundings.

• Understand what is meant by an ecosystem approach to environmental problem solving.

• Recognize that different geographic regions have somewhat different environmental problems, but the process for resolving them is often the same and involves compromise.

PAST PRESENT FUTURE

1983 Five percent of the colleges and universities in the United States offered undergraduate degrees in environmental science or environmental studies.

1983 There were 450 computers linked to the Internet.

1983 The gross world product (GWP) was 26.5 (trillion 2000 US$).

2003 Sixty-two percent of U.S. colleges and universities offered undergraduate degrees in environmental science or environmental studies.

2003 The number of computers linked to the Internet had increased to 162,344,723.

2003 The GWP had grown to 46.8 (trillion 2000 US$).

The Field of Environmental Science

Environmental science is an interdisciplinary area of study that includes both applied and theoretical aspects of human impact on the world. Since humans are generally organized into groups, environmental science must deal with politics, social organization, economics, ethics, and philosophy. Thus, environmental science is a mixture of traditional science, individual and societal values, and political awareness. (See figure 1.1.)

Although environmental science as a field of study is evolving, it is rooted in the early history of civilization. Many ancient cultures expressed a reverence for the plants, animals, and geographic features that provided them with food, water, and transportation. These features are still appreciated by many modern people. Although the following quote from Henry David Thoreau (1817–1862) is over a century old, it is consistent with current environmental philosophy:

> I wish to speak a word for Nature, for absolute freedom and wildness, as contrasted with a freedom and culture merely civil . . . to regard man as an inhabitant, or a part and parcel of Nature, rather than a member of society.

The current interest in the state of the environment began with philosophers like Thoreau and scientists like Rachel Carson and received emphasis from the organization of the first Earth Day on April 22, 1970. Subsequent Earth Days reaffirmed this commitment. As a result of this continuing interest in the state of the world and how people both affect it and are affected by it, environmental science is now a standard course or program at many colleges. It is also included in the curriculum of high schools. Most of the concepts covered by environmental science courses had previously been taught in ecology, conservation, biology, or geography courses. Environmental science incorporates the scientific aspects of these courses with input from the social sciences, such as economics, sociology, and political science, creating a new interdisciplinary field.

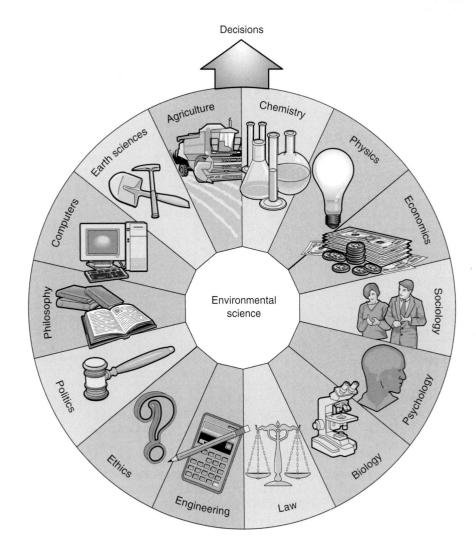

Figure 1.1 **Environmental Science** The field of environmental science involves an understanding of scientific principles, economic influences, and political action. Environmental decisions often involve compromise. A decision that may be supportable from a scientific or economic point of view may not be supportable from a political point of view without modification. Often political decisions relating to the environment may not be supported by economic analysis.

The Interrelated Nature of Environmental Problems

Environmental science is by nature an interdisciplinary field. The word *environmental* is usually understood to mean the surrounding conditions that affect people and other organisms. In a broader definition, **environment** is everything that affects an organism during its lifetime. In turn, all organisms including people affect many components in their environment. (See figure 1.1.) From a human perspective, environmental issues involve concerns about science, nature, health, employment, profits, politics, ethics, and economics.

Most social and political decisions are made with respect to political jurisdictions, but environmental problems do not necessarily coincide with these artificial political boundaries. For example, air pollution may involve several local units of government, several states or provinces,

Traditional Resource Use and Conflict Management in Keoladeo National Park, India—Science Versus Policy

Scientific knowledge and government policy do not always agree. The scientific community can advise governments but cannot insist that certain policies be adopted. This is true for international issues such as global warming as well as for local and regional issues such as national parks.

Keoladeo National Park is a small 2873 hectares; 7096 acres) artificial wetland system located near Bharatpur on the Ganges plain in India. The wetland was created in 1750 by local royalty to attract migratory birds for hunting. Today, over 350 bird species, including the endangered Siberian crane, inhabit the park seasonally. In 1982, Keoladeo was declared a national park.

Water buffalo, traditionally allowed to graze in the park, a water weed growing in the wetland, and the Siberian crane co-existed in a three-way relationship. The buffalo grazed on the weed, controlling its growth. Short-cropping of the weed by buffalo made it possible for the cranes to dig up the plant's tubers, one of their few food sources. In 1983, however, the India Wildlife Protection Act prohibited the grazing of buffalo in the park. As a result, the weed grew unchecked to maturity, thereby creating a physical barrier that prevented the cranes from accessing their main food source, which led to a dramatic decrease in the numbers of cranes in the park.

The Wildlife Protection Act was formulated and implemented without consultation with local scientists or local communities. A decade-long study, costing nearly US$ 1 million indicated that grazing buffalo were key to controlling growth of grasses and water weeds and, therefore, the Siberian crane and other bird populations. Local communities and scientists already knew this.

A three-week workshop in the late 1990s brought together the Indian federal authorities, regional scientists, and the local communities. The exercise sought to determine areas of agreement and disagreement between local communities and scientists concerning conservation objectives of the nationally adopted plans; identify new, potentially successful management options; and propose ways in which local scientists and community leaders could have input into the management of the park. A new, more cooperative management plan for the park resulted from the workshop, based on altering federal policy in preference to regional scientific knowledge.

---

and even different nations. The forest fires that raged in Mexico in 1998 had a severe impact on air quality in Texas. Air pollution generated in China affects air quality in western coastal states in the United States and in British Columbia, Canada. On a more local level, the air pollution problems in Juarez, Mexico, are also problems in El Paso, Texas. But the issue is more than air quality and human health. Lower wage rates and less strict environmental laws have influenced some U.S. industries to move to Mexico for economic advantages. Mexico and many other developing nations are struggling to improve their environmental image and need the money generated by foreign investment to improve the conditions and the environment in which their people live.

Air pollutants produced in the major industrial regions of the United States drift across the border into Canada, where acid rain damages lakes and forests. A longstanding dispute exists between the United States and Canada over this issue. Canada claims that the United States should be doing more to reduce emissions that cause acid rain, and the United States claims it is doing as much as it can. In another example, farmers who use water from the Colorado River for irrigation reduce the quality and quantity of water entering Mexico. This causes political friction between Mexico and the United States.

The issue of declining salmon stocks in the Pacific Northwest of the United States and British Columbia, Canada, is another example of political friction over a shared natural resource. It has been calculated that on the U.S. side of the salmon issue alone, there are five federal cabinet-level departments, two federal agencies, and five federal laws in question as well as numerous tribal treaties, commissions, and court decisions. All of this is in addition to many state-level departments, commissions, and rulings. If all of this were not sufficient, international bodies such as the United Nations and international treaties impact the fate of the salmon. Considering all this complexity, it is not surprising that the salmon is in such a dangerous status. (See figure 1.2.)

Because of all these political, economic, ethical, and scientific links, solving environmental problems is complicated.

Environmental problems seldom have simple solutions. However, international organizations, such as the International Joint Commission, have had major bearing on the quality of the environment over broad regions of the world.

The International Joint Commission was established in 1909, when the Boundary Waters Treaty was signed between the United States and Canada. The treaty was established in part to provide that the "boundary waters and waters flowing across the boundary shall not be polluted on either side to the injury of health or property of the other." The commission has been instrumental in identifying areas of concern and encouraging the cleanup of polluted sites that affect the quality of the Great Lakes and other boundary waters. In general, the two governments have listened to the commission's advice and have responded by initiating cleanup activities.

The first worldwide meeting of heads of state directed to concern for the environment took place at the Earth Summit, formally known as the United Nations Conference on Environment and Development (UNCED) in Rio de Janeiro

Harvest 1

Overfishing has contributed to the decline of many fish populations. Often this exploitation is caused by fishery managers trying to access harvestable hatchery salmon or other abundant fish in areas that contain depleted wild salmon populations.

Hatcheries 2

Hatchery fish that augment harvest levels can interbreed with wild fish, resulting in the loss of genetic diversity. Hatchery fish can also spread disease and compete with wild fish for food and habitat.

Habitat 3

Rural and Urban: Salmon face multiple, complex threats in the developed lower regions of watersheds. Problems include low water flows, pollution, degraded physical habitats, and migration barriers such as culverts.

Forests: Improper forest practices and road construction and maintenance are the biggest threat to salmon in the upper watershed. The Department of Natural Resources in Washington receives 12,000 applications for forest practices annually.

Hydropower 4

Dams can block fish migration to and from the ocean, kill fish passing through turbines, delay migration, and increase predation. Dams can also cause inadequate flow downstream.

There are 1,018 dams on Washington rivers. The Columbia River hosts 150 hydroelectric projects and 250 reservoirs – more than half the length of the river is blocked to salmon and steelhead.

Figure 1.2   The Four H's: Human Activities That Affect Wild Salmon Survival The interrelated nature of environmental problems is evident in the diminishing numbers of wild salmon in the Pacific Northwest of the United States and in British Columbia, Canada. This diagram portrays the plight of wild salmon in the state of Washington, but the issue is regionwide.

Source: Washington Department of Natural Resources Newsletter, Winter 1998, Department of Natural Resources, Olympia, Washington.

in 1992. Most countries have also signed agreements on **sustainable development** and biodiversity. (See the Global Perspective on page 11.) The policy statements on sustainable development at the UNCED were identified as Agenda 21. Agenda 21 is a comprehensive plan of action to be taken globally, nationally, and locally by organizations of the UN system, national governments, and major groups in every area in which humans impact the environment.

More than 178 governments at the 1992 conference adopted Agenda 21, the Rio Declaration on Environment and Development, and the Statement of Principles for the Sustainable Management of Forests to ensure effective follow-up of UNCED. The Commission on Sustainable Development (CSD) was created in 1993 to monitor and report on implementation of the agreements at the local, national, regional, and international levels. It was agreed that a five-year review of Earth

Summit progress would be made in 1997 by the United Nations General Assembly meeting in special session. The Fifty-fifth General Assembly session decided in 2000 that the CSD would serve as the central organizing body for the 2002 World Summit on Sustainable Development, which was held in Johannesburg, South Africa. The Fifty-fifth General Assembly session also noted that, because of globalization, external factors have become critical in determining the success or failure of developing countries in their national sustainable development efforts.

In 1997, representatives from 125 nations met in Kyoto, Japan, for the Third Conference of the United Nations Framework Convention on Climate Change. This conference, commonly referred to as the Kyoto Conference on Climate Change, resulted in commitments from the participating nations to reduce their overall emissions of six greenhouse gases (linked

to global warming) by at least 5 percent below 1990 levels and to do so between the years 2008 and 2012. The Kyoto Protocol, as the agreement was called, was viewed by many as one of the most important steps to date in environmental protection and international diplomacy. It may be years before we will know if all countries that signed these agreements will meet their commitments to environmental improvement, but they have at least stated their intention to do so.

The United Nations, through the United Nations Educational, Scientific, and Cultural Organization (UNESCO) and the United Nations Environment Programme (UNEP), has supported many environmental programs. A recent undertaking is the International Environmental Education Programme (IEEP). This program recognizes the need for both formal environmental education in schools and the informal education that occurs through the media and groups of interested citizens. Conferences on environmental education were first held during the 1970s and continue to the present.

An Ecosystem Approach

The natural world is organized into interrelated units called ecosystems. An **ecosystem** is a region in which the organisms and the physical environment form an interacting unit. Weather affects plants, plants use minerals in the soil and affect animals, animals spread plant seeds, plants secure the soil, and plants evaporate water, which affects weather.

One of the first individuals to provide a formalized, contemporary description of ecosystems was A. G. Tansley in 1935, when he stated that an ecosystem is a unit of vegetation which . . . includes not only the plants of which it is composed but the animals habitually associated with them, and also all the physical and chemical components of the immediate environment or habitat which together form a recognizable self-contained entity.

He later supplemented this with, "All parts of such an ecosystem may be considered as interacting." During the 1950s and 1960s, "ecology" as a discipline came into its own, generating a surge of efforts by understanding the Earth and its systems and processes in a more holistic way. Ecosystem science provided a framework for these efforts. Books such as *Gaia: A New Look at Life on Earth* (Lovelock, 1979) helped popularize this holistic approach.

While accepting that every living thing and all processes on Earth are interrelated and somehow interact, some naturally occurring demarcations in the real world can justifiably be considered as boundaries for the purposes of ecosystem-based management projects. Examples are lakes, deltas, islands, floodplains, watersheds separated by mountains, and many others. Such "natural" ecosystems occur on many scales. Examples of domesticated ecosystems include small towns, urban areas, and agro-ecosystems.

Large ecosystems always include smaller ones. A large watershed, for example, may include a number of lakes that can be locally managed within the context of the overall watershed. Isolated coral reefs may also form part of a much larger atoll system. Thus, defining an ecosystem boundary is more a matter of practical convenience and is usually related to easily identifiable demarcations. Nonetheless, it is important not to forget that a large quantity of materials moves back and forth across natural demarcations. In the case of mountain divides, for example, wildlife, seeds, hunters, slash-and-burn farmers, refugees, atmospheric pollutants, and other elements may be transitory parts of the ecosystem.

Ecosystems sometimes have fairly discrete boundaries, as is the case with a lake, island, or biosphere. Sometimes the boundaries are indistinct, as in the transition from grassland to desert. Grassland gradually becomes desert, depending on the historical pattern of rainfall in an area.

An ecosystem approach requires a look at the way the natural world is organized. Where do the rivers flow? What are the prevailing wind patterns? What are the typical plants and animals in the area? How does human activity affect nature? The task of an environmental scientist is to recognize and understand the natural in-teractions that take place and to integrate these with the uses humans must make of the natural world.

To illustrate the interrelated nature of environmental issues, we will look at several regions of North America and highlight some of the key features and issues of each.

Regional Environmental Concerns

No region is free of environmental concerns. Most regions tend to focus on specific, local environmental issues that apply directly to them. For example, protecting endangered species is a concern in many parts of the world. In the Pacific Northwest, for example, an endangered species known as the northern spotted owl depends on undisturbed mature forests for its survival. Development and logging may conflict with the survival of the owl. In most metropolitan areas, the problem of endangered species is purely historical, since the construction of cities has destroyed the previously existing ecosystem. Here we present a number of regional vignettes to illustrate the complexity and interrelatedness of environmental issues. (See figure 1.3.)

The Wilderness North

Much of Alaska and Northern Canada can be characterized as **wilderness**—areas with minimal human influence. Much of this land is owned by governments, not by individuals, so government policies have a large effect on what happens in these regions. These areas have important economic values in their trees, animals, scenery, and other natural resources. Exploitation of the region's natural resources involves significant trade-offs. Usually, a portion of the natural world is altered permanently, but the area altered is so small that many people consider it insignificant. Because of the severe climate, northern wilderness areas tend to be very sensitive to insults and take a long time to repair damage done by unwise exploitation. Mining, oil exploration, development of hydroelectric projects, and harvesting of timber all require roads and other human artifacts, involve the insertion of new technologies into native culture, and generate economic benefits.

In the past, many short-term political and economic decisions failed to look at long-term environmental implications. Today, however, people are concerned about these remaining wilderness areas. Politicians are more willing to look at the scientific and recreational values of wilderness as well as the economic value of exploitation.

Native people, who consider much of this region to be their land, have become increasingly sophisticated in negotiating with state, provincial, and federal governments to protect rights they feel they were granted in treaties. They are sensitive to changes in land use or government policy that would force changes in their traditional way of life.

Concerned citizens, business interests, and environmental activists have become increasingly sophisticated in influencing decisions made by government. The process of compromise is often difficult and does not always ensure wise decisions, but most governments now realize they must listen to the concerns of their citizens and balance economic benefits with social and cultural benefits. (See figure 1.4.)

The Agricultural Middle

The middle of the North American continent is dominated by intensive agriculture. This means that the original, natural ecosystems have been replaced by managed agricultural enterprise. It is important to understand that this area was at one time wilderness. Today, you would need to search very hard to find regions of true wilderness in Iowa, Indiana, or southern Manitoba. Some special areas have been set aside to preserve fragments of the original natural plant and animal associations, but most of the land has been converted to agriculture wherever practical.

The economic value generated by this use of a rich soil resource is tremendous, and most of the land is privately owned. Governments cannot easily control what

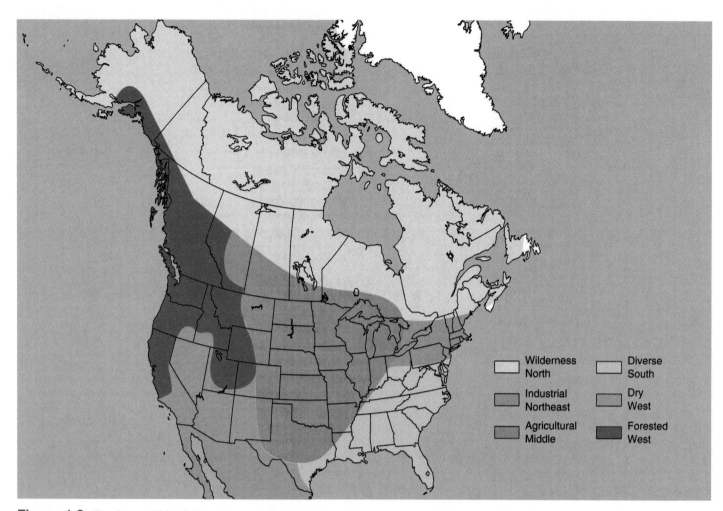

Figure 1.3 **Regions of North America** Because of natural features of the land and the uses people make of the land, different regions of North America face different kinds of environmental issues. In each region, people face a large number of specific issues, but certain kinds of issues are more important in some regions than others.

Legend:
- Wilderness North
- Industrial Northeast
- Agricultural Middle
- Diverse South
- Dry West
- Forested West

Walrus harvesting   A clear-cut forest   Grizzly bear fishing for salmon

Figure 1.4 **The Wilderness North** Protection of wilderness is a major issue in this region. The major points of conflict involve the government role in managing these lands and wildlife, the protection of the rights and beliefs of native people, and the desire of many to exploit the mineral and other resources of the region.

happens on these privately held lands. But governments indirectly encourage certain activities through departments of agriculture that encourage agricultural research, grant special subsidies to farmers in the form of guaranteed prices for their products and other special payments, and develop markets for products. Yet because the economic risks involved in farming are great, the number of farmers constantly declines. There are a number of reasons for farm failures, including drought, disease, lack of markets, increasing labor shortages, and fuel and equipment costs.

The Greater Yellowstone Ecosystem

In 1872, the U.S. government established Yellowstone National Park as the world's first national park. It was an expansive area that protected unique natural features such as geysers, hot springs, rivers, lakes, and mountains. It was also a preserve for many kinds of wildlife such as grizzly bears, elk, moose, and bison. At the time it was established, the park was thought to be of adequate size to protect the scenic resources and the wildlife. Since that time, the lands surrounding the park have been converted to a variety of uses, including cattle grazing, timber production, hunting, and mining.

Fortunately, most of the lands surrounding Yellowstone National Park and the adjacent Grant Teton National Park are still under government control as national forests, national wildlife refuges, and other state, local, or federal entities. Some of the park wildlife, particularly the grizzly bear and bison, often wander across the park boundaries. The grizzly in particular needs large regions of wilderness to survive as a species.

Many people assert that it is essential that these lands be integrated into a Greater Yellowstone Ecosystem management plan encompassing about 7.3 million hectares (18 million acres). The plan is based on more natural boundaries than the original boundaries established in 1872. This would require changes in the way much of the land surrounding Yellowstone is being used. The trade-offs are significant. Logging, mining, hunting, and grazing would be stopped or significantly reduced. This would result in a loss of jobs in those industries. Proponents argue that additional jobs would be created in the tourist and related service industries. The advantages, they argue, would be equal to or greater than the economic losses caused by stopping current uses. Individual and group decisions result in organizational policies and consumer behavior that support or weaken the ecosystem.

One of the major, nonpoint pollution sources (pollution that does not have an easily identified point of origin) is agriculture. Air pollution in the form of dust is an inevitable result of tilling the land. Soil erosion occurs when soil is exposed to wind and moving water and leads to siltation of rivers, impoundments, and lakes. Fertilizers and other agricultural chemicals blow or are washed from the areas where they are applied. Nutrients washed from the land enter rivers and lakes where they encourage the growth of algae, lowering water quality. The use of pesticides causes concern about human exposure, effects on wild animals that are accidentally exposed, and residues in foods produced.

Since many communities in this region rely on groundwater for drinking water, the use of fertilizers and pesticides, and their potential for entering the groundwater as a result of unwise or irresponsible use, is a consumer issue. In addition, many farmers use groundwater for irrigation, which lowers the water table and leaves less groundwater for other purposes.

Positive decisions regarding food choices are critical to reducing our collective ecological footprint. The manner in which our food is grown and how it is transported, packaged, and eventually purchased is crucial to environmental sustainability. Sustainable agriculture, a positive environmental trend, has blossomed within the last decade. Many farmers, both small and large, have converted to organic agriculture. Organic foods, once available only in local organic markets, are now available in larger conventional supermarkets as well.

Since the mid 1980s in the United States, many farmers have been taking an additional step toward sustainability by selling fresh vegetables and fruits directly to local community members. These community supported agriculture (CSA) farms exist in several states and total approximately 400. Typically, CSA farms are certified as organic or biodynamic, work to minimize pollution, and encourage land stewardship. California has 77 CSAs; three examples include Live Earth Farms and Two Small Farms in Watsonville and the University of California at Santa Cruz Farm and Garden CSA in Santa Cruz.

In an effort to stay in business and preserve their way of life, farmers must use modern technology. Careful use of these tools can reduce their impact; irresponsible use causes increased erosion, water pollution, and risk to humans. (See figure 1.5.)

The Dry West

Where rainfall is inadequate to support agriculture, ranching and raising livestock are possible. This is true in much of the drier portions of western North America. Because much of the land is of low economic value, most is still the property of government, which encourages its use by providing water for livestock and irrigation at minimal cost, offering low rates for grazing rights, and encouraging mining and other development.

Many people believe that government agencies have seriously mismanaged these lands. They assert that the agencies are controlled by special interest groups and powerful politicians sensitive to the demands of ranchers, that they subsidize ranchers by charging too little for grazing rights, and that they allow destructive overgrazing because of the economic needs of ranchers. Ranchers argue that they require access to government-owned land, cannot afford significantly increased grazing fees, and that changing government policies would destroy a way of life that is important to the regional economy.

Water is an extremely valuable resource in this region. It is needed for municipal use and for agriculture. Many areas, particularly the river valleys, have fertile soils that can be used for intensive agriculture. Cash crops such as cotton, fruits, and vegetables can be grown if water is available for irrigation. Because water tends to evaporate from the soil rapidly, long-term use of irrigated lands often results in the buildup of salts in the

GLOBAL PERSPECTIVE

Biodiversity, Human Welfare, and Economic Development

Biodiversity is fundamental to human welfare and economic development, and plays a critical role in meeting human needs by maintaining the ecological processes on which our survival depends. Broadscale ecological systems provide benefits such as clean air and fresh water that are needed by everyone, whether in urban or rural settings.

Biodiversity includes the full range of living organisms that people depend on for both direct and indirect uses. Direct benefits from biodiversity come from the supply of goods or products—such as food, timber, clothing materials, and medicine—that can be consumed or traded in exchange for other required or desired assets. Although all people depend on biodiversity to some extent, the poorest, especially the rural poor, most directly depend on the products of healthy ecosystems, harvesting wild plants and animals for their food, fuel, clothing, medicine, and shelter. Conserving biodiversity is therefore part of protecting the critical ecosystems that are essential for both environmental and economic sustainability.

Biodiversity also provides less tangible, indirect benefits that cannot be traded but underpin the natural and production systems central to human survival. Watershed protection, carbon storage, pollination, and nutrient recycling are all necessary environmental services. Genetic diversity and its associated information are used to create new crops or animal varieties and pharmaceuticals; modern agriculture, which depends on new genetic stock from natural ecological systems, is now a US$ 3 trillion global business. Biodiversity allows adaptation to take place through natural and artificial selection.

Many benefits of biodiversity do not rely on use. Biodiversity is closely linked with human cultural and spiritual values, nonuse benefits that are nonetheless powerful forces in many traditional cultures as well as in urbanized lives. For example, unique species and special landscapes provide aesthetic benefits that are important sources of revenue through economic activities such as tourism. Tourism based on an intact natural environment is rapidly becoming one of the leading sources of foreign exchange earnings in countries with high biodiversity.

Finally, other nonuse benefits of biodiversity, such as the capacity to adapt to future changes, risks, and uncertainties, cannot be captured by individuals, but are "owned" by society at local, regional, and global levels. (See chapter 12 for more information on biodiversity.)

A well-kept farm   Agricultural chemicals   Barges loaded with grain

Figure 1.5   **The Agricultural Middle**   The rich soil resource of this region has been converted to managed agricultural activity. The use of pesticides and fertilizer and exposure of the land to erosion cause concern about pollution of surface and groundwater. Most farmers still maintain that these practices are essential in modern agriculture and that they can be used safely and with minimal pollution.

soil, thus reducing fertility. Irrigation water flowing from fields is polluted by agricultural chemicals that make it unsuitable for other uses such as drinking. As cities in the region grow, an increasing conflict arises between urban dwellers who need water for drinking and other purposes, and ranchers and farmers who need the water for livestock and agriculture. Increased demand for water will re-

sult in shortages, and decisions will have to be made about who will ultimately get the water and at what price. If the urban areas get the water they want, some farmers and ranchers will go out of business. If the agricultural interests get the water, urban growth and development will have to be limited and expensive changes will have to be made to conserve domestic water use.

Because population density is low in most of this region, much of the land has a wilderness character. Increasingly, a conflict has developed between the economic management of the land for livestock production and the desire on the part of many to preserve the "wilderness." Designating an area as wilderness means that certain uses are no longer permitted. This offends individuals and groups who

Wilderness area

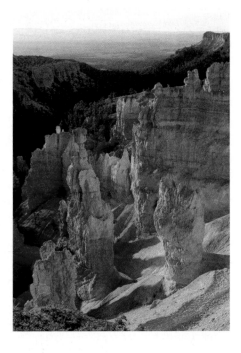

Irrigation water and electrical generation from Glen Canyon Dam

Overgrazed land

Bryce Canyon

Figure 1.6   **The Dry West** Water is a key issue in this region. Both city dwellers and rural ranchers and farmers need water, and conflict results when there is not enough water to satisfy the desires of all. In addition, much of the land in this region is owned by the government. This raises concerns about how the government manages the land and how government policy affects the people of the region.

have traditionally used the area for grazing, hunting, and other pursuits. A long history of use and abuse of this land by overgrazing, modification to encourage plants valuable for livestock, and the introduction of grasses for livestock has significantly altered the region so that it cannot truly be called wilderness. The low population density does, however, provide a remoteness and natural character that many seek to preserve. (See figure 1.6.)

The Forested West

The coastal areas and mountain ranges of the western United States and Canada receive sufficient rainfall for coniferous forests to dominate as vegetation. Since most of these areas are not suitable for farmland, they have been maintained as forests with some grazing activity in the more open forests. Governments and large commercial timber companies own large sections of these lands. Government forest managers (U.S. Forest Service, Bureau of Land Management, Environment Canada, and various state and provincial departments) historically have sold timber-

cutting rights at a loss and are thought by many to be too interested in the production of forest products at the expense of other, less tangible values. In 1993, the U.S. Forest Service was directed to stop below-cost timber sales.

This policy change has become a major issue in the old-growth forests of the Pacific Northwest, where timber interests maintain that they must have access to government-owned forests in order to remain in business. Many of these areas have significant wilderness, scenic, and recreational value. Environmental interests point out that it makes no sense to complain about the destruction of tropical rainforests in South America while North America makes plans to cut large areas of previously uncut, temperate rainforest. Are the intangible values of preserving an ancient forest ecosystem as important as the economic values provided by timber and jobs?

Environmental interests are concerned about the consequences logging would have on organisms that require mature, old-growth forests for their survival. Grizzly bear habitat in Alaska and British

Columbia could be altered significantly by logging; the northern spotted owl has become a symbol of the conflict between logging and preservation in Oregon and Washington; and preservation of coastal redwood forests has become an issue in northern California. (See figure 1.7.) The issue of government ownership of large areas of land and the policy of multiple uses of land is not new in the West.

The Great Lakes and Industrial Northeast

While much of the West and Central regions of North America are characterized by low population densities and small towns, major portions of the Great Lakes and Northeast are dominated by large metropolitan complexes that generate social and resource needs that are difficult to satisfy. Many of these older cities were formed around industrial centers that have declined, leaving behind poverty, environmental problems in abandoned industrial sites, and difficulties with solid waste disposal, air quality, and land-use priorities. Interspersed among the major metropoli-

Cut logs being hauled

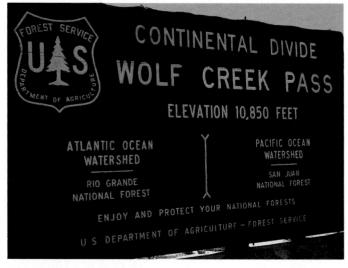

U.S. Forest Service ownership

Figure 1.7 **The Forested West** The cutting of forested areas for timber production destroys the previous ecosystem. Some see the trees as a valuable resource that provides jobs and building materials. Others see the forest ecosystem as a natural resource that should be preserved. In addition, government ownership of much of this land has generated considerable political debate about what the appropriate use of the land should be.

Native elk

tan areas are small towns, farmland, and forests.

One of the major resources of the region is water transport. The Great Lakes and eastern seacoast are extremely important to commerce; ships can travel throughout the area by way of the St. Lawrence Seaway and the Great Lakes through a series of locks and canals that bypass natural barriers. Because of the importance of shipping in this region, harbors have been constructed and waterways have been deepened by dredging. The waterways are maintained at considerable government expense.

One of the greatest problems associated with the industrial uses of the Great Lakes and East Coast is contamination of the water with toxic materials. In some cases, unthinking or unethical individuals have dumped toxins directly into the water. In other cases, small, accidental spills or leaks over long periods of time have contaminated the sediments in harbors and bays.

A major concern about these pollutants is that they bioaccumulate (see chapter 15) in the food chain. The concentrations of some chemicals in the fat tissue of top predators, such as lake trout and fish-eating birds, can be a million times higher than the concentration in the water. Because of this, government agencies have issued consumption advisories for some fish and shellfish in contaminated areas. Since many kinds of fish can swim great distances, advisories for the Great Lakes warn against eating certain fish taken anywhere within the lakes, not just from the site of contamination. Similarly, Chesapeake Bay has been subjected to years of thoughtless pollution, resulting in reduced fish and shellfish populations and advisories against consuming some organisms taken from the bay.

Water always generates considerable recreational value. Consequently, conflicts arise between those who want to use the water for industrial and shipping purposes and those who wish to use it for recreation. Due to the fact that so much of the

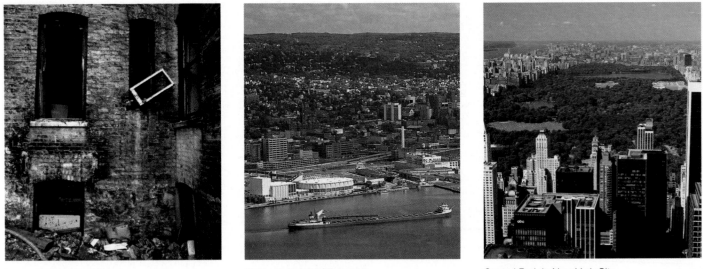

Inner-city decay in Chicago   Harbor in Duluth, Minnesota   Central Park in New York City

Figure 1.8   **The Great Lakes and the Industrial Northeast** Industry, waterways, and population centers are defining elements of this region. The historically extensive use of the Great Lakes and coastal areas of the Northeast for industry, because of the ease of providing water transportation, has resulted in many older cities with poor land-use practices. Rebuilding cities, providing recreational opportunities for urban dwellers, and repairing previous environmental damage are important issues. The water resources of the region provide transportation, recreation, and industrial opportunities.

North American population is concentrated in this region, the economic value of recreational use is extremely high. Consumer pressure is great to clean up contaminated sites and prevent the pollution of new ones. Contaminated areas do not enhance tourism or quality of life.

Most of these older, large cities had no plan to shape their growth. As a result, open space for people is limited and urban dwellers have few opportunities to interact with the natural world. Children who grow up in these cities often do not know that milk comes from a cow—they have never seen, smelled, or touched a cow. Consequently, urban people have difficulty understanding the feeling rural people have for the land. These urban dwellers may never have an opportunity to experience wilderness. Their major environmental priorities are cleaning up contaminated sites, providing more parks and recreation facilities, reducing air and wa-

ter pollution, and improving transportation. (See figure 1.8.)

The Diverse South

In many ways, the South is a microcosm of all the regions previously discussed. The petrochemical industry dominates the economies of Texas and Louisiana, and forestry and agriculture are significant elements of the economy in other parts of the region. Major metropolitan areas thrive, and much of the area is linked to the coast either directly or by the Mississippi River and its tributaries. The environmental issues faced in the South are as diverse as those in the other regions.

Some areas of the South (particularly Florida) have had extremely rapid population growth, which has led to groundwater problems, transportation problems, and concerns about regulating the rate of

growth. Growth means money to developers and investors, but it requires municipal services, which are the responsibility of local governments. Too many people and too much development also threaten remaining natural ecosystems.

Poverty has been a problem in many areas of the South. This creates a climate that encourages state and local governments to accept industrial development at the expense of other values. Often, jobs are more important than the environmental consequences of the jobs; low-paying jobs are better than no jobs.

The use of the coastline is of major concern in many parts of the South. The coast is a desirable place to live, which may encourage unwise development on barrier islands and in areas that are subject to flooding during severe weather. In addition, industrial activity along the coast has resulted in the loss of wetlands. (See figure 1.9.)

14   PART ONE   Interrelatedness

GLOBAL PERSPECTIVE

Constraints and Risks of a Regional Approach—
The Mekong River Delta Example

The Mekong River's watershed includes six of Southeast Asia's most affluent and poorest nations. In 1995, Cambodia, Lao People's Democratic Republic (Laos), Thailand, and Vietnam (the other two watershed countries being China and Myanmar) signed the Agreement on Cooperation for the Sustainable Development of the Mekong River Basin. Potential benefits of regional action include the establishment of a regional electricity grid and a regional growth plan. The agreement could also have environmental advantages by encouraging countries to consider the downstream effects of their activities.

Many demands are made on the river—to provide water for industrial and agricultural development, to sustain subsistence fishing, for transport, and to maintain delicate ecological and hydrological balances. Inevitably, there are conflicting demands on the resource and very different views as to how the water should (or should not) be used.

Disparities in the power and wealth of the countries involved pose problems in reaching solutions that are acceptable to all. This is espe-

cially true in determining how to balance economic development opportunities (such as dam construction) with the river's livelihood benefits to the 60 million people who live in the river basin. Around 30 percent of these people live below the poverty line and rely, for example, on the fish they catch from the river for their food. Similarly, the silt load that the river produces is crucial for intensive farming systems in Vietnam and Cambodia but less to elsewhere.

The Agreement on Cooperation specifies that the countries involved have neither the right to use nor to veto the use of the water of the Mekong, implying that consensus is necessary to move forward. The solution for the Mekong seems to be a balance: the careful selection and construction of environmentally "good" dams over "bad" ones, for example. But the financial and political clout of some countries may pose a threat even to this approach.

Mekong River Fact Sheet

Length	4200 km (2610 mi)
Countries	China, Myanmar, Thailand, Laos, Cambodia, and Vietnam
Basin population	60 million
Country population total	242 million
Per capita GDP	Varies from US$ 2565 (Thailand) to US$ 265 (Cambodia)
Uses	Irrigation, fisheries, power generation, transportation, industrial and domestic supply
Primary legal agreement	Agreement on the Cooperation for the Sustainable Development of the Mekong River Basin, Chiang Rai, Thailand, 5 April 1995 (Thailand, Laos, Cambodia, and Vietnam)
Institutional arrangements	Mekong River Commission (Thailand, Laos, Cambodia, and Vietnam)

Source: International Union for the Conservation of Nature (IUCN). Gland, Switzerland.

Miami metropolitan area

Everglades

Chemical plant on lower Mississippi

Figure 1.9 **The Diverse South** Poverty has been a historically important problem in the region. Often the creation of jobs was considered more important than the environmental consequences of those jobs. The use of coastal areas for industry has resulted in pollution of coastal waters. The heavy use of the Mississippi River for transportation and industry has caused pollution problems. In addition, the desirable climate in the South has resulted in intense pressure to develop new housing for those who want to move to the region. Unwise development of housing on fragile coastal sites has resulted in damage to buildings by storms and the actions of the oceans. This causes intense debate on land use.

Summary

Artificial political boundaries create difficulties in managing environmental problems because most environmental units, or ecosystems, do not coincide with political boundaries. Therefore, a regional approach to solving environmental problems, one that incorporates natural geographic units, is ideal. Each region of the world has certain environmental issues that are of primary concern because of the mix of population, resource use patterns, and culture.

Environmental problems become issues when there is disagreement. This inevitably leads to a confrontation between groups that have different views on the consequences of an environmental problem. Many social, economic, ethical, and scientific issues shape a person's opinions. The process of environmental decision making must account for all of these issues when seeking an acceptable compromise.

Environmental problems are people problems. They occur because the uses of natural resources, which some people feel are justified, result in a diminished environment for others in the region. Environmental problems are defined by the person who perceives the problem. When perceptions differ, conflict occurs. Environmental decisions inevitably involve economic consequences because someone is receiving value from the resources being used or someone perceives an economic loss because a use has been withdrawn.

- Some argue that economic consequences should not be important when making environmental decisions; others argue that economic considerations can resolve all environmental issues.

- Some argue that regulation is necessary to protect resources; others argue that regulation hinders valuable use of resources.

- Some consider nonhuman organisms as important as humans; others feel that humans have a primary place in nature.

- Some are against change; others recognize that change must occur if negative consequences are to be prevented.

- Some believe that environmental responsibility rests on each decision maker, whether at home, in the workplace, or in the community. Each hour and dollar the consumer spends involves environmental consequences. How do you feel about this statement?

With all these differing opinions, compromise is the only way to resolve the conflicts. The social institution of government must play a role. Economic evaluation is important. Recognition of the validity of opposing points of view is essential. The field of environmental science seeks to find that middle ground.

Issues – Analysis

The Fate of the Gray Wolf

Historically, large predators in the United States were seen as threats to game and livestock, and were trapped, poisoned, and shot. The federal government launched an aggressive program to eliminate predators early in the twentieth century that even targeted animals in the national parks. In 1926, Yellowstone's last wolf was killed, and by that time, wolves were seriously endangered throughout the western United States.

Since then, wolves have migrated from Canada into northwestern Montana, in the area of Glacier National Park. Conservationists have been advocating wolf introduction in Yellowstone and elsewhere since well before the passage of the Endangered Species Act (ESA) in 1973. But the law required the U.S. Fish and Wildlife Service (FWS) to create a recovery plan for the gray wolf, which was officially listed as endangered in 1973. The ensuing controversies were so acrimonious that years passed before any final decisions were made.

Public opinion in most of the United States seems to be strongly in favor of wolf reintroduction, while many local ranchers have remained vehemently opposed. Ranchers have expressed concern over the possibility of livestock loss on land adjacent to wolf habitat.

Thirty years after being added to the ESA list and nine years after being reintroduced to the Northern Rocky Mountains, the gray wolf population has grown to roughly 760 animals. In 2003, the FWS announced that this number, coupled with the identification of 30 breeding pairs in Montana, Idaho, and Wyoming, warranted delisting the wolf as endangered if the three states proposed management plans that would ensure maintenance of wolf populations.

In 2004, the FWS put the delisting of the gray wolf on hold after determining that the Wyoming proposal to manage its gray wolf population did not ensure maintenance of current population numbers. Wyoming had put forth a plan to establish a dual classification whereby some wolves would be protected from humans while others would be listed as predators subject to hunting. The FWS was unconvinced dual classification would sufficiently notify citizens as to which wolves are protected and which are subject to hunting. FWS feared that as a result, many protected wolves would be mistakenly killed, and thus the plan would not provide sufficient management control to ensure that the wolf population will remain above recovery levels.

Political fallout from the FWS determination was quickly apparent. The Wyoming governor was disappointed at the determination and charged election-year politics were to blame. The FWS director asserted that the governor and the state legislature should not have been surprised by the agency's determination, saying that the FWS had never endorsed the concept of dual classification. Environmental activist groups were pleased by the decision, stating that the wolf populations would be better protected.

While the fate of the gray wolf in Wyoming has been decided for the time being, the ultimate decision affecting its future is still to be determined. Subsequent FWS interpretations, political realignments, public pressure, and economic concerns will all factor into the next round of decisions.

- If you were to make the next decision regarding the gray wolf, how would you resolve the issue?
- Is the concept of dual classification workable within the Endangered Species Act and delisting?
- Why do you think that issues such as the gray wolf case are so complicated to resolve?

Key Terms

ecosystem 7
environment 5

environmental science 5
sustainable development 7

wilderness 8

Review Questions

1. Describe why finding solutions to environmental problems is so difficult. Do you think it has always been as complicated?
2. Describe what is meant by an ecosystem approach to environmental problem solving. Is this the right approach?
3. List two key environmental issues for each of the following regions: the wilderness North, the agricultural middle, the forested West, the dry West, the Great Lakes and industrial Northeast, and the South. How are the issues changing?
4. Define environment and ecosystem and provide examples of these terms from your region.
5. Describe how environmental conflicts are resolved.
6. Select a local environmental issue and write a short essay presenting all sides of the question. Is there a solution to this problem?

Critical Thinking Questions

1. Imagine you are a U.S. congressional representative from a western state and a new wilderness area is being proposed for your district. Who might contact you to influence your decision? What course of action would you take? Why?

2. How do you weigh in on the issue of jobs or the environment? What limits do you set on economic growth? Environmental protection?

3. Imagine you are an environmentalist in your area who is interested in local environmental issues. What kinds of issues might these be?

4. Imagine that you lived in the urban East and that you were an advocate of wilderness preservation. What disagreements might you have with residents of the wilderness North or the arid West? How would you justify your interest in wilderness preservation to these residents?

5. You are the superintendent of Yellowstone National Park and want to move to an ecosystem approach to managing the park. How might an ecosystem approach change the current park? How would you present your ideas to surrounding landowners?

6. Look at the issue of global warming from several different disciplinary perspectives—economics, climatology, sociology, political science, agronomy. What might be some questions that each discipline could contribute to our understanding of global warming?

Welcome to McGraw-Hill's Online Learning Center

Location: www.mhhe.com/enger10e

Mc Graw Hill

Interactive Exploration

Check out the website at **www.mhhe.com/enger10e** and click on the cover of this textbook for quizzing, career information, case studies, and hot links for the following topics:

General Environmental Sites

General Ecology Sites

Environmental and Ecological Organization Sites

History of Environmental Studies

Miscellaneous Environmental Resources

Introductory Materials and Governmental Sites

Science as a Process

Introductory Sites

Writing Papers and Study Tips

Glossaries and Dictionaries

Careers in Science

Utility and Organizational Sites

Global Ecology

Environmental Ethics

Objectives

After reading this chapter, you should be able to:

- Differentiate between ethics and morals.
- Define personal ethics.
- Explain the connection between material wealth and resource exploitation.
- Describe how industry exploits resources and consumes energy to produce goods.
- Describe how environmental leaders in industry are promoting more sustainable practices.
- Explain how corporate behavior is determined.
- Describe the influential power that corporations wield because of their size.
- Explain why governmental action was necessary to force all companies to meet environmental standards.
- Describe the factors associated with environmental justice.
- Describe what has been the general attitude of consumers and business toward the environment.
- Explain the relationship between economic growth and environmental degradation.
- List three conflicting attitudes toward nature.

Chapter Outline

PAST PRESENT FUTURE

1975 The world's population was about 4 billion people.	**2004** The world's population was over 6 billion people.
1975 The average person in a developing country ate 24 pounds of meat per year.	**1995** The average person in a developing country ate 51 pounds of meat per year.
1980 Forty-seven percent of children were malnourished worldwide.	**2000** Thirty-three percent of children were malnourished worldwide.
1990 The Oil Pollution Act was passed. Prior to that time, oil spills in the United States averaged 70,000 barrels per year.	**2004** Oil spills have been reduced 94 percent, to 4000 barrels per year.
1994 President Bill Clinton signed an executive order on environmental justice to assist minority and poor communities that suffered environmental damage.	**2003** The U.S. Commission on Civil Rights charged that several federal agencies had failed to implement the order.

Views of Nature

The most beautiful object I have ever seen in a photograph in all my life is the planet Earth seen from the distance of the moon, hanging in space, obviously alive. Although it seems at first glance to be made up of innumerable separate species of living things, on closer examination every one of its things, working parts, including us, is interdependently connected to all the other working parts. It is, to put it one way, the only truly closed ecosystem any of us know about.

—Lewis Thomas

There are no passengers on Spaceship Earth. We are all crew members.

—Buckminster Fuller

One of the marvels of recent technology is that we can see the Earth from the perspective of space, a blue sphere unique among all the planets in our solar system. (See figure 2.1.) Looking at ourselves from space, it becomes obvious, says ecologist William Clark of Harvard University, that only as a global species, "pooling our knowledge, coordinating our actions, and sharing what the planet has to offer—do we have any prospect for managing the planet's transformation along pathways of sustainable development."

Many people see little value in an undeveloped river and feel it is unreasonable to leave it flowing in a natural state. It could be argued that rivers throughout the world have been "controlled" to provide power, irrigation, and navigation at the expense of the natural world. It could also be argued that to not use these resources would be wasteful.

In the U.S. Pacific Northwest, there is a conflict over the value of old-growth forests. Economic interests want to use the forests for timber production and feel that to not do so would cause economic hardship. They argue that the trees are going to die anyway and they might as well be used for the betterment of the human community. Others feel that all the living things that make up the forest have a value we do not yet appreciate. Removing the trees would destroy something that took

Figure 2.1 **The Earth as Seen from Space** Political, geographical, and nationalistic differences among humans do not seem so important from this perspective. In reality, we all share the same "home."

hundreds of years to develop and may never be replaced.

Interactions between people and their environment are as old as human civilization. The problem of managing those interactions, however, has been transformed today by unprecedented increases in the rate, scale, and complexity of the interactions. At one time, pollution was viewed as a local, temporary event. Today, pollution may involve several countries—as with the concern over acid deposition in Europe and in North America—and will affect multiple generations. The debates over chemical and radioactive waste disposal are examples of the increasingly international nature of pollution. For example, many European countries are concerned about the transportation of radioactive and toxic wastes across their

borders. What were once straightforward confrontations between ecological preservation and economic growth now involve multiple linkages that blur the distinction between right and wrong. For example, the enhanced greenhouse effect is thought to result from energy consumption, agricultural practices, and climatic change.

Many people believe that we have entered an era characterized by global change that stems from the interdependence between human development and the environment. They argue that self-conscious, intelligent management of the earth is one of the greatest challenges facing humanity as we begin the twenty-first century. To meet this challenge, they believe, a new environmental ethic must evolve.

Environmental Ethics

Ethics is one branch of philosophy. Ethics seeks to define fundamentally what is right and what is wrong, regardless of cultural differences. For example, most cultures have a reverence for life and hold that all humans have a right to live. It is considered unethical to deprive an individual of life.

Morals differ somewhat from ethics because morals reflect the predominant feelings of a culture about ethical issues. For example, in almost all cultures, it is certainly unethical to kill someone; however, when a country declares war, most of its people accept the necessity of killing the enemy. Therefore, it is a moral thing to do even though ethics says that killing is wrong. No nation has ever declared an immoral war.

Environmental issues require a consideration of ethics and morals. For example, because there is enough food in the world to feed everyone adequately, it is unethical to allow some people to starve while others have more than enough. However, the predominant mood of those in the developed world is one of indifference. They don't feel morally bound to share what they have with others. In reality, this indifference says that it is permissible to allow people to starve. This moral stand is not consistent with a purely ethical one.

As we can see, ethics and morals are not always the same; thus, it is often difficult to clearly define what is right and what is wrong. Some individuals view the world's energy situation as serious and have reduced their consumption. Others do not believe there is a problem and so have not modified their energy use. Still others do not care what the situation is. They will use energy as long as it is available.

Other issues are population and pollution. Is it ethical to have more than two children when the world faces overpopulation? Should an industry persuade legislators to vote no on a particular bill because it might reduce profits, even though its passage would improve the environment? The stand we take on such issues often depends on our position. An industrial leader, for example, would probably not look upon pollution as negatively as someone who participates in outdoor activities. In fact, many business leaders view the behavior of active preservationists as immoral because it restricts growth and, in some cases, causes unemployment.

Most ethical questions are very complex. Ethical issues dealing with the environment are no different. It is important to explore environmental issues from several points of view before taking a stand.

When we take an ethical stand, we become open to attack from those who disagree with our stand. Often, individuals are portrayed as villains for pursuing a course of action they consider righteous.

Environmental ethics is a topic of applied ethics that examines the moral basis of environmental responsibility. In these environmentally conscious times, most people agree that we need to be environmentally responsible. Toxic waste contaminates groundwater, oil spills destroy shore lines, and fossil fuels produce carbon dioxide, thus adding to global warming. The goal of environmental ethics, then, is not to convince us that we should be concerned about the environment—many already are. Instead, environmental ethics focuses on the moral foundation of environmental responsibility and how far this responsibility extends. There are three primary theories of moral responsibility regarding the environment. Although each supports environmental responsibility, their approaches are different.

The first of these theories is **anthropocentric,** or human-centered. Environmental anthropocentrism is the view that all environmental responsibility is derived from human interests alone. The assumption here is that only human beings are morally significant organisms and have a direct moral standing. Since the environment is crucial to human well-being and human survival, we have a duty toward the environment, that is, a duty that is derived from human interests. This involves the duty to ensure that the Earth remains environmentally hospitable for supporting human life and that its beauty and resources are preserved so that human life on Earth continues to be pleasant.

Some have argued that our environmental duties are derived both from the immediate benefit that people receive from the environment and from the benefit that future generations of people will receive. But critics have maintained that since future generations of people do not yet exist, then, strictly speaking, they cannot have rights any more than a dead person can have rights. Nevertheless, both parties to this dispute acknowledge that environmental concern derives solely from human interests.

A second theory of moral responsibility to the environment is **biocentric.** According to the broadest form of the life-centered theory, all forms of life have an inherent right to exist. Some biocentric thinkers give species a hierarchy of values. Some, for example, believe we have greater responsibility to protect animal species than plant species. Others determine the rights of various species depending on the harm they do to humans. For example, they see nothing wrong in killing pest species such as rats or mosquitoes. Some go further and believe that each individual organism, not just each species, has a basic right to survive. Individuals who support the animal rights movement tend to place more value on the individuals of animal species than on plant species. Trying to decide what types of species or individuals should be protected from early extinction or death resulting from human activities is an ethical dilemma. It is hard to know where to draw the line and be ethically consistent.

The third approach to environmental responsibility, called **ecocentrism,** maintains that the environment deserves direct moral consideration and not one that is merely derived from human (and animal) interests. In ecocentrism, it is suggested that the environment has direct rights, that it qualifies for moral personhood, that it is deserving of a direct duty, and that it has inherent worth. The environment, by itself, is considered to be on a moral par with humans.

The position of ecocentrism is the view advocated by the ecologist and writer Aldo Leopold in his book *A Sand County Almanac* (1949). In response to the relentless destruction of the landscape, *A Sand County Almanac* redefined the

Preservation   Development   Conservation

Figure 2.2 **Environmental Attitudes** Individuals envision the same resources used differently.

relationship between human-kind and the Earth. Leopold devoted an entire chapter of his book to "The Land Ethic."

All ethics so far evolved rest upon a single premise: that the individual is a member of a community of inter-dependent parts. . . . The land ethic simply enlarges the boundaries of the community to include soils, waters, plants, and animals, or collectively the land . . . a land ethic changes the role of *Homo sapiens* from conqueror of the land-community to plain member and citizen of it. It implies respect for his fellow-members, and also respect for the community as such.

It is inconceivable to me that an ethical relation to land can exist without love, respect, and admiration for land, and a high regard for its value. By value, I of course mean something far broader than mere economic value; I mean value in the philosophical sense.

What Leopold put forth in "The Land Ethic" was viewed by many as a radical shift in how humans perceive themselves in relation to the environment. Originally we saw ourselves as conquerors of the land. Now, according to Leopold, we need to see ourselves as members of a community that also includes the land and the water.

Leopold also wrote that "a thing is right when it tends to preserve the integrity, stability, and beauty of the biotic community. It is wrong when it tends otherwise. . . . We abuse land because we regard it as a commodity belonging to us.

When we see land as a community to which we belong, we may begin to use it with love and respect."

As traditional political and nationalis-tic boundaries begin to fade or shift glob-ally, new variations of environmental thought and ethics are also evolving. Some of the new thoughts on environmen-tal ethics are founded on an awareness that humanity is part of nature and that na-ture's many parts are interdependent. In any natural community, the well-being of the individual and of each species is tied to the well-being of the whole. In a world increasingly without environmental bor-ders, nations, like individuals, should have a fundamental ethical responsibility to respect nature and to care for the Earth, protecting its life-support systems, bio-diversity, and beauty, and caring for the needs of other countries and future generations.

Environmental ethicists argue that to consider environmental protection as a "right" of the planet is a natural extension of the concept of human rights. Many also argue that an environmental ethic consid-ers one's actions toward the environment as a matter of right and wrong, rather than one of self-interest.

Environmental Attitudes

There are many different attitudes about the environment, most of which fall under one of three headings: (*a*) the develop-ment ethic, (*b*) the preservation ethic, and (*c*) the conservation ethic. Each of these

ethical positions has its own code of con-duct against which ecological morality may be measured. (See figure 2.2.)

The **development ethic** is based on individualism or egocentrism. It assumes that the human race is and should be the master of nature and that the Earth and its resources exist for our benefit and pleasure. This view is reinforced by the work ethic, which dictates that humans should be busy creating continual change and that things that are bigger, better, and faster represent "progress," which itself is good. This philosophy is strengthened by the idea that "if it can be done, it should be done" or that our actions and energies are best harnessed in creative work.

Examples of the development ethic abound. The notion that bigger is better is certainly not new to us, nor is the belief that if something can be done or built, it should be. The dream of upward mobility is embodied in this ethic. In some circles, questioning growth is considered almost unpatriotic. In the development ethic, nature has only in-strumental value; that is, the environment has value only insofar as human beings economically utilize it. Only in the past 50 to 100 years have the by-products and waste associated with development been considered.

The **preservation ethic** considers na-ture special in itself. Nature, it is argued, has intrinsic value or inherent worth apart from human appropriation. Preservation-ists have diverse reasons for wanting to preserve nature. Some hold an almost reli-gious belief regarding nature. They have a reverence for life and respect the right of

Naturalist Philosophers

The philosophy behind the environmental movement had its roots in the nineteenth century. Among many notable conservationist philosophers, several stand out: Ralph Waldo Emerson, Henry David Thoreau, John Muir, Aldo Leopold, and Rachel Carson.

In Emerson's first essay, *Nature,* published in 1836, he claimed that "behind nature, throughout nature, spirit is present." Emerson was an early critic of rampant economic development, and he sought to correct what he considered to be the social and spiritual errors of his time. In his *Journals,* published in 1840, Emerson stated that "a question which well deserves examination now is the Dangers of Commerce. This invasion of Nature by Trade with its Money, its Credit, its Steam, its Railroads, threatens to upset the balance of Man and Nature."

Henry David Thoreau was a naturalist who held beliefs similar to Emerson's. Thoreau's bias fell on the side of "truth in nature and wilderness over the deceits of urban civilization." The countryside around Concord, Massachusetts, fascinated and exhilarated him as much as the commercialism of the city depressed him. It was near Concord that Thoreau wrote his classic, *Walden,* which describes a year in which he lived in the country to have direct contact with nature's "essential facts of Life." In his later writings and journals, Thoreau summarized his feelings toward nature with prophetic vision:

> But most men, it seems to me, do not care for Nature and would sell their share in all her beauty, as long as they may live, for a stated sum—many for a glass of rum. Thank God, man cannot as yet fly, and lay waste the sky as well as the earth! We are safe on that side for the present. It is for the very reason that some do not care for these things that we need to continue to protect all from the vandalism of a few. (1861)

John Muir combined the intellectual ponderings of a philosopher with the hard-core, pragmatic characteristics of a leader. Muir believed that "wilderness mirrors divinity, nourishes humanity, and vivifies the spirit." Muir tried to convince people to leave the cities for a while to enjoy the wilderness. However, he felt that the wilderness was threatened. In the 1876 article entitled, "God's First Temples: How Shall We Preserve Our Forests?" published in the *Sacramento Record Union,* Muir argued that only government control could save California's finest sequoia groves from the "ravages of fools." In the early 1890s, Muir organized the Sierra Club to "explore, enjoy, and render accessible the mountain regions of the Pacific Coast" and to enlist the support of the government in preserving these areas. His actions in the West convinced the federal government to restrict development in the Yosemite Valley, which preserved its beauty for generations to come.

Aldo Leopold was another thinker as well as a doer in the early conservation field. As a philosopher, Leopold summed up his feelings in *A Sand County Almanac:*

> Wilderness is the raw material out of which man has hammered the artifact called civilization. No living man will see again the long grass prairie, where a sea of prairie flowers lapped at the stirrups of the pioneer. No living man will see again the virgin pineries of the Lake States, or the flatwoods of the coastal plain, or the giant hardwoods.

Ralph Waldo Emerson

Henry David Thoreau

John Muir

Aldo Leopold

Rachel Carson

Leopold founded the field of game management. In the 1920s, while serving in the Forest Service, he worked for the development of a wilderness policy and pioneered his concepts of game management. He wrote extensively in the *Bulletin* of the American Game Association and stated that the amount of space and the type of forage of a wildlife habitat determine the number of animals that can be supported in an area. Furthermore, he said that regulated hunting can maintain a proper balance of wildlife.

While most people talk about what's wrong with the way things are, few actually go ahead and change it. Rachel Carson ranks among those few. A distinguished naturalist and best-selling nature writer, Rachel Carson published in the *New Yorker* in 1960 a series of articles that generated widespread discussion about pesticides. In 1962, she published *Silent Spring,* which dramatized the potential dangers of pesticides to food, wildlife, and humans and eventually led to changes in pesticide use in the United States.

Although some technical details of her book have been shown to be in error by later research, her basic thesis that pesticides can contaminate and cause widespread damage to the ecosystem has been established. Unfortunately, Carson's early death from cancer came before her book was recognized as one of the most important events in the history of environmental awareness and action in the twentieth century.

all creatures to live, no matter what the social and economic costs.

During the nineteenth century, preservationists forthrightly gave ethical and spiritual reasons for protecting the natural world. John Muir condemned the "temple destroyers, devotees of ravaging commercialism" who, "instead of lifting their eyes to the God of the mountains, lift them to the Almighty dollar." This was not a call for better cost-benefit analysis: Muir described nature not as a commodity but as a companion. Nature is sacred, Muir held, whether or not resources are scarce.

Philosophers such as Emerson and Thoreau thought of nature as full of divinity. Walt Whitman celebrated a leaf of grass as no less than the "journeywork of the stars." "After you have exhausted what there is in business, politics, conviviality, love, and so on and found that none of these finally satisfy, or permanently wear—what remains? Nature remains," Whitman wrote. These philosophers thought of nature as a refuge from economic activity, not as a resource for it.

Some preservationists' interest in nature is primarily aesthetic or recreational. They believe that nature is beautiful and refreshing and should be available for picnics, hiking, camping, fishing, or just peace and quiet.

In addition to the religious and recreational preservationists, there are preservationists whose reasons are essentially scientific. They argue that the human species depends on and has much to learn from nature. Rare and endangered species and ecosystems, as well as the more common ones, must be preserved because of their known or assumed long-range, practical utility. In this view, natural diversity, variety, complexity, and wilderness are thought to be superior to humanized uniformity, simplicity, and domesticity. Scientific preservationists want to lock up not all the land but only what they consider important to future generations.

The third environmental ethic is referred to as the **conservation** or **management ethic.** It is related to the scientific preservationist view but extends the rational consideration to the entire Earth and for all time. It recognizes the desirability of decent living standards, but it works toward a balance of resource use and re-

source availability. The conservation ethic stresses a balance between total development and absolute preservation. It stresses that rapid and uncontrolled growth in population and economics is self-defeating in the long run. The goal of the conservation ethic is one people living together in one world, indefinitely.

Societal Environmental Ethics

Society is composed of a great variety of people with diverse viewpoints. This variety can be distilled into a set of ideas that reflect the prevailing attitudes of society. The collective attitudes can be analyzed from an ethical point of view. Western, developed societies have long acted as if the Earth has unlimited reserves of natural resources, an unlimited ability to assimilate wastes, and a limitless ability to accommodate unchecked growth.

The economic direction and rationale of developed nations have been that of continual growth. Unfortunately, this growth has not always been carefully planned or even desired. This "growth mania" has resulted in the use of our nonrenewable resources for comfortable homes, well-equipped hospitals, convenient transportation, fast-food outlets, VCRs, home computers, and battery-operated toys, among other things. In economic statistics, such "growth" measures out as "productivity." But the question arises, "What is enough?" Poor societies have too little, but rich societies never say, "Halt! We have enough." The Indian philosopher and statesman Mahatma Ghandi said, "The Earth provides enough to satisfy every person's need, but not every person's greed."

Growth, expansion, and domination remain the central sociocultural objectives of most advanced societies. **Economic growth** and **resource exploitation** are attitudes shared by developing societies. We continue to consume natural resources as if the supplies were never ending. All of this is reflected in our increasingly unstable relationship with the environment, which grows out of our tendency to take

from the "common good" without regard for the future.

This attitude is deeply embedded in the fabric of our society. Since the first settlers arrived in North America, nature has been considered an enemy. Frequently, the colonists expressed their relation to the wilderness in military terms. They viewed nature as an enemy to be "conquered," "subdued," or "vanquished" by a pioneer "army." Any qualms the pioneers may have felt about invading and exploiting the wilderness were justified by religious beliefs. They were driven by what they perceived to be a "moral imperative." This attitude toward nature is still popular today. Many view wilderness solely as underdeveloped land and see value in land only if it is farmed, built upon, or in some way developed. The notion that land and wilderness should be preserved is incomprehensible to some. The thought of purposely opting to not develop a resource is considered almost a sin.

Corporate Environmental Ethics

Many tasks of industry, such as procuring raw materials, manufacturing and marketing, and disposing of wastes, are in large part responsible for pollution. This is not because any industry or company has adopted pollution as a corporate policy, but because industries must consume energy and resources to make their products, and they must make a profit to exist. However, they have choices between using sustainable or nonsustainable manufacturing processes. When raw materials are processed, some waste (useless material) is inevitable. It is usually not possible to completely control the dispersal of all by-products of a manufacturing process. Also, some of the waste material may simply be useless.

For example, the food-service industry uses energy to prepare meals. Much of this energy is lost as waste heat. Smoke and odors are released into the atmosphere, and spoiled food items must be discarded.

What's in Your Backyard?

When asked, most Americans would say that they don't live in a polluted environment nor do they want to. This concept is called, "not in my backyard," or NIMBY. It comes from a long history of communities protesting the construction of hazardous waste facilities, landfills, nuclear power plants, sewage treatment facilities, or any other public or private facility that is viewed as unsightly or harmful. Unfortunately, many people don't know what sources of pollution are in their backyard. On the National Report Card on Environmental Knowledge, Attitudes, and Behavior, only 23 percent of Americans knew that water runoff from yards, streets, and farms was the most common source of water pollution. In contrast, 69 percent knew that the leading source of carbon monoxide pollution in the air was automobiles.

Where do you stand? Do you know what the level of pollution is in your backyard? If you knew, would you take action? How close do you live to a nuclear power plant, Superfund site, sewage treatment facility, or landfill? Find out the sources and levels of pollution in your neighborhood by typing your home zip code, county, or city into the Environmental Protection Agency's Envirofacts website (http://www.epa.gov/enviro/). Now compare these results to the Environmental Defense Fund's Scorecard website (http://www.scorecard.org/). How did you compare to the rest of the country? What are the most common sources of air and water pollution in your hometown? What agencies are working to protect the environment? Who are the major polluters? Were you surprised? Did the two websites agree or disagree? Share this information with three friends or family members and get their comments on what's in your backyard and what you can do about it.

The cost of controlling waste can be very important in determining a company's profit margin. Some innovative companies have made tremendous strides in reducing waste during manufacturing. For example, Ray Anderson of Interface Incorporated has led the way in greening the carpet industry by reducing the amount of waste produced by 75 percent. This measure has saved the company $209 million since 1995. **Corporations** are legal entities designed to operate at a profit, which is not in itself harmful. The corporation has no ethics, but the people who make up the corporation are faced with ethical decisions. Ethics are involved when a corporation cuts corners in production quality or waste disposal to maximize profit. The cheaper it is to produce an item, the greater the possible profit. It is cheaper in the short run to dump wastes into a river than to install a wastewater treatment facility, and it is cheaper in the short run to release wastes into the air than it is to trap them in filters. Many people consider such pollution unethical and immoral, but some corporations think of it as just one of the factors that determines **profitability.** (See figure 2.3.) Because stockholders expect an immediate return on their investment, corporations often make decisions based on short-term profitability rather than long-term benefit to society.

The amount of profit a corporation realizes determines how much it can expand. To expand continually, a corporation increases the demand for its products through advertising. The more it expands, the more power it attains. The more power it has, the greater its influence over decision makers who can create conditions favorable to its expansion plans. The process becomes a seemingly never-ending spiral.

Nations of the world must confront the problem of corporate irresponsibility toward the environment. In business, incorporation allows for the organization and concentration of wealth and power far surpassing that of individuals or partnerships. Some of the most important decisions affecting our environment are made not by governments or the public but by executives who wield massive corporate power. Often, these executives make only minimal concessions to the public interest, while they make every effort to maximize profits.

Business decisions and technological developments have increased the exploitation of natural resources. In addition, many political and legal institutions have generally supported the development of private enterprise. They have also defended and promoted private property rights rather than social and environmental concerns. Businesses and individuals typically use loopholes, political pressure, and the time-consuming nature of legal action to circumvent or delay compliance with social or environmental regulations.

Is industry becoming more environmentally concerned? Corporations have certainly made more frequent references to worldwide environmental issues over the past several years. Is such concern only rhetoric and social marketing, or is it the beginning of a new corporate ethic?

Reaction by the business community to the 1989 *Exxon Valdez* oil spill in Alaska is a good example of different ethical approaches to the same ecological disaster. (See figure 2.4.) The Oil Pollution Act (OPA) was passed in 1990 to reduce the environmental impact of future oil spills and has resulted in a 94 percent reduction in spills. One of the new regulations in OPA was that all large tankers must have double hulls or be phased out of service by 2010. To get around the law, however, many oil carriers shifted their oil transport operations to lightly regulated oil barges. This reduction in oil spill safety led to several barge oil spills, including one in January 1996 in Rhode Island's Moonstone Bay and another in March 1997 in Texas's Galveston Bay.

In stark contrast, another group of environmentalists, investors, and companies motivated by the *Valdez* spill and other environmental incidents formed the Coalition for Environmentally Responsible Economics (CERES) group in 1989 and created a set of 10 environmental standards by which their business practices could be measured called the CERES Principles (they were first called the Valdez Principles). CERES companies pledge to voluntarily go beyond the requirements set by the law to strive for environmental excellence through business practices that: (1) protect the biosphere, (2) sustainably use natural resources, (3) reduce and dispose of waste safely, (4) conserve energy, (5) minimize environmental risks through safe technologies, (6) reduce the use, manufacture, and sale of products and services that cause environmental damage, (7) restore environmental damage, (8) inform the public of any health, safety, or environmental conditions, (9) consider environmental policy in management decisions, and (10) report the results of an annual environmental audit to the public. Today, over 70 companies have publicly endorsed the CERES Principles, including 13 Fortune 500 firms. In addition, socially invested assets total $300 billion, and worldwide, CERES firms have won environmental awards from many organizations in recognition of their approach.

The CERES Principles are looked upon as a guide for corporate environmentalism. The goal, some argue, should be to make compliance with the CERES Principles a prerequisite for doing business.

In 1997, the Global Reporting Initiative (GRI) was established. Convened by CERES in partnership with the United National Environment Programme, the GRI incorporates the active participation of corporations, nongovernmental organizations, accountancy organizations, business associations, and other stakeholders from around the world. The mission of the GRI was to develop globally applicable guidelines for reporting on economic, environmental, and social performance, initially for corporations and eventually for any business, governmental, or nongovernmental organization.

The GRI's *Sustainability Reporting Guidelines* were released in draft form in 1999. The GRI guidelines represent the first global framework for comprehensive

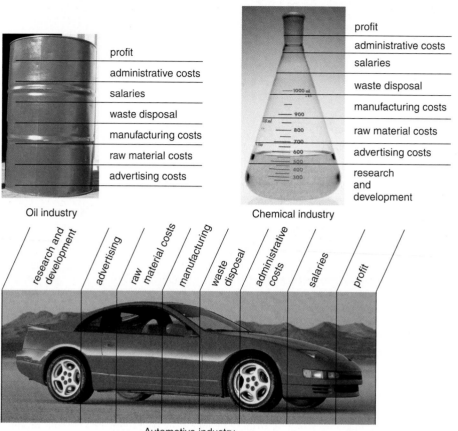

Figure 2.3 **Corporate Decision Making** Corporations must make a profit. When they look at pollution control, they view its cost like any other cost: any reductions in cost increase profits.

a.   b.

Figure 2.4 **Oil Spill Response** The 1989 oil spill in Alaska led to the development of the CERES Principles. (a) These waterfowl were victims of the spill. (b) The *Exxon Valdez*.

sustainability reporting, encompassing the "triple bottom line" of economic, environmental, and social issues.

Improved disclosure of sustainability information is an essential ingredient in the mix of approaches needed to meet the governance challenges in the globalizing economy. Today, at least 2000 companies around the world voluntarily report information on their economic, environmental, and social policies, practices, and performance. Yet, this information is generally inconsistent, incomplete, and unverified. Measurement and reporting practices vary widely according to industry, location, and regulatory requirements. The GRI's *Sustainability Reporting Guidelines* are designed to address some of these challenges. Currently, 392 organizations in 33 countries follow GRI's guidelines.

26   PART ONE   Interrelatedness

GLOBAL PERSPECTIVE

Chico Mendes and Extractive Reserves

Francisco "Chico" Alves Mendes Filho was born in 1944 in the western Brazilian Amazon. A second-generation rubber tapper, Chico was active in the rubber tappers' union for over 15 years. He and many other peasants made a living by extracting latex from rubber trees and selling it. Rubber tappers also collect and sell other natural products of the forest, such as Brazil nuts, fruits, and native medicines. Mendes was interested in preserving the portion of the Brazilian rainforest that provided their livelihood, and he supported the concept of "extractive reserves."

Extractive reserves involves setting aside land for rubber tappers, who would continue their traditional lifestyles and use the rainforest in its natural state for generations to come. This idea put Mendes in conflict with powerful people interested in clearing the rainforest to raise cattle. Most cattle-ranching operations show short-term economic gains but ultimately become uneconomical when land fertility declines from overgrazing.

In 1987, Mendes received two international environmental awards for his efforts in establishing extractive reserves. One was the Global 500 award from the United Nations Environment Programme; the other was Ted Turner's Better World Society Environment Award.

In 1988, Chico led the Xapuri Rural Workers Union in a winning effort to stop cattle rancher Darli Alves da Silva from deforesting an area the rubber tappers wanted to make into a reserve. On December 22,

1988, as he walked from his home, Chico Mendes was shot by members of a vigilante group who supported local ranchers. In 1990, Darli Alves da Silva and his son were convicted of the murder. Alves da Silva was sentenced to 19 years in prison but escaped in 1993, leading to a three-year exhaustive manhunt. Both men are now out of prison on a work release program, and the government has reopened the murder investigation because it is likely that there are additional suspects.

Before his death, Mendes had said, "I want to live to defend the Amazon." His life and death appear to have made a difference in the way the Brazilian rainforest is being used. In 1990, the Chico Mendes Extractive Reserve was established covering about 6 percent of the state of Acre in northwest Brazil.

In a recent poll in Acre, 75 percent of the people said they wanted "development without destruction of the forest," which is a substantial change from the 1970s emphasis on roads and ranching.

Today, colleagues of Chico Mendes hold political office and are in a position to shape the government's environmental laws in favor of sustainable forest management. The challenges they face are massive unemployment, low prices for rubber on the world market, illegal logging, and expanding agricultural interests. Chico Mendes is considered an environmental hero, and a number of agencies around the world have created a Chico Mendes Award to honor environmental activists like him.

Practicing an environmental ethic should not interfere with corporate and other social responsibilities or obligations, though this is not always the case. It must be integrated into overall systems of belief and coordinated with economic systems. Environmental advocates, in turn, need to consider others' objectives just as they demand that others consider environmental consequences in decision making. It makes little sense to preserve the environment if that objective produces national economic collapse. Nor does it make sense to maintain stable industrial productivity at the cost of breathable air, drinkable water, wildlife species, parks, and wilderness. But to maintain profitability, influence, and freedom, businesses must be sensitive to their impact on current and future citizens, not just in terms of the price and quality of the goods they produce but also in terms of public approval of their social and political influence. A

2000 Harris Poll, for example, found that 65 percent of Americans wanted increased government spending, either national or state, to address environmental problems, even if they had to pay higher taxes. In another poll, eight of 10 Americans said they would be willing to pay extra for a product packaged with recyclable materials.

In the middle 1990s, a concept emerged called **industrial ecology** that reflects the link between the economy and the environment. This concept argues that good ecology is also good economics and that alternatives exist for corporations to provide goods and services in ways that do not destroy the environment.

One of the most important elements of industrial ecology is that, as in biological systems, it accounts for waste. Dictionaries define *waste* as useless or worthless material. In nature, however, nothing is eternally discarded; in various ways, all materials are reused. In our industrial

world, discarding materials taken from the Earth at great cost is also generally unwise. Perhaps materials and products that are no longer in use should be termed *residues* rather than *wastes;* wastes are merely residues that our economy has not yet learned to use efficiently. A simpler way of saying this is to view a pollutant as a resource out of place. Such a statement forces us to view pollution and waste in a new way.

Environmental Justice

In 1982, a proposed PCBs landfill in Warren County, North Carolina, was protested by residents of an African-American community that was located close to the planned facility. Some observers asserted that local officials had been practicing a form of environmental racism, citing studies showing that hazardous waste treatment, storage, and disposal facilities were disproportionately located in areas occupied predominantly by minorities. Others argued that poor communities were often poorly represented in political circles and, therefore, were the easiest places in which to locate objectionable facilities. Eventually, the U.S. Environmental Protection Agency (EPA) granted the permits for this waste disposal site, which brought nationwide protests.

In response to protests stemming from this and similar incidents, President Bill Clinton issued an executive order (EO 12,898) in 1994 mandating that all federal agencies make environmental justice part of their mission. In 1998, the EPA defined **environmental justice** as fair treatment, meaning that "no group of people, including racial, ethnic, or socioeconomic groups, should bear a disproportionate share of the negative environmental consequences resulting from industrial, municipal, and commercial operations or the execution of federal, state, local, and tribal programs and policies." According to the EPA definition, deliberate discrimination need not be involved. Any action that affects protected groups disproportionately is in violation of EPA's rules. The difficulty arises in defining what to measure and what should be the standard of comparison.

As a first step in evaluating whether a group is unduly disadvantaged, a policy maker must consider who is affected. Most ethnic data relate to census tracts, zip codes, city boundaries, and counties. If the facility is to be located in a wealthy county but near the county border close to a poor community, how does the policy maker draw the line? Should prevailing winds be considered? Many industrial sites are located where land is cheap; people of low income may choose to live in those areas to minimize living expenses. How are these decisions weighed?

Another difficulty arises in determining whether and how particular groups will be disadvantaged. Landfills, chemical plants, and other industrial works bring benefits to some, although they may harm others. They create jobs, change land values, and generate revenues that are spent in the community. How do officials compare the benefits with the losses? How should potential health risks from a facility be compared with the overall health benefits that jobs and higher incomes bring?

The environmental justice movement is also occasionally referred to as "environmental equity"—which the EPA defines as equal protection from environmental hazards of all individuals, groups, or communities regardless of race, ethnicity, or economic status. Although environmental justice has many facets (e.g., legal, economic, and political), it may be approached appropriately in a variety of ways by the public and private sectors. In addition, the health community should naturally focus on the health aspect of environmental justice.

At its core, environmental justice means fairness. It speaks to the impartiality that should guide the application of laws designed to protect the health of human beings and the productivity of ecological systems on which all human activity, economic activity included, depends. It is emerging as an issue because studies show that certain groups of North Americans and citizens of other nations may suffer disproportionately from the effects of pollution.

Governments have established numerous laws, mandates, and directives to eliminate discrimination in housing, education, and employment, but few attempts have been made to address discriminatory environmental practices. In the United States, people of color have borne a disproportionate burden in the location of municipal landfills, incinerators, and hazardous-waste treatment, storage, and disposal facilities.

Hazardous waste sites and incinerators are not randomly located. While waste generation is correlated directly with per capita income, few toxic waste sites are located in affluent suburbs. Waste facilities are often located in communities that have high percentages of poor, elderly, young, and minority residents. Often such facilities are deliberately sited in these communities because they are seen as providing the least resistance and cheaper land costs.

Questions of environmental justice extend beyond the location of toxic waste sites. Exposure to harmful pesticides and other toxic agricultural substances is a major health issue among hired farm workers, the majority of whom are people of color. There is also concern that because some Native American communities consume much greater amounts of fish from certain areas such as the Great Lakes than does the general population, they are at greater risk for dietary exposure to toxic chemicals.

Historically, the environmental movement has been a concern of middle-class whites, but there is a growing level of activism by people of color. Minority participation has broadened the debate to include many issues that were being ignored. It has also forced a dialogue about race, class, discrimination, and equity. Minorities have pushed the plight of their communities to the forefront. They have also brought a new perspective to the environmental movement and will be a part of any future environmental agenda.

Environmental Justice Highlights

1982

National attention focused on a series of protests in the low-income, minority community of Warren County, North Carolina, over a landfill filled with PCB-contaminated soil from 14 other counties throughout the state. Over 500 people were arrested, including a congressman. The protest prompted the U.S. General Accounting Office (GAO) to initiate a study of hazardous waste landfills in eight southern states. The GAO study concluded that three out of every four landfills were located in minority communities.

1987

The United Church of Christ Commission on Racial Justice published a report showing that race was the most significant factor in the siting of toxic waste facilities throughout the nation. More than 60 percent of African-American and Hispanic people lived in a neighborhood near a hazardous waste site. A similar study by the *National Law Journal* found that polluters in minority communities paid 54 percent lower fines and the EPA took 20 percent longer to place toxic sites on the national priority action list.

1992

EPA created the Office of Environmental Justice to examine the issue of environmental justice in all agency policies and programs. EPA reported that low-income and minority communities were more likely to be exposed to lead, contaminated fish, air pollution, hazardous waste, and agricultural pesticides.

1993

Environmental justice became one of EPA's top priorities, and an independent advisory group, the National Environmental Justice Advisory Council, was formed from industry experts, activists, and officials.

1994

President Clinton signed an environmental justice executive order requiring all federal agencies to begin taking the issue into account. "Each Federal agency shall make achieving environmental justice part of its mission by identifying and addressing, as appropriate, disproportionately high and adverse human health or environmental effects of its programs, policies, and activities on minority populations and low-income populations." In a separate memo, the order called for enforcement through three federal statutes: the Civil Rights Act of 1964, the National Environmental Policy Act of 1969, and the Clean Air Act of 1972.

2000

EPA issued a memo outlining additional federal statutes in which environmental justice issues might be involved, including: the Resource Conservation and Recovery Act, the Clean Water Act, the Safe Drinking Water Act, and the Marine Protection, Research and Sanctuaries Act.

2001

A special task force was created within EPA to provide additional resources to investigate the backlog of environmental justice complaints.

2003

The U.S. Commission on Civil Rights, an independent group charged with monitoring federal civil rights enforcement, issued to Congress its report titled "Not in My Backyard," which found that several federal agencies (EPA, DOT, HUD, and DOI) have failed to fully implement the 1994 Environmental Justice Executive Order.

Individual Environmental Ethics

The environmental movement has effectively influenced public opinion and moved the business community toward an environmental ethic. The result of this changing view of business's responsibilities will complicate business decision making through this century. More complex environmental and safety demands by the public and a broadening of horizons on the part of business will be a dominant theme of corporate life during the next decade. As human populations and economic activity continue to grow, we are facing a number of environmental problems that threaten not only human health and the productivity of ecosystems but in some cases the very habitability of the globe.

If we are to respond to these problems successfully, our future environmental ethic must express itself in broader and more fundamental ways. We have to recognize that each of us is individually responsible for the quality of the environment we live in and that our personal actions affect environmental quality, for better or worse. The recognition of individual responsibility must then lead to changes in individual behavior. In other words, our environmental ethic must begin to express itself not only in national laws but also in subtle but profound changes in the ways we all live our daily lives.

Various public opinion polls conducted over the past decade have indicated that Americans think environmental problems can often be given a quick tech-

nological fix. The Roper polling organization has stated that, "They believe that cars, not drivers, pollute, so business should invent pollution-free autos. Coal utilities, not electricity consumers, pollute, so less environmentally dangerous generation methods should be found." It appears that many individuals want the environment cleaned up, but they do not want to make major lifestyle changes to make that happen.

Decisions and actions by individuals faced with ethical choices collectively determine the hopes and quality of life for everyone. As ecological knowledge and awareness begin to catch up with good intentions, people in all walks of life will need to live by an environmental ethic.

Do We Consume Too Much?

In 1994, when delegates from around the world gathered in Cairo for the International Conference on Population and Development, representatives from developing countries protested that a baby born in the United States will consume during its lifetime 20 times as much of the world's resources as an African or Indian baby. The problem for the world's environment, they argued, is overconsumption in the Northern Hemisphere, not just overpopulation in the Southern Hemisphere. Do we in the Northern Hemisphere consume too much?

North Americans, only 5 percent of the world's population, consume one-fourth of the world's oil. They use more water and own more cars than anybody else. They waste more food than most people in sub-Saharan Africa eat.

As the rest of the world becomes more like America (China now consumes almost half as much meat per capita as Americans do), will something vital—water, oil, food—simply run out?

Ever since he wrote a book called *The Population Bomb* in 1968, ecologist Paul Ehrlich has argued that the American lifestyle is driving the global ecosystem to the brink of collapse. But others, including

GLOBAL PERSPECTIVE

International Trade in Endangered Species

Among the countries of the world, plants and animals are bought and sold both legally and illegally. The *legal* trade in more than 350 million plants and animals has grown from earning $3 billion in the 1980s to more than $10 billion today. Examples include live animals for pets or medical research, and animal and plant products that are used in fashion or for food. In 1995, over 100 countries exported medicinal plants valued at $880 million. *Illegal* trade, by contrast, is estimated to earn over $5 billion to $8 billion a year, second only to smuggling in drugs or arms. Often rare or endangered species are targeted because their sales yield a premium price. For example, a single Asian bear gall bladder, which is used in traditional Asian medicine, can sell for up to $15,000.

Illegal trade directly affects the populations of more than 37,000 animal and plant species and represents a severe threat to their survival. The international groups that conduct this trading make a fortune by buying plants and animal products at extremely low prices at the source (often a developing country) and selling them for outrageous sums where demand is high. For example, it is estimated that each year 1.5 million caiman skins leave Brazil, Bolivia, and Paraguay, which results in a loss of millions of dollars for these countries, as well as a major threat to biodiversity. Most species of caimans (which are relatives of crocodiles and alligators) are either threatened or endangered.

The most important markets for both legal and illegal trade are found in the wealthy countries; the chief buyers for these wildlife products are in Japan, the United States, and the European Union. In the United States, legal trading accounts for more than $200 million, while illegal trading accounts for more than $300 million. Legal trade has resulted in the decline of species due to high demand for a species that exceeds its reproductive ability, slow response by governments to a shift in marketing, or uneven enforcement of existing laws to protect species. For example, high demand for caviar (sturgeon eggs) has resulted in overfishing of the Caspian Sea, and harvests have fallen severely to less than 1000 tons in the 1990s from 20,000 tons in the 1970s.

Illegal trade squanders natural resources and is one of the most dangerous forms of wildlife and biosphere destruction. It does serious damage to countries in the Southern Hemisphere that are rich in biodiversity, as it yields no lasting profits for them and only enriches the intermediaries. It pays no taxes or customs duty. The returns for the poachers are limited, because what they receive for illegally capturing or gathering macaws, tigers, crocodiles, or orchids is very little compared to what is paid for them in the wealthy countries. Such trading threatens the sustainability of the ecosystems where the species live and perpetuates inequalities between the wealthy consumer countries and the poorer producer countries, which hardly gain anything from the trade.

Many animals are sacrificed for one specific product. But in the traffic of live specimens, the mortality is very high, both at the moment of capture and during shipping; many animals die so that a few can reach their destination alive. Traditional, more respectful forms of hunting and gathering are abandoned; poachers often cut down trees to reach the highest nests, taking males and females indiscriminately.

In 1973, after much debating and pressure by scientific and non-governmental organizations, the Convention on International Trade in Endangered Animal and Plant Species (CITES) was signed in Washington, D.C., by 21 Western countries. Today, there are 164 member countries. The chief aim of CITES is to prevent illegal international trading in endangered species that are divided into three categories:

- Appendix 1 lists species in which trading is not allowed due to their imminent danger of extinction.
- Appendix 2 lists species in which trading is allowed under rigid scientific control, including those listed in Appendix 1 born in captivity.
- Appendix 3 lists species for which there are no general restrictions on trade but which have endangered populations in certain specific countries.

The convention foresees that each country should pass its own legislation to support and enforce the treaty's final provisions. This means that protection can vary from one member country to another. As with any law enforcement, it is not easy. Both legal and illegal trade continues to grow because the demand is there. In the United States alone, over 7 million people own a pet bird and 4 million own a snake, turtle, or iguana. Have you ever encountered illegally traded plants or wildlife? Have you ever asked when purchasing fish for an aquarium or perhaps a pet bird where they came from? Would you purchase a plant, fish, or perhaps jewelry with coral in it if you thought that it was traded illegally?

---

the economist Julian Simon, have argued that Ehrlich couldn't be more wrong. It is not resources that limit economic growth and lifestyles, Simon has insisted, but human ingenuity.

In 1980, the two wagered money on their competing world views. They picked something easily measurable—the value of metals—to put their theories to the test. Ehrlich predicted that world economic growth would make copper, chrome, nickel, tin, and tungsten scarcer and thus drive the prices up. Simon figured human ingenuity would overcome scarcity and that the prices would go down. Ehrlich lost. By 1990, all five metals had decreased in value. Ehrlich claimed it was a result of a global recession that had reduced industrial demand for raw metals. But Simon argued that the metals decreased in price because superior materials such as plastics, fiber optics, and ceramics had been developed to replace them. The Ehrlich-Simon argument is actually an old one, and, despite the outcome of their bet, it remains unsettled. What do you think? With the question of consumption in mind, let's look at how consumption could affect several areas in the future—food, nature, oil, and water.

Food

Two centuries ago, Thomas Malthus declared that worldwide famine was inevitable as human population growth outpaced food production. In 1972, a group of scholars known as the Club of Rome predicted much the same thing for the waning years of the twentieth century. It did not happen because—so far, at least—human ingenuity has outpaced population growth.

Fertilizers, pesticides, and high-yield crops have more than doubled world food production in the past 40 years. The reason 840 million people go hungry today and 6 million children under the age of five die each year is not that there is not enough food in the world but that they cannot afford to buy it. In fact, in 2000, the world reached the historic landmark of there being the same number of overweight people as those that were malnourished. It is argued that it is not a lack of resources that makes a people poor today but rather bad government. Consider, for example, Angola, a resource-rich country too wracked by civil war to exploit its wealth, and Russia, comparable to the United States in natural resources and intellectual capital but impoverished by the legacy of communism. Norman Borlaug, who won the Nobel peace prize in 1970 for his role in developing high-yield crops, predicts that genetic engineering and other new technologies will keep food production ahead of population increases over the next half century. New technologies, however, are not free from controversy. The Mexican government recently confirmed that genetically modified corn has escaped into native corn populations, and the European Union ended a five-year ban on U.S. imports by requiring labeling of all foods containing more than 0.9 percent materials from genetically modified organisms.

Most experts—even those who have doubts about genetic engineering—agree that enough food can be produced for the world in the twenty-first century. Whether everybody will get a fair share is much less certain.

Nature

As more people around the world achieve the American dream, they will consume more resources and generate more pollution. Tropical rainforests will be cut, and wilderness entombed under pavement. Mighty rivers like the Yangtze and Nile, already dammed and diverted, will become even more canal-like. As the new century progresses, fewer and fewer of us will live on the land. Half of humanity will live in "megacities" like Tokyo and Sao Paulo, Brazil—cities of 12 million to 25 million people. Untamed nature will exist only in scattered remnants, preserved like artifacts in a museum. We will increasingly live in a world of our own making.

Oil

If everybody on Earth consumed as much oil as the average American, the world's known reserves would be gone in a decade. Even at current rates of consumption, known reserves would not last through the current century. Experts, however, are not worried. New technologies, they say, will avert a global energy crisis.

Already, oil companies have developed cheaper ways to find oil and extract it from the ground, effectively extending the supply into the twenty-second century. Still, there is a finite amount of oil on the planet, and someday it will be gone. Even before that happens, concerns about global warming may compel the world to stop burning so much fossil fuel.

The energy industry is preparing for that day by investing in technologies that will replace fossil fuels. Solar, nuclear, and wind power are all possibilities, but many experts say the most likely candidate is the fuel cell. A fuel cell is essentially a hydrogen-powered battery that produces no pollution. Its only by-product is water. And since hydrogen is the most abundant element in the universe, supply should not be a problem. This all depends, however, on the technology being developed.

Water

The world of the future may not need oil, but without water, humanity could not last more than a few days. Right now, humans use about half the planet's accessible supply of renewable, fresh water—the supply regenerated each year and available for human use. A simple doubling of agricultural production with no efficiency improvements would push that fraction to about 85 percent. Unlike fossil fuels, which could eventually be replaced by other energy sources, there is no substitute for water.

Technologies such as desalination, which removes salt from seawater, can be used in rare circumstances. But removing salt from water takes a lot of energy and is expensive. In the Persian Gulf, one place that uses desalination, wealth makes it possible. It has been said that in the Persian Gulf, they turn "oil into water."

Some regions have already reached their water limits, with massive dams and aqueducts diverting almost every drop of water for human use. In the U.S. Southwest, the diversion is so complete that by the time the Colorado River reaches its mouth in the Sea of Cortez, it has no water in it. Much of Los Angeles's water comes from more than 300 kilometers (186 miles) away.

More than any other resource, water may limit the expansion of American consumerism during the next century. "In the next century," World Bank Vice President Ismail Serageldin predicted a few years ago, "wars will be fought over water."

The Unknown

How many people will be able to live the American way of life 50 years from now? The question is impossible to answer, many experts say, because of one major unknown, global warming. Estimates of the effect of global warming vary; however, most experts do expect some change in the Earth's average surface temperature. A warming at the lower end of the range might be barely noticeable, perhaps bringing rain instead of snow a few days each winter. But a warming at the farther end of the range could be catastrophic, shifting agricultural regions, threatening species with extinction, and pushing tropical diseases into areas where they are currently unknown. Glaciers would melt and ocean waters would expand, flooding heavily populated, low-lying places like Florida, the Netherlands, and Bangladesh.

If that happened, food production would almost certainly decline. Hundreds of millions of people would be driven from their homes by famine, flood, and drought. Billions more would be hard pressed to maintain their current lifestyles, much less aspire to an American standard of living. It has been stated that—never mind Botswana and Bangladesh, Cambodia and Cameroon—in 2050, perhaps even people in the United States may not be able to live the way Americans do today. What are your thoughts? Is there an ethical argument to the question of consumption in the United States? Do we consume too much?

Global Environmental Ethics

In 1990, Noel Brown, the director of the United Nations North American Environmental Programme, stated:

Suddenly and rather uniquely the world appears to be saying the same thing. We are approaching what I have termed a consensual moment in history, where suddenly from most quarters we get a sense that the world community is now agreeing that the environment has become a matter of global priority and action.

This new sense of urgency and common cause about the environment is leading to unprecedented cooperation in some areas. Despite their political differences, Arab, Israeli, Russian, and American environmental professionals have been working together for several years. Ecological degradation in any nation almost inevitably impinges on the quality of life in others. For years, acid rain has been a major irritant in relations between the United States and Canada. Drought in Africa and deforestation in Haiti have resulted in waves of refugees. From the Nile to the Rio Grande, conflicts flare over water rights. The growing megacities of the Third World are time bombs of civil unrest.

Much of the current environmental crisis is rooted in and exacerbated by the widening gap between rich and poor nations. Industrialized countries contain only 20 percent of the world's population, yet they control 80 percent of the world's goods and create most of its pollution. The developing countries are hardest hit by overpopulation, malnutrition, and disease. As these nations struggle to catch up with the developed world and improve the quality of life for their people, a vicious circle begins: Their efforts at rapid in-

dustrialization poison their cities, while their attempts to boost agricultural production often result in the destruction of their forests and the depletion of their soils, which lead to greater poverty. (See figure 2.5.)

Perhaps one of the most important questions for the future is, "Will the nations of the world be able to set aside their political differences to work toward a global environmental course of action?" The United Nations Conference on Human Environment held in Stockholm, Sweden, in 1972 was a step in the right direction. Out of that international conference was born the UN Environment Programme, a separate department of the United Nations that deals with environmental issues. A second world environmental conference was held in 1992 in Brazil. It followed up the Stockholm conference with many new international initiatives. A major world conference on climate change was held in Kyoto, Japan, in 1997. (See Global Perspective: The Kyoto Protocol in chapter 17.) Through organizations and conferences such as these, nations can work together to solve common environmental problems. Several major international treaties are listed in Table 2.1.

 Table 2.1 Major International Environmental Treaties and Their Impact

Treaty	Year	What This Treaty Accomplished
Convention on International Trade in Endangered Species of Wild Fauna and Flora (CITES)	1973	Regulated trade of endangered species. Today, 164 countries have signed, and trade in 30,000 plant and animals species is monitored. Enforcement varies by country; however, not one CITES-listed species has gone extinct since the treaty was fully initiated in 1975. See chapter 12 or the CITES website: http://www.cites.org
International Tropical Timber Agreement (ITTO)	1983	Created the International Tropical Timber Organization to set trade and conservation policies for tropical forest products. ITTO has 58 member organizations, which represent 90% of global timber trade and 80% of tropical forest resources. See the ITTO website: http://www.itto.or.jp/live/index.jsp
Montreal Protocol on Substances That Deplete the Ozone Layer	1987	Controlled the production and consumption of chemicals that cause ozone depletion in the atmosphere. (Ozone shields the Earth from harmful ultraviolet radiation.) This treaty successfully phased out production of these chemicals and promoted production of alternatives. See chapter 17 or the Montreal Protocol website: http://www.afeas.org/montreal_protocol.html.
UN Framework Convention on Climate Change, Council of the Parties 3: Kyoto Protocol	1992, 1997	These two treaties resulted in significant debate on the scientific evidence for global climate change and its environmental impact. The Kyoto Protocol is a strategy for reducing gas emissions that contribute to climate change. Eighty-four countries have signed the treaty; however, the United States and Russia have not. See chapter 17 or the United Nations website: http://unfccc.int/resource/convkp.html

Figure 2.5 **Lifestyle and Environmental Impact** Significant differences in lifestyles and their environmental impact exist between the rich and poor nations of the world. What would be the environmental impact on the Earth if the citizens of China and India and other less-developed countries enjoyed the standard of living in North America? Can we deny them that opportunity?

At the individual level, people have begun to respond to increased awareness of global environmental change by altering their values, beliefs, and actions. Changes in individual behavior are necessary but are not enough. As a global species, we are changing the planet. By pooling our knowledge, coordinating our actions, and sharing what the planet has to offer, we can achieve a global environmental ethic.

The Gray Whales of Neah Bay

In the mid-1990s, when gray whales were removed from the endangered species list, the Makah Tribe of Neah Bay in the state of Washington expressed the desire to resume their ancient tradition of hunting them. Through the U.S. government, they petitioned the International Whaling Commission for permission to do so. When news of this request reached the general public, it was met with controversy; some people expressed support for the Makah, while many others reacted with disapproval or anger.

The Makah were granted the legal right to hunt whales in the 1855 Treaty of Neah Bay, in which the tribe ceded the majority of their land at Cape Flattery to the U.S. government. Their fishing, sealing, and whaling rights were explicitly provided in the treaty. However, the tribe ceased their ceremonial whale hunts in the 1920s, when gray whale populations dwindled to near extinction. It was with the recovery of the gray whale populations from 17,000 to 26,000 animals that the Makah sought to begin whaling once again.

The controversy surrounding the decision to approve whaling is multifaceted. The tribe argues that their culture has disintegrated since the cessation of whaling; the last two generations have not known the discipline and pride that come with ceremonial hunting and have nutritionally suffered from the lack of a traditional seafood and sea mammal meat diet. Restoring the tradition of whale hunting, they claim, will result in the restoration of their culture.

Meanwhile, environmentalists concerned about the killing of gray whales argue that the tribe's intended hunting techniques are not traditional, thereby discrediting the Makah's arguments of cultural preservation. Historically, the whale would have been killed with repeated thrusts of a spear, causing internal bleeding and ultimate death. In lieu of this, the Makah, in consultation with the International Whaling Commission, intended to use a rifleshot following the initial harpooning, causing the immediate death of the whale. While this method is more humane than its traditional counterpart, antiwhaling groups argue that this deviation from tradition reveals that the Makah are not, in fact, concerned with restoring culture but intend to sell the meat for economic gain. The Makah assert that they are committed to using the meat solely for ceremonial and subsistence purposes. Some activists also present an ethical side of the argument, as they contend that it is morally unacceptable to kill whales due to their demonstrated intelligence.

International policy is also entangled in this controversy. The International Whaling Commission granted a yearly quota of five whales to the Makah. This quota is accommodated under the 620 gray whales currently allotted for a five-year time frame to the indigenous people of Chukotka, Russia. Also, some environmentalists allege that the Makah have received funding from Japan and Norway to support their campaign for whaling. Both of these countries host commercial whaling industries, and allowing the tribe to resume hunting, environmentalists argue, would set an important precedent for the Japanese and Norwegians to argue for "cultural subsistence" whaling also.

Numerous national entities are involved in the policy behind this debate as well. Compiling the agreements put forth by the International Whaling Commission and those reached between the National Oceanic and Atmospheric Administration and the Makah tribal council, the Northwest Indian Fisheries Commission put forth a management plan that explicitly defines quotas, hunting techniques, area restrictions, and monitoring and enforcement regulations. Additionally, the National Marine

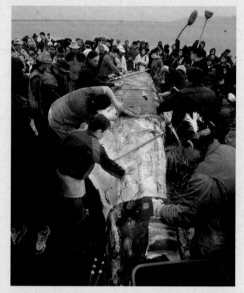

Makah Indian whalers strip a gray whale of its flesh in Neah Bay, Washington.

Fisheries Service has played an active part in defining sustainable populations of whales and other marine animals. The Makah are subject to strict regulation and would be under careful surveillance if whaling commenced.

In the fall of 1998, the Makah began their first attempt at harvesting a gray whale in the waters of Neah Bay. The tribe met with animated opposition, as citizens and activists protested on the shore of the hunting site and the Sea Shepherd Conservation Society sailed their fleet of vessels into the bay to shield whales from the hunting canoes. The presence of local and national media brought international attention to the dispute. Although numerous attempts were made, no whale was taken at the time.

The tribe made a second attempt in May 1999 and again met strong opposition, the intensity of which led to the arrests of numerous activists. However, on May 17, the hunting team succeeded in harpooning and fatally shooting a young female gray whale. Before any more whales were harvested, the case was brought before the U.S. Circuit Court of Appeals. In 2002, the court ruled that under the Marine Mammal Protection Act, further study of the environmental impact of hunting was warranted, and it banned any further hunting. In 2004, a federal appeals court upheld the ban. The Makah can appeal to the U.S. Supreme Court or complete another environmental review with public comment on the hunt.

What do you think? Should the U.S. government honor its treaty with the Makah despite other conservation legislation? Is it morally wrong to hunt whales in the first place? Is "cultural subsistence" a valid argument for taking a whale? What stance should the international community take?

For more information, visit these websites:

www.makah.com

http://www.nwifc.wa.gov/whaling/

http://www.nwr.noaa.gov/factshet/mak-inf.pdf

People of different cultures view their place in the world from different perspectives. Among the things that shape their views are religious understandings, economic pressures, geographic location, and fundamental knowledge of nature. Because of this diversity of backgrounds, different cultures put different values on the natural world and the individual organisms that compose it.

Three prevailing attitudes toward nature are the development ethic, which assumes that nature is for people to use for their own purposes; the preservation ethic, which assumes that nature has value in itself and should not be disturbed; and the conservation ethic, which recognizes that we will use nature but that it should be used in a sustainable manner.

Ethical issues can be examined at several levels. Growth and exploitation have been the prevailing priorities of our society and individual consumers for generations. This does not mean that everyone in society has the same opinions, but the general attitude has been one of development rather than preservation. Most individual environmental decisions have actually been economic decisions, and the rationale has been that if a resource is available for use, it should be used.

Corporate ethics are even more strongly influenced by economics. Corporations exist to make a profit. Any way that they can reduce costs makes them more profitable. Unfortunately, pollution, and exploitation of rare resources may be costly to individuals or society while being profitable to corporations. In addition, corporations wield tremendous economic power and can sway public opinion and political will. Many corporations have begun to openly acknowledge their responsibilities to carefully examine their impact on the natural world.

Society and corporations are composed of individuals. An increasing sensitivity of individual citizens to environmental concerns can change the political and economic climate for society and corporations. However, people often do not have a clear idea of what should be done and often do not act in a way that supports their stated beliefs.

Global environmental concerns have become more important. The world is getting "smaller" and more interrelated. As more people are added to the world's population each year, competition increases for the resources needed to live a decent life. An environmental disaster is no longer a local problem but affects us globally. The increasing economic difference between rich and poor nations affects the global environment, since the poor aspire to have what the rich take for granted. All peoples and nations need to work together to solve environmental problems.

Issues – Analysis

Environmental Dissent:
Is Ecoterrorism Justified?

At the first Earth Day in 1970, 20 million people from grassroots organizations throughout the United States came together in many different ways to voice their concern about the declining status of the environment. Since that first protest and celebration, environmental activists have used many creative strategies to bring about change, including letter writing, sit-ins, media campaigns, voting, lobbying, legislation, and land preservation. Many politicians, including former Vice President Al Gore, have asserted that greater political freedom can directly be tied to improved environmental conditions. Civil disobedience is also one of the strategies that has been used to call attention to environmental problems. For example, 25-year-old Julia Butterfly Hill recently descended to the ground after living in a 600-year-old redwood tree for more than two years to protest the logging of ancient forests on private land in California by Pacific Lumber. One of the results of the media attention created by her tree-sitting and the hard work of many in government and industry was the creation of a 10,000-acre public preserve that cost $480 million.

Environmental protest is taking on a new meaning, however, as radical ecoterrorist groups increasingly use violence to halt activity that they believe destroys or exploits the natural environment. Such special-interest terrorism is a unique form of violence in which a small group uses direct action to force a change in public policy. For example, in the name of environmental protection, ecoterrorist groups have burned, bombed, and sabotaged government offices, suburban subdivisions, ski resorts, logging operations, mink farms, restaurants, fur companies, universities, agricultural sites, and animal research laboratories. In 1995, following the Oklahoma City bombing, the Federal Bureau of Investigations established teams aimed at combating domestic terrorism. In 2002, the FBI testified before Congress that two ecoterrorist groups alone, the Earth Liberation Front (ELF) and the Animal Liberation Front (ALF),

had committed over 600 criminal acts in the United States costing more than $43 million in a six-year period.

Earth Liberation Front is believed to have been founded in Brighton, England, in 1992 by Earth First! members who felt violent and criminal acts such as bombing, arson, and tree-spiking should be used to protest environmental issues. Tree-spiking is the practice of driving a spike or metal object into tree trunks to destroy logging equipment when the trees are harvested. ELF's mission is to ensure that all forms of life have access to the basic necessities of life. It holds that anyone participating in earning wealth by "threatening the ability of all life on the planet to exist" should be stopped. ELF joined forces with the Animal Liberation Front, a radical animal rights group whose mission it is to discourage any acts that "harm any animal, human, or nonhuman," in 1993. Jointly, ELF and ALF are credited with raiding and burning the Bureau of Land Management wild horse corrals in Oregon and burning the U.S. Department of Agriculture Animal Damage Control Building in Washington. ELF alone is credited with arsons in Colorado, Oregon, Washington, Michigan, Indiana, and New York. Both groups use extensive and sophisticated surveillance systems, operate in strict secrecy, and leave little, if any, evidence at the scene. Federal, state, and local agencies have collaborated to arrest and prosecute several ELF members for such crimes as arson, tree-spiking, extortion, destruction of property, and attacks on businesses that work with animals.

- How should we as a society protect the rights of all life on the planet to have access to basic necessities? Do we have a moral obligation to do so?

- Is this method of environmental protest effective? Does it result in protection?

- What are the ethical issues involved in using violent methods compared to other approaches for environmental conservation?

Key Terms

Review Questions

1. How does personal wealth relate to ethics? Can you provide personal examples?

2. Why do industries pollute?

3. Why would normal economic forces work against pollution control? Do you feel that this is changing?

4. Is it reasonable to expect a totally unpolluted environment? Why or why not?

5. What has been the dominant societal attitude toward resource use?

6. Describe the differences between development, preservation, and conservation ethics. Must there always be conflict among these ethics?

7. What is a major motivating force of corporate management?

8. Why do decision makers view the actions of corporations differently from the way they view the actions of individuals?

Critical Thinking Questions

1. Using the definitions of moral and ethical judgment as presented in the text, identify at least two moral and ethical responses each to the issue of global climate change. What values, beliefs, and perspectives are at the root of these judgments?

2. What are our responsibilities to future generations regarding the environment? What values, beliefs, and perspectives lead you to think and act the way you do with regard to the environment?

3. Compare and contrast the three approaches to environmental ethics outlined in the text. Which is closest to your own? Why? How does, and how could, your ethical stance influence your actions?

4. The text makes the point that up until recently, humans have believed, almost universally, in unchecked growth as a positive good. Now, at the beginning of the twenty-first century, some are beginning to question this belief. What values, beliefs, and perspectives might these critics have? Describe some ways these critics might be received in a developing country. Why?

5. Imagine you are a business executive who wants to pursue an environmental policy for your company that limits pollution and uses fewer raw materials but would cost more. What might be the discussion at your next board of directors meeting? How would you respond to your board of directors and shareholders? Why?

6. In 1997, Ojibwa Indians in northern Wisconsin sat on railroad tracks to block from crossing their reservation a shipment of sulfu-ric acid that was headed for a controversial injection copper mine in northern Michigan. Try to put yourself in their position. What values, beliefs, and perspectives might have contributed to this action? Now put yourself in the position of the copper miners in northern Michigan. How might these copper miners have responded? What values, beliefs, and perspectives contribute to their action?

7. Read the Environmental Close-Up about environmental philosophy. Do you feel there is too much talk about environmental problems and not enough action? Or too little talk? Or some other problem? Please describe your position on this and your reasons for thinking the way you do.

8. Imagine yourself in the position of a person who lives on a poor Native American reservation that is contemplating building a storage facility for nuclear waste. What preconceptions, values, beliefs, or contextual perspectives might you bring to the issue? What might you propose as a course of action for yourself and for others? Why?

9. Consider environmental ethics issues in the year 2025. At the rate consumers, corporations, and governments are responding to environmental concerns, what quality-of-life consequences do you project for the year 2025? How will your health, lifestyle, income, employment, and community be affected?

Welcome to McGraw-Hill's Online Learning Center

Location: www.mhhe.com/enger10e

McGraw Hill

Interactive Exploration

Check out the website at **www.mhhe.com/enger10e** and click on the cover of this textbook for quizzing, career information, case studies, and hot links for the following topics:

Environmental Ethics

Environmental Philosophy

History of Environmental Studies

Environmental and Ecological Organization Sites

Extinction Issues

Environmental Policy, Law, and Planning

The Ultimate Ecological Answers

Individual Contributions to Environmental Issues

Environmental Organizations

3

Risk and Cost:
Elements of Decision Making

Chapter Outline

Objectives

After reading this chapter, you should be able to:

- Describe why the analysis of risk has become an important tool in environmental decision making.
- Understand the difference between risk assessment and risk management.
- Describe the issues involved in risk management.
- Understand the difference between true and perceived risks.
- Define what an economic good or service is.
- Understand the relationship between the available supply of a commodity or service and its price.
- Understand how and why cost-benefit analysis is used.
- Understand the concept of sustainable development.
- Understand environmental external costs and the economics of pollution prevention.
- Understand market approaches to solving environmental problems.

PAST PRESENT FUTURE

1980 The Superfund law was passed to bring about cleanup of abandoned toxic waste sites.

1983 Average U.S. price for kwh of electricity was 6.2¢

1983 Thirty percent of Americans smoked.

1988 The average fleet passenger and light truck mileage was 22.1 miles per gallon.

1983 Nineteen percent of the U.S. population always used seat belts.

2003 A total of 44,418 toxic waste sites had been evaluated. Seventy-five percent of the sites have been removed from the list of active sites.

2003 Average U.S. price for kwh of electricity was 8.5¢. The increase since 1983 is less than the rate of inflation.

2003 Twenty-three percent of Americans smoked.

2003 The average fleet passenger and light truck mileage was 20.4 miles per gallon.

2003 Eighty-one percent of the U.S. population always used seat belts.

Risk and Economics

Most decisions in life involve an analysis of two factors: risk and cost. We commonly ask such questions as "How likely is it that someone will be hurt?" and "What is the cost of this course of action?" Furthermore, these two factors are often interrelated. When we make economic decisions, we may be risking our hard-earned money. Risky decisions that lead to physical harm are often reduced to economic terms when medical care costs or legal fees are incurred. Environmental decision making is no different. Most environmental decisions involve finding a balance between the perceived cost of enduring the risk and the economic cost of eliminating the conditions that pose the risk. If a new air-pollution regulation is proposed to reduce the public's exposure to a chemical that is thought to cause disease in a small percentage of the exposed public, industry will be sure to point out that it will cost a considerable amount of money to put these controls in place and will reduce profitability. Citizens also may point out that their tax money will have to support a larger governmental bureaucracy to ensure that the regulations are followed. On the other side, advocates will point out the new regulations will lead to reduced risk of illness and reduced health care costs for people who live in areas affected by the pollutant.

Characterizing Risk

Risk is the probability that a condition or action will lead to an injury, damage, or loss. When we consider any activity or situation that poses a risk, we generally think about three factors: the probability of a bad outcome, the consequences of a bad outcome, and the cost of dealing with a bad outcome. **Probability** is a mathematical statement about how likely it is that something will happen. Probability is often stated in terms like "The probability of developing a particular illness is 1 in 10,000," or "The likelihood of winning the lottery is 1 in 5 million." It is important to make a distinction between *probability* and *possibility*. When we say

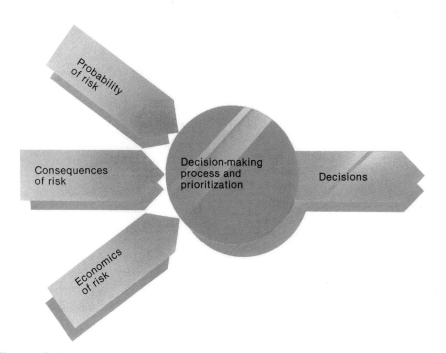

Figure 3.1 **Decision-Making Process** The assessment, cost, and consequences of risks are all important to the decision-making process.

something is *possible,* we are just saying that it could occur. It is a very inexact term. *Probability* specifically defines in mathematical terms how likely it is that a *possible* event will occur.

The *consequences* of a bad outcome resulting from the acceptance of a risk may be minor or catastrophic. For example, ammonia is a common household product. Exposure to ammonia will result in 100 percent of people reacting with watery eyes and other symptoms. The probability of an exposure and the probability of an adverse effect are high; however, the consequences are not severe, and there are no lasting effects after the person recovers. Therefore, we are willing to use ammonia in our homes and accept the high probability of an exposure. By contrast, if a large dam were to fail, it would cause extensive property damage and the deaths of thousands of people downstream. Because the consequences of a failure are high, we insist on very high engineering standards so that the probability of a failure is extremely low.

One of the consequences of accepting a risk is the *economic cost* of dealing with bad outcomes. If people become ill or are injured, health care costs are likely to be associated with the acceptance of the risk.

If a dam fails and a flood occurs downstream, there will be loss of life and property, which ultimately is converted to an economic cost. The assessment and management of risk involve an understanding of the probability and the consequences of decisions. (See figure 3.1.)

Risk Assessment

Environmental **risk assessment** is the use of facts and assumptions to estimate the probability of harm to human health or the environment that may result from particular management decisions. An environmental risk assessment process provides environmental decision makers with an orderly, clearly stated, and consistent way to deal with scientific issues when evaluating whether a risk exists, the magnitude of the risk, and the consequences of the negative outcome of accepting the risk.

Calculating the risk to humans of a particular activity, chemical, technology, or policy is difficult and several tools are used to help clarify the risk. If a situation is well-known, scientists use probabilities based on past experience to estimate risks. For example, the risk of developing black lung disease from coal dust in mines is well established, and people can be informed of the risks involved and actions that can reduce the risk.

There are also environmental risks that do not directly affect human health. If human activities cause the extinction of species, there is a negative environmental impact, although direct human impact may not be obvious. Similarly, unwise policy decisions may lead to the unsustainable harvest of forest products, fish, wildlife, or other resources that will deplete the resource for future generations.

To estimate the risks associated with new technologies or policies for which there is no established history, models must be used. One common method for modeling the risk of chemical exposure to human health is to expose animals to known quantities of a chemical to gain some insight into how dangerous a material or situation may be. However, a rat or rabbit may not react in the same way as a human. Therefore, animal studies are only indicators of human risk.

In other situations, the impact of a new policy initiative may be modeled using computer simulations. For example, in an attempt to understand the risks associated with global climate change, complex computer models of climate have been used to assess the effects resulting from current energy policy, which contributes to climate change. In the final analysis, most risk assessments are statistical statements that are estimates of the probability of negative effects, as in the examples listed in table 3.1. Such estimates typically are modified to ensure that a lack of complete knowledge does not result in an underestimation of the risk. People may be more or less sensitive to the effects of certain chemicals than the laboratory animals studied. Also, people vary in their sensitivity to compounds. Thus, what may present no risk to one person may be a high risk to others. Persons with breathing difficulties are more likely to be adversely affected by high levels of air pollutants than are healthy individuals. In addition, the estimate of human risk is based on extrapolation from animal tests in which high, chronic doses are used. Human exposure is likely to be lower or infrequent.

Because of all these uncertainties, government regulators have decided to err on the side of safety to protect the public

Table 3.1 Estimates of Selected Environmental Causes of Death

Risk Factor	Approximate Lifetime Risk of Death (per 1000)
Smoking 1–2 packs of cigarettes per day	38–175
Having 200 chest X rays per year	7–30
Driving a motor vehicle	17
Eating one 8-ounce meal per week of Great Lakes salmon with 1984 contaminant levels	11–12
Eating one 8-ounce meal per week of Great Lakes salmon with 1987 contaminant levels	3–6
Breathing air in U.S. urban areas at 1980 contaminant levels	0.1–6
Recreational boating	3.5
Drinking one 12-ounce beer per day	1–2
Recreational hunting	1.5
Complications from insect bites and stings	0.014

Source: Data from Indiana State Department of Health.

health. For example, the decisions to continue registration of pesticides, to list substances as hazardous air pollutants under the Clean Air Act, and to regulate water contaminants under the Safe Drinking Water Act set conditions of use and allowable exposure limits that provide a large margin of safety. Thus, if animal studies show an effect from the presence of a chemical at a certain dose, the allowable dose for humans is set at a lower level. This approach has been criticized by those who say it carries protection to the extreme, usually at the expense of industry. Conversely, others suggest that this method of setting regulations often underestimates the risks to humans of chronic, low-level exposures.

Risk assessment is also being used to help set regulatory priorities and support regulatory action. Those chemicals, technologies, or situations that have the highest potential to cause damage to health or the environment receive attention first, while those perceived as having minor impacts receive less immediate attention. Medical waste is perceived as high-risk, and laws have been enacted to minimize the risk, while the risk associated with the use of fertilizer on lawns is considered minimal and is not regulated.

Many of the most important threats to human health and the environment are highly uncertain. In addition to quantifying risk, a risk assessment process can

state the uncertainty associated with alternative approaches to dealing with environmental issues. This can help institutions determine research priorities and plan in a way that is consistent with scientific and public concern for environmental protection.

Risk Management

Risk management is a decision-making process that involves weighing policy alternatives and selecting the most appropriate regulatory action by integrating the results of risk assessment with engineering data and with social, economic, and political concerns. The purpose of risk management is to reduce the probability or magnitude of a negative outcome. This process involves understanding the probability and consequences of the risk and the factors that contribute to increasing or decreasing the risk. For example, automobile accidents are a leading cause of accidental death. Recognizing that the probability is high that a person will be involved in an automobile accident leads to management of the risk so that the consequences of the accident are minimized. Some management activities are designed to reduce the number of accidents. Traffic lights, warning signs, speed limits, and laws against drunk driving are all designed to reduce the number of accidents. Other activities are designed to reduce the trauma to

What's in a Number?

Risk values are often stated as a number. When the risk concern is cancer, the risk number represents the probability of additional cancer cases occurring. For example, such an estimate for pollutant X might be expressed as 1×10^{-6}, or simply 10^{-6}. This number can also be written as 0.000001, or one in a million—meaning one additional case of cancer projected in a population of 1 million people exposed to a certain level of pollutant X over their lifetimes. Similarly, 5×10^{-7}, or 0.0000005, or five in *10 million,* indicates a potential risk of five additional cancer cases in a population of 10 million people exposed to a certain level of the pollutant. These numbers signify additional cases above what normally occurs in the general population. The normal rate is referred to as the background cancer incidence. American Cancer Society statistics indicate that the background cancer incidence in the general population is one in three over a lifetime. (One-third of the population will develop some form of cancer during their lifetime.)

If the effect associated with pollutant X is a health effect other than cancer, such as neurotoxicity (nerve damage) or birth defects, then numbers are typically given as the levels of exposure below which no harm is estimated to occur. This often takes the form of a reference dose (RfD). A reference dose is typically expressed in terms of milligrams (of pollutant) per kilogram of body weight per day; for example, 0.004 mg/kg/day. A reference dose typically has a large uncertainty associated with it. It may be too high or too low by several orders of magnitude (i.e., multiples of 10).

An important point to remember is that numbers by themselves don't tell the whole story. For instance, a cancer risk value of 10^{-6} for the "average exposed person" is not the same thing as a cancer risk of 10^{-6} for a "most exposed individual" (perhaps someone exposed from living or working in a highly contaminated area), even though the numbers are identical. It's important to know the difference. Omitting the qualifier "average" or "most exposed" incompletely describes the risk and could result in an inappropriate assessment of the risk. A numerical estimate is only as good as the quality of the data it is based on. You need to ask yourself the following kinds of questions: How much data are there to support the risk assessment? Do the data include human epidemiological as well as animal studies? Do the laboratory studies include data on more than one species? If multiple species were tested, did they all respond similarly to the test substance? Are there pieces of information that you would like to have but do not? What assumptions underlie the risk assessment? What is the overall confidence level in the risk assessment? All of these qualitative considerations are essential when deciding how confident you are that the "numbers" used to characterize a risk are meaningful.

Source: Data from *EPA Journal.*

---

people who are involved in accidents. Air bags, seat belts, and car designs that absorb the energy of an impact are examples. A risk management plan includes:

1. Evaluating the scientific information regarding various kinds of risks
2. Deciding how much risk is acceptable
3. Deciding which risks should be given the highest priority
4. Deciding where the greatest benefit would be realized by spending limited funds
5. Deciding how the plan will be enforced and monitored

The process of developing a risk management plan begins with an evaluation of the scientific evidence that quantifies the magnitude of a risk. The scientific basis can be thought of as a kind of problem definition. Science determines that some threat or hazard exists but does not specify which risks are most important. With environmental concerns such as hazardous waste, climate change, ozone depletion, and acid rain, the scientific basis

for regulatory decisions is often controversial. Hazardous substances are tested on animals. Are animal tests appropriate for determining impacts on humans? There is no easy answer to this question. Dealing with climate change, ozone depletion, and acid rain require projecting into the future and estimating the magnitude of future effects. Will the sea level rise? How many additional skin cancers will be caused by depletion of the ozone layer? How many lakes will become acidified? Estimates from equally reputable sources vary widely. Which ones do we believe? For example, it is a fact that dioxin is a highly toxic material known to cause cancer in laboratory animals. It is also very difficult to prove that human exposure to dioxin has led to the development of cancer, although high exposures have resulted in acne in exposed workers.

From a risk management standpoint, whether one is dealing with a site-specific situation or a national standard, the deciding question ultimately is: What degree of risk is acceptable? In general, we are not talking about a "zero risk" standard but

rather the concept of **negligible risk:** At what point is there really no significant health or environmental risk? At what point is there an adequate safety margin to protect public health and the environment?

Once the scientific evidence has been evaluated, it is possible to integrate economic and political factors to determine how much risk is acceptable and to prioritize the assignment of economic and personnel resources to solving the problems. This is why problem definition is so important. Defining the problem helps to determine the rest of the policy process (making rules, passing laws, or issuing statements) and the appropriate enforcement actions.

Even after a policy has been developed and regulations have been put in place, however, there is often still controversy. For example, some observers believe specific chemicals such as herbicides pose many threats that need to be addressed. Others believe these chemicals pose little threat; instead, they see scare tactics and government regulations as unnecessary attacks on businesses. The commercial logging of forests poses risks of soil erosion and the loss of resident animal species. The timber industry sees these risks as minimal, while many environmentalists consider the risks unacceptable. These and similar disagreements are often serious public-relations problems for both government and business because most of the public have a poor understanding of the risks they accept daily.

True and Perceived Risks

Risk estimates by "experts" and by the "public" on many environmental problems differ significantly. Almost every daily activity—driving, walking, or working—involves some element of risk. (See table 3.2.) People often overestimate the frequency and seriousness of dramatic, sensational, well-publicized causes of death and underestimate the risks from more familiar causes that claim lives one by one.

This discrepancy and the reasons for it are extremely important because the public generally does not trust experts to make important risk decisions alone. The public generally perceives involuntary risks, such as nuclear power plants or nuclear weapons, as greater than voluntary risks, such as drinking alcohol or smoking. In addition, the public perceives newer technologies, such as genetic engineering or toxic-waste incinerators, as greater risks than more familiar technologies, such as automobiles and dams. Many people are afraid of flying for fear of crashing; however, motor vehicle accidents account for a far greater number of deaths—over 40,000 in the United States each year, compared to less than a thousand from plane crashes.

One of the most profound dilemmas facing decision makers and public health scientists is how to address the discrepancy between the scientific and public perceptions of environmental risks. Numerous studies have shown that in the last 20 years, environmental hazards truly affecting health status in the country are not those receiving the highest attention, whether measured by public opinion polls, news coverage, congressional actions, or government expenditures. (See figure 3.2.)

Table 3.2 Causes of Accidental Death in the United States, 2001

Cause	Number
Motor vehicle accidents (cars, trucks, buses)	42,443
Falls	15,019
Accidental poisoning	14,078
Unspecified accidents	7218
Suffocation	5555
Fires	3423
Drowning	3281
Aircraft accident	777
Lightning	50

Source: National Center for Health Statistics.

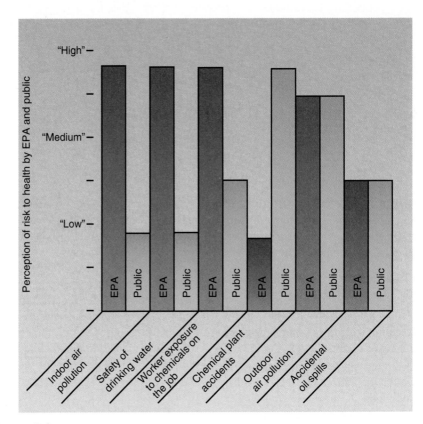

Figure 3.2 **Perception of Risk** Professional regulators and the public do not always agree on what risks are.

Indoor air pollution, in its various forms, receives relatively little attention compared with outdoor sources and yet probably accounts for as much, if not more, poor health. Hazardous waste dumps, on the other hand, which are difficult to associate with any measurable ill health, attract much attention and resources. The same chemicals in the form of common consumer products, such as household cleaners, pesticides, and fuel (gasoline), account for much more exposure and ill health and yet are comparatively of little concern to the public.

Several explanations exist for this difference in perception, the major ones relating to the fact that the public uses a number of criteria other than health risk to establish its concerns. However, this mismatch between real and perceived risks has significant consequences. In a world of finite financial resources, when money is used to reduce risks that have little measurable health impact, there is less to spend on interventions that address more significant risks.

Some researchers argue that the public is frequently misled by the politics of public health and environmental safety. This is understandable since many prominent people become involved in such issues and use their public image to encourage people to look at issues from a particular point of view.

Whatever the issue, it is hard to ignore the will of the people, particularly when sentiments are firmly held and not easily changed. A fundamental issue surfaces concerning the proper role of government and other organizations in a democracy when it comes to matters of risk. Should the government focus available resources and technology where they can have the greatest tangible impact on human and ecological well-being, or should it focus them on problems about which the public is most upset? What is the proper balance? For example, would adequate prenatal health care for all pregnant women have a greater effect on the health of children than removing asbestos from all school buildings?

Obviously, there are no clear answers to these questions. Experts and the public, however, are both beginning to realize that they each have something to offer to the debate. Many risk experts who have been accustomed to looking at numbers and probabilities are now conceding that a rationale exists for looking at risk in broader terms. At the same time, the public is being supplied with more data to enable them to make more informed judgments.

Throughout this discussion of risk assessment and management, we have made numerous references to costs and economics. It is not economically possible to eliminate all risk. As risk is eliminated, the cost of the product or service increases. Many environmental issues are difficult to evaluate from a purely economic point of view, but economics is one of the tools useful to analyzing any environmental problems.

Economics in an Environmental Context

Economics is the study of how people choose to use resources to produce goods and services and how these goods and services are distributed to the public. In other words, economics is an allocation process that determines the purposes to which resources are put. In many respects, environmental problems are primarily economic problems. While this may be an overstatement, it is not possible to view environmental issues outside the normal economic process that is central to our way of life. To appreciate the interplay between environmental issues and economics, it is important to have an understanding of some basic economic concepts.

Resources

Economists look at **resources** as the available supply of something that can be used. Classically, there are three kinds of resources: labor, capital, and land. Labor is commonly referred to as a human resource. Capital is anything that enables the efficient production of goods and services (technology and knowledge are examples). Land can be thought of as the natural resources of the planet. **Natural resources** are structures and processes that humans can use for their own purposes but cannot create. The agricultural productivity of the soil, rivers, minerals, forests, wildlife, and weather (wind, sunlight, rainfall) are all examples of natural resources. The landscape is also a natural resource, as we see in countries with a combination of mountainous terrain and high rainfall that can be used to generate hydroelectric power or those that have beautiful scenery or biotic resources that foster tourism.

Natural resources are usually categorized as either renewable or nonrenewable. **Renewable resources** can be formed or regenerated by natural processes. Soil, vegetation, animals, air, and water are renewable primarily because they naturally undergo processes that repair, regenerate, or cleanse them when their quality or quantity is reduced. Just because a resource is renewable, however, does not mean that it is inexhaustible. Overuse of renewable resources can result in their irreversible degradation. **Nonrenewable resources** are not replaced by natural processes, or the rate of replacement is so slow as to be ineffective. For example, iron ore, fossil fuels, and mountainous landscapes are nonrenewable on human timescales. Therefore, when nonrenewable resources are used up, they are gone, and a substitute must be found or we must do without.

Supply and Demand

An economic good or service can be defined as anything that is scarce. Scarcity exists whenever the demand for anything exceeds its supply. We live in a world of general scarcity, where resources are limited relative to the desires of humans to consume them. The mechanism by which resources are allocated involves the establishment of a price for a good or service. The price describes how we value goods and services and is set by the relationship between the supply of a good or service and society's demand for them.

The **supply** is the amount of a good or service people are willing to *sell* at a given price. **Demand** is the amount of a good or service that consumers are willing and able to *buy* at a given price. The **price** of a good or service is monetary value of a good or service. One of the important mechanisms that determines the price is the relationship between the supply and demand, which is often illustrated with a **supply/demand curve.** For any good or service, there is a constantly shifting relationship between supply, demand, and price. The price of a product or service reflects the strength of the demand for and the availability of the commodity. When demand exceeds supply, the price rises. The increase in price results in a chain of economic events. Price increases cause people to seek alternatives or decide not to use a product or service, which results in a lower quantity demanded.

For example, prices for recycled paper materials, such as old corrugated cardboard, fluctuate significantly based on their supply and demand. (See figure 3.3.) The supply of old corrugated cardboard does not vary much, because there are well-established recycling programs in place that capture over 70 percent of discarded corrugated cardboard. However, the demand fluctuates significantly depending on several factors. A primary factor that determines demand is the general state of the economy. When the economy is strong, people buy things and those things are typically shipped in corrugated cardboard containers. This results in an increase in demand and an increase in the price cardboard manufacturers are willing to pay for old corrugated cardboard. Conversely, when people are not buying things, less packaging is needed, demand falls, and the price falls as well. A second factor that determines demand is the strength of the export market. When other countries are buying old corrugated cardboard, less is available for the domestic market and the price increases. Finally, when the price of old corrugated cardboard approaches $125 per ton, cardboard producers can buy pulpwood at about the same price and begin to switch from using old corrugated cardboard to pulpwood.

Similarly, food production depends heavily on petroleum for the energy to plant, harvest, and transport food crops. In addition, petrochemicals are used to make fertilizer and chemical pest-control agents. If the demand for energy exceeds the supply, the price of petroleum increases. As petroleum prices rise, farmers reduce their petroleum use. Perhaps they farm less land or use less fertilizer or pesticide. Because farmers are using less energy, they will produce less food so that the supply of food decreases. Thus, an increase in petroleum prices results in an increase in food prices. As the prices of certain foods rise, consumers seek less costly foods.

When the supply of a commodity exceeds the demand, producers must lower their prices to get rid of the product, and eventually, some of the producers go out of business. Ironically, this happens to farmers when they have a series of good years. Production is high, prices fall, and some farmers go out of business.

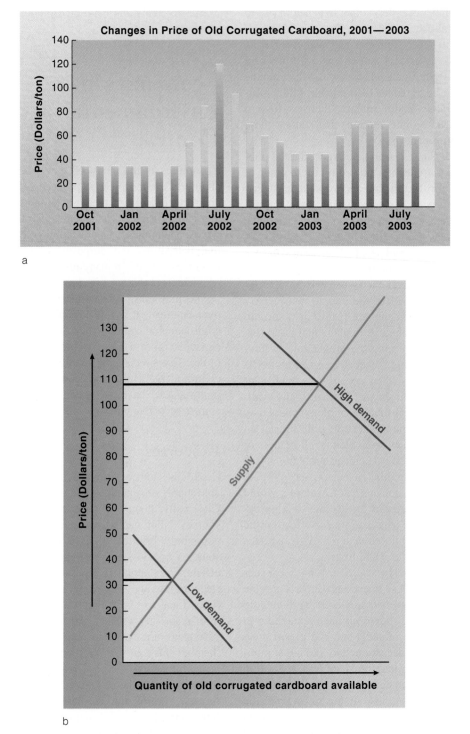

a

b

Figure 3.3   Supply and Demand for Old Corrugated Cardboard Graph *a* shows that the price for old corrugated cardboard varies considerably. The supply of old corrugated cardboard is relatively fixed, because about 70 percent is captured for recycling. However the demand varies. Graph *b* shows a typical supply/demand diagram. Demand for old corrugated cardboard is high when the U.S. economy is doing well or exports of old corrugated cardboard are high. Because demand is high the price is high. When the U.S. economy is not doing well or exports of old corrugated cardboard are low, the demand falls and so does the price.

Source: Price data from The Solid Waste Authority of Palm Beach County, Florida.

Assigning Value to Natural Resources

We assign value to natural resources based on our perception of their relative scarcity. We are willing to pay for goods or services we value highly and are unwilling to pay for things we think there is plenty of. For example, we will readily pay for a warm, safe place to live but would be offended if someone suggested that we pay for the air we breathe.

If a natural resource has always been rare, it is expensive. Pearls and precious metals are expensive because they have always been rare. If the supply of a resource is very large and the demand for it is low, the resource may be thought of as free. Sunlight, oceans, and air are often not even thought of as natural resources because their supply is so large. However, modern technologies have allowed us to exploit natural resources to a much greater extent than our ancestors were able to achieve and resources that were once considered limitless are now rare. For example, in the past, land and its covering of soil was considered a limitless natural resource, but as the population grew and the demand for food, lodging, and transportation increased, we began to realize that land is a finite, nonrenewable resource. The economic value of land is highest in metropolitan areas, where open land is unavailable. Unplanned, unwise, or inappropriate use can result in severe damage to the land and its soil. (See figure 3.4.)

Even renewable resources can be overexploited. If the overexploitation is severe and prolonged, the resource itself may be destroyed. For example, overharvesting of fish, wildlife, or forests can change the natural ecosystem so much that it cannot recover, and a resource that should have been renewable becomes a depleted nonrenewable resource.

Kinds of Environmental Costs

Air pollution, water pollution, plant and animal extinctions, depletion of a resource, and loss of scenic quality are all examples of the **environmental costs** of resource exploitation. Often environmen-

Figure 3.4 **Mismanagement of a Renewable Resource** Although soil is a renewable resource, extensive use can permanently damage it. Many of the world's deserts were formed or extended by unwise use of farmland. This photograph shows a once-productive farm now abandoned to the wind and sand because the soil was mistreated and allowed to erode.

tal costs are difficult to assess, since they are not easily converted to monetary values. In addition, since they may not be recognized immediately, environmental costs are often **deferred costs,** which must be paid at a later date. For example, when dams were built on the Colorado River to provide electric power and irrigation water, planners did not anticipate that the changes in the flow of the river would reduce habitat for endangered bird species, lead to the loss of fish species because the water is colder, and result in increased salinity in lower regions of the river. Soil erosion is another example of a deferred cost. The damage done by practices that increase soil erosion may not be felt immediately, but eventually as the amount of damage accumulates, the cost becomes obvious to future generations.

Many of the important environmental problems facing the world today arise because modern production techniques and consumption patterns transfer waste disposal, pollution, and health costs to society. Such expenses, whether they are measured in monetary terms or in diminished environmental quality, are borne by someone other than the individuals who use a resource. They are referred to as **external costs.** For example, when a logging operation removes so many trees

from a hillside that runoff from the hillside destroys streams and causes mudslides, the logging operation has transferred a cost to the public. Another example is the thousands of hazardous waste sites produced by industries that have since ceased to exist. The cleanup of these abandoned hazardous waste sites became the responsibility of government and the taxpayers. The entities that created the sites avoided paying for their cleanup. Similarly, when a new shopping complex is built, many additional external costs are paid for by the public and the municipality. Additional roads, police and fire protection, sewer and water services, runoff from parking lots, and pressure to convert the remaining adjacent land to shopping are all external costs typically borne by the taxpayer.

The extraction of mineral resources is a good example of the variety of environmental costs that accompany resource use. All mining operations involve the separation of the valuable mineral from the surrounding rock. The surrounding rock must then be disposed of in some way. These pieces of rock are usually piled on the surface of the Earth, where they are known as mine tailings and present an eyesore. It is difficult to get vegetation to grow on these deposits. Some mine tailings contain materials (such as asbestos, arsenic, lead, and radioactive materials) that can be harmful to humans and other living things.

Many types of mining operations require vast quantities of water for the extraction process. The quality of this water is degraded, so it is unsuitable for drinking, irrigation, or recreation. Since mining disturbs the natural vegetation in an area, water may carry soil particles into streams and cause erosion and siltation. Some mining operations, such as strip mining, rearrange the top layers of the soil, which lessens or eliminates its productivity for a long time. (See figure 3.5.) Strip mining has disturbed approximately 75,000 square kilometers (30,000 square miles) of U.S. land, an area equivalent to the state of Maine.

Probably most environmental costs have both deferred and external aspects. A good example of a problem that is both a deferred and an external cost is the damage caused by use of high-sulfur coal as an inexpensive way to produce electricity. The sulfur compounds released into the atmosphere resulted in acid rain that caused a decline in the growth of forests and damage to buildings and other structures. The damage accumulated over time so the cost of acid rain was a deferred cost. The cost of the damage was paid for by the public as less scenic vistas, by forest products industries with fewer trees to harvest, and by property owners as repair costs for buildings and other structures, so it was an external cost not paid for by the electric utilities directly.

Environmental costs also may include lost opportunities or values because the resource could not be used for another purpose. For example, if houses are built in a forested region, its possible use as a natural area for hiking or hunting is lost. Similarly, when land is converted to roads and parking lots, the opportunity is lost to use the land for farming or other purposes.

A primary environmental cost is pollution. **Pollution** is any addition of matter or energy that degrades the environment for humans and other organisms. When we think about pollution, however, we usually mean something that people produce in large enough quantities that it interferes with our health or well-being. Two primary factors that affect the

Figure 3.5   A Strip-Mining Operation It is easy to see the important impact a mine of this type has on the local environment. Unfortunately, many mining operations are located in areas that are also known for their scenic beauty.

amount of damage done by pollution are the size of the population and the development of technology that "invents" new forms of pollution.

When the human population was small and people lived in a simple manner, the wastes produced were biological and so dilute that they usually did not constitute a pollution problem. People used what was naturally available and did not manufacture many products. Humans, like any other animal, fit into their natural ecosystems. Their waste products were **biodegradable** materials that were broken down into simpler chemicals, such as water and carbon dioxide, by the action of decomposer organisms.

Human-initiated pollution became a problem when human populations became so concentrated that their waste materials could not be broken down as fast as they were produced. As the population increased, people began to congregate and establish villages, towns, and cities. The release of large amounts of smoke, biological waste, and trash faster than they could be absorbed and dispersed resulted in pollution, which led to unhealthy living conditions.

Throughout history, humans have sought to improve living conditions and

eliminate the misery caused by hunger and disease. In general, we rely on science and technology to improve our quality of life. While technological progress can improve the quality of life, it can also generate new sources of pollution. The development of the steam engine allowed machines to replace animal power and human labor but increased the amount of smoke and other pollutants in the air as well as the need for fuel. The modern chemical industry has produced many extremely valuable synthetic materials (plastics, pesticides, medicines), but it has also produced toxic pollutants.

It is not always easy to agree on what constitutes pollution. To some, the smell of a little wood smoke in the air is pleasant; others do not like the odor. A business may consider advertising signs valuable and necessary; others consider them to be visual pollution. Finally, it is important to recognize that it is impossible to eliminate all the negative effects produced by humans and our economic processes. The difficult question is to determine the levels of pollution that are acceptable. (See figure 3.6.)

As people recognize the significance of environmental costs, these costs are being converted to economic costs as more strict controls on pollution and environmental degradation are enforced. It takes money to clean up polluted water and air or to reclaim land that has been degraded, and the people who cause the damage should not be allowed to defer the cost or escape paying for the necessary cleanup or remediation.

Pollution-control costs include pollution costs and pollution-prevention costs. **Pollution costs** include such things as the private or public expenditures to correct pollution damage once pollution has occurred, the increased health costs because of pollution, and the loss of the use of public resources because of pollution. **Pollution-prevention costs** are those incurred either in the private sector or by government to prevent, either entirely or partially, the pollution that would otherwise result from some production or consumption activity. The cost incurred by local government to treat its sewage before releasing it into a river is a pollution-prevention cost; so is the cost incurred by an electric utility to prevent air pollution by installing new equipment.

Cost-Benefit Analysis

Because resources are limited and there are competing uses for most resources, it is essential that a process be used to help decide the most appropriate use of a scarce resource. **Cost-benefit analysis** is a formal quantitative method of assessing the costs and benefits of competing uses of a resource or solutions to a problem and deciding which is the most effective. It has long been the case in many developed countries that major projects, especially those undertaken by the government, require some form of cost-benefit analysis with respect to environmental impacts and regulations. In the United States, for example, such requirements were established by the National Environmental Policy Act of 1969, which mandates environmental impact statements for major government supported projects. Increasingly, similar analyses are required for projects supported by national and international lending institutions such as the World Bank.

People use cost-benefit analysis to determine whether a policy generates

Water pollution. This sign indicates it is unsafe to swim in this area because of high levels of bacteria.

Smoke. Smoke contains small particles that can cause lung problems.

Odors. Feedlots create an odor problem that many people find offensive.

Visual pollution. Unsightly surroundings are annoying but not hazardous to your health.

Health hazard

Smog. Smog that develops when air pollution is trapped is a serious health hazard.

Solvents. Solvents evaporate and cause localized pollution.

Thermal pollution. Cooling towers release heat into the atmosphere and can cause localized fog.

Annoyance

Litter. The presence of litter is unsightly but constitutes only a minor safety hazard.

Figure 3.6 **Examples of Pollution** There are many kinds of pollution. Some are major health concerns, whereas others merely annoy.

more social costs than social benefits and, if benefits outweigh costs, how much activity would obtain optimal results. Steps in cost-benefit analysis include:

1. Identification of the project to be evaluated
2. Determination of all impacts, favorable and unfavorable, present and future, on all of society

3. Determination of the value of those impacts, either directly through market values or indirectly through price estimates

4. Calculation of the net benefit, which is the total value of positive impacts less the total value of negative impacts

For example, the cost of reducing the amount of lead in drinking water in the United States to acceptable limits is estimated to be about $125 million a year. The benefits to the nation's health from such a program are estimated at nearly $1 billion per year. Thus, under a cost-benefit analysis, the program is economically sound. Table 3.3 gives examples of the kinds of costs and benefits involved in improving air quality. Although it is not a complete list, the table indicates the kinds of considerations that go into a cost-benefit analysis. Some of these are easy to measure in monetary terms; others are not.

Concerns About the Use of Cost-Benefit Analysis

Critics of cost-benefit analysis often raise the question of whether everything can be analyzed from an economic point of view. Some people argue that if the only measure of value is economic, many simple noneconomic values such as beauty or cleanliness can be justified only if they are given economic value. (See figure 3.7.)

There are clearly benefits to requiring such analysis. Although environmental issues must be considered at some point during project evaluation, efforts to do so are hampered by the difficulty of assigning specific value to environmental resources. In cases of Third World development projects, these already difficult environmental issues are made more difficult by cultural and socioeconomic differences. A less-developed country, for example, may be less inclined to insist on expensive emissions-treatment technology on a project that will provide jobs and economic development because it is unable to afford the treatment technology and it values the jobs very highly.

One particularly compelling critique of cost-benefit analysis is that for analysis to be applied to a specific policy, the analyst must decide which preferences

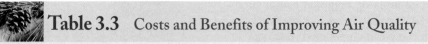

Table 3.3 Costs and Benefits of Improving Air Quality

Costs	Benefits
Installation and maintenance of new technology:	Reduced deaths and disease
Scrubbers on smokestacks	Fewer respiratory problems
Automobile emissions control	Reduced plant and animal damage
Redesign of industries and machines	Lower cleaning costs for industry and public
Additional energy costs to industry and public	More clear, sunny days; better visibility
Retraining of employees to use new technology	Less eye irritation
Costs associated with monitoring and enforcement	Fewer odor problems

count—that is, which preferences are the most important for cost-benefit analysis. In theory, cost-benefit analysis should count all benefits and costs associated with the policy under review, regardless of who benefits or bears the costs. In practice, however, this is not always done. For example, if a cost is spread thinly over a great many people, it may not be recognized as a cost at all. The cost of air pollution in many parts of the world could fall into such a category. Debates over how to count benefits and costs for future genera-

tions, inanimate objects such as rivers, and nonhumans, such as endangered species, are also common.

Comparing Economic and Ecological Systems

For most natural scientists, current crises such as biodiversity loss, climate change, and many other environmental problems are symptoms of an imbalance between the socioeconomic system and the natural world. While it is true that humans have

Figure 3.7 **Assigning Economic Values to Resource Use** The way we use resources is based on the perceived value of the resource. Not all people see the same value for a resource, and values are not always easy to measure.

always changed the natural world, it is also clear that this imprint is much greater now than anything experienced in the past. One reason for the profound effect of human activity on the natural world is the fact that there are so many of us.

One of the problems associated with matching economic processes with environmental resources is the great differences in the way economic systems and ecological systems function. The loss of biodiversity is an example that illustrates the conflicting frameworks of economics and ecology. Market decisions fail to account for the context of a species or the interconnections between resource quality and ecosystem functions. For example, from an economic point of view, the value of land used for beef production is measured according to its contribution to its economic output (beef). Yet long before economic output and the use value of land decreases, the diversity of grass varieties, microorganisms in the soil, or groundwater quality may be affected by intensive beef production. As long as yields are maintained, these environmental changes go unnoticed by economic measurements and are unimportant to land-use decisions. This does not need to be the case. It should be pointed out that in Zimbabwe and other African nations, some ranchers

now earn more money managing native species of wildlife for ecotourism in a biodiverse landscape than they would from raising cattle in a landscape with reduced biodiversity.

Another obvious difference between economics and ecology is the great difference in the time frame of markets and ecosystems. Many ecosystem processes take place over tens of thousands and even millions of years. The time frame for market decisions is short. It may be as short as minutes for stock trades or as long as a few years for the development and construction of a factory. Where U.S. economic policy is concerned, two- to four-year election cycles are the frame of reference. For investors and dividend earners, performance time frames of three months to one year are the rule.

Space or place is another issue. For ecosystems, place is critical. Take groundwater as an example. Soil quality, hydrogeological conditions, regional precipitation rates, plants that live in the region, and losses from evaporation, transpiration, and groundwater flow all contribute to the size and location of groundwater reservoirs. These capacities are not simply transferable from one location to another. For economic activities, place is increasingly irrelevant. Topogra-

phy, location, and function within a bioregion or local ecological features do not enter into economic calculations except as simple functions of transportation costs or comparative advantage. Production is transferable, and the preferred location is anywhere production costs are the lowest.

Another difference between economics and ecology is that they are measured in different units. The unifying measure of market economics is money. Progress is measured in monetary units that everyone uses and understands to some degree. Ecological systems are measured in physical units such as calories of energy, carbon dioxide absorption, centimeters of rainfall, or parts per million of nitrate contamination. Focusing only on the economic value of resources while ignoring environmental health may mask serious changes in environmental quality or function.

Common Property Resource Problems—The Tragedy of the Commons

Economists have stated that when everybody shares ownership of a resource, there is a strong tendency to overexploit and misuse that resource. Thus, common public ownership could be better described as effectively having no owner. The problems inherent in common ownership of resources were outlined by biologist Garrett Hardin in a classic essay entitled "The Tragedy of the Commons" (1968). The original "commons" were areas of pastureland in England that were provided free by the king to anyone who wished to graze cattle.

There are no problems on the commons as long as the number of animals is small in relation to the size of the pasture. From the point of view of each herder, however, the optimal strategy is to enlarge his or her herd as much as possible: If my animals do not eat the grass, someone else's will. Thus, the size of each herd grows, and the density of stock increases until the commons becomes overgrazed. The result is that everyone eventually loses as the animals die of starvation. The tragedy is that even though the eventual result should be perfectly clear, no one acts to avert disaster.

The ecosphere is one big commons stocked with air, water, and irreplaceable mineral resources—a "people's pasture," to be used in common, but it is a pasture with very real limits. Each nation attempts to extract as much from the commons as possible without regard to other countries. Furthermore, the United States and other industrial nations consume far more than their fair share of the total world resource harvest each year, much of it imported from less-developed nations.

One clear modern example of this problem involves the overharvest of marine organisms. Since no one owns the oceans, many countries feel they have the right to exploit the fisheries resources present. As in the case of grazing cattle, individuals seek to get as many fish as possible before someone else does. Currently, the UN estimates that nearly all of the marine fisheries of the world are being fished at or above capacity.

Finally, common ownership of land resources, such as parks and streets, is the source of other environmental problems. People who litter in public parks do not generally dump trash on their own property. The lack of enforceable property rights to commonly owned resources explains much of what economist John Ken-

Figure 3.8 **The Ocean Is a Common Property Resource** Since the oceans of the world are a shared resource that nobody owns, there is a tendency to use the resource unwisely. Growing populations in coastal areas lead to more marine pollution and destruction of coastal habitats. Many countries use the ocean as a dump for unwanted trash. Some 6.5 million metric tons of litter find their way into the sea each year.

neth Galbraith has termed "public squalor amid private affluence." Common ownership of the ocean makes it inexpensive for ships and oil drilling platforms to use the ocean as a dump for their wastes. (See figure 3.8.)

The tragedy of the commons also operates on an individual level. Most people are aware of air pollution, but they continue to drive their automobiles. Many families claim to need a second or third car. It is not that these people are antisocial; most would be willing to drive smaller or fewer cars if everyone else did, and they could get along with only one small car if public transport were adequate. But people frequently get "locked into" harmful situations, waiting for others to take the first step, and many unwittingly contribute to tragedies of the commons. After all, what harm can be done by the birth of one more child, the careless disposal of one more beer can, or the installation of one more air conditioner?

Using Economic Tools to Address Environmental Issues

The traditional way of dealing with environmental issues is to develop regulations that prohibit certain kinds of behavior.

This is often called a "command-and-control" approach. It has been very effective at reducing air and water pollution, protecting endangered species, and requiring that environmental concerns be addressed by environmental impact statements. However, there are also tools that use economic incentives to encourage environmental stewardship.

Subsidies

A **subsidy** is a gift from government to individuals or private enterprise to encourage actions considered important to the public interest. Subsidies may include consumer rebates for purchases of environmentally friendly goods, loans for businesses planning to implement environmental products, and other monetary incentives designed to reduce the costs of improving environmental performance.

Governments frequently subsidize agriculture, transportation, space technology, and communication. These gifts, whether loans, favorable tax situations, or direct grants, are all paid for by taxes on the public so in effect they are an external cost.

When subsidy programs have a clear purpose and are used for short periods to move to new ways of doing business, they can be very useful. Government payments to farmers that encourage them to permanently take highly erodable land from production reduces erosion and the buildup of sediment in local streams. The better water quality benefits fish, and the return of land to more natural vegetation benefits wildlife. Government programs that purchase the fishing boats of fishers who are displaced when fishing quotas are reduced are a form of subsidy to the fishing industry. The cost of government management of federal forests is a subsidy to the forest products industry, which can protect forest resources while ensuring a livelihood for loggers at the same time.

Subsidies are often used inappropriately, however, and when they are, they can lead to economic distortions. One of the effects of a subsidy is to keep the price of a good or service below its true market price. The actual cost of a subsidized good or service is higher than the subsidized market price because subsidy costs must be added to the market price to arrive at the product's true cost. Agricultural subsidies greatly distort the price of food. One common agricultural subsidy is a program that guarantees a price to a farmer for the products produced. If the market price is below the guaranteed price, the government buys the products at the guaranteed price or pays the farmer the difference between the market price and the guaranteed price. On average, U.S. farmers receive about 20 percent of their income from government payments. Other developed countries have programs that support their farmers in similar fashion. In addition, a huge bureaucracy is needed to manage the complex program. One of the unintended outcomes of such subsidies is that farmers are encouraged to produce more on less land. This results in the use of more fertilizer and pesticides that can damage the environment, and typically, there is overproduction of agricultural products.

Once subsidies become a part of the economic fabric of a country, they are very difficult to eliminate. In 1996, the U.S. Congress passed the Freedom to Farm Act, which eliminated many agricultural subsidies and was hailed as the end of agricultural subsidies. It did not work, however, and in 2002, a new federal farm bill abandoned the 1996 goal of reducing farm payments and authorized an 80 percent increase in expenditures.

Conversely, China has successfully reduced its subsidy for coal. In China, subsidy rates for coal declined from an estimated 61 percent in 1985 to 9 percent in 2000. Private mines now account for about half of all production, and some 80 percent of the coal is now sold at international prices. These reforms have had numerous benefits. Energy intensity in China has fallen by about 50 percent since 1980, and the government's total subsidy for fossil fuels fell from about $25 billion in 1990–91 to $9 billion in 2000.

The building of roads and bridges is a major part of the U.S. federal budget. In 2002, over $31 billion was allocated to the building and improvement of roads. This constitutes a subsidy for automobile transportation. Higher taxes on automobile use to cover the cost of building and repairing highways would encourage the use of more energy-efficient public transport.

Market-Based Instruments

With the growing interest in environmental protection during the past three decades, policy makers are examining new methods to reduce harm to the environment. One area of growing interest is market-based instruments. Market-based instruments provide an alternative to the common command-and-control legislation because they use economic forces and the ingenuity of entrepreneurs to achieve a high degree of environmental protection at a low cost. One of the benefits of market-based instruments is that they can be used to determine fair prices for environmental resources. Because of subsidies and external costs, many environmental resources are underpriced. Instead of inflexible, top-down government directives, market-based policies take advantage of price signals and give entrepreneurs the freedom to choose the solution most economically efficient for them. For example, a price can be established for pollution-causing activities. Companies are then allowed to decide for themselves how best to achieve the required level of environmental protection. To date, most of these market-based policies have been implemented in developed nations and in some rapidly growing developing nations. In virtually all cases, they have been introduced as supplements to, not substitutes for, traditional government regulations.

Several kinds of market-based instruments are currently in use:

Information programs provide consumers with information about the environmental consequences of purchasing decisions. Information about the environmental consequences of choices make clear to consumers that it is in their personal interest to change their decisions or behaviors. Examples include information tags on electric appliances that inform the public about the energy efficiency of the product, the mileage ratings of various automobiles, and labeling on pesticide products that describe safe use and disposal. Another type of program, such as the Toxic Release Inventory in the United States, discloses information on environmental releases by polluters. This provides corporations with incentives to improve their environmental performance to enhance their public image.

Tradable emissions permits give companies the right to emit specified quantities of pollutants. Companies that emit less than the specified amounts can sell their permits to other firms or "bank" them for future use. Thus, businesses responsible for pollution have an incentive to internalize the external cost they were previously imposing on society: If they clean up their pollution sources, they can realize a profit by selling their permit to pollute. Once a business recognizes the possibility of selling its permit, it sees that reducing pollution can have an economic benefit. The establishment of tradable sulfur dioxide pollution permits for coal-fired power plants has resulted in huge reductions in the amount of sulfur dioxide released.

Emission fees and taxes provide incentives for environmental improvement by making environmentally damaging activity or products more expensive. Businesses and individuals reduce their level of pollution wherever it is cheaper to reduce the pollution than to pay the charges. Emissions fees can be useful when pollution is coming from many small sources, such as vehicular emissions or agricultural runoff, where direct regulation or trading schemes are impractical. Taxes and fees contribute to government revenue and thus offset some of the indirect costs incurred by government to protect environmental resources.

Pollution Prevention Pays!

The philosophy of pollution prevention is that pollution should be prevented or reduced at the source whenever feasible. It is increasingly being shown that preventing pollution can cut business costs and thus increase profits. Pollution prevention, then, does make cents!

For example, several years ago, 3M's European chemical plant in Belgium switched from a polluting solvent to a safer but more expensive water-based substance to make the adhesive for its Scotch™ Brand Magic™ Tape. The switch was not made to satisfy any environmental law in Belgium or the European Union. 3M managers were complying with company policy to adopt the strictest pollution-control regulations that any of its subsidiaries is subject to—even in countries that have no pollution laws at all. Part of the policy is founded on corporate public relations, a response to growing customer demand for "green" products and environmentally responsible companies. But as many North American multinationals with similar global environmental policies are discovering, cleaning up waste, whether voluntarily or as required by law, can cut costs dramatically. Since 1975, 3M's "Pollution Prevention Pays" program—or 3P—has cut the company's air, water, and waste pollution by nearly 900,000 tons and saved the company almost $900 million. Less waste has meant less spending to comply with pollution control laws. But, in many cases, 3M actually has made money selling wastes it formerly hauled away. And because of recycling prompted by the 3P program, it has saved money by not having to buy as many raw materials.

AT&T followed a similar path. In 1990, it set voluntary goals for the company's 40 manufacturing and 2500 nonmanufacturing sites worldwide. According to its latest estimates, AT&T has (1) reduced toxic air emissions, many caused by solvents used in the manufacture of computer circuit boards, by 73 percent; (2) reduced emissions of chlorofluorocarbons—gases blamed for destroying the ozone layer in the Earth's atmosphere—by 76 percent; and (3) reduced manufacturing waste 39 percent.

Xerox Corporation had focused on recycling materials in its global environment efforts. It provides buyers of its copiers with free United Parcel Service pickup of used copier cartridges, which contain metal-alloy parts that otherwise would wind up in landfills. The cartridges and other parts are now cleaned and used to make new ones. Other recycled Xerox copier parts include power supplies, motors, paper transport systems, printed wiring boards, and metal rollers. In all, 1 million parts per year are remanufactured. The initial design and equipment investment was $10 million. Annual savings total $200 million.

Examples of Common Pollution Prevention Techniques

- Improved process control to use energy and materials more efficiently
- Improved catalysis or reactor design to reduce by-products, increase yield, and save energy in chemical processes
- Alternative processes (e.g., low- or no-chlorine pulping for paper)
- In-process material recovery (e.g., vapor recovery, water reuse, and heavy metals recovery)
- Alternatives to chlorofluorocarbons and other organic solvents
- High-efficiency paint and coating application
- Substitutes for heavy metals and other toxic substances
- Cleaner or alternative fuels and renewable energy
- Energy-efficient motors, lighting, heat exchangers, etc.
- Water conservation
- Improved "housekeeping" and maintenance in industry

Source: Data from *EPA Journal*.

In China, a pollution tax system is intended to raise revenue for investment in industrial pollution control, help pay for regulatory activities, and encourage enterprises to comply with emission and effluent standards. The system imposes noncompliance fees on discharges that exceed standards and assesses fines and other charges on violations of regulations.

In the Netherlands, an effluent tax on industrial wastewater has been viewed as successful. Especially among larger companies, the tax worked as an incentive to reduce pollution. In a survey of 150 larger companies, about two-thirds said the tax was the main factor in their decision to reduce discharges. As the volume of pollution from industrial sources dropped, rates were increased to cover the fixed costs of sewage water treatment plants. Rising rates are providing a further incentive for more companies to start purifying their sewage water.

Deposit-refund programs place a surcharge on the price of a product, which is refunded when the used product is returned for reuse or recycling. Deposit-refund schemes have been widely used to encourage recycling. In Japan, deposits are made for the return of bottles. In 2002, the German government imposed a deposit of 0.25 euros on drink cans and disposable glass and plastic (PET) bottles. Eleven states in the United States have similar laws, but so far Congress has been unwilling to pass a national bill.

Performance bonds are fees that are collected to ensure that proper care is taken to protect environmental resources. Some nations—including Indonesia, Malaysia, and Costa Rica—use performance bonds to ensure that reforestation takes place after timber harvesting. The United States also has used this approach to ensure that strip-mined lands

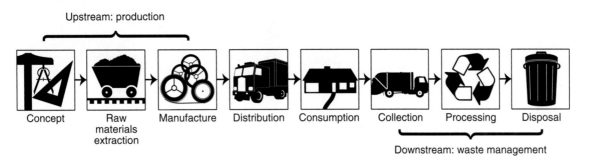

Figure 3.9 The Life Cycle of a Typical Product When life cycle analysis is undertaken, it is important to identify all the steps in the process—from obtaining raw materials, through the manufacturing process, to the final disposal of the item.

Source: *Environment*, vol. 39, no. 7, September 1997.

are reclaimed. Before a mining permit can be granted, a company must post a performance bond sufficient to cover the cost of reclaiming the site in the event the company does not complete reclamation. The bond is not fully released until all performance standards have been met and full reclamation of the site, including permanent revegetation, is successful—a five-year period in the East and Midwest and 10 years in the arid West. The bond can be partially released as various phases of reclamation are successfully completed.

Although each of these economic incentives can be effective by itself, they also can be used effectively in combination. For example, deposit-refund and tradable emissions programs work better if supported by information programs. Communities that adopted pay-by-the-bag systems of trash disposal had fewer problems if households were given adequate information well in advance. Environmental tax systems can incorporate emissions trading so that taxes can be levied on net emissions after trades.

Life Cycle Analysis and Extended Product Responsibility

Life cycle analysis is the process of assessing the environmental effects associated with the production, use, reuse, and disposal of a product over its entire useful life. Life cycle analysis can help us understand the full cost of new products and their associated technologies.

The various stages in the product chain include raw material acquisition, manufacturing processes, transportation, use by the consumer, and ultimately disposal of the used product. When this approach is used, it is possible to identify changes in product design and process technology that would reduce the ultimate environmental impact of the production, use, and disposal of the product. All factors along the product chain share responsibility for the life cycle environmental impacts of the product, from the upstream impacts inherent in selection of materials and impacts from the manufacturing process itself, to downstream impacts from the use and disposal of the product. (See figure 3.9.)

Because the relationships among industrial processes are complex, life cycle analysis requires understanding of material flows, resource reuse, and product substitution. Shifting to an approach that considers all resources, products, and waste as an interdependent system will take time, but governments can encourage this kind of thinking by establishing regulations that prevent industries from externalizing their pollution costs and providing economic incentives for those that use life cycle analysis in their product development and planning.

A logical extension of life cycle analysis is extended product responsibility. **Extended product responsibility** is the concept that the producer of a product is responsible for all the negative effects involved in its production, including the ultimate disposal of the product when its useful life is over. The logic behind extended product responsibility is that if manufacturers pay for the postconsumer impacts of products, they will design them differently to reduce waste.

Many people identify the German packaging ordinance as one of the first instances of extended product responsibility. Under the German packaging ordinance, consumers, retailers, and packaging manufacturers all share this responsibility, with the financial burden of waste management falling on the retailers and packaging manufacturers.

Although no national legislation in the United States mandates extended product responsibility, there are several instances in which manufacturers of specific products have implemented it. When several states passed legislation that required manufacturers of nickel-cadmium batteries to take back the worn-out batteries, manufacturers instituted a national take-back program. Kodak has established a program of taking back and recycling single-use cameras. The chemical industry has instituted a program known as Responsible Care®. The concept originated in Canada and has since spread to the 46 countries. The primary goals of Responsible Care® are to improve chemical processes and ensure safe production, transport, use, and disposal of the products of the industry.

Specific benefits of extended product responsibility include:

Cost savings result when manufacturers take back used products, because manufacturers recover valuable materials, reuse them, and save money.

Consideration of extended product responsibility has led to companies redesigning products to facilitate disassembly and recycling.

Georgia-Pacific Corporation: Recycled Urban Wood—A Case Study in Extended Product Responsibility

Georgia-Pacific manufactures particleboard from wood recovered from commercial disposal or general urban solid waste. The company has agreements with five recycling and processing companies that accept or collect wood at various sites. The wood is cleaned of contaminants and sent to a Georgia-Pacific particleboard manufacturing plant in Martell, California, or to other end users.

The project involves five stakeholder groups: (1) wood waste producers (e.g., operations involved with construction and demolition debris, waste from sawmills and lumber yards, and commercial wood waste from furniture manufacturing), (2) collection agents, (3) processors of wood waste, which make the waste into a product that can be reused, (4) transportation contractors, shippers, and haulers, and (5) end users (e.g., Georgia-Pacific's Martell plant). The project has a variety of goals, including increasing the availability of the wood supply for particleboard

production, contributing to Georgia-Pacific's goals of product stewardship, and contributing to California's mandated reduction in solid waste.

Business factors driving the project include the shortage of fiber for the particleboard plant, rising costs of landfilling, and mandated solid waste reductions. Currently, virgin fiber, a by-product from sawmills, is often less expensive than recovered fiber for use in particleboard. As wood becomes more scarce, however, the economics will reverse. Benefits include an expanded fiber supply in the Northwest United States.

Contamination is one of the most significant barriers to the wood recovery program. Often the collected wood is mixed in with metal, plastic, and paper, and must be cleared of these contaminants to be usable. The captured paper, plastic, and nonferrous metals are sent to a landfill. Wood by-products that cannot be used in particleboard processing are sold for use as animal bedding, playground cover, soil additive, and lawn or garden mulch.

There are more efficient environmental protections, since it is easier to design environmental safety into the product than to try to clean up the problems created by products after they have been dispersed to consumers.

Despite these benefits, obstacles to extended product responsibility exist. These include:

The cost of instituting extended product responsibility programs

The lack of information and tools to assess all impacts of the production, use, and disposal of a product

Difficulty in building relationships among individuals and institutions involved in different stages in the life cycle of a product

Hazardous waste regulations that require hazardous waste permits for collection and disposal of certain products

Antitrust laws that make it difficult for companies to cooperate

Nevertheless, extended product responsibility can be an important tool of industry and policy makers that can lead to

less costly and more flexible ways of dealing with the environmental costs of manufacturing and consuming goods.

Economics and Sustainable Development

Sustainable development has become an important policy priority for the world. The most commonly used definition of the term *sustainable development* is one that originated with the 1987 report, *Our Common Future,* by the World Commission on Environment and Development (known as the Bruntland Commission). It states that "**sustainable development** is development that meets the needs of the present without compromising the ability of future generations to meet their own needs." This definition reflects the dual societal objectives of economic development with environmental stewardship.

However, similar terms such as *sustainable growth* and *sustainable use* have been used interchangeably with *sustainable development,* as if their meanings

were the same. They are not. *Sustainable growth* is a contradiction in terms: Nothing physical can grow indefinitely. *Sustainable use* is applicable only to renewable resources: it means using them at rates within their capacity for renewal.

The concept of *sustainability* has gained usage because of increasing concern over exploitation of natural resources for economic development at the expense of environmental quality. Although disagreement exists as to the precise meaning of the term beyond respect for the quality of life of future generations, most definitions refer to the viability of natural resources and ecosystems over time and to maintenance of human living standards and economic growth.

A sustainable agricultural system, for example, can be defined as one that can indefinitely meet the demands for food and fiber at socially acceptable economic costs and environmental impacts.

Gaylord Nelson, the founder of the first Earth Day, lists five characteristics that define sustainability:

1. *Renewability:* A community must use renewable resources, such as water, topsoil, and energy sources no faster than they can replace themselves. The

rate of consumption of renewable resources cannot exceed the rate of regeneration.

2. *Substitution:* Whenever possible, a community should use renewable resources instead of nonrenewable resources. This can be difficult because of barriers to substitution. To be sustainable, a community has to make the transition before the nonrenewable resources become prohibitively scarce.

3. *Interdependence:* A sustainable community recognizes that it is a part of a larger system and that it cannot be sustainable unless the larger system is also sustainable. A sustainable community does not import resources in a way that impoverishes other communities, nor does it export its wastes in a way that pollutes other communities.

4. *Adaptability:* A sustainable community can absorb shocks and adapt to take advantage of new opportunities. This requires a diversified economy, educated citizens, and a spirit of solidarity. A sustainable community invests in and uses research and development.

5. *Institutional commitment:* A sustainable community adopts laws and political processes that mandate sustainability. Its economic system supports sustainable production and consumption. Its educational systems teach people to value and practice sustainable behavior.

Some people assume that a slowdown of economic growth is needed to prevent further deterioration of the environment. Whether or not a slowdown is necessary provokes sharp differences of opinion. One school of thought argues that economic growth is essential to finance the investments necessary to prevent pollution and to improve the environment by a better allocation of resources. Another school of thought, which is also progrowth, stresses the great potential of science and technology to solve problems and promotes technological advances as the way to solve environmental problems. Neither of these schools of thought sees any need for fundamental changes in the nature and

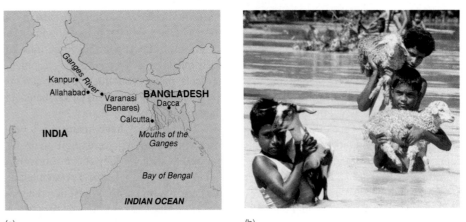

(a)   (b)

Figure 3.10 **Indian Deforestation Causes Floods in Bangladesh** (a) Because the Ganges River drains much of India and the country of Bangladesh is at the mouth of the river, deforestation and poor land use in India can result in (b) devastating floods in Bangladesh.

foundation of economic policy. Environmental issues are viewed mainly as a matter of setting priorities in the allocation of resources.

A newer school of economic thought believes that economic and environmental well-being are mutually reinforcing goals that must be pursued simultaneously if either one is to be reached. Economic growth will create its own ruin if it continues to undermine the healthy functioning of Earth's natural systems or to exhaust natural resources. It is also true that healthy economies are most likely to provide the necessary financial investments to support environmental protection. For this reason, one of the principal objectives of environmental policy must be to ensure a decent standard of living for all. The solution, at least in the broad scope, would be for a society to manage its economic growth in such a way as to do no irreparable damage to its environment.

If sustainable development is to become feasible, it will be necessary to transform our approach to economic policy. Historically, rapid exploitation of resources has provided only short-term economic growth, and the environmental consequences in some cases have been incurable. For example, 40 years ago, forests covered 30 percent of Ethiopia. Today, forests cover only 1 percent, and deserts are expanding. Trees once covered one-half of India; today, only 14 percent of the land is in forests. As the Indian trees and topsoil disappear, the citizens of

Bangladesh drown in India's runoff. (See figure 3.10.) One of the steps necessary to move economies toward sustainable development is to change the definition of gross national product (GNP) to include environmental improvement or decline.

Sustainable development is a worthy goal, but many changes are needed for the concept to be viable. One of these involves the transfer of modern, environmentally sound technology to developing nations. Tan Sri Razali, former chairman of the UN Commission on Sustainable Development, has stated that this transfer is the "key global action to sustainable development." Another major obstacle to sustainable development in many countries is a social structure that gives most of the nation's wealth to a tiny minority of its people. It has been said that a person who is worrying about his next meal is not going to listen to lectures on protecting the environment. What seem like some of the worst environmental outrages to residents in the Northern Hemisphere—cutting rainforests to make charcoal for sale as cooking fuel, for example—are often committed by people who have no other source of income.

The disparities that mark individual countries are also reflected in the planet as a whole. Most of the wealth is concentrated in the Northern Hemisphere. From the Southern Hemisphere's point of view, it is the rich world's growing consumption patterns—big cars, refrigerators, and climate-controlled shopping malls—that are the problem. The problem for the long term is that people in developing countries now want those consumer items that make life in the industrial world so comfortable—and these items use large amounts of energy and raw materials for their production and use. If the standard of living in China and India were to rise to that of Germany or the United States, the environmental impact on the planet would be significant.

Sustainable development requires choices based on values. To make intelligent choices, the public must have information about the way economic decisions affect the environment. A. W. Clausen, in his final address as president of the World Bank, noted the

> increasing awareness that environmental precautions are essential for continued economic development over the long run. Conservation, in its broadest sense, is not a luxury for people rich enough to vacation in scenic parks. It is not just a motherhood issue. Rather, the goal of economic growth itself dictates a serious and abiding concern for resource management.

High-income developed nations with high educational levels, such as the United States, Japan, and much of Europe, are in a position to promote sustainable development. They have the resources to invest in research and the technologies to implement research findings. Some believe that the world should not impose environmental protection standards on poorer nations without also helping them move into the economic mainstream.

Economics, Environment, and Developing Nations

As previously mentioned, the Earth's "natural capital," on which humankind depends for food, security, medicines, and machines, includes both nonrenewable resources such as minerals, oil, and mountains and renewable resources such as soil, sunlight, and biological diversity. Many countries in the developing world have resources that they wish to develop in order to improve the economic conditions of their inhabitants. To pay for development projects, many economically poor nations are forced to borrow money from banks in the developed world.

The debt they have incurred is a perverse incentive to overexploit their resources. Because they have borrowed money and creditors expect repayment of the debt, many countries must divert a major part of their gross domestic product to debt payment. In 2001, the debt in the developing nations had risen to over US $1900 billion, a figure equal to about half their collective gross national product. The burden of external debt is so great that many developing nations feel forced to overexploit their natural resources rather than manage them sustainably. The debt burdens have led to investments in programs absolutely necessary for immediate survival and in projects with safe, short-term returns. Environmental impacts are often neglected, because severely indebted countries feel they cannot afford to pay attention to environmental costs until other problems are resolved. This ignores the growing realization that environmental impacts frequently cause international problems.

One new method of helping manage a nation's debt crisis is referred to as debt-for-nature exchange. **Debt-for-nature exchanges** are an innovative mechanism for addressing the debt issue while encouraging investment in conservation and sustainable development. Three players are involved in debt-for-nature exchanges: the debtor nation, the creditor, and a third party interested in conservation initiatives. The exchange works as follows:

1. The conservation organization buys the debt from the creditor at a discount.
2. Although the creditor receives only partial payment of the initial loan, some return is better than a total loss.
3. The debtor country has the debt removed and is relieved of the huge burden of paying interest on the debt.
4. In exchange, the conservation organization requires the debtor country to spend money on appropriate conser-

vation and sustainable development projects.

Debt-for-nature exchanges originated in 1987, when a nonprofit organization, Conservation International, bought $650,000 of Bolivia's foreign debt in exchange for Bolivia's promise to establish a national park. By 2000, at least 16 debtor countries—in the Caribbean, Africa, Eastern Europe, and Latin America—had made similar deals with official and nongovernmental organizations. By 2000, nearly US $135 million of debt around the world had been purchased at a cost of some US $28 million, but the debtor countries spent the equivalent of US $72 million. This money was used to establish biosphere reserves and national parks, develop watershed protection programs, build inventories of endangered species, and develop environmental education.

The primary goal of debt-for-nature exchanges has not been debt reduction but the funding of natural-resource management investment. The contribution made by exchanges could increase, as in the case of the Dominican Republic, where 10 percent of the country's outstanding foreign commercial debt is to be redeemed by exchanges. Although eliminating the debt crisis alone is no guarantee of investment in environmentally sound projects, instruments such as debt-for-nature exchanges can, on a small scale, reduce the mismanagement of natural resources and encourage sustainable development.

Attitudes of banks in the industrialized nations also seem to be changing. For example, the World Bank, which lends money for Third World development projects, has long been criticized by environmental groups for backing large, ecologically unsound programs, such as a cattle-raising project in Botswana that led to overgrazing. During the past few years, however, the World Bank has been factoring environmental concerns into its programs. One product of this new approach is an environmental action plan for Madagascar. The 20-year plan, which has been drawn up jointly with the World Wildlife Fund, is aimed at heightening public awareness of environmental issues, setting up and managing protected areas, and encouraging sustainable development.

Another problem associated with resource exploitation in developing countries is the short-term exploitation of a country's resources by foreign corporations. Because poor countries do not have the financial means to develop their natural resources themselves, they often contract with corporations to do the development. Many of these corporations have no long-term commitment to the project and withdraw as soon as the project becomes unprofitable. For example, many forest resources have been logged with no interest in sustaining the resource. Companies are involved for the short-term economic gain and have little interest in the long-term economic needs of the people of the country.

Summary

Risk is the probability that a condition or action will lead to an injury, damage, or loss. Risk assessment is the use of facts and assumptions to estimate the probability of harm to human health or the environment that may result from exposures to pollutants, toxic agents, or management decisions. While it is difficult to calculate risks, risk assessment is used in risk management, which analyzes factors in decision making. The politics of risk management focus on the adequacy of scientific evidence, which is often open to divergent interpretations. In assessing risk, people frequently overestimate new and unfamiliar risks, while underestimating familiar ones.

To a large degree, environmental problems can be viewed as economic problems that revolve around decisions about how to use resources. Many environmental costs are deferred (paid at a later date) or external (paid by someone other than the entity that causes the problem). Pollution is a good example of both a deferred and an external cost. Cost-benefit analysis is concerned with whether a policy generates more social benefits than social costs. Criticism of cost-benefit analysis is based on the question of whether everything has an economic value. It has been argued that if economic thinking dominates society, then noneconomic values, such as beauty, can survive only if a monetary value is assigned to them. There is a strong tendency to overexploit and misuse resources that are shared by all. This concept was developed by Garrett Hardin in his essay "The Tragedy of the Commons."

Economic policies and concepts, such as supply and demand, and subsidies, play important roles in environmental decision making. The balance between the amount of a good or service available for purchase and the demand for that commodity determines the price. Subsidies are gifts from government to encourage desired behaviors. Recently, several kinds of market-based approaches have been developed to deal with the economic costs of environmental problems. These approaches include information programs, tradable emissions, emissions fees, deposit-refund programs, and performance bond programs. The goal of all these mechanisms is to introduce a profit motive for institutions and individuals to use resources wisely.

A newer school of economic thought is referred to as sustainable development. Sustainable development has been defined as actions that address the needs of the present without compromising the ability of future generations to meet their own needs. Sustainable development requires choices based on values. Economic concepts are also being applied to the debt-laden developing countries. One such approach is the debt-for-nature exchange. This program, which involves transferring loan payments for land that is later turned into parks and wildlife preserves, is gaining popularity.

Issues – Analysis

The Economics and Risks of Mercury Contamination

Mercury is a chemical element that is used in many industrial processes and products. Since it is a liquid metal, it is used in many kinds of electrical applications. Common uses of mercury include fluorescent light bulbs, mercury vapor lights, electrical tilt switches, and certain kinds of small batteries. It also is alloyed with silver and other metals to produce fillings in teeth. Elemental mercury by itself is poisonous, but so few people have direct access to elemental mercury that it constitutes a minor risk. Certain bacteria, however, are able to convert elemental mercury to methylmercury, which can be easily taken up by living things and stored in the body.

Some of the mercury available for conversion to methylmercury is from natural rock. Thus, some parts of the world naturally have high mercury levels compared to others. A common human-generated source of mercury is the combustion of coal in electric power plants and similar facilities. The mercury is released into the air and distributed over the landscape downwind from the power plant.

Contamination with methylmercury is a particular problem in aquatic food chains where fish at higher trophic levels accumulate high quantities of methylmercury in their tissues. Fish are an important source of omega-3 fatty acids that are associated with reductions in heart disease. If fish are declared unfit for human consumption because of methylmercury levels, there is a major economic impact on the people who catch and process fish. There can also be an increased health risk if people switch to eating other foods that do not have omega-3 fatty acids. One study of the risks and benefits of Alaskan natives eating fish determined that the risk to health from ingesting small amounts of methylmercury in fish was less than the risk associated with changing to a diet that included other sources of protein.

In 2000, under the Clinton administration, the EPA announced that it would require reductions in the amount of mercury released from coal-fired power plants. This would require expensive additions to power plants to capture the mercury so that it was not released into the atmosphere. The power industry protested that changes would be too expensive, and the program was never initiated. In December 2003, under the George W. Bush administration, the EPA announced a mercury emission trading proposal to deal with the problem. Under this proposal, coal-fired power plants will be issued permits to release specific amounts of mercury. If they do not release much mercury, they can sell their permit to other power plants that are not able to reduce the amount of mercury they release. Proponents point out that a similar emissions trading program for sulfur dioxide resulted in major reductions in sulfur dioxide release. Critics of the mercury emissions program say that the program is a concession to the power industry.

- Should power companies be able to buy rights to pollute?
- Should all pollution be prevented regardless of the cost, or should risk assessment and economic analysis be a part of the decision-making process?

Key Terms

Review Questions

1. How is risk assessment used in environmental decision making?
2. What is incorporated in a cost-benefit analysis? Develop a cost-benefit analysis for a local issue.
3. What are some of the concerns about the use of cost-benefit analysis in environmental decision making?
4. What concerns are associated with sustainable development?
5. What are some examples of external environmental costs?
6. Define what is meant by pollution-prevention costs.
7. Define the problem of common property resource ownership. Provide some examples.
8. Describe the concept of debt-for-nature exchanges.
9. Give examples of subsidies, market-based instruments, and life cycle analysis.
10. What kinds of risks are willingly accepted by people?
11. Give examples of renewable and nonrenewable resources.
12. Why are environmental costs often deferred costs?

Critical Thinking Questions

1. If you were a regulatory official, what kind of information would you require to make a decision about whether a certain chemical was "safe" or not? What level of risk would you deem acceptable for society? For yourself and your family?

2. Why do you suppose some carcinogenic agents, such as those in cigarettes, are so difficult to regulate?

3. Imagine you were assessing the risk of a new chemical plant being built along the Mississippi River in Louisiana. Identify some of the risks that you would want to assess. What kinds of data would you need to assess whether or not the risk was acceptable? Do you think that some risks are harder to quantify than others? Why?

4. Granting polluting industries or countries the right to buy and sell emissions permits is a controversial idea. Some argue that the market is the best way to limit pollution. Others argue that trade in permits allows polluting industries to continue to pollute and concentrates that pollution. What do you think?

5. Imagine you are an independent economist who is conducting a cost-benefit analysis of a hydroelectric project. What might be the costs of this project? The benefits? How would you quantify the costs of the project? The benefits? What kinds of costs and benefits might be hard to quantify or might be too tangential to the project to figure into the official estimates?

6. Do you think environmentalists should or should not stretch traditional cost-benefit analysis to include how development impacts the environment? What are the benefits to this? The risks?

7. Looking at your own life, what kinds of risks do you take? What kinds are you unwilling to take? What criteria do you use to make a decision about acceptable and unacceptable risk?

8. Is current worldwide growth and development sustainable? If there were less growth, what would be the effect on developing countries? How could we achieve a just distribution of resources and still limit growth?

9. Should our policies reflect an interest in preserving resources for future generations? If so, what level of resources should be preserved? What would you be willing to do without in order to save for the future?

PART II

Ecological Principles and Their Application

Hydrilla coverage in Lake Tohopekaliga in central Florida.

Imitating Mother Nature in a Florida Lake

Blase Maffia and Lisa Ganser

University of Miami

The lakes and river systems of central Florida have been greatly altered by human activity. Before urban growth led to the development of flood control devices such as reservoirs, locks, and canals, natural water systems maintained healthy plant and animal communities through periodic water-level fluctuations. During wet periods, sediments and "muck" (the general term for decomposing organic matter) were deposited on the shores of lakes and the banks of rivers or on floating organic islands called tussocks. During dry periods, the muck was exposed to sunlight, oxygen in the air, or freezing temperatures, which helped to oxidize and break it down. Subsequent periods of increased water flow resulted in a natural flushing action that carried the organic matter from the system.

Natural water fluctuations, however, do not occur when water levels are controlled by human means. This is the case for Lake Tohopekaliga (Lake "Toho" for short) located just south of Orlando, Florida. One of the largest lakes in Florida, Lake Toho is part of the 50-mile-long Kissimmee chain of lakes in central Florida. The lake covers about 8000 hectares (20,000 acres) and has 53 kilometers (33 miles) of shoreline. Lake Toho is home to over 150 species of plants and animals, and approximately 30 endangered species use its ecosystem. Forty percent of the lake is less than 1.5 meters (5 feet) deep, making Lake Toho one of Florida's best bass lakes for sportfishing.

Originally, Lake Toho had a hard, sandy bottom with clear water and a well-established community of aquatic plants, including Kissimmee grass, pondweed, eelgrass, and bulrushes. Later, dams and locks were built, overriding the natural flushing action of the river system. This event, coupled with the addition of nonpoint source nutrients such as nitrogen and phosphorus, has facilitated the growth of exotic species such as Eurasian milfoil, which outcompete the native plants.

To simulate the natural flushing action provided by seasonal fluctuations in water level and to maintain Lake Toho's health, the Florida Fish and Wildlife Conservation Commission recommended periodic *drawdowns* of the lake. During a drawdown, the water level is artificially lowered by allowing large amounts of water to pass the dams and other control structures. This action helps flush detrimental plant species and organic matter from the system. A drawdown exposes an estimated 5 million cubic meters (6.6 million cubic yards) of muck on the shoreline of the lake or on tussocks. Heavy equipment is used to dredge the lake bottom. Trucks then transport the muck to upland "disposal islands" and other approved disposal sites. Some nearby landowners use the muck to enhance the poor, sandy soil on their properties. Herbicide applications are necessary to control plant growth in any remaining muck on the shoreline or on tussocks.

This estimated $6 million project is intended to maintain Lake Toho in a healthy, muck-free state and preserve the environment. Thus far, the state, through the South Florida Water Management district (SFWM), has instituted drawdowns about every eight years, depending in part on the amount of rain, the degree of plant growth, and the funding available.

One of the primary goals of the periodic drawdowns is to maintain healthy sport fish and forage fish populations—and the plan seems to be working. Lake Toho's bass population is believed to be at its peak, with roughly 33,000 bass caught per year. A drawdown has the added benefit, from a fishing perspective, of concentrating fish populations so that it's easy to obtain a large catch from a limited area. Without management, however, the lake's fish populations would quickly decline.

We know that drawdowns will benefit the sport fish population, but how do they affect the native plants and wildlife? Regular drawdowns and dredging eliminate detrimental plants, but they also make it difficult to establish native aquatic plants that fish, wildlife, and waterfowl use for habitat. In addition, lowering Lake Toho displaces excess water south of the lake into other bodies in the Kissimmee chain of lakes. Unfortunately, this moves the problems of excess nutrients and exotic plants downriver as well.

Certainly, the improvement of Lake Toho for sportfishing reasons seems valid. After all, if there is no drawdown of Lake Toho, the fishery will decline. Direct results would include lost opportunities for fishing, boating, and tourism and a decline in the local economy. For the time being, humans are lending a helping hand in maintaining Lake Toho's ecosystem, but are they doing so for the right reasons?

What Do You Think?

1. How have we come to the point that we need to have drawdowns?

2. If you are a sportfisher, do the benefits outweigh the costs? If you are a taxpayer, do the benefits outweigh the costs? If government budgets are cut back, what type of priority would you give this project?

3. The risk of impact to downstream lakes (most notably, the largest, Lake Okeechobee) is a cost of the drawdown upstream. An environmental group tried to stop the drawdown of Lake Toho for this reason. A federal judge denied the motion to stop the drawdown. How would you fairly assess the needs of Lake Toho compared to those of the lakes downstream that receive the water released from Lake Toho?

4. A rare frog species depends on muck for part of its life cycle. Dredging the bottom sediments and muck disrupts the frog's life cycle. Which comes first—the bass or the frog? The local economy and its recreational fishing or a species' survival?

4 Interrelated Scientific Principles: Matter, Energy, and Environment

Chapter Outline

Objectives

After reading this chapter, you should be able to:

- Understand that science is usually reliable because information is gathered in a manner that requires impartial evaluation and continuous revision.

- Understand that matter is made up of atoms that have a specific subatomic structure of protons, neutrons, and electrons.

- Recognize that each element is made of atoms that have a specific number of protons and electrons and that isotopes of the same element may differ in the number of neutrons present.

- Recognize that atoms may be combined and held together by chemical bonds to produce molecules.

- Understand that rearranging chemical bonds results in chemical reactions and that these reactions are associated with energy changes.

- Recognize that matter may be solid, liquid, or gas, depending on the amount of kinetic energy contained by the molecules.

- Realize that energy can be neither created nor destroyed, but when energy is converted from one form to another, some energy is converted into a less useful form.

- Understand that energy can be of different qualities.

PAST PRESENT FUTURE

1982 The first transgenic (genetically modified) plants were created as laboratory specimens.

1990 The Human Genome Project (HGP) began with the goal of discovering all of the estimated 30,000 to 35,000 human genes and making them accessible for further biological study.

1983 Seven countries had appliance energy efficiency programs.

1983 World sales of compact fluorescent lamps were 32 million units.

2003 Transgenic crops were planted in over 60 million hectares (148 million acres) in 22 countries on six continents. The United States, Argentina, and Canada were the three leading countries in the amount of hectares planted.

2003 The Human Chromosome 14 Project was completed in January, and the HGP was completed in April.

2003 Forty-six countries have appliance energy efficiency programs.

2003 World sales of compact fluorescent lamps had grown to 709 million units.

Scientific Thinking

Since environmental science involves the analysis of data, it is useful to understand how scientists gather and evaluate information. It is also important to understand some chemical and physical principles as a background for evaluating environmental issues. An understanding of these scientific principles will also help you appreciate the ecological concepts in the chapters that follow.

The word *science* creates a variety of images in the mind. Some people feel that it is a powerful word and are threatened by it. Others are baffled by scientific topics and have developed an unrealistic belief that scientists are brilliant individuals who can solve any problem. For example, there are those who believe that the conservation of fossil fuels is unnecessary because scientists will soon "find" a replacement energy source. Similarly, many are convinced that if government really wanted to, it would allocate sufficient funds to allow scientists to find a cure for AIDS. Such images do not accurately portray what science is really like.

The Scientific Method

Science is a process used to solve problems or develop an understanding of nature that involves testing possible answers. Science is distinguished from other fields of study by how knowledge is acquired rather than by what is studied. The process has become known as the *scientific method*. The **scientific method** is a way of gaining information (facts) about the world by forming possible solutions to questions, followed by rigorous testing to determine if the proposed solutions are valid. Furthermore, when using the scientific method, scientists make several fundamental assumptions. They presume that:

1. There are specific causes for events observed in the natural world;
2. The causes can be identified;
3. There are general rules or patterns that can be used to describe what happens in nature;
4. An event that occurs repeatedly probably has the same cause;
5. What one person perceives can be perceived by others; and
6. The same fundamental rules of nature apply regardless of where and when they occur.

For example, we have all observed lightning associated with thunderstorms. According to the assumptions just stated, we should expect that there is an explanation that would account for all cases of lightning regardless of where or when they occur and that all people could make the same observations. We know from scientific observations and experiments that lightning is caused by a difference in electrical charge, that the behavior of lightning follows general rules that are the same as those seen with static electricity, and that all lightning that has been measured has the same cause wherever and whenever it occurred.

The scientific method requires a systematic search for information and a continual checking and rechecking to see if previous ideas are still supported by new information. If the new evidence is not supportive, scientists discard or change their original ideas. Scientific ideas undergo constant reevaluation, criticism, and modification. The scientific method involves several important identifiable components, including careful observation, asking questions about observed events, constructing and testing hypotheses, an openness to new information and ideas, and a willingness to submit one's ideas to the scrutiny of others. Underlying all of these activities is constant attention to accuracy and freedom from bias.

The scientific method is not, however, an inflexible series of steps that must be followed in a specific order. Figure 4.1 shows how these steps may be linked.

Observation

Scientific inquiry often begins with an observation that an event has occurred. An **observation** occurs when we use our senses (smell, sight, hearing, taste, touch) or an extension of our senses (microscope, tape recorder, X-ray machine, thermometer) to record an event. Observation is more than a casual awareness. You may hear a sound or see an image without really observing it. Do you know what music was being played in the shopping mall? You certainly heard it, but if you are unable to tell someone else what it was, you didn't "observe" it. If you had prepared yourself to observe the music being played, you would be able to identify it. When scientists talk about their observations, they are referring to careful, thoughtful recognition of an event—not just casual notice. Scientists train themselves to improve their observational skills, since careful observation is important in all parts of the scientific method.

Because many of the instruments used in scientific investigations are complicated, we might get the feeling that science is incredibly complex, when in reality, these sophisticated tools are being used simply to answer questions that are relatively easy to understand. For example, a microscope has several knobs to turn and a specially designed light source. It requires considerable skill to use properly, but it is essentially a fancy magnifying glass that allows small objects to be seen more clearly. The microscope has enabled scientists to answer some relatively fundamental questions such as: Are there living things in pond water? and Are living things made up of smaller subunits? Similarly, chemical tests allow us to determine the amount of specific materials dissolved in water, and a pH meter allows us to determine how acidic or basic a solution is. Both are simple activities, but if we are not familiar with the procedures, we might consider the processes hard to understand.

Questioning and Exploring

Observations often lead one to ask questions about the observations. Why did this event happen? Will it happen again in the same circumstances? Is it related to something else? Some questions may be simple speculation, but others may inspire you to further investigation. The formation of the questions is not as simple as it might seem, because the way the questions are asked will determine how you go about answering them. A question that is too broad or too complex may be impossible to answer; therefore, a great deal of effort is put into asking the question in the right

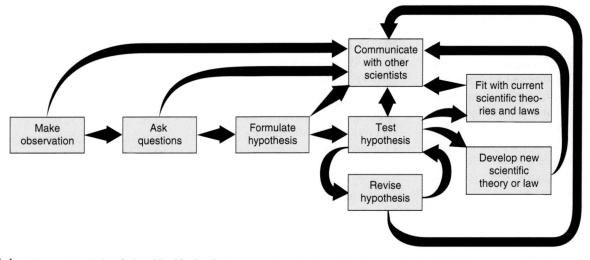

Figure 4.1 **Elements of the Scientific Method** The scientific method consists of several kinds of activities. Observation of a natural phenomenon is usually the first step. Observation often leads people to ask questions about the observation they have made or to try to determine why the event occurred. This questioning is typically followed by the construction of a hypothesis that explains why the phenomenon occurred. The hypothesis is then tested to see if it is supported. Often this involves experimentation. If the hypothesis is not substantiated, it is modified and tested in its new form. It is important at all times that others in the scientific community be informed by publishing observations of unusual events, their probable cause, and the results of experiments that test hypotheses. Occasionally, this method of inquiry leads to the development of theories that tie together many bits of information into broad statements that state why things happen in nature and serve to guide future thinking about a specific area of science. Scientific laws are similar broad statements that describe how things happen in nature.

way. In some situations, this can be the most time-consuming part of the scientific method; asking the right question is critical to how you look for answers. For example, you observe that robins eat the berries of many plants but avoid others. You could ask the following questions:

1. Do the robins dislike the flavor of some berries?
2. Will robins eat more of one kind of berry if given a choice between two kinds of berries?

The second question is obviously easier to answer.

Once a decision has been made about what question to ask, scientists *explore other sources of knowledge* to gain more information. Perhaps the question already has been answered by someone else or several possible answers already have been rejected. Knowing what others have already done saves one time. This process usually involves reading appropriate science publications, exploring information on the Internet, or contacting fellow scientists interested in the same field of study. Even if the particular question has not already been answered, scientific literature and other scientists can provide insights that may lead to a solution. After explor-

ing the appropriate literature, a decision is made about whether to continue to explore the question. If the scientist is still intrigued by the question, a formal hypothesis is constructed, and the process of inquiry continues at a different level.

Constructing Hypotheses

A **hypothesis** is a statement that provides a possible answer to a question or an explanation for an observation that can be tested. A good hypothesis must be logical, account for all the relevant information currently available, allow one to predict future events relating to the question being asked, and be testable. Furthermore, if one has the choice of several competing hypotheses, one should use the simplest hypothesis with the fewest assumptions. Just as deciding which questions to ask is often difficult, the formation of a hypothesis requires much critical thought and mental exploration. If the hypothesis does not account for all the observed facts in the situation, doubt will be cast on the work and perhaps eventually on the validity of the scientist's work. If a hypothesis is not testable or is not supported by the evidence, the explanation will be only hearsay and no more useful than mere speculation.

Keep in mind that a hypothesis is based on observations and information gained from other knowledgeable sources and predicts how an event will occur under specific circumstances. Scientists test the predictive ability of a hypothesis to see if the hypothesis is supported or is disproved. If you disprove the hypothesis, it is rejected, and a new hypothesis must be constructed. However, if you cannot disprove a hypothesis, it increases your confidence in the hypothesis, but it does not prove it to be true in all cases and for all time. Science always allows for the questioning of ideas and the substitution of new ones that more completely describe what is known at a particular point in time. It could be that an alternative hypothesis you haven't thought of explains the situation or that you have not made the appropriate observations to indicate that your hypothesis is wrong.

Testing Hypotheses

The test of a hypothesis can take several forms. It may simply involve the collection of pertinent information that already exists from a variety of sources. For example, if you visited a cemetery and observed from reading the tombstones that an unusually large number of people of different ages died in the same year, you could hypothesize that there was an epidemic of disease or a natural disaster that caused the deaths. Consulting historical newspaper accounts would be a good way to test this hypothesis.

In other cases, a hypothesis may be tested by simply making additional observations. For example, if you hypothesized that a certain species of bird used cavities in trees as places to build nests, you could observe many birds of the species and record the kinds of nests they built and where they built them.

Another common method for testing a hypothesis involves devising an experiment. An **experiment** is a re-creation of an event or occurrence in a way that enables a scientist to support or disprove a hypothesis. This can be difficult because a particular event may involve a great many separate happenings called **variables.** The best experimental design is a **controlled experiment** in which two groups differ in only one way. For example, tumors of the skin and liver occur in the fish that live in certain rivers (*observation*). This raises the question: What causes the tumors? Many people feel that the tumors are caused by toxic chemicals that have been released into the rivers by industrial plants (*hypothesis*). However, it is possible that the tumors are caused by a virus, by exposure to natural substances in the water, or are the result of genes present in the fish. How could an experiment be conducted to test the hypothesis that industrial contaminants cause the tumors? Fish could be collected from the river and placed in one of two groups. One group (the control group) would be raised in a container through which the normal river water passes. The second group (experimental group) would be raised in an identical container through which water from the industrial facility passes. There would need to be large numbers of fish in both groups. This kind of experiment is called a controlled experiment. If the fish in the experimental group have a significantly larger number of tumors than the control group, something in the water from the plant is the probable cause of the tumors. This is particularly true if the chemicals present in the water are already known to cause tumors. After the data have been

evaluated, the results of the experiment would be published.

The results of a well-designed experiment should be able to support or disprove a hypothesis. However, this does not always occur. Sometimes the results of an experiment are inconclusive. This means that a new experiment has to be conducted or that more information has to be collected. Often, it is necessary to have large amounts of information before a decision can be made about the validity of a hypothesis. The public often finds it difficult to understand why it is necessary to perform experiments on so many subjects or to repeat experiments again and again.

The concept of **reproducibility** is important to the scientific method. Because it is often not easy for scientists to eliminate unconscious bias, independent investigators must be able to reproduce the experiment to see if they get the same results. To do this, they must have a complete and accurate written document to work from. That means the scientists must publish the methods and results of their experiment. This process of publishing one's work for others to examine and criticize is one of the most important steps in the process of scientific discovery. If a hypothesis is supported by many experiments and by different investigators, it is considered reliable.

The Development of Theories and Laws

When broad consensus exists about an area of science, it is known as a theory or law. A **theory** is a widely accepted, plausible generalization about fundamental concepts in science that explains *why* things happen. An example of a scientific theory is the **kinetic molecular theory,** which states that all matter is made up of tiny, moving particles. As you can see, this is a very broad statement, and it is the result of years of observation, questioning, experimentation, and data analysis. Because we are so confident that the theory explains the nature of matter, we use this concept to explain why materials disperse in water or air, why materials change from solids to liquids, and why different chemicals can interact during chemical reactions.

Theories and hypotheses are different. A hypothesis provides a possible explanation for a specific question; a theory is a broad concept that shapes how scientists look at the world and how they frame their hypotheses. Because they are broad, unifying statements, there are few theories. However, just because a theory exists does not mean that testing stops. As scientists continue to gain new information, they may find exceptions to a theory or even in rare cases disprove a theory.

It is important to recognize that the word *theory* is often used in a much less restrictive sense. Often it is used incorrectly to describe a vague idea or a hunch. This is not a theory in the scientific sense. So when you see or hear the word *theory,* you must look at the context to see if the speaker or writer is referring to a theory in the scientific sense.

A **scientific law** is a uniform or constant fact of nature that describes *what* happens in nature. An example of a scientific law is the **law of conservation of mass,** which states that matter is not gained or lost during a chemical reaction. While laws describe what happens and theories describe why things happen, in a way, laws and theories are similar. They have both been examined repeatedly and are regarded as excellent predictors of how nature behaves.

Limitations of Science

Science is a powerful tool for developing an understanding of the natural world, but it cannot analyze international politics, decide if family-planning programs should be instituted, or evaluate the significance of a beautiful landscape. These tasks are beyond the scope of scientific investigation. This does not mean that scientists cannot comment on such issues. They often do. But they should not be regarded as more knowledgeable on these issues just because they are scientists. Scientists may know more about the scientific aspects of these issues, but they struggle with the same moral and ethical questions that face all people, and their judgments on these

Typical Household Chemicals

Modern society uses many different kinds of chemicals. A survey of a typical household would probably yield the following inorganic chemicals:

Common Name	Chemical Name	Use
Table salt	Sodium chloride, NaCl	Flavor
Saltpeter	Potassium nitrate, KNO_3	Preservative
Baking soda	Sodium bicarbonate, $NaHCO_3$	Leavening agent
Ammonia	Ammonia, NH_3	Disinfectant
Bleach	Sodium hypochlorite, NaHClO	Bleaching
Lye	Sodium hydroxide, NaOH	Drain cleaner

Other products we use contain mixtures of inorganic chemicals. Fertilizers are good examples. They usually contain a nitrate such as ammonium nitrate (NH_4NO_3), a phosphate compound (PO_4^{3+}), and potash, which is potassium oxide (K_2O).

In addition, we use a vast array of organic chemicals: ethyl alcohol in alcoholic beverages, acetic acid in vinegar, methyl alcohol for fuel, and cream of tartar (tartaric acid) for flavoring. We also use many complex mixtures of organic molecules in flavorings, pesticides, cleaners, and other applications.

Most of us know very little about the activities of the molecules we use. Many of them can be dangerous if used improperly. Fertilizer is poisonous, caustic soda can cause severe burns, and bleach or ammonia in high enough concentrations can damage skin or other tissues. The disposal of unused or unwanted household chemicals is a problem. Many of them should not just be dumped down the sink but should be disposed of in such a way that the material is converted to a harmless product or stored in a secure place. Unfortunately, most people do not know how to dispose of unwanted chemicals. For this reason, many manufacturers of household chemicals that have a potential to cause harm print statements on the containers explaining how to properly dispose of the unused product and the container. In addition, many communities have regular cleanup efforts for household hazardous waste, in which volunteers who know the contents of such products help determine how to dispose of them properly.

CARPET BEETLES — Thoroughly apply as a spot treatment. Spray along baseboards and edges of carpeting, under carpeting, rugs and furniture, in closets and on shelving, or wherever these insects are seen or suspected. FLEAS, BROWN DOG TICKS — Remove soiled bed bedding and clean thoroughly or destroy. Spray sleeping quarters of pets, along baseboards, windows, door frames, cracks and crevices, carpets, rugs, floors where these pests may be found. Put fresh bedding in pet quarters after spray has dried. DO NOT SPRAY ANIMALS. Pets should be treated with FLEA-B-GON® Flea Killer (aerosol) or ORTHO Pet Flea & Tick Spray Formula II (pump spray).
STORAGE: To store, rotate nozzle to closed position. Keep pesticide in original container. Do not put concentrate or dilute into food or drink containers. Avoid contamination of feed and foodstuffs. Store in a cool, dry place, preferably in a locked storage area.
DISPOSAL: PRODUCT — Partially filled bottle may be disposed of by securely wrapping original container in several layers of newspaper and discard in trash. CONTAINER — Do not reuse empty bottle. Rinse thoroughly before discarding in trash.
NOTICE: Buyer assumes all responsibility for safety and use not in accordance with directions.

Chevron Chemical Company © 1984
Ortho Consumer Products Division
P.O. Box 5047 San Ramon CA 94583-0947
Form 10152-N Product 5466 Made in U.S.A.
EPA Reg. No. 239-2490-AA
EPA Est. 239-IA-3

0   8
71549 01980 C

matters can be just as biased as anyone else's. Consequently, major differences of opinion often exist among lawmakers, regulatory agencies, special interest groups, and members of scientific organizations about the significance or value of specific scientific information.

It is important to differentiate between the scientific data collected and the opinions scientists have about what the data mean. Scientists form and state opinions that may not always be supported by fact, just as other people do. Equally reputable scientists commonly state opinions that are in direct contradiction. This is especially true in environmental science, where predictions about the future must be based on inadequate or fragmentary data. The issue of climate change (covered in chapter 17) is an example of this.

It is important to recognize that some scientific knowledge can be used to support both valid and invalid conclusions.

For example, the following statements are all factual.

1. Many of the kinds of chemicals used in modern agriculture are toxic to humans and other animals.

2. Small amounts of agricultural chemicals have been detected in some agricultural products.

3. Low levels of some toxic materials have been strongly linked to a variety of human illnesses.

This does not mean that all foods grown with the use of chemicals are less nutritious or are dangerous to health or that "organically grown" foods are necessarily more nutritious or healthful because they have been grown without agricultural chemicals. The idea that something that is artificial is necessarily bad and something natural is necessarily good is an oversimplification. After all, many plants such as tobacco, poison ivy, and rhubarb leaves naturally contain toxic materials, while the use of chemical fertilizers has contributed to the health and well-being of the human population because fertilizer use accounts for about one-third of the food grown in the world. However, it is appropriate to question if the use of agricultural chemicals is always necessary or if trace amounts of specific agricultural chemicals in food are dangerous. It is often easy to jump to conclusions or confuse fact with hypothesis, particularly when we generalize.

The Structure of Matter

Now that we have an appreciation for the methods of science, it is time to explore some basic information and theories about the structure and function of various kinds of matter. **Matter** is anything that takes up space and has mass. Air, water, trees, cement, and gold are all examples of matter. As stated earlier, the kinetic molecular theory is a central theory that describes the structure and activity of matter. This theory states that all matter is made up of tiny objects that are in constant motion. Although different kinds of matter have different properties, they all are similar in one fundamental way. They are all made up of one or more kinds of smaller subunits called atoms.

Atomic Structure

Atoms are fundamental subunits of matter. They in turn are made up of protons, neutrons, and electrons. There are 92 kinds of atoms found in nature. Each kind forms a specific type of matter known as

an **element.** Gold (Au), oxygen (O), and mercury (Hg) are examples of elements. All atoms have a central region known as a **nucleus,** which is composed of two kinds of relatively heavy particles: positively charged particles called **protons** and uncharged particles called **neutrons.** Surrounding the nucleus of the atom is a cloud of relatively lightweight, fast-moving, negatively charged particles called **electrons.** The atoms of each element differ from one another in the number of protons, neutrons, and electrons present. For example, a typical mercury atom contains 80 protons and 80 electrons; gold has 79 of each, and oxygen only eight of each. (See figure 4.2.) (Inside back cover contains a periodic table of the elements.) All atoms of an element always have the same number of protons and electrons, but the number of neutrons may vary from one atom to the next. Atoms of the same element that differ from one another in the number of neutrons they contain are called **isotopes.** For example, there are three isotopes of the element hydrogen. All atoms have one proton and one electron, but one isotope of hydrogen has no neutrons, one has one neutron, and one has two neutrons. These isotopes behave the same chemically but have different masses since they contain different numbers of neutrons. (See figure 4.3.)

The Molecular Nature of Matter

The kinetic molecular theory states that all matter is made of tiny particles that are in constant motion. However, there are several kinds of these tiny particles. In some instances, atoms act as individual particles. In other instances, atoms bond to one another chemically to form stable units called **molecules.** In still other cases, atoms or molecules may gain or lose electrons and thus become electrically charged particles called **ions.** Atoms or molecules that lose electrons are positively charged because they have more protons (+) than electrons (−). Those that gain electrons have more electrons (−) than protons (+) and are negatively charged. Oppositely charged ions are attracted to one another

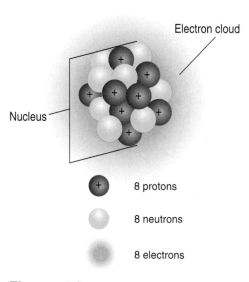

Figure 4.2 Diagrammatic Oxygen Atom Most oxygen atoms are composed of a nucleus containing eight positively charged protons and eight neutrons without charges. Eight negatively charged electrons move in a cloud around the nucleus.

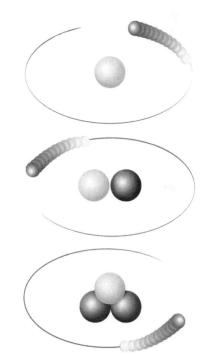

Figure 4.3 Isotopes of Hydrogen (a) The most common form of hydrogen is the isotope that is 1 AMU. It is composed of one proton and no neutrons. (b) The isotope deuterium is 2 AMU and has one proton and one neutron. (c) Tritium, 3 AMU, has two neutrons and one proton. Each of these isotopes of hydrogen also has one electron, but because the mass of an electron is so small, the electrons do not contribute significantly to the mass as measured in AMU. All three isotopes of hydrogen are found on Earth, but the most frequently occurring has 1 AMU and is commonly called hydrogen. Most scientists use the term *hydrogen* in a generic sense, that is, the term is not specific but might refer to any or all of these isotopes.

and may form stable units similar to molecules; however, they typically split into their individual ions when dissolved. For example, table salt (NaCl) is composed of sodium ions (Na^+) and chloride ions (Cl^-). It is a white, crystalline material when dry but separates into individual ions when placed in water. When two or more atoms or ions are bonded chemically, a new kind of matter called a **compound** is formed.

While only 92 kinds of atoms are commonly found, there are millions of ways atoms can be combined to form compounds. Water (H_2O), table sugar ($C_6H_{12}O_6$), table salt (NaCl), and methane gas (CH_4) are examples of compounds.

Many other kinds of matter are **mixtures,** variable combinations of atoms, ions, or molecules. Honey is a mixture of several sugars and water; concrete is a mixture of cement, sand, gravel, and reinforcing rods; and air is a mixture of several gases of which the most common are nitrogen and oxygen. Table 4.1 summarizes the various kinds of matter and the subunits of which they are composed.

Acids, Bases, and pH

Acids and bases are two classes of compounds that are of special interest. Their characteristics are determined by the nature of their chemical bonds. When acids are dissolved in water, hydrogen ions (H^+) are set free. A *hydrogen ion* is positive because it has lost its electron and now has only the positive charge of its proton. Therefore, a hydrogen ion is a proton. An **acid** is any compound that releases hydrogen ions (protons) in a solution. Some familiar examples of common acids are sulfuric acid (H_2SO_4) in automobile batteries and acetic acid (HCH_2COOH) in vinegar.

A **base** is the opposite of an acid in that it accepts hydrogen ions in solution. Many common bases release **hydroxide ions** (OH^-). This ion is composed of an oxygen atom and a hydrogen atom bonded together but with an additional electron. The hydroxide ion is negatively charged. It is a base because it is able to accept hydrogen ions in solution to

Table 4.1 Relationships Between the Kinds of Subunits Found in Matter

Category of Matter	Subunits	Characteristics
Subatomic particles	protons	Positively charged
		Located in nucleus of the atom
	neutrons	Have no charge
		Located in nucleus of the atom
	electrons	Negatively charged
		Located outside the nucleus of the atom
Elements	atoms	Atoms of an element are composed of specific arrangements of protons, neutrons, and electrons.
		Atoms of different elements differ in the number of protons, neutrons, and electrons present.
Compounds	molecules or ions	Compounds are composed of two or more atoms or ions chemically bonded together.
		Different compounds contain specific atoms or ions in specific proportions.
Mixtures	atoms, molecules, or ions	The molecular particles in mixtures are not chemically bonded to each other.
		The number of each kind of molecular particle present is variable.

form water ($H^+ + OH^- \rightarrow H_2O$). A very strong base often used in oven cleaners is sodium hydroxide (NaOH). Sometimes people refer to a base as an alkali and solutions that are basic as alkaline solutions.

The concentration of an acid or base solution is given by a number called its **pH.** The pH scale is a measure of hydrogen ion concentration. However, the pH scale is different from what you might expect. First, it is a reciprocal scale, which means that the lower the pH, the greater the number of hydrogen ions present. Second, the scale is logarithmic, which means that a difference between two consecutive pHs is really a difference of 10 times. For example, a pH of 7 indicates that the solution is neutral and has an equal number of H^+ ions and OH^- ions, but a pH of 6 means that the solution has 10 times more hydrogen ions than it would at a pH of 7. As the number of hydrogen ions in the solution increases, the pH gets smaller. A number higher than seven indicates that the solution has more OH^- than H^+. As the number of hydroxide ions increases, the pH gets larger. The higher the pH, the more concentrated the hydroxide ions. (See figure 4.4.)

Inorganic and Organic Matter

Inorganic and organic matter are usually distinguished from one another by one fact: organic matter consists of molecules that contain carbon atoms that are usually bonded to form chains or rings. Consequently, organic molecules can be very large. Many different kinds of organic compounds exist. Inorganic compounds generally consist of small molecules and combinations of ions, and relatively few kinds exist. All living things contain molecules of organic compounds. They must either be able to manufacture organic compounds from inorganic compounds or to modify organic compounds they obtain from eating organic material. Typically, chemical bonds in organic molecules contain a large amount of chemical energy that can be released when the bonds are broken and new inorganic compounds are produced. Salt, water, metals, sand, and oxygen are examples of inorganic matter. Sugars, proteins, and fats are examples of organic compounds that are produced and used by living things. Natural gas, oil, and coal are all examples of organic substances that were originally produced by living things but have been modified by geologic processes.

Chemical Reactions

When atoms or ions combine to form compounds, they are held together by chemical bonds, as in the water molecule in figure 4.5. **Chemical bonds** are attractive forces between atoms resulting from the interaction of their electrons. Each chemical bond contains a certain amount of energy. When chemical bonds are broken or formed, a chemical reaction occurs. During chemical reactions, the amount of energy within the chemical bonds changes. If the chemical bonds in the new compounds have less chemical energy than the previous compounds, some of the energy may be released as heat and light. These kinds of reactions are called **exothermic reactions.** In other cases, the newly formed chemical bonds contain more energy than was present in the compounds from which they were formed. Such reactions are called **endothermic reactions.** For such a reaction to occur, the additional energy must come from an external source.

A common example of an exothermic reaction is the burning of natural gas. The primary ingredient in natural gas is the compound methane. When methane and oxygen are mixed together and a small amount of energy is used to start the reaction, the chemical bonds in the methane and oxygen (reactants) are rearranged to form two different compounds, carbon dioxide and water (products). In this kind of reaction, the products have less energy than the reactants. The leftover energy is released as light and heat. (See figure 4.6.) In every reaction, the amount of energy in the reactants and in the products can be compared and the differences accounted for by energy loss or gain. Even energy-yielding reactions usually need an input of energy to get the reaction started. This initial input of energy is called **activation energy.** In certain cases, the amount of activation energy required to start the reaction can be reduced by the use of a catalyst. A **catalyst** is a substance that alters the rate of a reaction, but the catalyst itself is not consumed or altered in the process. Catalysts are used in catalytic converters, which are attached to automobile exhaust

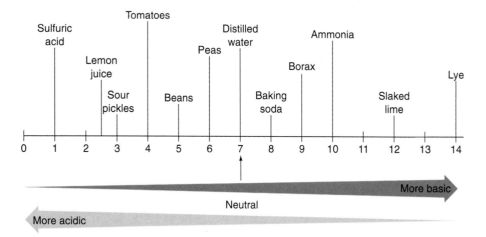

Figure 4.4 **The pH Scale** The concentration of acid is greatest when the pH is lowest. As the pH increases, the concentration of base increases. At a pH of 7.0, the concentrations of H^+ and OH^- are equal. We usually say as the pH gets smaller, the solution becomes more acidic. As the pH gets larger, the solution becomes more basic or alkaline.

Figure 4.5 **Water Molecule** A water molecule consists of an atom of oxygen bonded to two atoms of hydrogen. The molecule is not symmetrical. The hydrogen atoms are not on opposite sides of the oxygen atom.

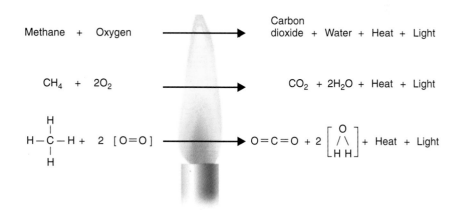

Figure 4.6 **A Chemical Reaction** When methane is burned, chemical bonds are rearranged, and the excess chemical bond energy is released as light and heat. The same atoms are present in the reactants as in the products, but they are bonded in different ways, resulting in molecules of new substances.

systems. The purpose of the catalytic converter is to bring about more complete burning of the fuel, thus resulting in less air pollution. Most of the materials that are not completely burned by the engine require high temperatures to react further; with the presence of catalysts, these reactions can occur at lower temperatures.

Endothermic reactions require an input of energy in order to occur. For example, nitrogen gas and oxygen gas can be combined to form nitrous oxide ($O_2 + N_2$ + heat $\rightarrow$ 2NO). Another important endothermic reaction is the process of photosynthesis, which is discussed in the section on "Chemical Reactions in Living Things."

Chemical Reactions in Living Things

Living things are constructed of cells that are themselves made up of both inorganic and organic matter in very specific arrangements. The chemical reactions that occur in living things are regulated by protein molecules called **enzymes** that function to reduce the activation energy needed to start the reactions. Enzymes are important since the high temperatures required to start these reactions without enzymes would destroy living organisms. Many enzymes are arranged in such a way that they cooperate in controlling a chain of reactions, as in photosynthesis and respiration.

Photosynthesis is the process plants use to convert inorganic material into organic matter, with the assistance of light energy. Light energy enables the smaller inorganic molecules (water and carbon dioxide) to be converted into organic sugar molecules. In the process, molecular oxygen is released.

$$6CO_2 + 6H_2O \xrightarrow[\text{in chloroplasts}]{\text{light energy}} C_6H_{12}O_6 + 6O_2$$

Carbon dioxide + Water $\rightarrow$ Sugar + Oxygen

Molecules of the green pigment, chlorophyll, are found in cellular structures called chloroplasts. Chlorophyll is responsible for trapping the sunlight energy needed in the process of photosynthesis. Therefore, photosynthesis takes place in the green portions of the plant, usually the leaves. (See figure 4.7.) The organic molecules produced as a result of photosynthesis can be used as an energy source by the plants and by organisms that eat the plants.

Respiration involves the use of atmospheric oxygen to break down large, organic molecules (sugars, fats, and pro-

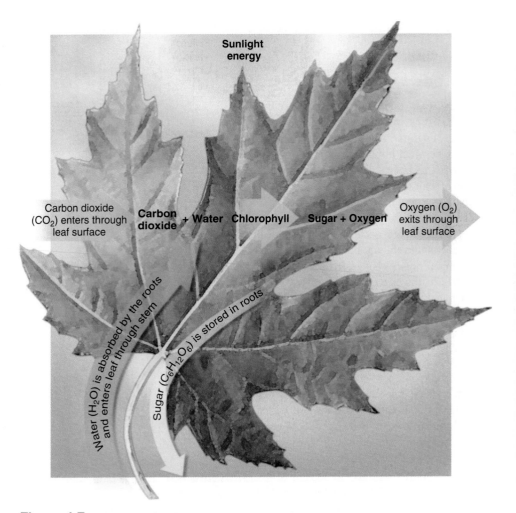

Figure 4.7 **Photosynthesis** This reaction is an example of one that requires an input of energy (sunlight) to combine low-energy molecules (CO_2 and H_2O) to form sugar ($C_6H_{12}O_6$) with a greater amount of chemical bond energy. Molecular oxygen (O_2) is also produced.

teins) into smaller, inorganic molecules (carbon dioxide and water). This process releases energy the organisms can use.

$$C_6H_{12}O_6 + 6O_2 \longrightarrow 6CO_2 + 6H_2O + \text{Energy}$$

Sugar + Oxygen $\longrightarrow$ Carbon dioxide + Water + Energy

(See figure 4.8.) All organisms must carry on some form of respiration, since all organisms need a source of energy to maintain life.

Energy Principles

The "Chemical Reactions in Living Things" section started out with a description of matter, yet it used the concept of energy to describe chemical bonds, chemical reactions, and molecular motion. That is because energy and matter are inseparable. It is difficult to describe one without the other. **Energy** is the ability to do work. Work is done when an object is moved over a distance. This occurs even at the molecular level.

Kinds of Energy

There are several kinds of energy. Heat, light, electricity, and chemical energy are common forms. The energy contained by moving objects is called **kinetic energy.** The moving molecules in air have kinetic energy, as does water running downhill or a dog chasing a ball. In contrast, **potential energy** is the energy matter has because of its position. The water behind a dam has potential energy by virtue of its elevated

PART TWO   Ecological Principles and Their Application

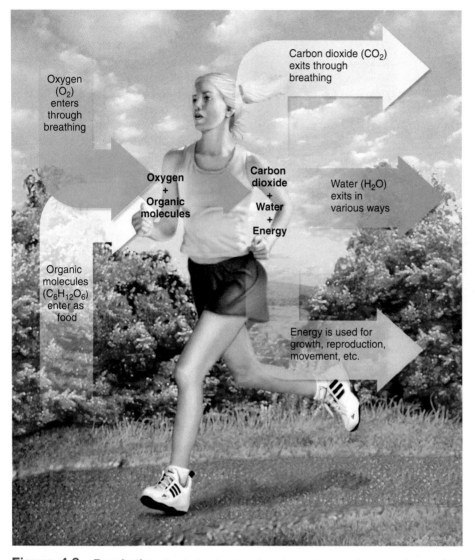

Oxygen
(O₂)
enters
through
breathing

Carbon dioxide (CO₂)
exits through
breathing

**Oxygen
+
Organic
molecules**

**Carbon
dioxide
+
Water
+
Energy**

Water (H₂O)
exits in
various ways

Organic
molecules
($C_6H_{12}O_6$)
enter as
food

Energy is used for
growth, reproduction,
movement, etc.

Figure 4.8 **Respiration** Respiration involves the release of energy from organic molecules when they react with oxygen. In addition to providing energy in a usable form, respiration produces carbon dioxide and water.

position. (See figure 4.9.) An electron moved to a position farther from the nucleus has increased potential energy due to the increased distance between the electron and the nucleus.

States of Matter

Depending on the amount of energy present, matter can occur in three common states: solid, liquid, or gas. The physical nature of matter changes when a change occurs in the amount of kinetic energy its molecular particles contain, but the chemical nature of matter and the kinds of chemical reactions it will undergo remain the same. For example, water vapor, liquid water, and ice all have the same chemical composition but differ in the arrangement and activity of their molecules. The amount of kinetic energy molecules have determines how rapidly they move. (See figure 4.10.) In solids, the molecular particles have comparatively little energy, and they vibrate in place very close to one another. In liquids, the particles have more energy, are farther apart from one another, and will roll, tumble, and flow over each other. In gases, the molecular particles move very rapidly and are very far apart. All that is necessary to change the physical nature of a substance is an energy change. Heat energy must be added or removed.

When two forms of matter have different temperatures, heat energy will flow from the one with the higher temperature to the one with the lower temperature. The temperature of the cooler matter increases while that of the warmer matter decreases. You experience this whenever you touch a cold or hot object. This is referred to as a **sensible heat** transfer. When heat energy is used to change the state of matter from solid to liquid at its melting point or liquid to gas at its boiling point, heat is transferred, but the temperature of the matter does not change. This is called a **latent heat** transfer. You have experienced this effect when water evaporates from your skin. Your body supplies the heat necessary to convert liquid water to water vapor. While the temperature of the water does not change, the physical state of the water does. The heat that was transferred to the water caused it to evaporate, and your body cooled. When substances change from gas to liquid at the boiling point or liquid to solid at the freezing point, there is a corresponding release of heat energy without a change in temperature.

First and Second Laws of Thermodynamics

Energy can exist in several different forms, and it is possible to convert one form of energy into another. However, the total amount of energy remains constant. The **first law of thermodynamics** states that energy can neither be created nor destroyed; it can only be changed from one form into another. From a human perspective, some forms of energy are more useful than others. We tend to make extensive use of electrical energy for a variety of purposes, but there is very little electrical energy present in nature. Therefore, we convert other forms of energy into electrical energy. When converting energy from one form to another, some of the useful energy is lost. This is the **second law of thermodynamics.** The energy that cannot be used to do useful work is called **entropy.** Therefore, another way to state the second law of thermodynamics is to say that when energy is converted from one form to another, entropy increases. An alternative way to look at the idea of entropy is to say that entropy is a measure of

Figure 4.9 **Kinetic and Potential Energy** Kinetic and potential energy are interconvertible. The potential energy possessed by the water behind a dam is converted to kinetic energy as the water flows to a lower level.

disorder and that the amount of disorder (entropy) typically increases when energy conversions take place. Obviously, it is possible to generate greater order in a system (living things are a good example of highly ordered things), but when things become more highly ordered, the disorder of the surroundings must increase. For example, all living things release heat to their surroundings. It is important to understand that when energy is converted from one form to another, there is no loss of *total* energy, but there is a loss of *useful* energy. For example, coal, which contains chemical energy, can be burned in a power plant to produce electrical energy. The heat from the burning coal is used to heat water to form steam, which turns turbines that generate electricity. At each step in the process, some heat energy is lost from the system. Therefore, the amount of useful energy (electricity) coming from the plant is much less than the total amount of chemical energy present in the coal that was burned. (See figure 4.11.)

Within the universe, energy is being converted from one form to another continuously. Stars are converting nuclear energy into heat and light. Animals are converting the chemical potential energy found in food into kinetic energy that allows them to move. Plants are converting sunlight energy into the chemical bond energy of sugar molecules. In each of these cases, some energy is produced that is not able to do useful work. This is generally in the form of heat lost to the surroundings.

Environmental Implications of Energy Flow

The heat produced when energy conversions occur is dissipated throughout the universe. This is a common experience. All machines and living things that manipulate energy release heat. It is also true that organized matter tends to become more disordered unless an external source of energy is available to maintain the ordered arrangement. Houses fall into ruin, automobiles rust, and appliances wear out unless work is done to maintain them. In reality, all of these phenomena involve the loss of heat.

Gas (water vapor)
1. Molecules have high kinetic energy.
2. Molecules are far apart.
3. Molecules have little attraction to one another.
4. Molecules are able to exchange places.

Liquid (water)
1. Molecules have moderate kinetic energy.
2. Molecules are close together.
3. Molecules are attracted to one another.
4. Molecules are able to exchange places.

Solid (ice)
1. Molecules have low kinetic energy.
2. Molecules are close together.
3. Molecules are attracted to one another.
4. Molecules vibrate in place—they do not exchange places.

Figure 4.10 **States of Matter** Matter exists in one of three states, depending on the amount of kinetic energy the molecules have. The higher the amount of energy, the greater the distance between molecules and the greater their degree of freedom of movement.

The organisms that decompose the wood in our houses release heat. The chemical reaction that causes rust releases heat. Friction, caused by the movement of parts of a machine against each other, generates heat and causes the parts to wear.

Ultimately, orderly arrangements of matter, such as clothing, automobiles, or living organisms, become disordered. There is an increase in entropy. Eventually, nonliving objects wear out and living things die and decompose. This process of becoming more disordered coincides with the constant flow of energy toward a dilute form of heat. This dissipated, low-quality heat has little value to us, since we are unable to use it.

It is important to understand that some forms of energy are more useful to us than others. Some forms, such as electrical energy, are of high quality because they can be easily used to perform a variety of useful actions. Other forms, such as the heat in the water of the ocean, are of low quality because we are not able to use them for useful purposes. Although the total *quantity* of heat energy in the ocean is much greater than the total amount of electrical energy in the world, little useful work can be done with the heat energy in the ocean because it is of *low quality*. Therefore, it is not as valuable as other forms of energy that can be used to do work for us.

The reason the heat of the ocean is of little value is related to the small temperature difference between two sources of heat. When two objects differ in temperature, heat will flow from the warmer to the cooler object. The greater the temperature difference, the more useful the work that can be done. For example, fossil-fuel power plants burn fuel to heat water and convert it to steam. High-temperature steam enters the turbine, while cold cooling water condenses the steam as it leaves the turbine. This steep temperature gradient also provides a steep pressure gradient as heat energy flows from the steam to the cold water, which causes a turbine to turn, which generates electricity. Because the average temperature of the ocean is not high, and it is difficult to find another object that has a greatly lower temperature than the ocean, it is difficult to use the huge heat content of the ocean to do useful work for us.

Figure 4.11   Second Law of Thermodynamics Whenever energy is converted from one form to another, some of the useful energy is lost, usually in the form of heat. The conversion of fuel to electricity produces heat, which is lost to the atmosphere. As the electricity moves through the wire, resistance generates some additional heat. When the electricity is converted to light in a light bulb, heat is produced as well. All of these steps produce low-quality heat in accordance with the second law of thermodynamics.

These quantitative and qualitative factors are also evident in the energy expended by a stream as the water runs downhill. The steeper the slope, the greater the amount of energy expended per kilometer of its length. If there is no point along the stream where the slope is very steep, the stream has low-quality energy, because the energy is dissipated along the entire length of the stream. To make this a high-quality (concentrated) source of energy, the water must be dammed so that it will drop a long distance at one point. This means that it will give up much of its energy over a short distance. With damming, the *quantity* of energy has not changed but the *quality* has. Whenever low-quality energy is converted to high-quality energy, work must be done.

It is important to understand that energy that is of low quality from our point of view may still have significance to the world in which we live. For example, the distribution of heat energy in the ocean tends to moderate the temperature of coastal climates, contributes to weather patterns, and causes ocean currents that are extremely important in many ways. It is also important to recognize that we can sometimes figure new ways to convert low-quality energy to high-quality energy. For example, it is possible to use the waste heat from power plants to heat cities if the power plants are located in the cities. Scientists have recently made major improvements in wind turbines and photovoltaic cells that allow us to economically convert to low-quality light or wind to high-quality electricity.

From the point of view of ecological systems, organisms such as plants do photosynthesis and are able to convert low-quality light energy to high-quality chemical energy in the organic molecules they produce. Eventually, they will use this stored energy for their needs, or it will be used by some other organism that has eaten the plant. In accordance with the second law of thermodynamics, all organisms, including humans, are in the process of converting high-quality energy into low-quality energy. Waste heat is produced when the chemical-bond energy in food is converted into the energy needed to move, grow, or respond. The process of releasing chemical-bond energy from food by organisms is known as cellular respiration. From an energy point of view, it is comparable to the process of **combustion,** which is the burning of fuel to obtain heat, light, or some other form of useful energy. The efficiency of cellular respiration is relatively high. About 40 percent of the energy contained in food is released in a useful form. The rest is dissipated as low-quality heat. Table 4.2 lists the efficiencies of many common energy conversion systems.

An unfortunate consequence of energy conversion is pollution. The heat lost from most energy conversions is a pollutant. The wear of the brakes used to stop

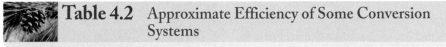

Table 4.2 Approximate Efficiency of Some Conversion Systems

Energy Conversion System	% Efficiency*
Electric motor	95
Hydroelectric power plant	85
Home oil furnace	65
Jet engine	55
Steam-power plant	47
Diesel engine	45
Fluorescent lamp	40
Automobile engine (gasoline)	25
Incandescent lamp	10

*Efficiency with which the energy of the power or fuel source is converted to a useful form.
Source: Data from several sources.

cars results in pollution. The emissions from power plants pollute. All of these are examples of the effect of the second law of thermodynamics. If each individual on Earth used less energy, there would be less waste heat and other forms of pollution that result from energy conversion. The amount of energy in the universe is limited. Only a small portion of that energy is of high quality. The use of high-quality energy decreases the amount of useful energy available, as more low-quality heat is generated. All life and all activities are subject to these important physical principles described by the first and second laws of thermodynamics.

Summary

Science is a method of gathering and organizing information. It involves observation, asking questions, exploring alternative sources of information, hypothesis formation, the testing of hypotheses, and publication of the results for others to evaluate. A hypothesis is a logical prediction about how things work that must account for all the known information and be testable. The process of science attempts to be careful, unbiased, and reliable in the way information is collected and evaluated. This often involves conducting experiments to test the validity of a hypothesis. If a hypothesis is continually supported by the addition of new facts, it may be incorporated into a theory. A theory is a broadly written, widely accepted generalization that ties together large bodies of information and explains why things happen. Similarly, a law is a broad statement that describes what happens in nature.

The fundamental unit of matter is the atom, which is made up of protons and neutrons in the nucleus surrounded by a cloud of moving electrons. The number of protons for any one type of atom is constant, but the number of neutrons in different atoms of the same type of atom may vary. The number of electrons is equal to the number of protons. Protons have a positive charge, neutrons lack a charge, and electrons have a negative charge.

Molecules are units made of a combination of two or more atoms bonded to one another. Chemical bonds are physical at-

tractions between atoms resulting from the interaction of their electrons. When chemical bonds are broken or formed, a chemical reaction occurs, and the amount of energy within the chemical bonds is changed. Chemical reactions require activation energy to get the reaction started.

Matter that is composed of only one kind of atom is known as an element. Matter that is composed of small units containing different kinds of atoms bonded in specific ratios is known as a compound. An atom or molecule that has gained or lost electrons so that it has an electric charge is known as an ion.

Matter can occur in three states: solid, liquid, and gas. These three differ in the amount of energy the molecular units contain and the distance between the units. Kinetic energy is the energy contained by moving objects. Potential energy is the energy an object has because of its position.

The first law of thermodynamics states that the amount of energy in the universe is constant, that energy can neither be created nor destroyed. The second law of thermodynamics states that when energy is converted from one form to another, some of the useful energy is lost (entropy increases). Some forms of energy are more useful than others. The quality of the energy determines how much useful work can be accomplished by expending the energy. Low-temperature heat sources are of poor quality, since they cannot be used to do useful work.

Issues – Analysis

Biofuels Technology

When Rudolf Diesel designed his prototype diesel engine nearly a century ago, he ran it on peanut oil. He envisioned that diesel engines would operate on a variety of vegetable oils. But when petroleum-based diesel fuel hit the marketplace, it was cheap, reasonably efficient, and readily available, and it quickly became the diesel fuel of choice.

Today, however, due to the issue of diversifying U.S. fuel resources and concerns about the potential of global climate change, the impetus is greater to develop biodiesel as an alternative to petroleum diesel. Since biodiesel is made from feedstocks that take carbon dioxide out of the atmosphere, using biodiesel adds no net carbon dioxide to the atmosphere, which is important to reducing the buildup of greenhouse gases.

When Henry Ford first designed his Model T automobile in 1908, he expected ethanol, made from renewable resources, to be the major fuel used. From 1920 to 1924, the Standard Oil Company marketed a 25 percent by volume absolute ethanol in gasoline in the Baltimore area, but high corn prices combined with storage and transportation difficulties terminated the project. During the 1930s, more than 2000 service stations in the Midwest sold a mixture of gasoline and ethanol made from corn and called it gasohol. Low petroleum prices during the 1940s virtually eliminated gasohol production.

In 1979, ethanol-gasoline blends were reintroduced to the U.S. market when oil supply disruptions in the Middle East became a national security issue and Americans were waiting in long lines to buy gasoline. The American Oil Company and several other oil companies began to market ethanol blends as a "gasoline extender" and an octane enhancer.

The U.S. Clean Air Act Amendments of 1990 mandated the sale of oxygenated fuels in areas of the country with unhealthy levels of carbon monoxide. Since then, demand has been strong for ethanol as an oxygenate blended with gasoline. E10 (a blend containing 10 percent ethanol) has been the most commonly distributed. Blends of E85 and E95 have been successfully tested in North America in government fleet vehicles and urban buses.

As the demand for ethanol increases, other biomass resources, such as agricultural and forestry wastes, municipal solid wastes, industrial wastes, and crops grown solely for energy purposes, will be used to make ethanol. Research activities over the past 20 years have developed technology to convert these feedstocks to ethanol.

Converting biomass to bioethanol involves four basic steps:

1. Producing biomass results in the fixing of atmospheric carbon dioxide into organic carbon.
2. Converting this biomass to a usable fermentation feedstock (typically, some form of sugar) can be achieved using a variety of process technologies.
3. Fermenting sugar by using yeasts produces ethanol in a relatively dilute aqueous solution that is probably the oldest form of biotechnology developed by humankind.
4. Processing the dilute mixture yields fuel grade ethanol and by-products that can be used to produce other fuels, chemicals, heat, and electricity.

- Are there benefits to producing and using biofuels, aside from the transportation fuel supply issue?
- Is it realistic to expect that biofuels will replace a significant amount of petroleum-based fuels?
- What are the negative aspects of using biofuels?

CO_2 → Produce biomass → Convert and/or process biomass to fermentation feedstock → Ferment biomass intermediates to ethanol → Recover ethanol and by-products

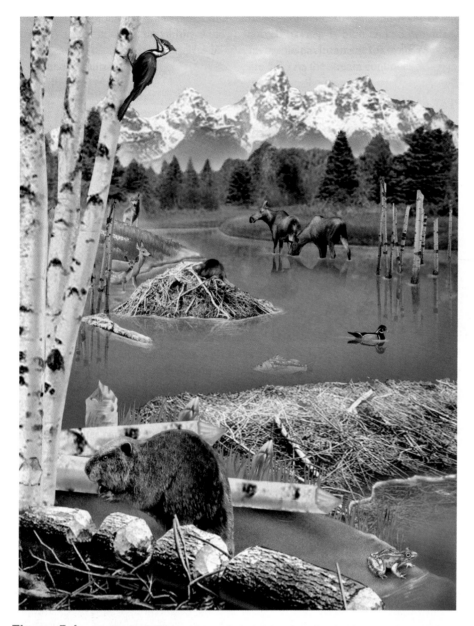

Figure 5.4 **Ecological Niche** The ecological niche of an organism is a complex set of interactions between an organism and its surroundings, which includes all of the ways an organism influences its surroundings as well as all of the ways the organism is affected by its environment. A beaver's ecological niche includes building dams and flooding forested areas, killing trees, providing habitat for ducks and other animals, serving as food for predators, and many other effects.

Figure 5.5 **The Niche of a Dandelion** A dandelion is a familiar plant that commonly invades disturbed sites because it produces many seeds that are blown easily to new areas. It serves as food to various herbivores, supplies nectar to bees, and can regrow quickly from its root if its leaves are removed.

velop an understanding of the processes that led to this situation. Furthermore, since the mechanisms that result in adaptation occur within a species, we need to understand the nature of a species.

Genes, Populations, and Species

We can look at an organism from several points of view. We can consider an individual, groups of individuals of the same kind, or groups that are distinct from other groups. This leads us to discuss three interrelated concepts. **Genes** are distinct pieces of DNA that determine the characteristics an individual displays. There are genes for structures such as leaf shape or feather color, behaviors such as cricket chirps or migratory activity, and physiological processes such as photosynthesis or muscular contractions. Each individual has a particular set of genes. A **population** is considered to be all the organisms of the same kind found within a specific geographic region. The individuals of a population will have very similar sets of genes, although there will be some individual variation. Because some genetic difference exists among individuals in a population, a population contains more kinds of genes than any individual within the population. Reproduction also takes place among individuals in a population so that genes are passed from one generation to the next. The concept of a species is an extension of this thinking about genes,

sunny, disturbed site, including lawns. Since it is a plant, one major aspect of its niche is the ability to carry on photosynthesis and grow. Dandelions need direct sunlight to grow successfully. Mowing lawns helps provide just the right conditions for dandelions, since the vegetation is never allowed to get so tall that dandelions are shaded. Many kinds of animals, including some humans, use the plant for food. The young leaves may be eaten in a salad, and the blossoms can be used to

make dandelion wine. Bees visit the flowers regularly to obtain nectar and pollen.

The Role of Natural Selection and Evolution

Since organisms generally are well adapted to their surroundings and fill a particular niche, it is important that we de-

Figure 5.10 **Commensalism** Remoras hitch a ride on sharks and feed on the scraps of food lost by the sharks. This is a benefit to the remoras. The sharks do not appear to be affected by the presence of the remoras.

The root-fungus associations are called **mycorrhizae.** The fungus obtains organic molecules from the roots of the plant, and the branched nature of the fungus assists the plant in obtaining nutrients such as phosphates and nitrates. In many cases, it is clear that the relationship is obligatory.

Some Relationships Are Difficult to Categorize

Sometimes it is not easy to categorize the relationships that organisms have with each other. For example, it is not always easy to say whether a relationship is a predator-prey relationship or a host-parasite relationship. How would you classify a mosquito or a tick? Both of these animals require blood meals to live and reproduce. They don't kill and eat their prey. Neither do they live in or on a host for a long period of time. This question points out the difficulty encountered when we try to place all kinds of organism interactions into a few categories. However, we can eliminate this problem if we call them temporary parasites or blood predators.

Another relationship that doesn't fit well is the relationship that certain birds such as cowbirds and European cuckoos have with other birds. Cowbirds and European cuckoos do not build nests but lay their eggs in the nests of other species of birds, who are left to care for a foster nestling at the expense of their own nestlings, who generally die. This situation is usually called nest parasitism or brood parasitism. (See figure 5.12.)

Figure 5.11 **Mutualism** The growths on the roots of this plant contain beneficial bacteria that make nitrogen available to the plant. The relationship is also beneficial to the bacteria, since the bacteria obtain necessary raw materials from the plant. It is a mutually beneficial relationship.

What about grazing animals? Are they predators or parasites on the plants that they eat? Sometimes they kill the plant they eat, while at other times they simply remove part of the plant and the rest continues to grow. In either case, the plant has been harmed by the interaction and the grazer has benefited.

There are also mutualistic relationships that do not require permanent contact between the participants in the relationship. Bees and the flowering plants they pollinate both benefit from their interactions. The bees obtain pollen and nectar for food and the plants are pollinated. But the active part of the relationship involves only a part of the life of any plant, and the bees are not restricted to any one species of plant for the food. They must actually switch to different flowers at different times of the year.

Human Interaction—A Different Look

In terms of our ability to modify ecosystems, humans are the dominant organisms on Earth. Our niche is very broad and we

Figure 5.12 **Nest (Brood) Parasitism** This red-eyed vireo is feeding the nestling of a brown-headed cowbird. A female cowbird laid its egg in the vireo's nest. The vireo is harmed because it is not raising its own young, and the cowbird benefits because it did not need to expend energy to build and defend a nest or collect food for its own young.

interact in many ways with the organisms with which we share the planet. If we examine our activities, we can see that we have complicated interactions with other organisms, and these interactions can be placed into the same categories we use to describe relationships between nonhuman organisms.

- Predator—Humans throughout the world use animals as food. Some actually kill animals themselves, while others rely on employees of slaughterhouses to do the killing for them.

- Herbivore—Humans rely on many kinds of plants as their primary source of food.

  Grazing and browsing involve consuming part of a living plant without killing the entire plant. We graze or browse parts of plants such as asparagus, rhubarb, lettuce, broccoli, and many other plants.

  Foraging involves searching for food that is available from nature. Hunter-gatherer peoples spend a great deal of time foraging for

Environmental Close-Up

Human Health and Exotic Species

Monkeypox is a communicable viral disease that occurs in several kinds of mammals, primarily in rainforest regions of central and west Africa. It is similar to smallpox but generally a much milder form of disease. Discovered in 1958 in Africa, monkeypox was first reported in prairie dogs and humans in the United States in May 2003. Individuals began contracting monkeypox primarily as a result of contact with prairie dogs that had contracted monkeypox from diseased imported African rodents. Investigations implicated a shipment of imported animals from Ghana that arrived in Texas on April 9, 2003. This shipment contained approximately 800 small mammals of nine different species, including six genera of African rodents, which came in contact with prairie dogs.

In June 2003, the Centers for Disease Control (CDC) implemented an immediate embargo on the importation of all rodents from Africa. As a public health measure, the Food and Drug Administration (FDA), along with the CDC, prohibited the transportation in interstate commerce or sale or offering for sale or any other type of commercial or public distribution, including release into the environment, of prairie dogs and the following rodents from Africa: tree squirrels, rope squirrels, dormice,

Gambian giant pouched rats, brush-tailed porcupines, and striped mice. The 2003 embargo was replaced by an interim final rule, with further restrictions by the FDA in November 2003.

The initial outbreak of monkeypox linked to exotic animals caused 37 confirmed, 12 probable, and 22 suspect human cases in the United States. While these numbers do not appear to be that great, there is no way of knowing how widespread such an outbreak could be. Emerging infectious diseases that originate in animals, such as monkeypox, plague, and West Nile virus, do pose a threat to public health in North America.

In early 2003, an outbreak of a new viral disease known as severe acute respiratory syndrome (SARS) originated in China. It is a variation of a coronavirus, a class of virus commonly associated with the common cold, but it causes severe symptoms and, if untreated, may result in death. In June 2003, the SARS virus was isolated from an animal known as the masked palm civet. This animal is used for food in China and is the likely origin of this new human disease. In January 2004, the government of China ordered the killing of thousands of masked palm civets when a new case of SARS was detected in that country.

---

edible plant materials such as roots, fruits, and seeds. Even individuals from sophisticated cultures engage in foraging activities when they pick wild berries, mushrooms, or asparagus.

Scavenger—Scavenging involves finding and consuming animals that are already dead. Our distant ancestors were probably actively engaged in scavenging by seizing the kills of more efficient carnivores. Even today in places where protein-rich food is in short supply, any recently dead animal is a valued food source. And many states in the United States have laws that allow people to take animals killed by collisions with automobiles.

Commensalism—Humans find themselves on both sides of commensal relationships. Many kinds of organisms use our homes as places to live without affecting us. Birds may nest on our buildings, spiders build webs in our windows, and rats may live under our decks. Other animals benefit from the animals we accidentally kill along our highways. We also derive benefit from organ-

isms without affecting them, such as when we are able to get out of the hot sun by sitting under the shade of a tree or when we rely on decomposers to decay wastes.

Parasitism—Although humans do not live in or on other living things, we do engage in relationships that are parasitic in nature. In some African cultures, blood is drawn from cows and mixed with milk to serve as food. Maple syrup is made by "bleeding" maple trees. Similarly, humans tap rubber trees to obtain sap that is used to make rubber. Many human activities regularly rob animals. Honeybees are robbed of their honey and chickens are robbed of their eggs.

Mutualism—Humans have many mutualistic relationships with plants and animals. Our domesticated plants and animals rely on us for support and nutrition, and we extract from them payment in the form of companionship, food, or other valuable resources.

Competition—Humans are in competition with all other organisms on Earth. As we convert land and

aquatic resources to our uses, we deprive other organisms of what they need to survive. When humans hunt and kill large grazing animals such as bison or gazelles, we are in direct competition with other predators such as wolves or leopards. Because of our technological superiority and our huge population, we usually win in the game of competition.

Community and Ecosystem Interactions

Thus far, we have discussed specific ways in which individual organisms interact with one another and with their physical surroundings. However, often it is useful to look at ecological relationships from a broader perspective. Two concepts that focus on relationships that involve many different kinds of interactions are community and ecosystem. A **community** is an assemblage of all the interacting populations of different species of organisms in an area. Some species play minor roles, while others play major roles, but all are part of the community. For example, the grasses of the prairie have a major role since they carry on photosynthesis and provide food and shelter for the animals that live in the area. Grasshoppers, prairie dogs, and bison are important consumers of grass. Meadowlarks consume many kinds of insects, and though they are a conspicuous and colorful part of the prairie scene, they have a relatively minor role and little to do with maintaining a prairie community. Bacteria and fungi in the soil break down the bodies of dead plants and animals and provide nutrients to plants.

Communities consist of interacting populations of different species, but these species interact with their physical world as well. An **ecosystem** is a defined space in which interactions take place between a community, with all its complex interrelationships, and the physical environment. The physical world has a major impact on what kinds of plants and animals can live in an area. We do not expect to see a banana tree in the Arctic or a walrus in the Mississippi River. Banana trees are adapted to warm, moist, tropical areas, and walruses require cold ocean waters. Some ecosystems, such as grasslands and certain kinds of forests, are shaped by periodic fires. The kind of soil and the amount of moisture also influence the kinds of organisms found in an area.

While it is easy to see that the physical environment places limitations on the kinds of organisms that can live in an area, it is also important to recognize that organisms impact their physical surroundings. Trees break the force of the wind, grazing animals form paths, and earthworms create holes that aerate the soil. While the concepts of community and ecosystem are closely related, an ecosystem is a broader concept because it involves physical as well as biological processes.

Every system has parts that are related to one another in specific ways. A bicycle has wheels, a frame, handlebars, brakes, pedals, and a seat. These parts must be organized in a certain way or the system known as a bicycle will not function. Similarly, ecosystems have parts that must be organized in specific ways or the systems will not operate. To more fully develop the concept of ecosystem, we will look at ecosystems from three points of view: the major roles played by organisms, the way energy is utilized within ecosystems, and the way atoms are cycled from one organism to another.

Major Roles of Organisms in Ecosystems

Several categories of organisms are found in any ecosystem. **Producers** are organisms that are able to use sources of energy to make complex, organic molecules from the simple inorganic substances in their environment. In nearly all ecosystems, energy is supplied by the sun, and organisms such as plants, algae, and tiny aquatic organisms called phytoplankton use light energy to carry on photosynthesis. Since producers are the only organisms in an ecosystem that can trap energy and make new organic material from inorganic material, all other organisms rely on producers as a source of food, either directly or indirectly. These other organisms are called **consumers** because they consume organic matter to provide themselves with energy and the organic molecules necessary to build their own bodies. An important part of their role is the process of respiration in which they break down organic matter to inorganic matter.

However, some consumers have significantly different roles from others. **Primary consumers,** also known as **herbivores,** are animals that eat producers (plants or phytoplankton) as a source of food. Herbivores, such as leaf-eating insects and seed-eating birds, are usually quite numerous in ecosystems, where they serve as food for the next organisms in the chain. **Secondary consumers** or **carnivores** are animals that eat other animals. Secondary consumers can be further subdivided into categories based on what kind of prey they capture and eat. Some carnivores, such as ladybird beetles, primarily eat herbivores, such as aphids; others, such as eagles, primarily eat fish that are themselves carnivores. While these are interesting conceptual distinctions, most carnivores will eat any animal they can capture and kill. In addition, many animals, called **omnivores,** include both plants and animals in their diet. Even animals that are considered to be carnivores (foxes, bears) regularly include large amounts of plant material in their diets. Conversely, animals often thought of as herbivores (mice, squirrels, seed-eating birds) regularly consume animals as a source of food.

A final category of consumer is the decomposer. **Decomposers** are organisms that use nonliving organic matter as a source of energy and raw materials to build their bodies. Whenever an organism sheds a part of itself, excretes waste products, or dies, it provides a source of food for decomposers. Since decomposers carry on respiration, they are extremely important in recycling matter by converting organic matter to inorganic material. Many small animals, fungi, and bacteria fill this niche. (See table 5.1.)

Table 5.1 Roles in an Ecosystem

Category	Major Role or Action	Examples
Producer	Converts simple inorganic molecules into organic molecules by the process of photosynthesis	Trees, flowers, grasses, ferns, mosses, algae
Consumer	Uses organic matter as a source of food	Animals, fungi, bacteria
Herbivore	Eats plants directly	Grasshopper, elk, human vegetarian
Carnivore	Kills and eats animals	Wolf, pike, dragonfly
Omnivore	Eats both plants and animals	Rats, raccoons, most humans
Scavenger	Eats meat but often gets it from animals that died by accident or illness, or were killed by other animals	Coyote, vulture, blowflies
Parasite	Lives in or on another living organism and gets food from it	Tapeworm, many bacteria, some insects
Decomposer	Returns organic material to inorganic material; completes recycling of atoms	Fungi, bacteria, some insects and worms

Keystone Species

Ecosystems typically consist of many different species interacting with each other and their physical surroundings. However, some species have more central roles than others. In recognition of this idea, ecologists have developed the concept of keystone species. A **keystone species** is one that has a critical role to play in the maintenance of specific ecosystems. In prairie ecosystems, grazing animals are extremely important in maintaining the mix of species typical of a grassland. Without the many influences of the grazers, the nature of the prairie changes. A study of the American tallgrass prairie indicated that when bison are present, they increase the biodiversity of the site. Bison typically eat grasses and, therefore, allow smaller plant species that would normally be shaded by tall grasses to be successful. In ungrazed plots, the tall grasses become the dominant vegetation and biodiversity decreases. Bison dig depressions in the soil, called wallows, to provide themselves with dust or mud with which they can coat themselves. These wallows retain many species of plants that typically live in disturbed areas. Bison urine has also been shown to be an important source of nitrogen for the plants.

The activities of bison even affect the extent and impact of fire, another important feature of grassland ecosystems. Since bison prefer to feed on recently burned sites and revisit these sites several times throughout the year, they tend to create a patchwork of grazed and ungrazed areas. The grazed areas are less likely to be able to sustain a fire, and fires likely will be more prevalent in ungrazed patches.

The concept of keystone species has also been studied in marine ecosystems. The relationship between sea urchins, sea otters, and kelp forests suggests that sea otters are a keystone species. Sea otters eat sea urchins, which eat kelp. A reduction in the number of otters results in an increase in the number of sea urchins. Increased numbers of sea urchins lead to heavy grazing of the kelp by sea urchins. When the amount of kelp is severely reduced, fish and many other animals that live within the kelp beds lose their habitat and biodiversity is significantly reduced.

The concept of keystone species is useful to ecologists and resource managers because it helps them to realize that all species cannot be treated equally. Some species have pivotal roles, and their elimination or severe reduction can significantly alter ecosystems. In some cases, the loss of a keystone species can result in the permanent modification of an ecosystem into something considerably different from the original mix of species.

Energy Flow Through Ecosystems

An ecosystem is a stable, self-regulating unit. This does not mean that an ecosystem is unchanging. The organisms within it are growing, reproducing, dying, and decaying. In addition, an ecosystem must have a continuous input of energy to remain its stability. The only significant source of energy for most ecosystems is sunlight. Producers are the only organisms that are capable of trapping solar energy through the process of photosynthesis and making it available to the ecosystem. The energy is stored in the form of chemical bonds in large organic molecules such as carbohydrates (sugars, starches), fats, and proteins. The energy stored in the molecules of producers is transferred to other organisms when the producers are eaten. Each step in the flow of energy through an ecosystem is known as a **trophic level.** Producers (plants, algae, phytoplankton) constitute the first trophic level, and herbivores constitute the second trophic level. Carnivores that eat herbivores are the third trophic level, and carnivores that eat other carnivores are the fourth trophic level. Omnivores, parasites, and scavengers occupy different trophic levels, depending on what they happen to be eating at the time. If we eat a piece of steak, we are at the third trophic level; if we eat celery, we are at the second trophic level. (See figure 5.13.)

The second law of thermodynamics states that whenever energy is converted from one form to another, some of the energy is converted to a nonuseful form (typically, low-quality heat). Thus, there is always less useful energy following an energy conversion. When energy passes

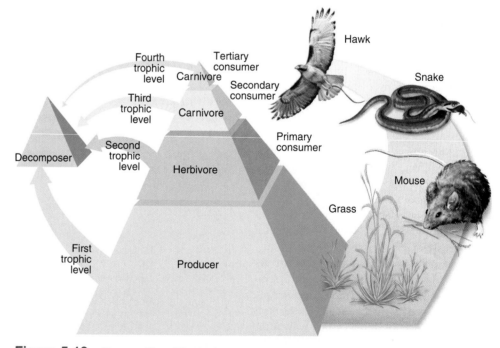

Figure 5.13 **Energy Flow Through an Ecosystem** As energy flows through an ecosystem, it passes through several levels known as trophic levels. Each trophic level contains a certain amount of energy. Each time energy flows to another trophic level, approximately 90 percent of the useful energy is lost, usually as heat to the surroundings. Therefore, in most ecosystems, higher trophic levels contain less energy and fewer organisms.

from one trophic level to the next, there is less useful energy left with each successive trophic level. This loss of low-quality heat is dissipated to the surroundings and warms the air, water, or soil. In addition to this loss of heat, organisms must expend energy to maintain their own life processes. It takes energy to chew food, defend nests, walk to waterholes, or produce and raise offspring. Therefore, the amount of energy contained in higher trophic levels is considerably less than that at lower levels. Approximately 90 percent of the useful energy is lost with each transfer to the next highest trophic level. So in any ecosystem, the amount of energy contained in the herbivore trophic level is only about 10 percent of the energy contained in the producer trophic level. The amount of energy at the third trophic level is approximately 1 percent of that found in the first trophic level.

Because it is difficult to actually measure the amount of energy contained in each trophic level, ecologists often use other measures to approximate the relationship between the amounts of energy at each level. One of these is the biomass. The **biomass** is the weight of living material in a trophic level. It is often possible in a simple ecosystem to collect and weigh all the producers, herbivores, and carnivores. The weights often show the same 90 percent loss from one trophic level to the next as happens with the amount of energy.

Food Chains and Food Webs

The passage of energy from one trophic level to the next as a result of one organism consuming another is known as a **food chain.** For example, willow trees grow well in very moist soil, perhaps near a pond. The trees' leaves capture sunlight and convert carbon dioxide and water into sugars and other organic molecules. The leaves serve as a food source for insects, such as caterpillars and leaf beetles, that have chewing mouth parts and a digestive system adapted to plant food. Some of these insects are eaten by spiders, which fall from the trees into the pond below, where they are consumed by a frog. As the frog swims from one lily pad to another, a large bass consumes the frog. A human

may use an artificial frog as a lure to entice the bass from its hiding place. A fish dinner is the final step in this chain of events that began with the leaves of a willow tree. (See figure 5.14.) This food chain has six trophic levels. Each organism occupies a specific niche and has special abilities that fit it for its niche, and each organism in the food chain is involved in converting energy and matter from one form to another.

Some food chains rely on a constant supply of small pieces of dead organic material coming from situations where photosynthesis is taking place. The small bits of nonliving organic material are called **detritus.** Detritus food chains are found in a variety of situations. The bottoms of the deep lakes and oceans are too dark for photosynthesis. The animals and decomposers that live there rely on a steady rain of small bits of organic matter from the upper layers of the water where photosynthesis does take place. Similarly, in most streams, leaves and other organic debris serve as the major source of organic material and energy. A sewage treatment plant is also a detritus food chain in which particles and dissolved organic matter are constantly supplied to a series of bacteria and protozoa that use this material for food.

In another example, the soil on a forest floor receives leaves, which fuel a detritus food chain. In detritus food chains, a mixture of insects, crustaceans, worms, bacteria, and fungi cooperate in the breakdown of the large pieces of organic matter, while at the same time feeding on one another. When a leaf dies and falls to the forest floor, it is colonized by bacteria and fungi, which begin the breakdown process. An earthworm will also feed on the leaf and at the same time consume the bacteria and fungi. If that earthworm is eaten by a bird, it becomes part of a larger food chain that includes material from both a detritus food chain and a photosynthesis-driven food chain. When several food chains overlap and intersect, they make up a **food web.** (See figure 5.15.) This diagram is typical of the kinds of interactions that take place in a community. Each organism is likely to be a food source for several other kinds of organisms. Even the simplest food webs are complex.

Figure 5.14 **Food Chain** As one organism feeds on another organism, energy flows through the series. This is called a food chain.

Notice in the upper-left-hand corner of figure 5.15 that the Cooper's and sharp-shinned hawks use many different kinds of birds as a source of food. These hawks

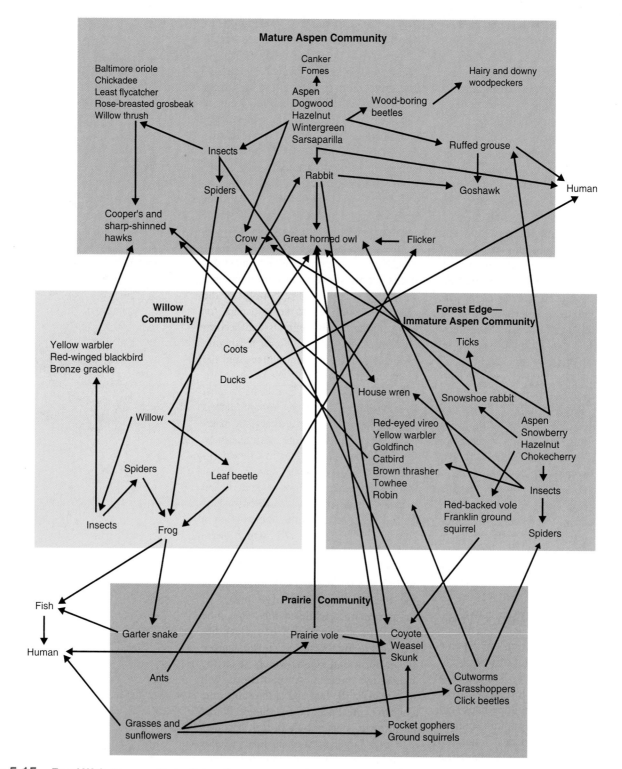

Figure 5.15 **Food Web** The many kinds of interactions among organisms in an ecosystem constitute a food web. In this network of interactions, several organisms would be affected if one key organism were reduced in number. Look at the rabbit in the mature aspen community and note how many organisms use it as food.

Source: R.D. Bird, "Biotic Communities of the Aspen Parkland of Central Canada," *Ecology*, 11 (April 1930): 410.

Contaminants in the Food Chain of Fish from the Great Lakes

As organisms eat, molecules move through the food chain from one organism to another. Most molecules are simply recycled through nutrient pathways because they can be broken down by decay organisms. However, humans have invented a large number of organic molecules that are not easily decomposed, and these tend to remain in food chains and accumulate in higher concentrations in organisms that are at higher trophic levels.

The Great Lakes area developed as an industrial center because the lakes provided an efficient way to move raw materials and products and because water was important for manufacturing processes. In the past, many of these industries released heavy metals and organic molecules into the water as an accidental by-product of the manufacturing process or because it was a cheap way to get rid of unwanted material. Many of the organic molecules are products of modern organic synthesis and, therefore, are not something that bacteria and fungi are able to decay. Abnormally high concentrations of inorganic materials in the environment can also lead to abnormally high concentrations in organisms.

Approximately 500 different organic compounds that scientists think are contaminants have been identified in the bodies of fish from the Great Lakes. Most are present in extremely small amounts and probably do not represent a serious hazard, but others are present in high enough concentrations to cause public health officials to be concerned. Since these materials do not break down, fish tend to accumulate more of these toxic materials in their bodies as they get older. Furthermore, carnivorous fish that are feeding at higher trophic levels accumulate more of these compounds. It just so happens that most of the fish that people catch and eat are carnivores, which tend to have the highest concentrations of contaminants.

If people place themselves in this food chain by eating the contaminated fish, they will tend to accumulate the contaminants in their bodies, which could have health effects. It is not possible to check every fish caught to see if it is fit to eat. The cost of doing so would be on the order of several hundred to several thousands of dollars per individual fish, depending on which contaminants are looked for. Therefore, the states and provinces surrounding the Great Lakes have developed advisory statements to help people avoid unsafe fish. While there are some minor differences among state and provincial governments, the following advisories are typical.

1. Some species of fish, such as carp and catfish, feed on the bottom and tend to accumulate contaminants that are in the bottom sediments. In many areas of the Great Lakes, people are advised not to eat these fish.

2. Larger fish have generally consumed more food and have had an opportunity to accumulate more contaminants. Therefore, people are advised to eat only smaller fish.

3. Since many of the organic contaminants are fat soluble, removal of the fat or cooking in such a way that the fat is allowed to separate from the flesh is also advised in many areas.

4. Because the amount of contamination a person is exposed to is directly related to the number of fish a person eats, people are advised to limit the number of fish consumed, and women of childbearing age and young children are often advised to not eat the fish at all.

The lip cancer on this bullhead indicates carcinogens in Wisconsin's Fox River. In some tributaries of the Great Lakes, fish cancer rates may reach 84 percent.

---

fit into several food chains. If one source of prey is in short supply, they can switch to something else without too much trouble. These kinds of complex food webs tend to be more stable than simple food chains with few cross-links.

Nutrient Cycles in Ecosystems— Biogeochemical Cycles

All matter is made up of atoms. These atoms are cycled between the living and nonliving portions of an ecosystem. The activities involved in the cycling of atoms include biological, geological, and chemical processes. Therefore, these nutrient cycles are often called **biogeochemical cycles.** Some atoms are more common in living things than are others. Carbon, nitrogen, oxygen, hydrogen, and phosphorus are found in important organic molecules such as proteins, DNA, carbohydrates, and fats, which are found in all kinds of living things. Organic molecules contain large numbers of carbon atoms attached to one another. These organic molecules are initially manufactured from inorganic molecules by the activities of producers and are transferred from one living organism to another in food chains. The processes of respiration and decay ultimately break down the complex organic molecules of organisms and convert them to simpler, inorganic constituents that are returned to the abiotic environment. In this section, we will look at the flow of three kinds of atoms within communities and between the biotic and abiotic portions of an ecosystem: carbon, nitrogen, and phosphorus.

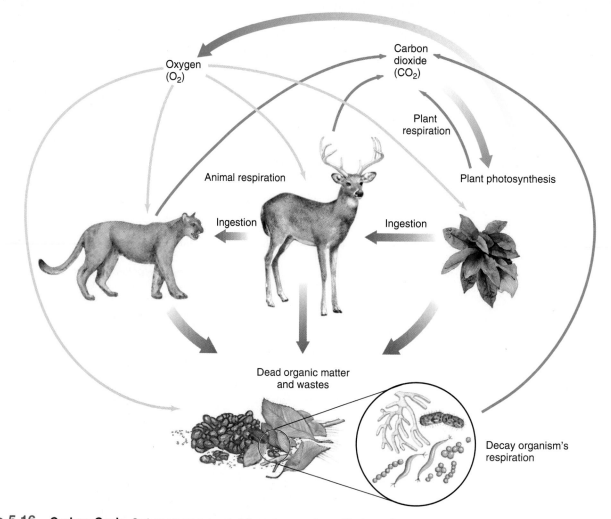

Figure 5.16 **Carbon Cycle** Carbon atoms are cycled through ecosystems. Plants can incorporate carbon atoms from carbon dioxide into organic molecules when they carry on photosynthesis. The carbon-containing organic molecules are passed to animals when they eat plants or other animals. Organic wastes or dead organisms are consumed by decay organisms. All organisms, plants, animals, and decomposers return carbon atoms to the atmosphere when they carry on respiration. Oxygen atoms are being cycled at the same time that carbon atoms are being cycled.

Carbon Cycle

All living things are composed of organic molecules that contain atoms of the element carbon. The **carbon cycle** includes the processes and pathways involved in capturing inorganic carbon-containing molecules, converting them into organic molecules that are used by organisms, and the ultimate release of inorganic carbon molecules back to the abiotic environment. (See figure 5.16.) Carbon and oxygen combine to form the molecule carbon dioxide (CO_2), which is present in small quantities as a gas in the atmosphere and dissolved in water. During photosynthesis, carbon dioxide from the atmosphere is taken into the leaves of plants where it is combined with hydrogen from water molecules (H_2O), which are absorbed from the soil by the roots and transported to the leaves. Many kinds of aquatic organisms such as algae and some bacteria also perform photosynthesis but absorb carbon dioxide and water molecules from the water in which they live. The energy needed to perform photosynthesis is provided by sunlight. As a result of photosynthesis, complex organic molecules such as carbohydrates (sugars) are formed. Oxygen molecules (O_2) are released into the atmosphere or water during the process of photosynthesis because water molecules are split to provide hydrogen atoms necessary to manufacture carbohydrate molecules. The remaining oxygen is released as a waste product of photosynthesis. In this process, light energy is converted to chemical-bond energy in organic molecules, such as sugar. Plants and other producer organisms use these sugars for growth and to provide energy for other necessary processes.

Herbivores can use these complex organic molecules as food. When an herbivore eats plants or algae, it breaks down the complex organic molecules into simpler organic molecular building blocks, which can be reassembled into the specific organic molecules that are part of its chemical structure. The carbon atom, which was once part of an organic molecule in a producer, is now part of an organic molecule in an herbivore. Nearly all organisms also carry on the process of respiration, in which oxygen from the atmosphere is used to break down large organic molecules into carbon dioxide and water. Much of the chemical-bond energy is released by respiration and is lost as heat, but the remainder is used by the herbivore for movement, growth, and other activities.

In similar fashion, when an herbivore is eaten by a carnivore, some of the carbon-containing molecules of the herbivore become incorporated into the body of the carnivore. The remaining organic molecules are broken down in the process of respiration to obtain energy, and carbon dioxide and water are released.

The organic molecules contained in animal waste products and dead organisms are acted upon by decomposers that use these organic materials as a source of food. The decay process of decomposers involves respiration and releases carbon dioxide and water so that naturally occurring organic molecules are typically recycled.

In the carbon cycle, all organisms require organic molecules for their survival and must either manufacture them or consume them. Photosynthetic organisms (producers) capture inorganic carbon in the form of carbon dioxide molecules and manufacture organic molecules. Nearly all organisms, including plants, carry on respiration, in which organic molecules are broken down to provide energy and inorganic carbon dioxide is released. The same carbon atoms are used over and over again. In fact, you are not exactly the same person today that you were yesterday. Some of your carbon atoms are different. Furthermore, those carbon atoms have been involved in many other kinds of living things over the past several billion years. Some of them were temporary residents in dinosaurs, extinct trees, or insects, but at this instant, they are part of you.

Although this is a terrestrial example, it is important to recognize that this same cycle operates in aquatic systems as well. Carbon dioxide dissolves in water and, therefore, is available for aquatic plants and algae to use in the process of photo-synthesis. When the consumers in the food web use molecules of oxygen that are dissolved in water to carry on respiration, carbon dioxide is released back into the water.

Fossil fuels (coal, oil, and natural gas) are part of the carbon cycle as well. At one time, these materials were organic molecules in the bodies of living organisms. The organisms were buried and the organic compounds in their bodies were modified by geologic forces. Thus, the carbon atoms present in fossil fuels were removed temporarily from the active, short-term carbon cycle. When we burn fossil fuels, the carbon reenters the active carbon cycle.

Nitrogen Cycle

Another very important nutrient cycle, the **nitrogen cycle,** involves the cycling of nitrogen atoms between the abiotic and biotic components and among the organisms in an ecosystem. Seventy-eight percent of the gas in the air we breathe is made up of molecules of nitrogen gas (N_2). However, the two nitrogen atoms are bound very tightly to each other, and very few organisms are able to use nitrogen in this form. Since plants and other producers are at the base of nearly all food chains, they must make new nitrogen-containing molecules, such as proteins and DNA. Plants and other producers are unable to use the nitrogen in the atmosphere and must get it in the form of nitrate (NO_3^-) or ammonia (NH_3). Because atmospheric nitrogen is not usable by plants, nitrogen-containing compounds are often in short supply and the availability of nitrogen is often a factor that limits the growth of plants. The primary way in which plants obtain nitrogen compounds they can use is with the help of bacteria that live in the soil.

Bacteria, called **nitrogen-fixing bacteria,** are able to convert the nitrogen gas (N_2) that enters the soil into ammonia that plants can use. Certain kinds of these bacteria live freely in the soil and are called **free-living nitrogen-fixing bacteria.** Others, known as **symbiotic nitrogen-fixing bacteria,** have a mutualistic relationship with certain plants and live in nodules in the roots of plants known as legumes (peas, beans, and clover) and certain trees such as alders. Some grasses and evergreen trees appear to have a similar relationship with certain root fungi that seem to improve the nitrogen-fixing capacity of the plant.

Once plants and other producers have nitrogen available in a form they can use, they can construct proteins, DNA, and other important nitrogen-containing organic molecules. When herbivores eat plants, the plant protein molecules are broken down to smaller building blocks called amino acids. These amino acids are then reassembled to form proteins typical for the herbivore. This same process is repeated throughout the food chain.

Bacteria and other types of decay organisms are involved in the nitrogen cycle also. Dead organisms and their waste products contain molecules, such as proteins, urea, and uric acid, that contain nitrogen. Decomposers break down these nitrogen-containing organic molecules, releasing ammonia, which can be used directly by many kinds of plants. Still other kinds of soil bacteria called **nitrifying bacteria** are able to convert ammonia to nitrite, which can be converted to nitrate. Plants can use nitrate as a source of nitrogen for synthesis of nitrogen-containing organic molecules.

Finally, bacteria known as **denitrifying bacteria** are, under conditions where oxygen is absent, able to convert nitrite to nitrogen gas (N_2), which is ultimately released into the atmosphere. These nitrogen atoms can reenter the cycle with the aid of nitrogen-fixing bacteria. (See figure 5.17.)

Although a cyclic pattern is present in both the carbon cycle and the nitrogen cycle, the nitrogen cycle shows two significant differences. First, most of the difficult chemical conversions are made by bacteria and other microorganisms. Without the activities of bacteria, little nitrogen would be available and the world would be a very different place. Second, although nitrogen enters organisms by way of nitrogen-fixing bacteria and returns to the atmosphere through the actions of denitrifying bacteria, there is a secondary loop in the cycle that recycles nitrogen compounds directly from dead organisms and wastes directly back to producers.

In naturally occurring soil, nitrogen is often a limiting factor of plant growth. To

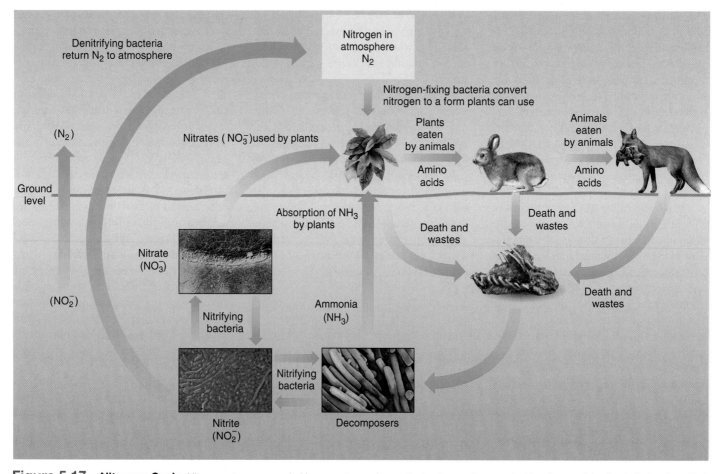

Figure 5.17 **Nitrogen Cycle** Nitrogen atoms are cycled in ecosystems. Atmospheric nitrogen is converted by nitrogen-fixing bacteria to a form that plants can use to make protein and other compounds. Proteins are passed to other organisms when one organism is eaten by another. Dead organisms and waste products are acted on by decay organisms to form ammonia, which may be reused by plants or converted to other nitrogen compounds by other kinds of bacteria. Denitrifying bacteria are able to convert inorganic nitrogen compounds into atmospheric nitrogen.

increase yields, farmers provide extra sources of nitrogen in several ways. Inorganic fertilizers are a primary method of increasing the nitrogen available. These fertilizers may contain ammonia, nitrate, or both.

Since the manufacture of nitrogen fertilizer requires a large amount of energy, fertilizer is expensive. Therefore, farmers use alternative methods to supply nitrogen and reduce their cost of production. Several different techniques are effective. Farmers can alternate nitrogen-yielding crops such as soybeans with nitrogen-demanding crops such as corn. Since soybeans are legumes that have symbiotic nitrogen-fixing bacteria in their roots, if soybeans are planted one year, the excess nitrogen left in the soil can be used by the corn plants grown the next year. Some farmers even plant alternating strips of soybeans and corn in the same field. A slightly different technique involves growing a nitrogen-fixing crop for a short period of time and then plowing the crop into the soil and letting the organic matter decompose. The ammonia released by decomposition serves as fertilizer to the crop that follows. This is often referred to as green manure. Farmers can also add nitrogen to the soil by spreading manure from animal production operations or dairy farms on the field and relying on the soil bacteria to decompose the organic matter and release the nitrogen for plant use.

Phosphorus Cycle

Phosphorus is another kind of element common in the structure of living things. It is present in many important biological molecules such as DNA and in the membrane structure of cells. In addition, the bones and teeth of animals contain significant quantities of phosphorus. The phosphorus cycle differs from the carbon and nitrogen cycles in one important respect. Phosphorus is not present in the atmosphere as a gas. The ultimate source of phosphorus atoms is rock. In nature, new phosphorus compounds are released by the erosion of rock and become dissolved in water. Plants use the dissolved phosphorus compounds to construct the molecules they need. Animals obtain the phosphorus they need when they consume plants or other animals. When an organism dies or excretes waste products, decomposer organisms recycle the phosphorus compounds back into the soil. Phosphorus compounds that are dissolved in water are ultimately precipitated as deposits. Geologic processes elevate these deposits and expose them to erosion, thus making these deposits available to organisms. Waste products of animals often have significant amounts of phosphorus. In places where large numbers of seabirds or bats congregate for hundreds of years, the thickness of their droppings (called guano) can be a significant source of phosphorus for fertilizer. (See figure 5.18.) In many soils, phosphorus is in short supply and must be provided to crop plants to get maximum yields. Phosphorus is also in short supply in aquatic ecosystems.

Fertilizers usually contain nitrogen, phosphorus, and potassium compounds. The numbers on a fertilizer bag indicate the percentage of each in the fertilizer. For example, a 6–24–24 fertilizer has 6 percent nitrogen, 24 percent phosphorus, and 24 percent potassium compounds. In addition to carbon, nitrogen, and phosphorus, potassium and other elements are cycled within ecosystems. In an agriculture ecosystem, these elements are removed when the crop is harvested. Therefore, farmers must not only return the nitrogen, phosphorus, and potassium, they must also analyze for other less prominent elements and add them to their fertilizer mixture as well. Aquatic ecosystems are also sensitive to nutrient levels. High levels of nitrates or phosphorus compounds often result in rapid growth of aquatic producers. In aquaculture, such as that used to raise catfish, fertilizer is added to the body of water to stimulate the production of algae which is the base of many aquatic food chains.

Human Impact on Nutrient Cycles

To appreciate how ecosystems function, it is important to have an understanding of how elements such as carbon, nitrogen, and phosphorus flow through them. When we look at these cycles from a global perspective, it is apparent that humans have significantly altered them in many ways. Two activities have caused significant changes in the carbon cycle: burning fossil fuels and converting forests to agricultural land. Fossil fuels are carbon-containing molecules produced when organisms were fossilized. Burning fossil fuels (coal, oil,

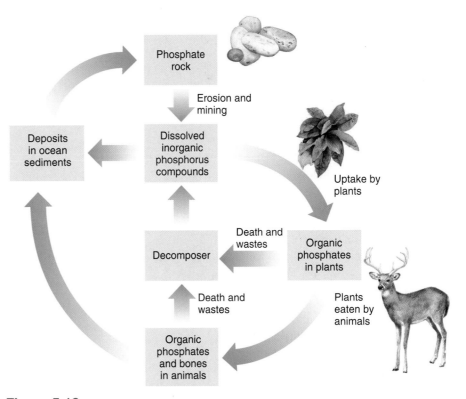

Figure 5.18 **Phosphorus Cycle** The source of phosphorus is rock that, when dissolved, provides the phosphate used by plants and animals.

and natural gas) releases large amounts of carbon dioxide into the atmosphere. The conversion of forest ecosystems that tend to store carbon for long periods to agricultural ecosystems that store carbon only temporarily has also disrupted the natural carbon cycle. Less carbon is being stored in the bodies of large, long-lived plants such as trees. One consequence of these actions is that the amount of carbon dioxide in the atmosphere has been increasing steadily since humans began to use fossil fuels extensively. It has become clear that increasing carbon dioxide is causing changes in the climate of the world, and many nations are seeking to reduce energy use and prevent deforestation. This topic will be discussed in more detail in chapter 17, which deals with air pollution.

The burning of fossil fuels has also altered the nitrogen cycle. When fossil fuels are burned, the oxygen and nitrogen in the air are heated to high temperatures and a variety of nitrogen-containing compounds are produced. These compounds are used by plants as nutrients for growth. Many people suggest that these sources of nitrogen, along with that provided by fer-

tilizers, have doubled the amount of nitrogen available today as compared to preindustrial times.

Fertilizer is used in agriculture to increase the growth of crops. These nutrients are intended to become incorporated into the bodies of the plants and animals that we raise for food. However, if too much nitrogen or phosphorus is applied as fertilizer or if they are applied at the wrong time, much of this fertilizer is carried into aquatic ecosystems. In addition, raising large numbers of animals for food in concentrated settings results in huge amounts of animal waste that contain nitrogen and phosphorus compounds that often enter local water sources. This addition of nitrogen and phosphorus to aquatic ecosystems is particularly significant since aquatic ecosystems normally are starved for these nutrients. The presence of large amounts of these nutrients in either freshwater or saltwater results in increased rates of growth of bacteria, algae, and aquatic plants. Increases in the number of these organisms can have many different effects. Many algae are toxic, and when their numbers increase significantly, fish are killed and incidents of human poisoning occur. An increase in the number of plants and algae in aquatic ecosystems also can lead to low oxygen concentrations in the water. When these organisms die, decomposers use oxygen from the water as they break down the dead organic matter. This lowers the oxygen concentration and many organisms die. For example, each summer a major "dead zone" develops in the Gulf of Mexico off the mouth of the Mississippi River. In 2002 it covered 22,000 square kilometers (8,600 square miles). This "dead zone" contains few fish and bottom-dwelling organisms. It is caused by low oxygen levels brought about by the rapid growth of algae and bacteria in the nutrient-rich waters. The nutrients can be traced to the extensive use of fertilizer in the major farming areas of the central United States, areas drained by the Mississippi River and its tributaries. Thus, fertilizer use in the agricultural center of the United States results in the death of fish in the Gulf of Mexico. (See figure 5.19.)

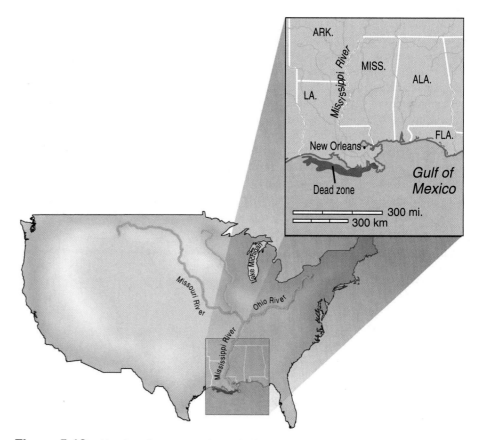

Figure 5.19 **Nutrient Impact on Aquatic Organisms** The Mississippi River drainage system carries water from the agricultural center of the United States to the Gulf of Mexico. Extensive use of fertilizer for agriculture results in nitrogen and phosphorus compounds being carried to the Gulf of Mexico. When the aquatic organisms that grew as a result of the fertilizer die, a major region of the Gulf of Mexico has oxygen concentrations too low to support most kinds of life.

Summary

Everything that affects an organism during its lifetime is collectively known as its environment. The environment of an organism can be divided into biotic (living) and abiotic (nonliving) components.

The space an organism occupies is known as its habitat, and the role it plays in its environment is known as its niche. The niche of a species is the result of natural selection directing the adaptation of the species to a specific set of environmental conditions.

Organisms interact with one another in a variety of ways. Predators kill and eat prey. Organisms that have the same needs compete with one another and do mutual harm, but one is usually harmed less and survives. Symbiotic relationships are those in which organisms live in physical contact with one another. Parasites live in or on another organism and derive benefit from the relationship, harming the host in the process. Commensal organisms derive benefit from another organism but

do not harm the host. Mutualistic organisms both derive benefit from their relationship.

A community is the biotic portion of an ecosystem that is a set of interacting populations of organisms. Those organisms and their abiotic environment constitute an ecosystem. In an ecosystem, energy is trapped by producers and flows from producers through various trophic levels of consumers (herbivores, carnivores, omnivores, and decomposers). About 90 percent of the energy is lost as it passes from one trophic level to the next. This means that the amount of biomass at higher trophic levels is usually much less than that at lower trophic levels. The sequence of organisms through which energy flows is known as a food chain. Several interconnecting food chains constitute a food web.

The flow of atoms through an ecosystem involves all the organisms in the community. The carbon, nitrogen, and phosphorus cycles are examples of how these materials are cycled in ecosystems.

PART TWO   Ecological Principles and Their Application

Issues – Analysis

Phosphate Mining in Nauru

The mining of phosphate on the island of Nauru, located in a remote corner of the Pacific Ocean, has devastated the island environmentally and has created financial, legal, and cultural problems for the islanders. The phosphate deposits found on Nauru and a few other Pacific Islands are a combination of limestone and guano from nesting seabirds that have accumulated for thousands of years. Nauru's phosphate is the only resource with which the island can sustain an economy. Nauru produces about 2 million metric tons of phosphate per year. Most of this is exported to Australia, where it is used as fertilizer.

Phosphate mining on Nauru generally occurs in the interior of the island. The phosphate is actually a composite of two materials that have combined and solidified over time: decayed oceanic microorganisms and bird droppings. These elements, intertwined with the coral and limestone that forms the island, and extraction of the phosphate have left behind deep pits and tall pillars, some as high as 35 meters (115 feet). This creates a moonlike scene, which contributes to the uninhabitable atmosphere. Four-fifths of the island is a barren wasteland.

As a result of the mining, the vast majority of soil and vegetation has been stripped away. This prevents agriculture from taking place and makes it very difficult for a viable ecosystem to establish itself and flourish. In addition, the combination of a pillar-and-pit landscape and the loss of vegetation creates a very hot interior, such that rising hot air prevents rain clouds from settling over the island. This contributes to frequent droughts on the island.

The mining that has taken place on Nauru for the past 90 years has had a physical toll on the islanders. Because of the lack of soil and vegetation, the Nauruans have been forced to import nearly all of their food. The result of eating processed, fatty foods has been an increase in the occurrence of high blood pressure, diabetes, and obesity. These problems have led to a decrease in the life expectancy of the islanders, which is between 50 and 60 years.

Nauru's problems are becoming increasingly acute, as the phosphate on the island has been nearly exhausted and mining has virtually ceased. Thus, the government of Nauru is looking into the question of responsibility for the ecological disaster and seeking ways to rehabilitate the island.

The Nauruans live on a strip of land along the coast, and with the population expanding, they need more living space. The population has

increased from 2000 in 1968 to over 6000 by 2004. What the island also needs is new construction, most notably a hospital, schools, and government buildings. This development can only occur in the central part of the island, which is currently a wasteland of limestone and coral.

Several ideas have been discussed regarding the future of the island. One solution is to crush the pillars and import topsoil, humus, and other nutrients, thus beginning the long process of rebuilding the ecosystem. This would be very expensive and could take more than 30 years. Creating an area for agriculture is paramount, and the island, for reasons of at least minimal self-sufficiency, must consider a water filter, fish and pig farms, and tree plantations. Another option is more drastic. It calls for the total removal of the population to another island. This solution is considered because the present island of Nauru is so damaged, and there are no other economically viable industries except the phosphate, which is now almost gone. This rather hopeless situation leads many to believe that evacuating the Nauruans to some other Pacific island is the only choice.

- Do you think the island can be rehabilitated?
- Should the countries that benefited from the phosphate mining be expected to rehabilitate the island?
- Since the cost of rehabilitating the island would be huge, is evacuating the island the only viable option?

Key Terms

abiotic factors *79*
biogeochemical cycles *95*
biomass *93*
biotic factors *79*
carbon cycle *96*
carnivore *91*
coevolution *85*
commensalism *88*
community *90*
competition *86*
competitive exclusion principle *87*
consumer *91*
decomposer *91*
denitrifying bacteria *97*
detritus *93*
ecology *79*
ecosystem *91*
ectoparasite *88*
endoparasite *88*

environment *79*
evolution *84*
extinction *85*
food chain *93*
food web *93*
free-living nitrogen-fixing bacteria *97*
genes *82*
habitat *80*
herbivore *91*
host *87*
interspecific competition *87*
intraspecific competition *87*
keystone species *92*
limiting factor *80*
mutualism *88*
mycorrhizae *89*
natural selection *84*
niche *80*
nitrifying bacteria *97*

nitrogen cycle *97*
nitrogen-fixing bacteria *97*
omnivore *91*
parasite *87*
parasitism *87*
population *82*
predation *86*
predator *86*
prey *86*
primary consumer *91*
producer *91*
range of tolerance *80*
secondary consumer *91*
speciation *85*
species *83*
symbiosis *87*
symbiotic nitrogen-fixing bacteria *97*
trophic level *92*
vector *88*

Review Questions

1. Define *environment*.
2. Describe, in detail, the niche of a human.
3. How is natural selection related to the concept of niche?
4. List five predators and their prey organisms.
5. How is an ecosystem different from a community?
6. Humans raising cattle for food is what kind of relationship?
7. Give examples of organisms that are herbivores, carnivores, and omnivores.
8. What are some different trophic levels in an ecosystem?
9. Describe the carbon cycle, the nitrogen cycle, and the phosphorus cycle.
10. Analyze an aquarium as an ecosystem. Identify the major abiotic and biotic factors. List members of the producer, primary consumer, secondary consumer, and decomposer trophic levels.

Critical Thinking Questions

1. Ecologists and political scientists look at habitat destruction differently. Consider the Environmental Close-Up about habitat conservation plans and the political, economic, and scientific issues that surround conservation plans. Identify some perspectives each discipline has to contribute to our understanding of habitat destruction. What values does each place on the ideas of the other discipline? What do you think about the issue of creating conservation plans? Protecting habitat from destruction? Why?

2. Even before humans entered the scene, many species of plants and animals were extinct and new ones had developed. Why are we even concerned about endangered species, given the fact that species have always come and gone?

3. Concentrations of industrial chemicals are high in some species of fish, high enough to call for an advisory to limit the number of fish a person should eat within a given period of time. Many of these chemicals are thought to cause cancer, but cancer is an effect that is often not felt for decades after exposure. How do scientists decide how many fish can be safely eaten? Is there any "safe" level? What evidence would convince you that there is danger? How could you tell?

4. You notice that after using pesticides on your farm field that the number of insects declines for a year. The next year, though, they come back and you need to reapply the pesticide. This time, though, there is less of an effect on the insect population. A third application in another year has even less of an effect. What is your hypothesis about what is happening here? Design an experiment that tests your hypothesis.

Interactive Exploration

Check out the website at **www.mhhe.com/enger10e** and click on the cover of this textbook for quizzing, career information, case studies, and hot links for the following topics:

General Ecology Sites

Animal Population Ecology

Evolution

Natural Selection

Community Ecology

Competition

Parasitism, Predation, Herbivory

Mutualism

Field Methods for Studies of Populations

Field Methods for Studies of Ecosystems

Species Abundance, Diversity, and Complexity

Food Webs

Nutrient Cycling

Extinction Issues

Biodiversity

Trees of Life

6

Kinds of Ecosystems and Communities

Chapter Outline

Objectives

After reading this chapter, you should be able to:

• Recognize the difference between primary and secondary succession.

• Describe the process of succession from pioneer to climax community in both terrestrial and aquatic situations.

• Associate typical plants and animals with the various terrestrial biomes.

• Recognize the physical environmental factors that determine the kind of climax community that will develop.

• Differentiate the forest biomes that develop based on temperature and rainfall.

• Describe the various kinds of aquatic ecosystems and the factors that determine their characteristics.

PAST PRESENT FUTURE

1970 Ten percent of marine fisheries were either fished to capacity or overfished.

1988 There were only 59 Transboundary Protected Park Areas, mainly in Europe and North America.

1992 The Convention on Biological Diversity is passed

2003 Seventy percent of marine fisheries were either fished to capacity or overfished.

2003 The number of Transboundary Protected Park Areas had grown to 159 in all regions of the world, including 36 sites in Africa and 30 sites in Asia.

2003 More than 180 countries are signatories of the convention.

Succession

Ecosystems are dynamic, changing units. On a daily basis, plants grow and die, animals feed on plants and on one another, and decomposers recycle the chemical elements that make up the biotic portion of any ecosystem. Abiotic factors (such as temperature, rainfall, intensity of sunlight, and seasonality) also have a major influence on the kind of community that will be established. Since all organisms are linked together in a community, any change in the community affects many organisms within it. Certain conditions within a community are keys to the kinds of organisms that are found associated with one another. Grasshoppers need grass for food, robins need trees to build nests, and herons need shallow water to find food. Each organism has specific requirements that must be met in the community, or it will not survive.

Over long time periods, it is possible to see trends in the way the structure of a community changes and to recognize that climate greatly influences the kind of community that becomes established in an area. Generally, this series of changes eventually results in a relatively long-lasting, stable combination of species that is self-perpetuating. The concept that communities proceed through a series of recognizable, predictable changes in structure over time is called **succession.** The relatively stable, long-lasting community that is the result of succession is called a **climax community.** As ecologists have studied the process of succession, they have come to recognize that the process is not always as predictable as once was thought. However, to begin to understand the process of succession, it often is valuable to look at general models before looking at the exceptions. In the traditional view of succession, the kind of climax community that develops is primarily determined by climate. Some communities will be forests, while others will be grassland or deserts. Succession occurs because the activities of organisms cause changes in their surroundings that make the environment suitable for other kinds of organisms. When new species become established, they compete with the original inhabitants. In some cases, original species may be replaced completely. In other cases, early species may not be replaced but may become less numerous as invading species take a dominant role. Slowly, over time, it is recognized that a significantly different community has become established. Several factors determine the pace and direction of the successional process. For example, climate, locally available seed sources, frequent disturbance, and invasions of organisms from outside the area all greatly affect the course of succession. Ecologists have traditionally recognized two kinds of succession. **Primary succession** is a successional progression that begins with a total lack of organisms and bare mineral surfaces or water. Such conditions can occur when volcanic activity causes lava flows or glaciers scrape away the organisms and soil. Similarly, a lowering of sea level exposes new surfaces for colonization by terrestrial organisms. Primary succession often takes an extremely long time, since there is no soil and few readily available nutrients for plants to use for growth. **Secondary succession** is much more commonly observed and generally proceeds more rapidly, because it begins with the destruction or disturbance of an existing ecosystem. Fire, flood, windstorm, or human activity can destroy or disturb a community of organisms. Usually, however, there is at least some soil and often there are seeds or roots from which plants can begin growing almost immediately.

Primary Succession

Primary succession can begin on a bare rock surface, pure sand, or standing water. Since succession on rock and sand is somewhat different from that which occurs with watery situations, we deal with them separately. We discuss terrestrial succession first.

Terrestrial Primary Succession

Several factors determine the rate of succession and the kind of climax community that will develop in an area. The kind of substrate (rock, sand, clay) will greatly affect the kind of soil that will develop. The kinds of spores, seeds, or other reproductive structures will determine the species available to colonize the area. The climate will determine the species that will be able to live in an area and how rapidly they will grow. The rate of growth will determine how quickly organic matter will accumulate in the soil. The kind of substrate, climate, and amount of organic matter will influence the amount of water available for plant growth. Let's look at a specific example of how these factors are interrelated in an example of primary succession from bare mineral surfaces.

Bare rock or sand is a very inhospitable place for organisms to live. The temperature changes drastically, there is no soil, there is little moisture, the organisms are exposed to the damaging effect of the wind, few nutrients are available, and few places are available for organisms to attach themselves or hide. However, wind-blown spores or other tiny reproductive units of a few kinds of organisms can become established and survive in even this inhospitable environment. This collection of organisms is known as the **pioneer community** because it is the first to colonize bare rock. (See figure 6.1.)

Figure 6.1 Pioneer Organism The lichen growing on this rock is able to accumulate bits of debris, carry on photosynthesis, and aid in breaking down the rock. All of these activities contribute to the formation of a thin layer of soil, which is necessary for plant growth in the early stages of succession.

| Bare rock | Lichens | Small annual plants, lichens | Perennial herbs, grasses | Grasses, shrubs, shade-intolerant trees | Shade-tolerant trees |

Pioneer stages   Intermediate stages   Climax community

Hundreds of years

Figure 6.2 **Primary Succession on Land** The formation of soil is a major step in primary succession. Until soil is formed, the area is unable to support large amounts of vegetation, which modify the harsh environment. Once soil formation begins, the site proceeds through an orderly series of stages toward a climax community.

A common kind of organism in this initial community is something called a lichen. Lichens are actually mutualistic relationships between two kinds of organisms: algae or bacteria that carry on photosynthesis and fungi that attach to the rock surface and retain water. The growth and development of lichens is often a slow process. It may take lichens 100 years to grow as large as a dinner plate. Lichens are the producers in this simple ecosystem, and many tiny consumer organisms may be found associated with lichens. Some feed on the lichen and many use it as a place of shelter, since even a drizzle is like a torrential rain for a microscopic animal. Since lichens are firmly attached to rock surfaces, they also tend to accumulate bits of airborne debris and store small amounts of water that would otherwise blow away or run off the rock surface. Acids produced by the lichen tend to cause the breakdown of the rock substrate into smaller particles. This fragmentation of rock, aided by physical and chemical weathering processes, along with the trapping of debris and the contribution of organic matter by the death of lichens and other organisms, ultimately leads to the accumulation of a very thin layer of soil.

This thin layer of soil is the key to the next stage in the successional process. The layer can retain some water and support some fungi, certain small worms, insects, bacteria, protozoa, and perhaps a few tiny annual plants that live for only one year but produce flowers and seeds that fall to the soil and germinate the following growing season. Many of these initial organisms or their reproductive structures are very tiny and will arrive as a result of wind and rain. As these organisms grow, reproduce, and die, they contribute additional organic material for the soil-building process, and the soil layer increases in thickness and is better able to retain water. This stage, which is dominated by annual plants, eliminates the lichen community because the plants are taller and shade the lichens, depriving them of sunlight. This annual plant stage is itself replaced by a community of small perennial grasses and herbs. The perennial grasses and herbs are often replaced by larger perennial woody shrubs, which are often replaced by larger trees that require lots of sunlight, which often are replaced in turn by trees that can tolerate shade. Sun-loving (shade-intolerant) trees are replaced by shade-tolerant trees because the seedlings of shade-intolerant trees cannot grow in the shade of their parents, while seedlings of shade-tolerant trees can. Eventually, a relatively stable, long-lasting, complex, and interrelated climax community of plants, animals, fungi, and bacteria is produced. Each step in this process from pioneer community to climax community is called a **successional stage,** or **seral stage,** and the entire sequence of stages—from pioneer community to climax community—is called a **sere.** (See figure 6.2.)

Although in this example we have described a successional process that began with a lichen pioneer community and ended with a climax forest, it is important to recognize that the process of succession can stop at any point along this continuum. In certain extreme climates, lichen communities may last for hundreds of years and must be considered climax communities. Others reach a grass-herb stage

Tens to hundreds of years

Figure 6.3   Primary Succession from a Pond to a Wet Meadow A shallow pond will fill slowly with organic matter from producers in the pond. Eventually, a wet soil will form and grasses will become established. In many areas, this will be succeeded by a climax forest.

and proceed no further. The specific kind of climax community produced depends on such things as climate and soil type, which are discussed in greater detail later in this chapter. However, when successional communities are compared to climax communities, climax communities show certain characteristics.

1. Climax communities maintain their mix of species for a long time, while successional communities are temporary.
2. Climax communities tend to have many specialized ecological niches, while successional communities tend to have more generalized niches.
3. Climax communities tend to have many more kinds of organisms and kinds of interactions among organisms than do successional communities.
4. Climax communities tend to recycle nutrients and maintain a relatively constant biomass, while successional communities tend to accumulate large amounts of new material.

The general trend in succession is toward increasing complexity and more efficient use of matter and energy compared

to the successional communities that preceded them.

Aquatic Primary Succession

The principal concepts of land succession can be applied to aquatic ecosystems. Except for the oceans, most aquatic ecosystems are considered temporary. Certainly, some are going to be around for thousands of years, but eventually they will disappear and be replaced by terrestrial ecosystems as a result of normal successional processes. All aquatic ecosystems receive a continuous input of soil particles and organic matter from surrounding land, which results in the gradual filling in of shallow bodies of water such as ponds and lakes.

In deep portions of lakes and ponds, only floating plants and algae can exist. However, as sediment accumulates, the water depth becomes less, and it becomes possible for certain species of submerged plants to establish their roots in the sediments of the bottom of shallow bodies of water. They carry on photosynthesis, which results in a further accumulation of organic matter. These plants also tend to trap sediments that flow into the pond or

lake from streams or rivers, resulting in a further decrease in water depth. Eventually, as the water becomes shallower, emergent plants become established. They have leaves that float on the surface of the water or project into the air. The network of roots and stems below the surface of the water results in the accumulation of more material, and the water depth decreases as material accumulates on the bottom. As the process continues, a wet soil is formed and grasses and other plants that can live in wet soil become established. This successional stage is often called a wet meadow. The activities of plants tend to draw moisture from the soil, and, as more organic matter is added to the top layer of the soil, it becomes somewhat drier. Once this occurs, the stage is set for a typical terrestrial successional series of changes, eventually resulting in a climax community typical for the climate of the area. (See figure 6.3.)

Since the shallower portions of most lakes and ponds are at the shore, it is often possible to see the various stages in aquatic succession from the shore. In the central, deeper portions of the lake, there are only floating plants and algae. As we approach the shore, we first find submerged plants such as *Elodea* and algal mats, then emergent vegetation such as water lilies and cattails, then grasses and sedges that can tolerate wet soil, and on the shore, the beginnings of a typical terrestrial succession resulting in the climax community typical for the area.

In many northern ponds and lakes, sphagnum moss forms thick, floating mats. These mats may allow certain plants that can tolerate wet soil conditions to become established. The roots of the plants bind the mat together and establish a floating bog, which may contain small trees and shrubs as well as many other smaller, flowering plants. (See figure 6.4.) Someone walking on such a mat would recognize that the entire system is floating when they noticed

Figure 6.4 Floating Bog In many northern regions, sphagnum moss forms a floating mat that can be colonized by plants that tolerate wet soils. A network of roots ties the mat together to form a floating community.

the trees sway or when they stepped through a weak zone in the mat and sunk to their hips in water. Eventually, these bogs will become increasingly dry and the normal climax vegetation for the area will succeed the more temporary bog stage.

Secondary Succession

The same processes and activities drive both primary and secondary succession. The major difference is that secondary succession occurs when an existing community is destroyed but much of the soil and some of the organisms remain. A forest fire, a flood, or the conversion of a natural ecosystem to agriculture may be the cause. Since much of the soil remains and many of the nutrients necessary for plant growth are still available, the process of succession can advance more rapidly than primary succession. In addition, because some plants and other organisms may survive the disturbance and continue to grow and others will survive as roots or seeds, they can quickly reestablish themselves in the area. Furthermore, undamaged communities adjacent to the disturbed area can serve as sources of seeds and animals that migrate into the disturbed area. Thus, the new climax community is likely to resemble the one that was destroyed. Figure 6.5 shows old field succession—the typical secondary succession found on abandoned farmland in the southeastern United States.

Similarly, when beavers flood an area, the existing terrestrial community is replaced by an aquatic ecosystem. As the area behind the dam fills in with sediment and organic matter, it goes through a series of changes that may include floating plants, submerged plants, emergent plants, and wet meadow stages, but it eventually returns to the typical climax community for the area. (See figure 6.6.)

Many kinds of communities exist only as successional stages and are continually reestablished following disturbances. Many kinds of woodlands along rivers exist only where floods remove vegetation, allowing specific species to become established on the disturbed flood plain. Some kinds of forest and shrub communities exist only if fire occasionally destroys the mature forest. Windstorms such as hurricanes are also important in causing openings in forests that allow the establishment of certain kinds of plant communities.

Modern Concepts of Succession and Climax

The discussion of the nature of succession and climax communities in the "Succession" section is an oversimplification of the true nature of the process. Some historical perspective will help to clarify how ecologists have changed their concept of successional change. When European

Mature oak/hickory forest destroyed	Farmland abandoned	Annual plants	Grasses and biennial herbs	Perennial herbs and shrubs begin to replace grasses and biennials	Pines begin to replace shrubs	Young oak and hickory trees begin to grow	Pines die and are replaced by mature oak and hickory trees	Mature oak/hickory forest
		1–2 years	3–4 years	4–15 years	5-15 years	10–30 years	50–75 years	

Figure 6.5 **Secondary Succession on Land** A plowed field in the southeastern United States shows a parade of changes over time, involving plant and animal associations. The general pattern is for annual weeds to be replaced by grasses and other perennial herbs, which are replaced by shrubs, which are replaced by trees. As the plant species change, so do the animal species.

explorers traveled across the North American continent, they saw huge expanses of land dominated by specific types of communities: hardwood forests in the east, evergreen forests in the north, grasslands in central North America, and deserts in the southwest. These regional communities came to be considered the steady-state or normal situation for those parts of the world. When ecologists began to explore the way in which ecosystems developed over time, they began to think of these ecosystems as the end point or climax of a long journey, beginning with the formation of soil and its colonization by a variety of plants and other organisms.

As settlers removed the original forests or grasslands and converted the land to farming, the original "climax" community was destroyed. Eventually, as poor farming practices destroyed the soil, many farms were abandoned and the land was allowed to return to its "original" condition. This secondary succession often resulted in forests that resembled those that had been destroyed. However, in most cases, these successional forests contained fewer species and in some cases were entirely different kinds of communities from the originals. These new stable communities were also called climax communities, but they were not the same as the original climax communities.

In addition, the introduction of species from Europe and other parts of the world changed the mix of organisms that might colonize an area. Many grasses and herbs that were introduced either on purpose or accidentally have become well established. Today, some communities are dominated by these introduced species. Even diseases have altered the nature of climax communities. Chestnut blight and Dutch elm disease have removed tree species that were at one time dominant species in certain plant communities.

Ecologists began to recognize that there was no fixed, predetermined community for each part of the world and began to modify the way they looked at the concept of climax communities. The concept today is a more plastic one. It is still used to talk about a stable stage following a period of change, but ecologists no longer feel that land will eventually return to a "preordained" climax condition. They have also recognized in recent years that the type of climax community that develops depends on many factors other than simply climate. One of these is the availability of seeds to colonize new areas. Some seeds may lie dormant in the soil for a decade or more, while others may be carried to an area by wind, water, or animals. Two areas with very similar climate and soil characteristics may develop very different successional and "climax" communities because of the seeds that were present in the area when the lands were released from agriculture. Furthermore, we need to recognize that the only thing that differentiates a "climax" community from a successional one is the time scale over which change occurs. "Climax" communities do not change as rapidly as successional ones. However, all communities are eventually replaced, as were the swamps that produced coal deposits, the preglacial forests of Europe and North America, and the pine forests of the northeastern United States. Many human activities alter the nature of the successional process. Agricultural practices obviously modify the original community to allow for the raising of crops. However, several other management practices have also significantly altered communities. Regular

Figure 6.6 **Secondary Succession from a Beaver Pond** A colony of beavers can dam up streams and kill trees by the flooding that occurs and by using trees for food. Once the site is abandoned, it will slowly return to the original forest community by a process of succession.

CHAPTER 6 Kinds of Ecosystems and Communities

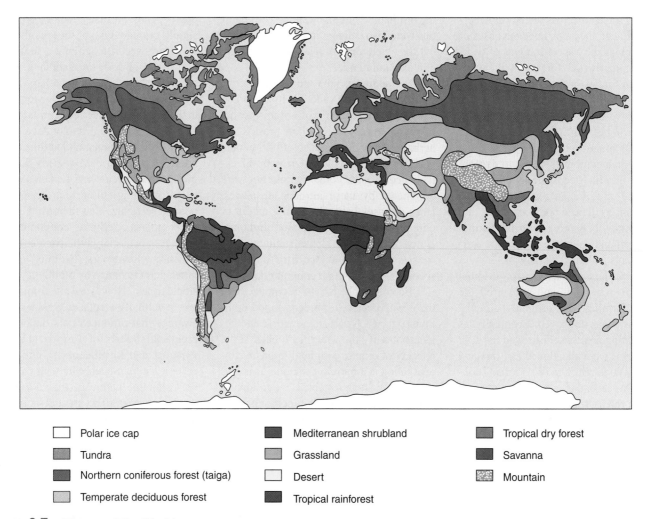

☐ Polar ice cap	■ Mediterranean shrubland	■ Tropical dry forest
■ Tundra	■ Grassland	■ Savanna
■ Northern coniferous forest (taiga)	☐ Desert	▨ Mountain
■ Temperate deciduous forest	■ Tropical rainforest	

Figure 6.7 **Biomes of the World** Although most biomes are named for a major type of vegetation, each includes a specialized group of animals adapted to the plants and the biome's climatic conditions.

logging returns a forest to an earlier stage of succession. The suppression of fire in many forests has also changed the mix of organisms present. When fire is suppressed, those plants that are killed by regular fires become more common and those that are able to resist fire become less common. Changing the amount of water present will also change the kind of community. Draining an area makes it less suitable for the original inhabitants and more suitable for those that live in drier settings. Similarly, irrigation and flooding increase the amount of water present and change the kinds of organisms that can live in an area.

So what should we do with these concepts of succession and climax communities? Although the climax concept embraces a false notion that there is a specific end point to succession, it is still important to recognize that there is an identifiable, predictable pattern of change during succession and that later stages in succession are more stable and longer lasting than early stages. Whether we call a specific community of organisms a climax community is not really important.

Biomes: Major Types of Terrestrial Climax Communities

Biomes are terrestrial climax communities with wide geographic distribution. (See figure 6.7.) Although the concept of biomes is useful for discussing general patterns and processes, it is important to recognize that when different communities within a particular biome are examined, there will be variations in the exact species present. However, in broad terms, the general structure of the ecosystem and the kinds of niches and habitats present are similar. Two primary nonbiological factors have major impacts on the kind of climax community that develops in any part of the world: precipitation and temperature. Several aspects of precipitation are important: the total amount of precipitation per year, the form in which it arrives (rain, snow, sleet), and its seasonal distribution. Precipitation may be evenly spaced throughout the year, or it may be concentrated at particular times so that there are wet and dry seasons.

The temperature patterns are also important and can vary considerably in

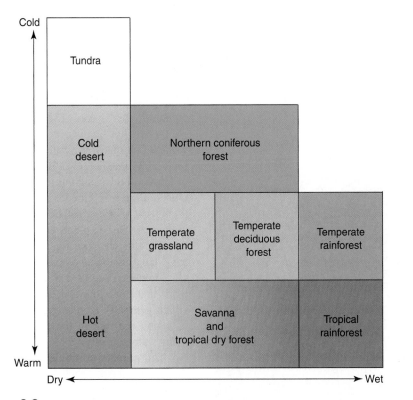

Figure 6.8 **Influence of Precipitation and Temperature on Vegetation** Temperature and moisture are two major factors that influence the kind of vegetation that can occur in an area. Areas with low moisture and low temperatures produce tundra; areas with high moisture and freezing temperatures during part of the year produce deciduous or coniferous forests; dry areas produce deserts; moderate amounts of rainfall or seasonal rainfall support grasslands or savannas; and areas with high rainfall and high temperatures support tropical rainforests.

different parts of the world. Tropical areas have warm, relatively unchanging temperatures throughout the year. Areas near the poles have long winters with extremely cold temperatures and relatively short, cool summers. Other areas are more evenly divided between cold and warm periods of the year. (See figure 6.8.)

Although temperature and precipitation are of primary importance, several other factors may influence the kind of climax community present. Periodic fires are important in maintaining some grassland and shrub climax communities because the fires prevent the establishment of larger, woody species. Some parts of the world have frequent, strong winds that prevent the establishment of trees and cause rapid drying of the soil. The type of soil present is also very important. Sandy soils tend to dry out quickly and may not allow the establishment of more water-demanding species such as trees, while extremely wet soils may allow only certain species of trees to grow. Obviously, the kinds of organisms currently living in

the area are also important, since their offspring will be the ones available to colonize a new area.

The Effect of Elevation on Climate and Vegetation

The distribution of terrestrial ecosystems is primarily related to precipitation and temperature. The temperature is warmest near the equator and becomes cooler toward the poles. Similarly, as the height above sea level increases, the average temperature decreases. This means that even at the equator, it is possible to have cold temperatures on the peaks of tall mountains. As one proceeds from sea level to the tops of mountains, it is possible to pass through a series of biomes that are similar to what would be encountered as one traveled from the equator to the North Pole. (See figure 6.9.)

In the next sections, we will look at the major biomes of the world and highlight the abiotic and biotic features typical of each biome.

Desert

A lack of water is the primary factor that determines that an area will be a desert. **Deserts** are areas that generally average less than 25 centimeters (10 inches) of precipitation per year. (See figure 6.10.) When and how precipitation arrives is quite variable in different deserts. Some deserts receive most of the moisture as snow or rain in the winter months, while in others rain comes in the form of thundershowers at infrequent intervals. If rain comes as heavy thundershowers, much of the water does not sink into the ground but runs off into gullies. Also, since the rate of evaporation is high, plant growth and flowering usually coincide with the periods when moisture is available. Deserts are also likely to be windy. We often think of deserts as hot, dry wastelands devoid of life. However, many deserts are quite cool during a major part of the year. Certainly, the Sahara Desert and the deserts of the southwestern United States and Mexico are hot during much of the year, but the desert areas of the northwestern United States and the Gobi Desert in Central Asia can be extremely cold during winter months and have relatively cool summers. Furthermore, the temperature can vary greatly during a 24-hour period. Since deserts receive little rainfall, it is logical that most will have infrequent cloud cover. With no clouds to block out the sun, during the day the soil surface and the air above it tend to heat up rapidly. After the sun has set, the absence of an insulating layer of clouds allows heat energy to be reradiated from the Earth, and the area cools off rapidly. Cool to cold nights are typical even in "hot" deserts, especially during the winter months.

Another misconception about deserts is that few species of organisms live in the desert. There are many species, but they typically have low numbers of individuals. For example, a conspicuous feature of deserts is the dispersed nature of the plants. There is a significant amount of space between them. Similarly, animals do not have large, dense populations. However, those species that are present are specially adapted to survive in dry, often hot environments. For example, water evaporates from the surfaces of leaves. As an adaptation to this condition, many

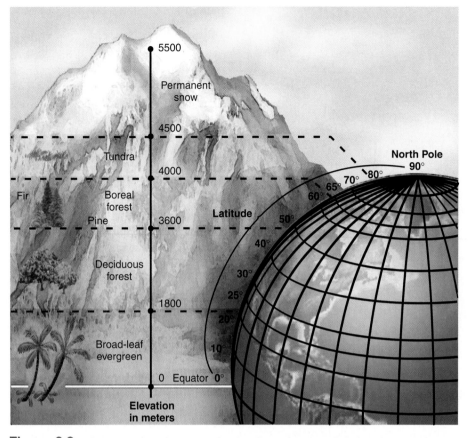

Figure 6.9   Relationship Between Height above Sea Level, Latitude, and Vegetation As one travels up a mountain, the climate changes. The higher the elevation, the cooler the climate. Even in the tropics, tall mountains can have snow on the top. Thus, it is possible to experience the same change in vegetation by traveling up a mountain as one would experience traveling from the equator to the North Pole.

desert plants have very small leaves that allow them to conserve water. Some even lose their leaves entirely during the driest part of the year. Some, such as cactus, have the ability to store water in their spongy bodies or their roots for use during drier periods. Other plants have parts or seeds that lie dormant until the rains come. Then they grow rapidly, reproduce, and die, or become dormant until the next rains. Even the perennial plants are tied to the infrequent rains. During these times, the plants are most likely to produce flowers and reproduce. Many desert plants are spiny. The spines discourage large animals from eating the leaves and young twigs.

The desert has many kinds of animals. However, they are often overlooked because their populations are low, numerous species are of small size, and many are inactive during the hot part of the day. They also aren't seen in large, conspicuous groups. Many insects, lizards, snakes, small mammals, grazing mammals, carnivorous mammals, and birds are common in desert areas. All of the animals that live in deserts are able to survive with a minimal amount of water. Some receive nearly all of their water from the moisture in the food they eat. They generally have an outer skin or cuticle that resists water loss, so they lose little water by evaporation. They also have physiological adaptations, such as extremely efficient kidneys, that allow them to retain water. They often limit their activities to the cooler part of the day (the evening) and may spend considerable amounts of time in underground burrows during the day, which allows them to avoid extreme temperatures and to conserve water.

Grassland

Grasslands, also known as **prairies** or **steppes,** are widely distributed over temperate parts of the world. As with deserts, the major factor that contributes to the establishment of a grassland is the amount of available moisture. Grasslands generally receive between 25 and 75 centimeters (10 to 30 inches) of precipitation per year. These areas are windy with hot summers and cold-to-mild winters. In many grasslands, fire is an important force in preventing the invasion of trees and releasing nutrients from dead plants to the soil. Grasses make up 60 to 90 percent of the vegetation. Many other kinds of flowering plants are interspersed with the grasses. (See figure 6.11.) Typically, the grasses and other plants are very close together, and their roots form a network that binds the soil together. Trees, which generally require greater amounts of water, are rare in these areas except along watercourses.

The primary consumers are animals that eat the grasses, such as large herds of migratory, grazing mammals such as bison, wildebeests, wild horses, and various kinds of sheep, cattle, and goats. While the grazers are important as consumers of the grasses, they also supply fertilizer from their dung and discourage invasion by woody species of plants because they eat the young shoots.

In addition to grazing mammals, many kinds of insects, including grasshoppers and other herbivorous insects, dung beetles (which feed on the dung of grazing animals), and several kinds of flies are common. Some of these flies bite to obtain blood. Others lay their eggs in the dung of large mammals. Some feed on dead animals and lay their eggs in carcasses. Small herbivorous mammals, such as mice and ground squirrels, are also common. Birds are often associated with grazing mammals. They eat the insects stirred up by the mammals or feed on the insects that bite them. Other birds feed on seeds and other plant parts. Reptiles (snakes and lizards) and other carnivores such as coyotes, foxes, and hawks feed on small mammals and insects.

Most of the moist grasslands of the world have been converted to agriculture, since the rich, deep soil that developed as a result of the activities of centuries of soil building is useful for growing cultivated grasses such as corn (maize) and wheat.

City: Cairo, Egypt
Latitude: 29° 52′N
Altitude: 116 m (381 ft.)
Yearly precipitation: 1.8 cm (0.7 in.)

Climate name: Hot desert
Other cities with similar climates:
   Mecca, Karachi

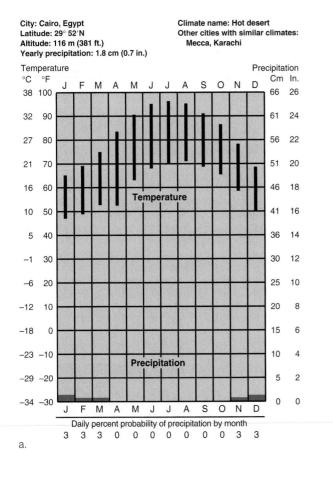

Daily percent probability of precipitation by month
3 3 3 0 0 0 0 0 0 0 3 3

a.

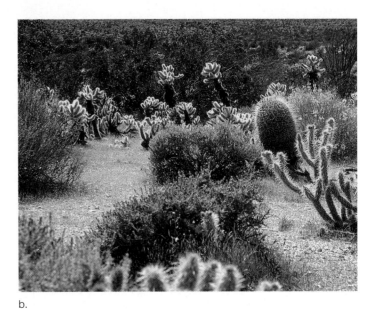

b.

Figure 6.10   Desert (a) Climagraph for Cairo, Egypt. (b) The desert receives less than 25 centimeters (10 inches) of precipitation per year, yet it teems with life. Cactus, sagebrush, lichens, snakes, small mammals, birds, and insects inhabit the desert. Because daytime temperatures are often high, most animals are active only at night, when the air temperature drops significantly. Cool deserts also exist in many parts of the world, where rainfall is low but temperatures are not high.

City: Tehran, Iran
Latitude: 35° 41′N
Altitude: 1220 m (4002 ft.)
Yearly precipitation: 26 cm (10.1 in.)

Climate name: Midaltitude dryland
Other cities with similar climates:
   Salt Lake City, Ankara

Daily percent probability of precipitation by month
13 14 16 10 6 3 3 0 0 3 10 13

a.

b.

Figure 6.11   Grassland (a) Climagraph for Tehran, Iran. (b) Grasses are better able to withstand low water levels than are trees. Therefore, in areas that have moderate rainfall, grasses are the dominant plants.

Grassland Succession

Because there are many kinds of grasslands, it is difficult to generalize about how succession takes place in these areas. Most grasslands in North America have been heavily influenced by agriculture and the grazing of domesticated animals. The grasslands reestablished in these areas may be quite different from the original ecosystem. However, there appear to be several stages typically involved in grassland succession.

After land is abandoned from cultivation, a short period of one to three years elapses in which the field is dominated by annual broad-leaf weeds. In this respect, grassland succession is like deciduous forest succession. The next stage varies in length (10 or more years) and is dominated by annual grasses. Usually, in these early stages, the soil is in poor condition, lacking organic matter and nutrients. After several years, the soil fertility increases as organic material accumulates from the death and decay of annual grasses. This leads to the next stage in development, perennial grasses. Eventually, a mature grassland develops as prairie flowers invade the area and become interspersed with the grasses. In general, throughout this sequence, the soil becomes more fertile and of higher quality.

Because so much of the original North American grassland has been used for agriculture, when the land is allowed to return to a prairie, there may not be seeds of all of the original plants native to the area. Thus, the grassland that results from secondary succession may not be exactly like the original; some species may be missing. Consequently, in many managed restorations of prairies, seeds that are no longer available in the local soils are introduced from other sources.

The low amount of rainfall and the fires typical of grasslands generally cause the successional process to stop at this point. However, if more water becomes available or if fire is prevented, woody trees may invade moist sites.

Actively farmed   *Recently abandoned*   *Several years of succession*

The drier grasslands have been converted to the raising of domesticated grazers such as cattle, sheep, and goats. Therefore, little undisturbed grassland is left, and those fragments that remain need to be preserved as refuges for the grassland species that once occupied huge portions of the globe.

Savanna

Tropical parts of Africa, South America, and Australia have extensive grasslands spotted with occasional trees or patches of trees. (See figure 6.12.) This kind of a biome is often called a **savanna.** Although savannas receive 50 to 150 centimeters (20 to 60 inches) of rain per year, the rain is not distributed evenly throughout the year. Typically, a period of heavy rainfall is followed by a prolonged drought. This results in a very seasonally structured ecosystem. The plants and animals time their reproductive activities to coincide with the rainy period, when limiting factors are least confining. The predominant plants are grasses, but many drought-resistant, flat-topped, thorny trees are common. As with grasslands, fire is a common feature of the savanna, and the trees present are resistant to fire damage. Many of these trees are particularly important because they are legumes that are involved in nitrogen fixation. They also provide shade and nesting sites for animals. As with grasslands, the predominant mammals are the grazers. Wallabies in Australia, wildebeests, zebras, elephants, and various species of antelope in Africa, and capybaras (rodents) in South America are examples. In Africa, the large herds of grazing animals provide food for many different kinds of large carnivores (lions, hyenas, leopards). Many kinds of rodents, birds, insects, and reptiles are associated with this biome. Among the insects, mound-building termites are particularly common.

Mediterranean Shrublands (Chaparral)

The **Mediterranean shrublands** are located near an ocean and have wet, cool winters and hot, dry summers. Rainfall is 40 to 100 centimeters (15 to 40 inches) per year. As the name implies, this biome is typical of the Mediterranean coast and is also found in coastal southern California, the southern tip of Africa, a portion of the west coast of Chile, and southern Australia. The vegetation is dominated by woody shrubs that are adapted to withstand the hot, dry summer. (See figure 6.13.) Often the plants are dormant during the summer. Fire is a common feature of

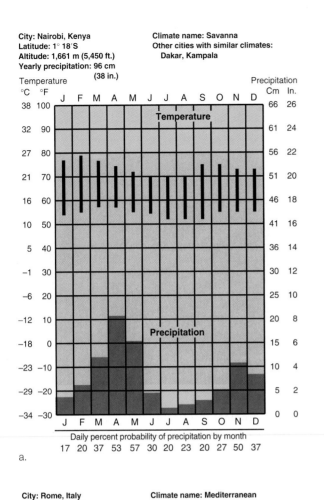

City: Nairobi, Kenya
Latitude: 1° 18′S
Altitude: 1,661 m (5,450 ft.)
Yearly precipitation: 96 cm
(38 in.)

Climate name: Savanna
Other cities with similar climates:
Dakar, Kampala

Temperature °C °F — Precipitation Cm In.

Temperature

Precipitation

Daily percent probability of precipitation by month
17 20 37 53 57 30 20 23 20 27 50 37

a.

b.

Figure 6.12 **Savanna** (a) Climagraph for Nairobi, Kenya. (b) Savannas develop in tropical areas that have seasonal rainfall. They typically have grasses as the dominant vegetation with drought- and fire-resistant trees scattered through the area.

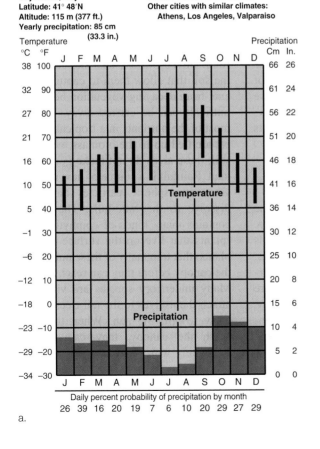

City: Rome, Italy
Latitude: 41° 48′N
Altitude: 115 m (377 ft.)
Yearly precipitation: 85 cm
(33.3 in.)

Climate name: Mediterranean
Other cities with similar climates:
Athens, Los Angeles, Valparaiso

Temperature °C °F — Precipitation Cm In.

Temperature

Precipitation

Daily percent probability of precipitation by month
26 39 16 20 19 7 6 10 20 29 27 29

a.

b.

Figure 6.13 **Mediterranean Shrubland** (a) Climagraph for Rome, Italy. (b) Mediterranean shrublands are characterized by a period of winter rains and a dry, hot summer. The dominant plants are drought-resistant, woody shrubs.

Protecting Old-Growth Temperate Rainforests of the Pacific Northwest

The coastal areas of northern California, Oregon, Washington, British Columbia, and southern Alaska have an unusual set of environmental conditions that support a special kind of forest, a temperate rainforest. The prevailing winds from the west bring moisture-laden air to the coast. As this air meets the coastal mountains and is forced to rise, it cools and the moisture falls as rain or snow. Most of these areas receive 200 or more centimeters (80 or more inches) of precipitation per year. This abundance of water, along with fertile soil and mild temperatures, results in a luxuriant growth of plants.

Sitka spruce, Douglas fir, and western hemlock are typical evergreen coniferous trees. Undisturbed (old-growth) forests of this region have trees as old as 800 years that are almost as tall as the length of a football field. Deciduous trees of various kinds (red alder, bigleaf maple, black cottonwood) grow in places where they can get enough light. All trees are covered with mosses, ferns, and other plants that grow on their surface. The dominant color is green, since most surfaces have something growing on them.

When a tree dies and falls to the ground, it rots in place and often serves as a site for the establishment of new trees. This is such a common feature of the forest that the fallen, rotting trees are called nurse trees. The fallen trees also serve as a food source for a variety of insects, which are food for a variety of other animals. Several endangered or threatened animals, such as the northern spotted owl, the marbled murrelet (a seabird), and the Roosevelt elk, are dependent on undisturbed forest for their survival.

Because of the rich resource of trees, 90 percent of the original temperate rainforest has been logged. What remains has become a source of controversy. Some maintain that it should be protected as a remnant of the original forest of the region, just as small patches of prairie and eastern woodland have been preserved in other parts of North America. Others point out that the intact old-growth forest must be preserved to protect endangered species. Some maintain that since the trees are old and will die in the near future, they should be harvested for their timber value and the jobs the logging will provide. Others counter that dying and dead trees are important to the maintenance of this unique ecosystem. The fate of this unusual ecosystem is not likely to be resolved without legislation or legal action.

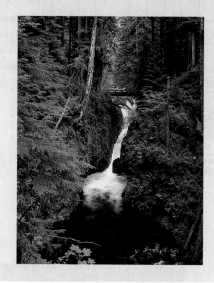

and can last as long as six months. Typically, the soil freezes during the winter. Precipitation ranges between 25 and 100 centimeters (10 to 40 inches) per year. However, the climate is typically humid because there is a great deal of snowmelt in the spring and generally low temperatures reduce evaporation. The landscape is typically dotted with lakes, ponds, and bogs.

Conifers such as spruces, firs, and larches are the most common trees in these areas. These trees are specifically adapted to winter conditions. Winter is relatively dry as far as the trees are concerned because the moisture falls as snow and stays above the soil until it melts in the spring. The needle-shaped leaves are adapted to prevent water loss; in addition, the larches lose their needles in the fall. The branches of these trees are flexible, allowing them to bend under a load of snow so that the snow slides off the pyramid-shaped trees without greatly damaging them. As with the temperate deciduous forest, many of the inhabitants of this biome are temporarily active during the summer. Most birds are migratory and feed on the abundant summer insect population, which is not available during the long, cold winter. A few birds, such as woodpeckers, owls, and grouse, are permanent residents. Typical mammals are deer, caribou, moose, wolves, weasels, mice, snowshoe hares, and squirrels. Because of the cold, few reptiles and amphibians live in this biome.

Tundra

North of the taiga is the **tundra,** a biome that lacks trees and has a permanently frozen subsurface soil. This frozen soil layer is known as **permafrost.** (See figure 6.18.) Because of the permanently frozen soil and extremely cold, windy climate (up to 10 months of winter), no trees can live in the area. Although the amount of precipitation is similar to that in some deserts—less than 25 centimeters (10 inches) per year—the short summer is generally wet because the winter snows

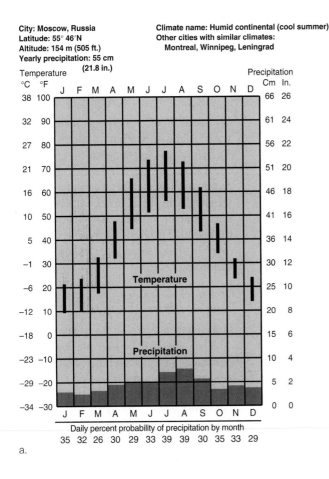

City: Moscow, Russia
Latitude: 55° 46′N
Altitude: 154 m (505 ft.)
Yearly precipitation: 55 cm (21.8 in.)

Climate name: Humid continental (cool summer)
Other cities with similar climates: Montreal, Winnipeg, Leningrad

Daily percent probability of precipitation by month
35 32 26 30 29 33 39 39 30 35 33 29

a.

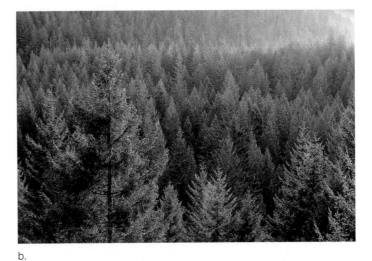

b.

Figure 6.17 **Taiga, Northern Coniferous Forest, or Boreal Forest** (a) Climagraph for Moscow. (b) The taiga, northern coniferous forest, or boreal forest occurs in areas with long winters and heavy snowfall. The trees have adapted to these conditions and provide food and shelter for the animals that live there.

City: Fairbanks, Alaska
Latitude: 64° 51′N
Altitude: 134 m (440 ft.)
Yearly precipitation: 31.5 cm (12.4 in.)

Climate name: Subarctic tundra
Other cities with similar climates: Yellowknife, Yakutsk

Daily percent probability of precipitation by month
32 21 19 13 29 33 42 48 33 35 33 29

a.

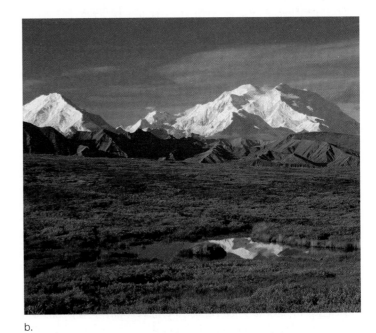

b.

Figure 6.18 **Tundra** (a) Climagraph for Fairbanks, Alaska. (b) In the northern latitudes and on the tops of some mountains, the growing season is short and plants grow very slowly. Trees are unable to live in these extremely cold areas, in part because there is a permanently frozen layer of soil beneath the surface, known as the permafrost. Because growth is so slow, damage to the tundra can still be seen generations later.

melt in the spring and summer temperatures are usually less than 10°C (50°F), which reduces the evaporation rate. Since the permafrost does not let the water sink into the soil, waterlogged soils and many shallow ponds and pools are present. Many waterfowl such as ducks and geese migrate to the tundra in the spring; there, they mate and raise their young during the summer before migrating south in the fall. When the top few centimeters (inches) of the soil thaw, many plants (grasses, dwarf birch, dwarf willow) and lichens, such as reindeer moss, grow. The plants are short, usually less than 20 centimeters (8 inches) tall.

Clouds of insects are common during the summer and serve as food for migratory birds. Permanent resident birds are the ptarmigan and snowy owl. No reptiles or amphibians survive in this extreme climate. A few hardy mammals such as musk oxen, caribou (reindeer), arctic hare, and lemmings can survive by feeding on the grasses and other plants that grow during the short, cool summer. Arctic foxes, wolves, and owls are the primary predators in this region. Because of the very short growing season, damage to this kind of ecosystem is slow to heal, so the land must be handled with great care.

Scattered patches of tundralike communities also are found on mountaintops throughout the world. These are known as **alpine tundra.** Although the general appearance of the alpine tundra is similar to true tundra, many of the species of plants and animals are different. Many of the birds and large mammals migrate up to the alpine tundra during the summer and return to lower elevations as the weather turns cold.

Major Aquatic Ecosystems

Terrestrial biomes are determined by the amount and kind of precipitation and by temperatures. Other factors, such as soil type and wind, also play a part. Aquatic ecosystems also are shaped by key environmental factors. Several important factors are the ability of the sun's rays to penetrate the water, the depth of the water, the nature of the bottom substrate, the water temperature, and the amount of dissolved salts.

Marine Ecosystems

An important determiner of the nature of aquatic ecosystems is the amount of salt dissolved in the water. Those that have little dissolved salt are called **freshwater ecosystems,** and those that have a high salt content are called **marine ecosystems.**

Pelagic Marine Ecosystems

In the open ocean, many kinds of organisms float or swim actively. Crustaceans, fish, and whales swim actively as they pursue food. Organisms that are not attached to the bottom are called **pelagic** organisms, and the ecosystem they are a part of is called a **pelagic ecosystem.** The term **plankton** is used to describe aquatic organisms that are so small and weakly swimming that they are simply carried by currents. As with all ecosystems, the organisms at the bottom of the energy pyramid carry on photosynthesis. The planktonic organisms that carry on photosynthesis are called **phytoplankton.** In the open ocean, a majority of these organisms are small, microscopic, floating algae and bacteria. The upper layer of the ocean, where the sun's rays penetrate, is known as the **euphotic zone.** It is in this euphotic zone where phytoplankton are most common. The thickness of the euphotic zone varies with the degree of clarity of the water but in clear water can be up to 150 meters (500 feet) in depth. Small, weakly swimming animals of many kinds, known as **zooplankton,** feed on the phytoplankton. Zooplankton are often located at a greater depth in the ocean than the phytoplankton but migrate upward at night and feed on the large population of phytoplankton. The zooplankton are in turn eaten by larger animals such as fish and larger shrimp, which are eaten by larger fish such as salmon, tuna, sharks, and mackerel. (See figure 6.19.)

A major factor that influences the nature of a marine community is the kind and amount of material dissolved in the water. Of particular importance is the amount of dissolved, inorganic nutrients available to the organisms carrying on photosynthesis. Phosphorus, nitrogen, and carbon are all required for the construction of new living material. In water, these are often in short supply. Therefore, the most productive aquatic ecosystems are those in which these essential nutrients are most common. These areas include places in oceans where currents bring up nutrients that have settled to the bottom and places where rivers deposit their load of suspended and dissolved materials.

Benthic Marine Ecosystems

Organisms that live on the ocean bottom, whether attached or not, are known as **benthic** organisms, and the ecosystem of which they are a part is called a **benthic ecosystem.** Some fish, clams, oysters, various crustaceans, sponges, sea anemones, and many other kinds of organisms live on the bottom. In shallow water, sunlight can penetrate to the bottom, and a variety of attached photosynthetic organisms commonly called seaweeds are common. Since they are attached and some, such as kelp, can grow to very large size, many other bottom-dwelling organisms, such as sea urchins, worms, and fish, are associated with them.

The substrate is very important in determining the kind of benthic community that develops. Sand tends to shift and move, making it difficult for large plants or algae to become established, although some clams, burrowing worms, and small crustaceans find sand to be a suitable habitat. Clams filter water and obtain plankton and detritus or burrow through the sand, feeding on other inhabitants. Mud may provide suitable habitats for some kinds of rooted plants, such as mangrove trees or sea grasses. Although mud usually contains little oxygen, it still may be inhabited by a variety of burrowing organisms that feed by filtering the water above them or that feed on other animals in the mud. Rocky surfaces in the ocean provide a good substrate for many kinds of large algae. Associated with this profuse growth of algae is a

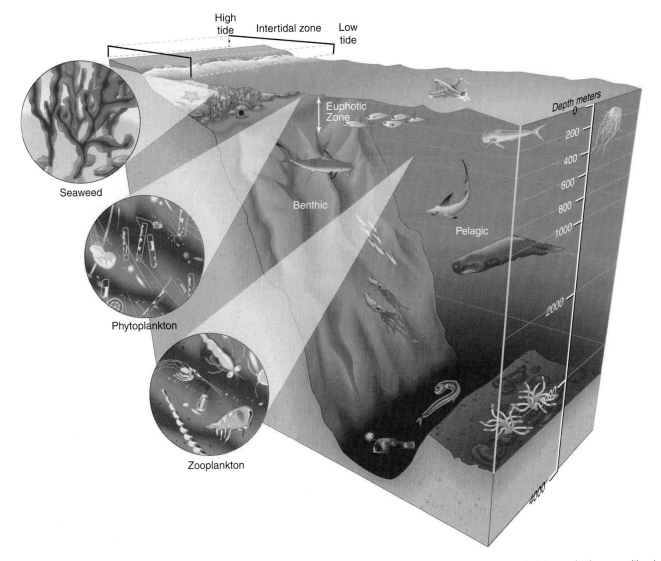

High tide | Intertidal zone | Low tide

Euphotic Zone

Seaweed

Phytoplankton

Zooplankton

Benthic

Pelagic

Depth meters
0
200
400
600
800
1000
2000
4000

Figure 6.19 **Marine Ecosystems** All of the photosynthetic activity of the ocean occurs in the shallow water called the euphotic zone, either by attached algae near the shore or by minute phytoplankton in the upper levels of the open ocean. Consumers are either free-swimming pelagic organisms or benthic organisms that live on the bottom. Small animals that feed on phytoplankton are known as zooplankton.

large variety of animals. (See figure 6.20.)

Temperature also has an impact on the kind of benthic community established. Some communities, such as coral reefs or mangrove swamps, are found only in areas where the water is warm. **Coral reef ecosystems** are produced by coral animals that build cup-shaped external skeletons around themselves. Corals protrude from their skeletons to capture food and expose themselves to the sun. Exposure to sunlight is important because corals contain single-celled algae within their bodies. These algae carry on photosynthesis and provide both themselves and

the coral animals with the nutrients necessary for growth. This mutualistic relationship between algae and coral is the basis for a very productive community of organisms. The skeletons of the corals provide a surface upon which many other kinds of animals live. Some of these animals feed on corals directly, while others feed on small plankton and bits of algae that establish themselves among the coral organisms. Many kinds of fish, crustaceans, sponges, clams, and snails are members of coral reef ecosystems.

Because they require warm water, coral ecosystems are found only near the equator. Coral ecosystems also require

shallow, clear water since the algae must have ample sunlight to carry on photosynthesis. Coral reefs are considered one of the most productive ecosystems on Earth. (See figure 6.21.)

Mangrove swamp ecosystems occupy a region near the shore. The dominant organisms are special kinds of trees that are able to tolerate the high salt content of the ocean. In areas where the water is shallow and wave action is not too great, the trees can become established. They have long seeds that float in the water. When the seeds become trapped in mud, they take root. The trees can excrete salt from their leaves. They also have

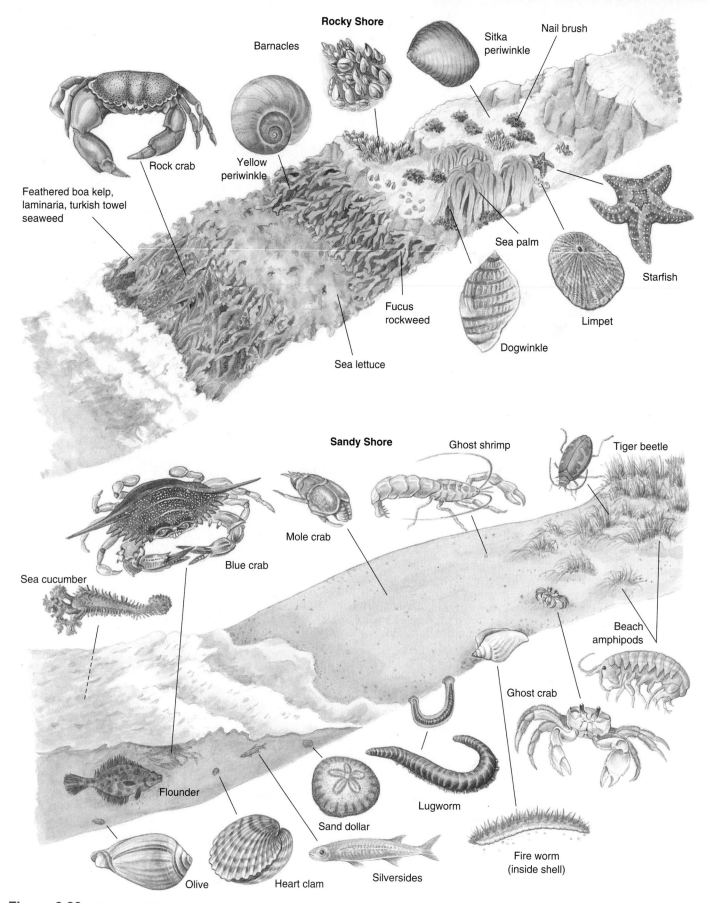

Figure 6.20 **Types of Shores** The kind of substrate determines the kind of organisms that can live near the shore. Rocks provide areas for attachment that sands do not, since sands are constantly shifting. Muds usually have little oxygen in them; therefore, the organisms that live there must be adapted to those kinds of conditions.

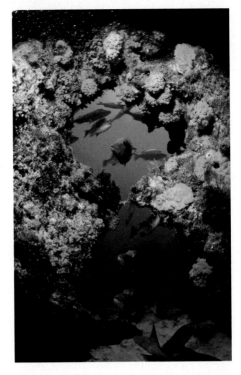

Figure 6.21 **Coral Reef** Corals are small sea animals that secrete external skeletons. They have a mutualistic relationship with certain algae, which allows both kinds of organisms to be very successful. The skeletal material serves as a substrate upon which many other kinds of organisms live.

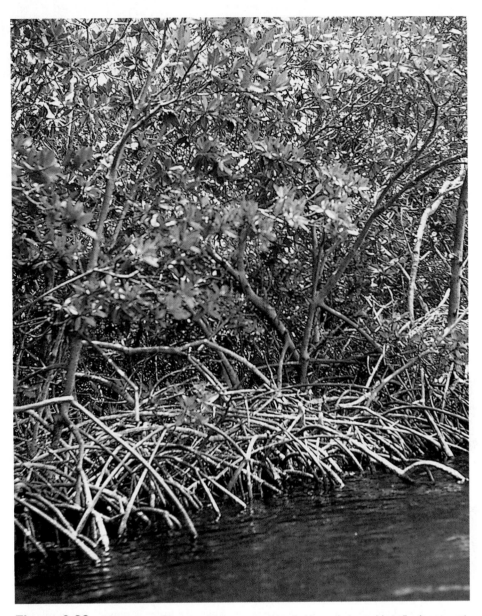

Figure 6.22 **Mangrove Swamp** Mangroves are tropical trees that are able to live in very wet, salty muds found along the ocean shore. Since they are able to trap additional sediment, they tend to extend farther seaward as they reproduce.

extensively developed roots that extend above the water, where they can obtain oxygen and prop up the plant. The trees trap sediment and provide places for oysters, crabs, jellyfish, sponges, and fish to live. The trapping of sediment and the continual extension of mangroves into shallow areas result in the development of a terrestrial ecosystem in what was once shallow ocean. Mangroves are found in south Florida, the Caribbean, Southeast Asia, Africa, and other parts of the world where tropical mudflats occur. (See figure 6.22.)

At great depths in the ocean is a benthic ecosystem that must rely on a continuous rain of organic matter from the euphotic zone. These areas are known as abyssal areas, and the ecosystem is known as an **abyssal ecosystem.** No light penetrates to this region, and the amount of food available is limited. Essentially, all of the organisms in this environment are scavengers that feed on whatever drifts their way. Many of the animals are small

and generate light that they use for finding or attracting food.

Estuaries

An **estuary** is a special category of aquatic ecosystem that consists of shallow, partially enclosed areas where freshwater enters the ocean. The saltiness of the water in the estuary changes with tides and the flow of water from rivers. The organisms that live here are specially adapted to this set of physical conditions, and the number of species is less than in the ocean or in

freshwater. Estuaries are particularly productive ecosystems because of the large amounts of nutrients introduced into the basin from the rivers that run into them. This is further enhanced by the fact that the shallow water allows light to penetrate to most of the water in the basin. Phytoplankton and attached algae and plants are able to use the sunlight and the nutrients for rapid growth. This photosynthetic activity supports many kinds of organisms in the estuary. Estuaries are especially important as nursery sites for fish and crustaceans such as flounder and shrimp.

Environmental Close-Up

Non-native Invasive Aquatic Plants

The term *non-native invasive species* refers to plants that spread aggressively and are not native to an area. All ecosystems are susceptible to invasion by these plants. These invasions can lead to the loss of biological diversity by eliminating habitat for native plants and can have a wide variety of other impacts. Prolific growth of plants such as Eurasian water milfoil (*Myriophyllum spicatum*) can clog irrigation canals, hydroelectric systems, and navigation waterways. Flooding can occur due to obstructed drainage systems. Boating and swimming are impacted and become less desirable, and aesthetics of lakes are often impaired. Dense beds of plants alter water chemistry, temperature, and habitat structure, leading to changes in invertebrate and fish communities. Dense growth may also be a major source of organic matter to the water body, and decomposition of these plants may deplete dissolved oxygen. This may lead to problems with odor, color, and taste in the water. Depletion of dissolved oxygen may also cause fish kills.

Lake margins are also vulnerable to invasion by non-native plants. Invasive species such as purple loosestrife (*Lythrum salicaria*) and reed canarygrass (*Phalaris arundinaceae*) may create large, single-species stands, which reduce diversity and eliminate natural sources of food and cover essential to wetland wildlife.

Invasive species have been introduced into North America in a variety of ways. Purple loosestrife was sold as an ornamental landscaping plant. Plants that plague lakes include a number of species introduced by the aquarium industry, such as Brazilian elodea (*Egeria densa*) and hydrilla (*Hydrilla verticillata*). Because of their attractive flowers, gardeners and landscapers have planted water hyacinth (*Eichornia crassipes*) and white water lily (*Nymphaea odorata*). Some aquatic species that are now considered undesirable were introduced for "restoration" purposes. The U.S. Department of Agriculture has promoted invasive species, such as reed canarygrass as soil binders and forage plants. Agricultural canals, drains, and irrigation systems have further spread aquatic invaders in the western United States. Once introduced into an area, invasive aquatic plants may spread from lake to lake on boats or be carried by birds or other animals.

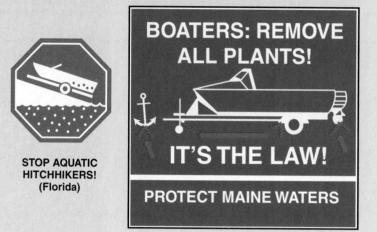

STOP AQUATIC HITCHHIKERS! (Florida)

States that have programs aimed at preventing the introduction of nonnative plants to public waters include Florida (left) and Maine (right).

The adults enter these productive, sheltered areas to reproduce and then return to the ocean. The young spend their early life in the estuary and eventually leave as they get larger and are more able to survive in the ocean. Estuaries also trap sediment. This activity tends to prevent many kinds of pollutants from reaching the ocean and also results in the gradual filling in of the estuary, which may eventually become a salt marsh and then part of a terrestrial ecosystem.

Freshwater Ecosystems

Freshwater ecosystems differ from marine ecosystems in several ways. The amount of salt present is much less, the temperature of the water can change greatly, the water is in the process of moving to the ocean, oxygen can often be in short supply, and the organisms that inhabit freshwater systems are different.

Freshwater ecosystems can be divided into two categories: those in which the water is relatively stationary, such as lakes, ponds, and reservoirs, and those in which the water is running downhill, such as streams and rivers.

Lakes and Ponds

Large lakes have many of the same characteristics as the ocean. If the lake is deep, there is a euphotic zone at the top, with many kinds of phytoplankton, and zooplankton that feed on the phytoplankton. Small fish feed on the zooplankton and are in turn eaten by larger fish. The species of organisms found in freshwater lakes are different from those found in the ocean, but the roles played are similar, so the same terminology is used.

Along the shore and in the shallower parts of lakes, many kinds of flowering plants are rooted in the bottom. Some have leaves that float on the surface or protrude above the water and are called **emergent plants.** Cattails, bulrushes, arrowhead plants, and water lilies are examples. Rooted plants that stay submerged below the surface of the water are called **submerged plants.** *Elodea* and *Chara* are examples.

Many kinds of freshwater algae also grow in the shallow water, where they may appear as mats on the bottom or attached to vegetation and other objects in the water. Associated with the plants and algae are a large number of different kinds of animals. Fish, crayfish, clams, and many kinds of aquatic insects are common inhabitants of this mixture of plants and algae. This region, with rooted vegetation, is known as the **littoral zone,** and the portion of the lake that does not have rooted vegetation is called the **limnetic zone.** (See figure 6.23.)

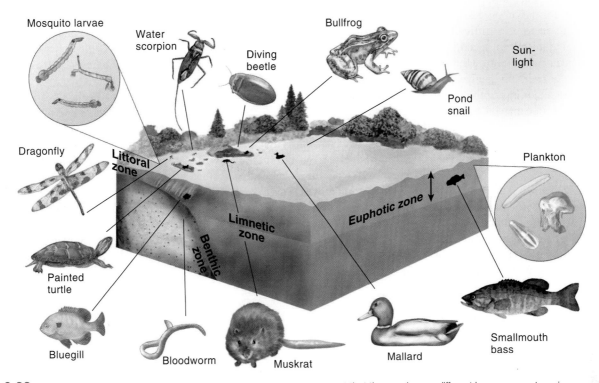

Figure 6.23 **Lake Ecosystem** Lakes are similar in structure to oceans except that the species are different because most marine organisms cannot live in freshwater. Insects are common organisms in freshwater lakes, as are many kinds of fish, zooplankton, and phytoplankton.

The productivity of the lake is determined by several factors. Temperature is important, since cold temperatures tend to reduce the amount of photosynthesis. Water depth is important because shallow lakes will have light penetrating to the lake bottom, and therefore, photosynthesis can occur throughout the entire water column. Shallow lakes also tend to be warmer as a result of the warming effects of the sun's rays. A third factor that influences the productivity of lakes is the amount of nutrients present. This is primarily determined by the rivers and streams that carry nutrients to the lake. River systems that run through areas that donate many nutrients will carry the nutrients to the lakes. Farming and construction expose soil and release nutrients, as do other human activities such as depositing sewage into streams and lakes. Deep, clear, cold, nutrient-poor lakes are low in productivity and are called **oligotrophic lakes.** Shallow, murky, warm, nutrient-rich lakes are called **eutrophic lakes.**

Although the water molecule (H_2O) has oxygen as part of its structure, this oxygen is not available to organisms. The oxygen that they need is dissolved molecular oxygen (O_2), which enters water from the air or when it is released as a result of photosynthesis by aquatic plants. When water tumbles over rocks in a stream or crashes on the shore as a result of wave action, air and water mix, which allows more oxygen to dissolve in the water.

The dissolved oxygen content of the water is important since the quantity of oxygen determines the kinds of organisms that can inhabit the lake. When organic molecules enter water, they are broken down by bacteria and fungi. These decomposer organisms use oxygen from the water as they perform respiration. The amount of oxygen used by decomposers to break down a specific amount of organic matter is called the **biochemical oxygen demand (BOD).** Organic materials enter aquatic ecosystems in several ways. The organisms that live in the water produce the metabolic wastes. When organisms that live in or near water die or shed parts, their organic matter is contributed to the water. The amount of nutrients entering the water is also important, since the algae and plants whose growth is stimulated will

eventually die and their decomposition will reduce oxygen concentration. Many bodies of water experience a reduced oxygen level during the winter, when producers die. The amount and kinds of organic matter determine, in part, how much oxygen is left to be used by other organisms, such as fish, crustaceans, and snails. Many lakes may experience periods when oxygen is low, resulting in the death of fish and other organisms. Human activity often influences the health of bodies of water, because we tend to introduce nutrients from agriculture and organic wastes from a variety of industrial, agricultural, and municipal sources. These topics are discussed in greater depth in chapter 16.

Streams and Rivers

Streams and rivers are a second category of freshwater ecosystem. Since the water is moving, planktonic organisms are less important than are attached organisms. Most algae grow attached to rocks and other objects on the bottom. This collection of attached algae, animals, and fungi is called the **periphyton.** Since the water is shallow, light can penetrate easily to the bottom (except for large or extremely muddy rivers). Even so, it is difficult for photosynthetic organisms to accumulate the nutrients necessary for growth, and most streams are not very productive. As a matter of fact, the major input of nutrients is from organic matter that falls into the stream from terrestrial sources. These are primarily the leaves from trees and other vegetation, as well as the bodies of living and dead insects.

Within the stream is a community of organisms that are specifically adapted to use the debris as a source of food. Bacteria and fungi colonize the organic matter, and many kinds of insects shred and eat this organic matter along with the fungi and

bacteria living on it. The feces (intestinal wastes) of these insects and the tiny particles produced during the eating process become food for other insects that build nets to capture the tiny bits of organic matter that drift their way. These insects are in turn eaten by carnivorous insects and fish.

Organisms in larger rivers and muddy streams, which have less light penetration, rely in large part on the food that drifts their way from the many streams that empty into the river. These larger rivers tend to be warmer and to have slower-moving water. Consequently, the amount of oxygen is usually less, and the species of plants and animals change. Any additional organic matter added to the river

system adds to the BOD, further reducing the oxygen in the water. Plants may become established along the river bank and contribute to the ecosystem by carrying on photosynthesis and providing hiding places for animals.

Just as estuaries are a bridge between freshwater and marine ecosystems, swamps and marshes are a transition between aquatic and terrestrial ecosystems. **Swamps** are wetlands that contain trees that are able to live in places that are either permanently flooded or flooded for a major part of the year. **Marshes** are wetlands that are dominated by grasses and reeds. Many swamps and marshes are successional states that eventually become totally terrestrial communities.

Summary

Ecosystems change as one kind of organism replaces another in a process called succession. Ultimately, a relatively stable stage is reached, called the climax community. Succession may begin with bare rock or water, in which case it is called primary succession, or may occur when the original ecosystem is destroyed, in which case it is called secondary succession. The stages that lead to the climax are called successional stages.

Major regional terrestrial climax communities are called biomes. The primary determiners of the kinds of biomes that develop are the amount and yearly distribution of rainfall and the yearly temperature cycle. Major biomes are desert, grassland, savanna, Mediterranean shrublands, tropical dry forest, tropical rainforest, temperate deciduous forest, taiga, and tundra. Each has a particular set of organisms that is adapted to the climatic conditions typical for the area. As one proceeds up a mountainside, it is possible to witness the same kind of change in biomes that occurs if one were to travel from the equator to the North Pole.

Aquatic ecosystems can be divided into marine (saltwater) and freshwater ecosystems. In the ocean, some organisms live in open water and are called pelagic organisms. Light penetrates only the upper layer of water; therefore, this region is called the euphotic zone. Tiny photosynthetic organisms that float near the surface are called phytoplankton. They are eaten by small animals known as zooplankton, which in turn are eaten by fish and other larger organisms.

The kind of material that makes up the shore determines the mixture of organisms that live there. Rocky shores provide surfaces for organisms to attach; sandy shores do not. Muddy shores are often poor in oxygen, but marshes and swamps may develop in these areas. Coral reefs are tropical marine ecosystems dominated by coral animals. Mangrove swamps are tropical marine shoreline ecosystems dominated by trees. Estuaries occur where freshwater streams and rivers enter the ocean. They are usually shallow, very productive areas. Many marine organisms use estuaries for reproduction.

Insects are common in freshwater and absent in marine systems. Lakes show a structure similar to that of the ocean, but the species are different. Deep, cold-water lakes with poor productivity are called oligotrophic, while shallow, warm-water, highly productive lakes are called eutrophic. Streams differ from lakes in that most of the organic matter present in them falls into them from the surrounding land. Thus, organisms in streams are highly sensitive to the land uses that occur near the streams.

Issues – Analysis

Ecosystem Loss in North America

North America contains a variety of species and ecosystems, including temperate rainforests, grasslands, wetlands, deserts, and more. Species in the United States include grizzly bears, spotted owls, ghost-faced bats, horned puffins, and redwood trees, as only a few examples.

As in the tropics, North America's storehouse of biodiversity is being threatened. As of May 2002, the U.S. Fish and Wildlife Service and the National Marine Fisheries Service combined had listed 1231 species (496 animal species and 735 plant species) as endangered or threatened in the United States. Hundreds of other species are being considered as possible additions to the list. According to the Nature Conservancy, one-third of all U.S. plant and animal species are in need of protection. Many freshwater fishes and wetland species such as mussels, crayfish and amphibians are particularly vulnerable. Nearly 500 species in the United States may be nearly extinct.

Canada's endangered species list included 353 species as of May 2002. Among them are the wolverine, killer whale, eastern barn owl, western rattlesnake, tailed frog, white-throated swift, peregrine falcon, and whooping crane. Many of Canada's ecosystems are also in danger. According to the Canadian Nature Federation, 240 hectares (593 acres) of wildlife habitat are converted or fragmented every hour in Canada, and habitat destruction threatens more than 80 percent of Canada's endangered species with extinction.

Mexico's rich biodiversity is also being lost. Home to nearly 10 percent of the world's terrestrial species, Mexico has a high number of endemic species, the richest diversity of reptiles and cacti, and the second richest diversity of mammals in the world. But almost half of Mexico's 25 million hectares (62 million acres) of tropical dry and humid forests have been cleared for agriculture and grazing, leaving only 10 percent in stable condition. More than 50 percent of Mexico is dry coastal sage scrub or desert, and overgrazing and human-caused fires have degraded much of this land.

Consider the following facts compiled by the World Wildlife Fund:

North American Ecosystem	Percent of Ecosystem Lost
Original North American tallgrass prairie	More than 99 percent transformed
Original primary forest in the 48 contiguous United States	More than 95 percent lost
Midwest oak savanna	More than 98 percent altered
Old-growth forest in the Pacific Northwest	About 90 percent cleared
Wild or scenic rivers in the United States	Between 90 percent and 98 percent degraded
Coastal sage scrub in the United States	Between 70 percent and 90 percent disturbed
Original wetlands in the United States	More than 50 percent drained and filled

- Can you give examples of ecosystem lost in your area?
- What were the circumstances that led to the loss in your area?
- Was there an alternative to the loss of the ecosystem?
- Were any endangered or threatened species affected?

Source: President's Committee of Advisors on Science and Technology, Biodiversity: Connecting with the Tapestry of Life. Washington, D.C., 2002.

Key Terms

abyssal ecosystem *125*
alpine tundra *121*
benthic *122*
benthic ecosystem *122*
biochemical oxygen demand (BOD) *127*
biome *110*
boreal forest *120*
climax community *105*
coral reef ecosystem *123*
desert *111*
emergent plants *126*
estuary *125*
euphotic zone *121*
eutrophic lake *127*
freshwater ecosystem *121*
grassland *112*

limnetic zone *126*
littoral zone *126*
mangrove swamp ecosystem *123*
marine ecosystem *121*
marsh *128*
Mediterranean shrublands *114*
northern coniferous forest *120*
oligotrophic lake *127*
pelagic *121*
pelagic ecosystem *121*
periphyton *127*
permafrost *120*
phytoplankton *121*
pioneer community *105*
plankton *121*
prairie *112*

primary succession *105*
savanna *114*
secondary succession *105*
seral stage *106*
sere *106*
steppe *112*
submerged plants *126*
succession *105*
successional stage *106*
swamp *128*
taiga *120*
temperate deciduous forest *119*
tropical dry forest *116*
tropical rainforest *116*
tundra *120*
zooplankton *122*

Review Questions

1. Describe the process of succession. How does primary succession differ from secondary succession?
2. How does a climax community differ from a successional community?
3. List three characteristics typical of each of the following biomes: tropical rainforest, desert, tundra, taiga, savanna, Mediterranean shrublands, tropical dry forest, grassland, and temperate deciduous forest.
4. What two primary factors determine the kind of terrestrial biome that will develop in an area?
5. How does height above sea level affect the kind of biome present?
6. What areas of the ocean are the most productive?
7. How does the nature of the substrate affect the kinds of organisms found at the shore?
8. What is the role of each of the following organisms in a marine ecosystem: phytoplankton, zooplankton, algae, coral animals, and fish?
9. List three differences between freshwater and marine ecosystems.
10. What is an estuary? Why are estuaries important?

Critical Thinking Questions

1. Does the concept of a "climax community" make sense? Why or why not?
2. What do you think about restoring ecosystems that have been degraded by human activity? Should it be done or not? Why? Who should pay for this reconstruction?
3. Identify the biome in which you live. What environmental factors are instrumental in maintaining this biome? What is the current health of your biome? What are the current threats to its health? How might your biome have looked 100, 1000, 10,000 years ago?
4. Imagine you are a conservation biologist who is being asked by local residents what the likely environmental outcomes of development would be in the tropical rainforest in which they live. What would you tell them? Why do you give them this evaluation? What evidence can you cite for your claims?
5. The text says that 90 percent of the old-growth temperate rainforest in the Pacific Northwest has been logged. What to do with the remaining 10 percent is still a question. Some say it should be logged, and others say it should be preserved. What values, beliefs, and perspectives are held by each side? What is your ethic regarding logging old-growth in this area? What values, beliefs, and perspectives do you hold regarding this issue?
6. Much of the old-growth forest in the United States has been logged, economic gains have been realized, and second-growth forests have become established. This is not the case in the tropical rainforests, although they are being lost at alarming rates. Should developed countries, which have already "cashed in" on their resources, have anything to say about what is happening in developing countries? Why do you think the way you do?

Welcome to McGraw-Hill's Online Learning Center

Location: www.mhhe.com/enger10e

Interactive Exploration

Check out the website at www.mhhe.com/enger10e and click on the cover of this textbook for quizzing, career information, case studies, and hot links for the following topics:

Field Methods for Studies of Ecosystems

Succession and Stability

Biomes and Environmental Habitats

Deserts

Savannah

Grasslands

Rangelands and Land Use Issues

Tropical Rainforests

Tropical Rainforests and Land Use Issues

Tropical Forests and Extinction

Temperate Forests

Nontropical Forests and Land Use Issues

Taiga

Tundra

Marine Habitats

Marine Ecology

The Marine Pelagic Zone

The Marine Deep Sea Zone

Shallow Subtidal Communities

Estuaries

Mangroves

Coral Reefs

Freshwater Habitats

Population Principles

Objectives

After reading this chapter, you should be able to:

- Understand that birthrate and death rate are both important in determining the population growth rate.
- Define the following characteristics of a population: natality, mortality, sex ratio, age distribution, biotic potential, and spatial distribution.
- Explain the significance of biotic potential to the rate of population growth.
- Describe the lag, exponential growth, and stable equilibrium phases of a population growth curve. Explain why each of these stages occurs.
- Describe how limiting factors determine the carrying capacity for a population.
- List the four categories of limiting factors.
- Describe the death phase of a population growth curve typical of some kinds of populations.
- Recognize that humans are subject to the same forces of environmental resistance as are other organisms.
- Understand the implications of overreproduction.
- Understand that the human population is still growing rapidly.
- Explain how human population growth is influenced by social, theological, philosophical, and political thinking.

Chapter Outline

Population Characteristics
   Natality and Mortality
   Sex Ratio and Age Distribution
   Population Density and Spatial Distribution
   Summary of Factors That Influence Population
     Growth Rates
A Population Growth Curve
Carrying Capacity
Reproductive Strategies and Population
   Fluctuations
Human Population Growth
   Available Raw Materials
   Available Energy

Waste Disposal
Interaction with Other Organisms
Social Factors Influence Human Population
Ultimate Size Limitation
Issues—Analysis: *The Lesser Snow Goose—A Population Problem*

Environmental Close-Up
Population Growth of Invading Species, p. 138

Global Perspective
Managing Elephant Populations—Harvest or Birth Control?, p. 143

PAST PRESENT FUTURE

1995–96 Thirty-one wolves were released in Yellowstone National Park.

1985 Zebra mussels were introduced into Lake St. Clair near Detroit, Michigan.

1983 The human population was about 4.68 billion.

1983 There were 75 whooping cranes in North America.

2004 There were about 170 wolves in Yellowstone National Park.

2004 Zebra mussels are found in rivers and lakes in over 20 eastern states and two Canadian provinces.

2004 The human population is about 6.4 billion.

2003 There were 189 whooping cranes in North America.

Population Characteristics

A **population** can be defined as a group of individuals of the same species inhabiting an area. Just as individuals within a population are recognizable, different populations of the same species have specific characteristics that distinguish them from one another. Some important ways in which populations differ include natality (birthrate), mortality (death rate), sex ratio, age distribution, growth rates, density, and spatial distribution.

Natality and Mortality

Natality refers to the number of individuals added to the population through reproduction over a particular time period. There are two ways in which new individual organisms are produced—asexual reproduction and sexual reproduction. Bacteria and other tiny organisms reproduce primarily asexually when they divide to form new individuals that are identical to the original parent organism. Even plants and many kinds of

animals, such as sponges, jellyfish, and many kinds of worms, reproduce asexually by dividing into two parts or by budding off small portions of themselves that become independent individuals. Even some insects and lizards have a special kind of asexual reproduction in which the females lay unfertilized eggs that are genetically identical to the female. However, most species have some stage in their life cycle in which they reproduce sexually. In plant populations, sexual reproduction results in the production of numerous seeds, but the seeds must land in appropriate soil conditions before they will germinate to produce a new individual. Animal species also typically produce large numbers of offspring as a result of sexual reproduction. In human populations, natality is usually described in terms of the **birthrate,** the number of individuals born per 1000 individuals per year. For example, if a population of 2000 individuals produced 20 offspring during one year, the birthrate would be 10 per thousand per year. The natality for most species is typically quite high. Most species produce many more offspring than are needed to replace the parents.

It is important to recognize that the growth of a population is not determined by the birthrate (natality) alone. **Mortality,** the number of deaths in a population over a particular time period, is also important. For most species, mortality rates are very high, particularly among the younger individuals. For example, of all the seeds that plants produce, very few will result in a mature plant that itself will produce offspring. Many seeds are eaten by animals, some do not germinate because they never find proper soil conditions, and those that germinate must compete with other organisms for nutrients and sunlight.

In human population studies, mortality is usually discussed in terms of the **death rate,** the number of people who die per 1000 individuals per year. Compared to the high mortality of the young of most species, the infant death rate of long-lived animals such as humans is relatively low. In order for the size of a population to grow, the number of individuals added by reproduction must be greater than the number leaving it by dying. (See figure 7.1)

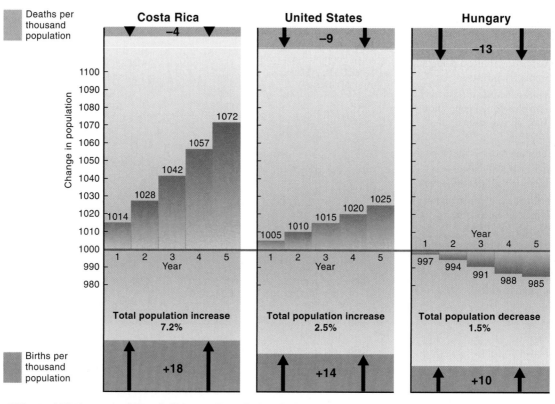

Figure 7.1   **Effect of Birthrate and Death Rate on Population Size** For a population to grow, the birthrate must exceed the death rate for a period of time. These three human populations illustrate how the combined effects of births and deaths would change population size if birthrates and death rates were maintained for a five-year period.

Source: Data from World Population Data Sheet 2003, Population Reference Bureau, Inc., Washington, D.C.

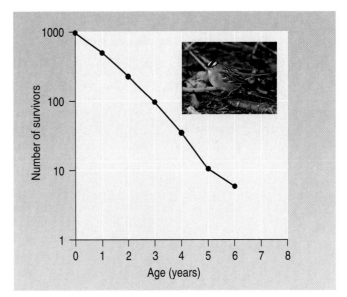

Figure 7.2 **Types of Survivorship Curves** (a) The Dall sheep is a large mammal that produces relatively few young. Most of the young survive, and survival is high until individuals reach old age, when they are more susceptible to predation and disease. (b) The curve shown for the white-crowned sparrow is typical of that for many kinds of birds. After a period of high mortality among the young, the mortality rate is about equal for all ages of adult birds. (c) Many small animals and plants, such as the Mediterranean shrub *Cleome droserifolia,* produce enormous numbers of offspring. Mortality is very high in the younger individuals, and few individuals reach old age.

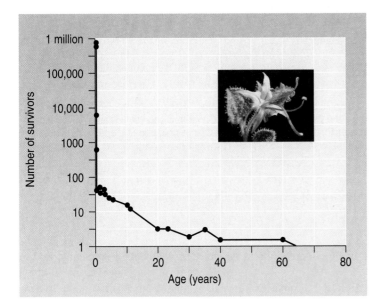

The **population growth rate** is the birthrate minus the death rate. In human population studies, the population growth rate is usually expressed as a percentage of the total population. For example, in the United States, the birthrate is 15 births per thousand individuals in the population. The death rate is 9 per thousand. The difference between the two is 6 per thousand, which is equal to an annual population increase of 0.6 percent (6/1000).

Another way to view mortality is to view how likely it is that an offspring will survive to a specific age. One way of visualizing this is with a survivorship curve. A **survivorship curve** shows the proportion of individuals likely to survive to each age. While each species is different, three general types of survivorship curves can be recognized: species that have high mortality among their young, species in which mortality is evenly spread over all age groups, and species in which survival is high until old age, when mortality is high. Figure 7.2 gives examples of species that fit these three general categories.

Sex Ratio and Age Distribution

The population growth rate is greatly influenced by sex ratio and age distribution of the population. The **sex ratio** refers to the relative numbers of males and females. (Many kinds of organisms, such as earthworms and most plants, have both kinds of sex organs in the same body; sex ratio has no meaning for these species.) The number of females is very important, since they ultimately determine the number of offspring produced in the population. In polygamous species, one male may mate with many females. Therefore, the number of males is less important to the population growth rate than the number of females. In monogamous species, a male and female pair up, mate, and raise their young together. Unpaired females are

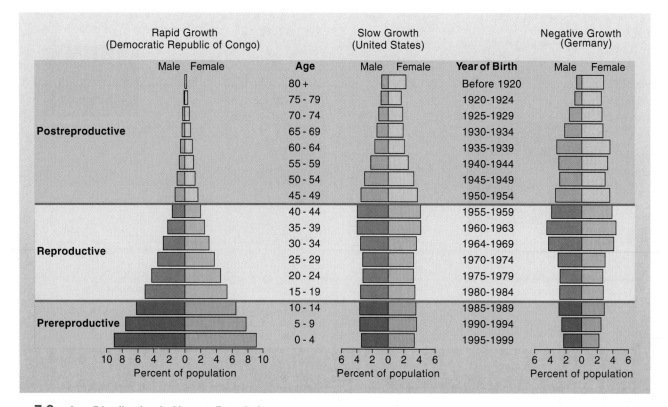

Figure 7.3 **Age Distribution in Human Populations** The relative numbers of individuals in each of the three categories (prereproductive, reproductive, and postreproductive) are good clues to the future growth of a population. The Democratic Republic of Congo has a large number of young individuals who will become reproducing adults. Therefore, this population is likely to grow rapidly. The United States has a large proportion of reproductive individuals and a moderate number of prereproductive individuals. Therefore, this population is likely to grow slowly. Germany has a declining number of reproductive individuals and a very small number of prereproductive individuals. Therefore, its population has begun to decline.

Source: Data from Population Reference Bureau.

not likely to be fertilized and raise young. Even if an unpaired female is fertilized, she will be less successful in raising young.

It is typical in most species that the sex ratio is about 1:1 (one female to one male). However, there are populations in which this is not true. In populations of many species of game animals, the males are shot (have a higher mortality) and the females are not. This results in an uneven sex ratio in which the females outnumber the males. In many social insect populations (bees, ants, and wasps), the number of females greatly exceeds the number of males at all times, though most of the females are sterile. In humans, about 106 males are born for every 100 females. However, in the United States, by the time people reach their mid-twenties, a higher death rate for males has equalized the sex ratio. The higher male death rate continues into old age, when women outnumber men.

The **age distribution,** the number of individuals of each age in the population, greatly influences the population growth rate. As you can see in figure 7.3, some are prereproductive juveniles, some are reproducing adults, and some are postreproductive adults. The age distribution greatly influences the population growth rate. If the population has a large number of prereproductive juveniles, it would be expected to grow in the future as the young become sexually mature. If the majority of a population is made up of reproducing adults, the population should be growing. If the population is made up of old individuals whose reproductive success is low, the population is likely to fall.

Many species, particularly those that have short life spans, have age distributions that change significantly during the course of a year. Species typically produce their young during specific parts of the year. Annual plants (those that live for only one year) produce seeds that germinate in the spring or following a rainy period of the year. Therefore, during one part of the year, most of the individuals are prereproductive. As time passes, nearly all of those that survive become reproducing adults and produce seeds. Later in the year, they all die. A similar pattern is seen in many insects that emerge from eggs as larvae, transform into adults, mate and lay eggs, and die. Other animals typically produce their young at a time when food is available. For many birds, this is during the spring and summer. Many herbivores produce their young in the spring or following rains, when plants begin to grow. Thus, there is a surge in the number of prereproductive individuals at specific times of the year.

In species that live a long time, it is possible for a population to have an age distribution in which the proportion of individuals in these three categories is relatively constant. Since mortality is generally higher among young individuals, such populations typically have more prereproductive individuals than reproductive individuals and more reproductive individuals than postreproductive individuals.

Human populations exhibit several types of age distribution. (See figure 7.3.) The Democratic Republic of Congo's population has a large prereproductive and reproductive component. This means that it will continue to increase rapidly for some time. The United States has a very large reproductive component with a declining number of prereproductive individuals. Eventually, if there were no immigration, the U.S. population would begin to decline if current trends in birthrates and death rates continued. Germany has an age distribution with high postreproductive and low prereproductive portions of the population. With low numbers of prereproductive individuals entering their reproductive years, the population of Germany has begun to decline.

Population Density and Spatial Distribution

Because of such factors as soil type, quality of habitat, and availability of water, organisms normally are distributed unevenly. Some populations have many individuals clustered into a small space, while other populations of the same species may be widely dispersed. **Population density** is the number of organisms per unit area. For example, fruitfly populations are very dense around a source of rotting fruit, while they are rare in other places. Similarly, humans are often clustered into dense concentrations we call cities, with lower densities in rural areas. When the population density is too great, all individuals within the population are injured because they compete severely with each other for necessary resources. Plants may compete for water, soil nutrients, or sunlight. Animals may compete for food, shelter, or nesting sites. In animal populations, overcrowding might

cause some individuals to explore and migrate into new areas. This movement from densely populated locations to new areas is called **dispersal.** It relieves the overcrowded conditions in the home area and, at the same time, increases the population in the places to which they migrate. Often, it is juvenile individuals that relieve overcrowding by leaving. The pressure to migrate from a population (**emigration**) may be a result of seasonal reproduction leading to a rapid increase in population size or environmental changes that intensify competition among members of the same species. For example, as water holes dry up, competition for water increases, and many desert birds emigrate to areas where water is still available.

The organisms that leave one population often become members of a different population. This migration into an area (**immigration**) may introduce characteristics that were not in the population originally. One of the concerns about genetically modified plants is that when they are introduced into a region, they may transfer genes to wild relatives of the crop plant. When Europeans immigrated to North America, they brought genetic and cultural characteristics that had a tremendous impact on the existing Native American population. Among other things, Europeans brought diseases that were foreign to the Native Americans. These diseases increased the death rate and lowered the birthrate of Native Americans, resulting in a sharp decrease in the size of their populations.

Summary of Factors That Influence Population Growth Rates

Populations have an inherent tendency to increase in size. However, as we have just seen, many factors influence the rate at which a population can grow. At the simplest level, the rate of increase is determined by subtracting the number of individuals leaving the population from the number entering. Individuals leave the population either by death or emigration. Individuals enter the population by birth or immigration. Birthrates and death rates are influenced by several factors, includ-

ing the number of females in the population and their age. In addition, the density of a population may encourage individuals to leave because of intense competition for a limited supply of resources.

A Population Growth Curve

Sex ratios and age distributions directly influence the rate of reproduction within a population. However, each species also has an inherent reproductive capacity, or **biotic potential,** which is its biological ability to produce offspring. Some species, such as apple trees, may produce thousands of offspring (seeds) per year, while others, such as pigs or geese, may produce 10 to 12 young per year. (See figure 7.4.) Some large animals, such as bears or elephants, may produce one young every two to three years. Although there are large differences among species, generally, adults produce many more offspring during their lifetimes than are needed to replace themselves when they die. However, most of the young die, so only a few survive to become reproductive adults themselves.

Because most species have a high biotic potential, there is a natural tendency for populations to increase. If we consider a hypothetical situation in which mortality is not a factor, we could have the following situation. If two mice produced four offspring and they all lived, eventually they would produce offspring of their own, while their parents continued to reproduce as well. Under these conditions, the population will grow exponentially. (Exponential growth results in the number of individuals in each succeeding generation being a multiple of the previous generation; for example, 2, 4, 8, 16, 32, etc.). While populations cannot grow exponentially forever, they often have an exponential period of growth.

Population growth often follows a particular pattern, consisting of a lag phase, an exponential growth phase, and a stable equilibrium phase. Figure 7.5 shows a typical population growth curve. During the first portion of the curve,

Apples   Geese   Pigs

Figure 7.4   **Biotic Potential** The ability of a species to reproduce greatly exceeds the number necessary to replace those who die. Here are some examples of the prodigious reproductive abilities of some species.

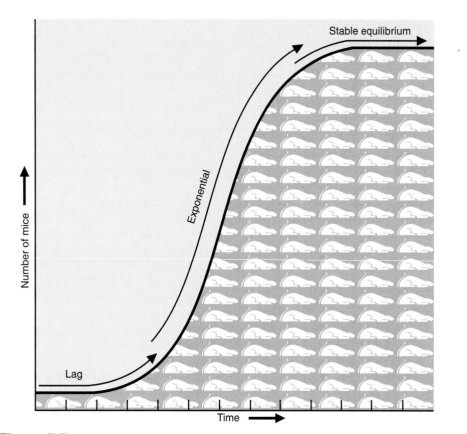

Figure 7.5   **A Typical Population Growth Curve** In this mouse population, there is little growth during the lag phase. During the exponential growth phase, the population rises rapidly as increasing numbers of individuals reach reproductive age. Eventually, the population reaches a stable equilibrium phase, during which the birthrate equals the death rate.

phase (log phase). This growth will continue for as long as the birthrate exceeds the death rate. Eventually, however, the death rate and the birthrate will come to equal one another, and the population will stop growing and reach a relatively stable population size. This stage is known as the **stable equilibrium phase.**

It is important to recognize that although the size of the population may not be changing, the individuals are changing. As new individuals enter by birth or immigration, others leave by death or emigration. For most organisms, the first indication that a population is entering a stable equilibrium phase is an increase in the death rate. A decline in the birthrate may also contribute to the stabilizing of population size. Usually, this occurs after an increase in the death rate. To understand why populations cannot grow continuously, it is necessary to discuss the concept of carrying capacity.

Carrying Capacity

The **carrying capacity** of an area is the number of individuals of a species that can be maintained in an area, over time, without harming the habitat. The concept of carrying capacity is usually applied to relatively long-lasting habitats and is helpful when we examine why populations stabilize. However, nothing is permanent, and as the habitats change, because of disturbance or succession, the carrying capacity for a species changes also. Seasonal changes also influence the numbers of in-

known as the **lag phase,** the population grows very slowly because there are few births, since the process of reproduction and growth of offspring takes time. Organisms must mature into adults before they can reproduce. When the offspring begin to mate and have young, the parents may be producing a second set of offspring. Since more organisms now are reproducing, the population begins to increase at an accelerating rate. This stage is known as the **exponential growth** dividuals that can be supported in an area. The number of individuals that can be supported during the summer may be much larger than can be supported in winter. Indeed, some animals (such as birds) use certain habitats only during the summer and migrate in the fall when living conditions become difficult. The combination of factors that sets the carrying capacity for an area is called **environmental resistance.** (See figure 7.6.)

When a particular condition or factor can be identified as a key component that limits the size of a population, it is identified as a **limiting factor.** For most populations, four categories of limiting factors are recognized as components of environmental resistance that set the carrying capacity. These are: (1) the availability of raw materials, (2) the availability of energy, (3) the accumulation of waste products and their means of disposal, and (4) interactions among organisms.

Raw materials come in many forms. For example, plants need nitrogen and magnesium from the soil as raw materials for the manufacture of chlorophyll. If these minerals are not present in sufficient quantities, the plant population cannot increase. The application of fertilizers is a way of preventing certain raw materials from being a limiting factor. In effect, the carrying capacity has been increased because this limiting factor has been removed. A carrying capacity still exists, but it is set at a new level, and some new primary limiting factor will emerge. Perhaps it will be the amount of water, the number of insects that feed on the plants, or competition for sunlight. Animals also require certain minerals as raw material, which they obtain in their diet. They may also require materials for building nests or to serve as observation sites.

Energy sources are important to all organisms. Plants require energy in the form of sunlight for photosynthesis, so the amount of light can be a limiting factor for many plants. When small plants are in the shade of trees, they often do not grow well and have small populations because they do not receive enough sunlight. Animals require energy in the form of the food they eat, and if food is scarce, they die.

The accumulation of waste products is not normally a limiting factor for plants,

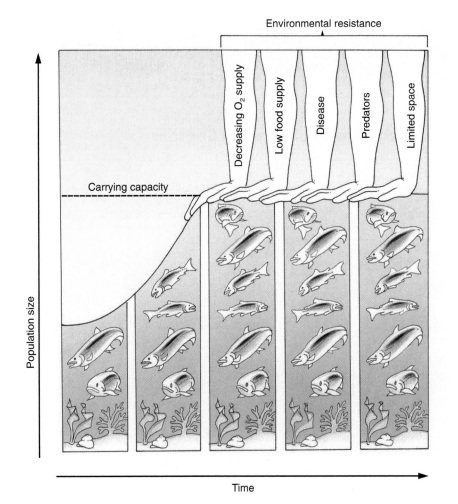

Figure 7.6 **Carrying Capacity** A number of factors in the environment, such as oxygen supply, food supply, diseases, predators, and space, determine the number of organisms that can survive in a given area—the carrying capacity of that area. The environmental factors that limit populations are known collectively as environmental resistance.

since they produce few wastes, but it can be for other kinds of organisms. Bacteria, other tiny organisms, and many kinds of aquatic organisms that live in small ecosystems such as puddles, pools, or aquariums may be limited by wastes. When a small number of a species of bacterium are placed on a petri plate with nutrient agar (a jellylike material containing food substances), the population growth follows a curve shown in figure 7.7. As expected, it begins with a lag phase, continues through an exponential growth phase, and eventually levels off in a stable equilibrium phase. However, in this small, enclosed space, there is no way to get rid of the toxic waste products, which accumulate, eventually killing the bacteria.

This decline in population size is known as the **death phase.** When a population decreases rapidly, it is said to crash.

Interactions among organisms are also important in determining population size. Since white-tailed deer and cottontail rabbits eat the twigs of many species of small trees, they have a limiting effect on the size of some tree populations. Many aquatic organisms produce waste products that build up to toxic levels and result in the death of fish. At the same time, decomposer organisms that break down these toxic waste materials may prevent the negative impact of the toxins. Parasites and predators cause the premature death of individuals, thus limiting the size of the population.

Environmental Close-Up

Population Growth of Invading Species

When a new species is introduced into an area suitable for its survival, it often has great potential to increase its population size, since it may not have natural enemies to keep mortality high. New species may also be able to compete favorably for available resources and lower the populations of native species. A typical population growth curve starts with a few individuals being released. Once established, the population will increase in number and expand its range. The zebra mussel, for example, is thought to have entered the Great Lakes about 1985. Today, it is found throughout all the Great Lakes and has been introduced into the Mississippi River and its tributaries, where it has been discovered as far south as New Orleans. Similarly, the gypsy moth has spread from its place of original release to much of the forested land of the Midwest and is still expanding its range. Another invading species is the kudzu vine, which has become a pest in many areas of the southern United States.

Eventually, the invading species occupies all the habitat suitable to it and the population stabilizes. Dandelions and starlings are introduced species that are no longer expanding their range. They have simply become a normal part of the biology of North America.

Kudzu vine

Dandelion

Zebra mussels

Gypsy moths

Starlings

PART TWO   Ecological Principles and Their Application

Some studies indicate that populations can be controlled by interaction among individuals within the population. A study of laboratory rats shows that crowding causes a breakdown in normal social behavior, which leads to fewer births and increased deaths. The changes observed include abnormal mating behavior, decreased litter size, fewer litters per year, lack of maternal care, and increased aggression in some rats or withdrawal in others. Thus, limiting factors can reduce birthrates as well as increase death rates. Many other kinds of animals have shown similar reductions in breeding success when population densities were high.

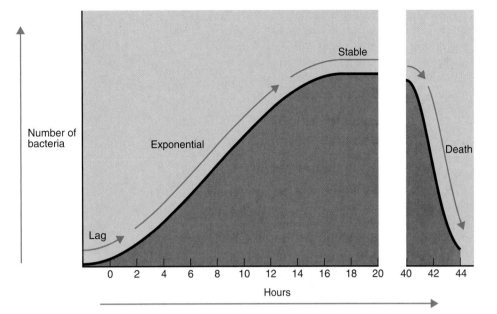

Figure 7.7 **A Bacterial Growth Curve** The initial change in population size follows a typical population growth curve until waste products become lethal. The buildup of waste products lowers the carrying capacity. When a population begins to decline, it enters the death phase.

Reproductive Strategies and Population Fluctuations

So far, we have talked about population growth as if all organisms reach a stable population when they reach the carrying capacity. That is an appropriate way to begin to understand population changes, but the real world is much more complicated. Species can be divided into two broad categories based on their reproductive strategies. **K-strategists** are organisms that typically reach a stable population as the population reaches the carrying capacity. K-strategists usually occupy relatively stable environments and tend to be large organisms that have relatively long lives, produce few offspring, and provide care for their offspring. Their reproductive strategy is to invest a great deal of energy in producing a few offspring that have a good chance of living to reproduce. Deer, lions, and swans are examples of this kind of organism. Humans generally produce single offspring, and even in countries with high infant mortality, 80 percent of the children survive beyond one year of age, and the majority of these will reach adulthood. Generally, populations of K-strategists are controlled by density-dependent limiting factors. **Density-dependent limiting factors** are those that become more severe as the size of the population increases. For

example, there is a carrying capacity that limits the size of a hawk population. Hawks feed on mice, snakes, and small birds. As the size of the hawk population increases, the competition among hawks for available food becomes more severe. The increased competition for food is a density-dependent limiting factor that leads to less food for the young in the nest. Therefore, many of the young die, and population growth rate slows as the carrying capacity for the area is reached.

The **r-strategist** is typically an organism that exploits unstable environments. Typically, r-strategists are small organisms that have a short life, produce many offspring, and do not reach a carrying capacity. Examples are grasshoppers, gypsy moths, and some mice. The reproductive strategy of r-strategists is to expend large amounts of energy producing many offspring but to provide limited care (often none) for them. Consequently, there is high mortality among the young. For example, one female oyster may produce a million eggs, but few of them ever find suitable places to attach themselves and grow. Typically, these populations are limited by **density-independent limiting factors** in which the size of the population has nothing to do with the limiting factor. Typical density-independent limiting factors include changing weather conditions that kill large numbers of organisms, habitat loss such as when a pond dries up or fire destroys a forest, or events such as a deep snow or flood that buries sources of food and leads to the death of entire populations. The population size of r-strategists is likely to fluctuate wildly. They reproduce rapidly, and the size of the population increases until some density-independent factor causes the population to crash; then they begin the cycle all over again. (The letters *K* and *r* in *K-strategist* and *r-strategist* come from a mathematical equation in which *K* represents the carrying capacity of the environment and *r* represents the biotic potential of the species.)

Since humans are K-strategists, it may be difficult for us to appreciate that the r-strategy can be viable from an evolutionary point of view. There is no carrying capacity for temporary resources. Resources that are present only for a short time can be exploited most effectively if many individuals of one species monopolize the resource, while denying other species access to it. Rapid reproduction can place a species in a position to compete against other species that are not able to increase numbers as rapidly. Obviously, most of the individuals will die, but not before they have left some offspring or resistant stages that will be capable of exploiting the resource should it become available again.

Even K-strategists, however, have some fluctuations in their population size for a variety of reasons. One reason is that even in relatively stable ecosystems, there will be variations from year to year. Floods, droughts, fires, extreme cold, and similar events may affect the carrying capacity of an area, thus causing fluctuations in population size. Epidemic disease or increased predation may also lead to populations that vary in size from year to year. Many endangered species have reduced populations because their normal environment has been altered either naturally or as a result of human activity. (See chapter 12 on biodiversity issues.)

In northern regions of the world, many kinds of animals show distinct population cycles—periods of relatively large populations followed by periods of small populations. This is generally thought to be the case because the ecosystems are relatively simple, with few kinds of organisms affecting one another. Many of these cycles are quite regular. Biologists have been studying these cycles since the 1920s, and have developed several theories about why northern populations cycle. One idea is that heavy feeding by large populations of herbivores causes the plants to produce increased amounts of chemicals, which taste bad or are toxic. A second thought is that when an herbivore population is large that many different predators shift to eating them and the herbivore populations crashes. Another idea is that interactions between a prey organism and a specialized predator naturally lead to population cycles. The length of the population cycle depends on the reproductive biology of the prey and their predators.

A recent study of the population biology of the collared lemming (*Dicrostonyx groenlandicus*) on Greenland illustrates the population interactions between lemmings and four different predators. Lemmings have a very high biotic potential. They produce two to three litters of young per year. Their population is held in check

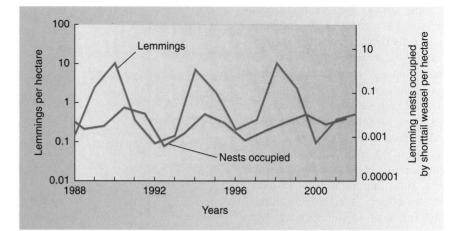

Figure 7.8 **Population Cycles** In many northern regions of the world, population cycles are common. In the case of collared lemmings and shorttail weasels on Greenland, interactions between the two populations result in population cycles of about four years. The graph shows the population of lemmings per hectare and the number of lemming nests occupied by weasels per hectare.

Source: Data from O. Gilg, I. Hamski, and B. Sittler, "Cyclic Dynamics in a Simple Predator-Prey Community," *Transpol'air* Online Magazine, www.transpolair.com.

by four different predators. Three of these predators—the snowy owl, arctic fox, and longtailed skua (a bird that resembles a gull)—are generalist predators whose consumption of lemmings is directly related to the size of the lemming population. They constitute a density-dependent limiting factor for the lemming population. When lemming numbers are low, these predators seek other prey. For example, the snowy owl often migrates to other regions when lemming numbers are low in a particular region. The fourth predator is the shorttail weasel (*Mustela ermina*), a specialist predator on lemmings. The weasels are much more dependent on lemmings for food than the other predators. Since the weasels mate once a year, their populations increase at a slower rate than those of the lemmings. As weasel populations increase, however, they eventually become large enough that they drive down the lemming populations. The resulting decrease in lemmings leads to a decline in the number of weasels, which allows greater survival of lemmings, which ultimately leads to another cycle of increased weasel numbers. (See figure 7.8.)

Although local human populations often show fluctuations, the worldwide human population has increased continually for the past several hundred years. Humans have been able to reduce environmental resistance by eliminating competing organisms, increasing food production, and controlling disease organisms.

Human Population Growth

The human population growth curve has a long lag phase followed by a sharply rising exponential growth phase that is still rapidly increasing. (See figure 7.9.) A major reason for the continuing increase in the size of the human population is that the human species has lowered its death rate. When various countries reduce environmental resistance by increasing food production or controlling disease, they share this technology throughout the world. Developed countries send health care personnel to all parts of the globe to improve the quality of life for people in less-developed countries. Physicians offer advice on nutrition, and engineers develop wastewater treatment systems. Improved sanitary facilities in India and Indonesia, for example, decreased deaths caused by cholera. These advancements tend to reduce death rates while birthrates remain high. Thus, the size of the human population increases rapidly.

Let us examine the human population situation from a different perspective. The world population is currently

PART TWO Ecological Principles and Their Application

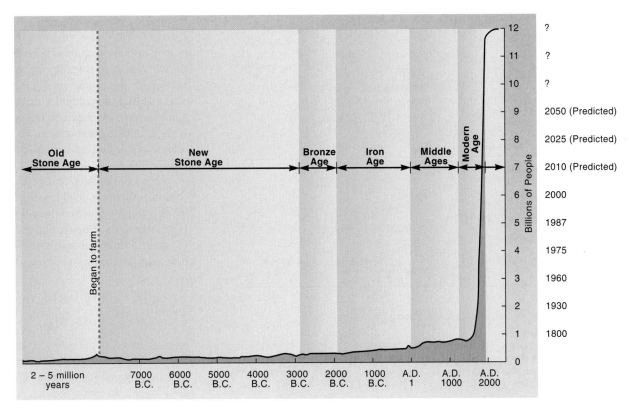

Figure 7.9   The Historical Human Population Curve From A.D. 1800 to A.D. 1930, the number of humans doubled (from 1 billion to 2 billion) and then doubled again by 1975 (4 billion) and could double again (8 billion) by the year 2025. How long can this pattern continue before the Earth's ultimate carrying capacity is reached?

Source: Data from Jean Van Det Tak, et al., "Our Population Predicament: A New Look," *Population Bulletin*, vol. 34, no. 5 (December 1979), Population Reference Bureau, Washington, D.C.; and more recent data from the Population Reference Bureau.

increasing at an annual rate of 1.3 percent. That may not seem like much, but even at 1.3 percent, the population is growing rapidly. It can be difficult to comprehend the impact of a 1 or 2 percent annual increase. Remember that a growth rate in any population compounds itself, since many of the additional individuals eventually reproduce, thus adding more individuals. One way to look at this growth is to determine how much time is needed to double the population. This is a valuable method because most of us can appreciate what life would be like if the number of people in our locality were doubled, particularly if the doubling were to occur within our lifetime.

Figure 7.10 shows the relationship between the rate of annual increase for the human population and the number of years it would take to double the population if that rate were to continue. The doubling time for the human population is easily calculated by dividing the number 70 by the annual rate of increase. Thus, at a 1 percent rate of annual increase, the population will double in 70 years (70/1). At a 2 percent rate of annual increase, the human population will double in 35 years (70/2). The current worldwide rate of annual increase of about 1.3 percent will double the world human population in about 54 years.

What does this very rapid rate of growth mean to the human species? As a species, humans are subject to the same limiting factors as all other species. We cannot increase beyond our ability to acquire raw materials and energy and safely dispose of our wastes. We also must remember that interactions with other species and with other humans will help determine our carrying capacity. Furthermore, when we think about the human population and carrying capacity, we need to distinguish between the biological carrying capacity, which describes how many people the Earth can support, and a cultural carrying capacity, which describes how many people the Earth can support with a reasonable standard of living. Re-

gardless of which of these two concepts we are thinking about, the same four basic factors are involved. Let us look at these four factors in more detail.

Available Raw Materials

Many of us think of raw materials simply as the amount of food available. However, we have become increasingly dependent on technology, and our lifestyles are directly tied to our use of other kinds of resources, such as irrigation water, genetic research, and antibiotics. Food production is becoming a limiting factor for some segments of the world's human population. Malnutrition is a serious problem in many parts of the world because sufficient food is not available. Currently, about 1 billion people (one-sixth of the world's population) suffer from a lack of adequate food. Chapter 8 deals in greater detail with the problems of food production and distribution and their relationship to human population growth.

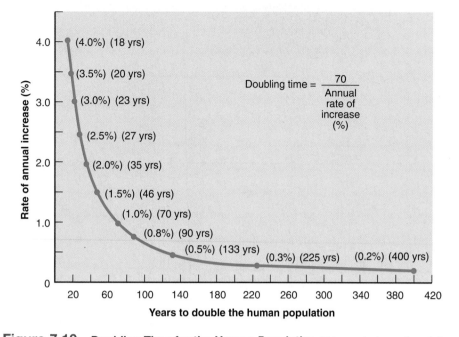

$$\text{Doubling time} = \frac{70}{\text{Annual rate of increase (\%)}}$$

Figure 7.10 Doubling Time for the Human Population This graph shows the relationship between the rate of annual increase in percent and doubling time. A population growth rate of 1 percent per year would result in the doubling of the population in about 70 years. A population growth rate of 3 percent per year would result in a population doubling in about 23 years.

Available Energy

The second factor, available energy, involves problems similar to those of raw materials. Essentially, all species on Earth are ultimately dependent on sunlight for their energy. Currently, the world's human population depends on fossil fuels to raise food, modify the environment, and move from place to place. When energy prices increase, much of the world's population is placed in jeopardy because incomes are not sufficient to pay the increased costs for energy and other essentials. New, less disruptive methods of harnessing this energy must be developed to support an increasing population. Increases in the efficiency with which energy is used could reduce demands on fossil fuels. In addition, the development of more efficient solar and wind energy conversion systems could reduce the need for fossil fuels. (See chapter 10.)

Waste Disposal

Waste disposal is the third factor determining the carrying capacity for humans. Most pollution is, in reality, the waste product of human activity. Lack of adequate sewage treatment and safe drinking water causes large numbers of deaths each year. Some people are convinced that disregard for the quality of our environment will be a major limiting factor. In any case, it makes good sense to control pollution and to work toward cleaning our environment. Chapters 16–19 deal with aspects of pollution and how pollution can be reduced.

Interaction with Other Organisms

The fourth factor that determines the carrying capacity of a species is interaction with other organisms. We need to become aware that we are not the only species of importance. When we convert land to meet our needs, we displace other species from their habitats. Many of these displaced organisms are not able to compete with us successfully and must migrate or become extinct. Unfortunately, as humans expand their domain, the areas available to these displaced organisms become rarer. Parks and natural areas have become tiny refuges for the plants and animals that once occupied vast expanses of land. If these refuges fall to the developer's bulldozer or are converted to agri-

cultural use, many organisms will become extinct. What today seems like an unimportant organism, one that we could easily do without, may someday be seen as an important link to our very survival. It is also important to recognize that many organisms provide services that we enjoy without thinking about them. Forest trees release water and moderate temperature changes, bees and other insects pollinate crops, insect predators eat pests, and decomposers recycle dead organisms. All of these activities illustrate how we rely on other organisms. Eliminating the services of these valuable organisms would be detrimental to our way of life.

Social Factors Influence Human Population

Human survival depends on interaction and cooperation with other humans. Current technology and medical knowledge are available to control human population growth and to improve the health of the people of the world. Why, then, does the population continue to increase, and why do large numbers of people continue to live in poverty, suffer from preventable diseases, and endure malnutrition? Humans are social animals who have freedom of choice and frequently do not do what is considered "best" from an unemotional, uninvolved, biological point of view. People make decisions based on history, social situations, ethical and religious considerations, and personal desires. The biggest obstacles to controlling human population are not biological but are the province of philosophers, theologians, politicians, and sociologists. People in all fields need to understand that the cause of the population problem has both biological and social components if they are to successfully develop strategies for addressing it. Chapter 8 will cover many of these social, political, and ethical topics in greater detail.

Ultimate Size Limitation

The human population is subject to the same biological constraints as other species of organisms. We can say with certainty that our population will ultimately reach its carrying capacity and

GLOBAL PERSPECTIVE

Managing Elephant Populations—Harvest or Birth Control?

During the 1980s, uncontrolled hunting of elephants for ivory and food had reduced the African elephant (*Loxodonta africana*) population from 1.3 million to 650,000. In addition, expansion of agriculture into areas that had been excellent elephant habitat compounded the problem because it reduced the amount of land available to elephants. Many conservation organizations became alarmed by the rapid decline in the numbers of elephants and supported a ban on the export of ivory. In 1989, the Convention on International Trade in Endangered Species of Wild Fauna and Flora (CITES) banned all international trade in elephant products, of which ivory is the most important. The ban worked. The price of ivory fell and poaching became unprofitable. Although the total population of elephants in Africa continued to fall to between 400,000 and 660,000 by the year 2003, the decline is not true for all parts of Africa. Some southern African countries actually have experienced an elephant population explosion that threatens to destroy the limited habitat available to elephants and increasingly results in conflicts between farmers and elephants.

The African elephant requires huge amounts of food (150 to 250 kilograms per day [330 to 550 pounds per day]) to sustain its large body. The animals strip bark from trees, uproot the trees, and eat large quantities of grass. Where humans and elephants share the same habitat, elephants can do great damage to crops. Because elephants have been increasingly confined to national parks and nature preserves by agricultural expansion, two natural options for relieving population pressure (migration or starvation) have not been available. Migration from the park results in increased agricultural damage and risk of injury to farmers, and increasing populations can cause irreparable damage to the protected habitat they occupy as they seek food. If they are allowed to starve, public pressure from wildlife groups will be enormous, and the park managers will be condemned. If the populations are to remain healthy, not destroy their habitat, and not come into conflict with farming communities, the size of the population must be controlled. There are only two ways to do that—increase the death rate (culling) or decrease the birthrate (sterilization or birth control).

Because they have growing elephant populations, several southern African countries, such as Botswana, Namibia, Zimbabwe, and South Africa, have argued that they should be allowed to manage their elephant populations as a natural resource in the same way that other large game animals (deer, elk, caribou) are managed. The sustainable harvest of surplus elephants would be wise use of a natural resource that would provide income to the local people as well as a much needed source of protein in this region. Furthermore, the income from harvesting elephants could be used to provide additional funding for park management. Those who oppose harvesting maintain that killing the animals for trophies or ivory will take many older individuals from the herd. It is well known that older elephants serve as leaders. Selectively killing older animals, therefore, could remove important memories that could disrupt traditional migration patterns and other important behaviors necessary to their survival.

In 1997, CITES approved the sale of 60 metric tons of ivory from Botswana, Namibia, and Zimbabwe to Japan. The ivory had been stockpiled from previous times and was not to be from killing of elephants. Following the sale, conservation organizations reported an increase in poaching of elephants and unlawful sale of ivory. In 2000, several coun-

tries again petitioned CITES to approve additional sales of ivory. Their request was not approved because of concerns about poaching and monitoring of illegal sales. In 2003, CITES approved the sale of 66 metric tons of stockpiled ivory from South Africa, Namibia, and Botswana. The revenue from the sale during 2004 is to be used to improve protection for wildlife reserves.

Animal rights groups see these sales as a prelude to a return to the practice of culling elephants from herds that are too large. Therefore, the Humane Society of the United States has advocated the use of birth control to reduce the birthrate and solve the elephant population problem. Several methods have been tested and they can work. However, critics of this approach suggest that preventing female elephants from conceiving for long periods of time may disrupt the normal social structure of the herd. Under most circumstances, a female that is pregnant will not come into heat again for two years. If females do not conceive, they will come into heat approximately every 15 weeks, and there will be more females in heat than is normal throughout the year. Since the mating activity of elephants is disruptive of their usual pattern of activity, some people are concerned that the normal social structure of the herd will be disrupted. Furthermore, a herd usually includes various ages of immature individuals. The younger individuals learn from older siblings or cousins. The use of birth control could disrupt the typical spacing of pregnancies and interfere with the normal process of "educating" young elephants. What looks like a simple problem to solve has resulted in two diametrically opposed camps: those who advocate increased deaths by harvesting and those who advocate decreased births by manipulating births. Neither is "natural," but something must be done or the survival of elephants could be threatened.

stabilize. There is disagreement about how many people can exist when the carrying capacity is reached. Some people suggest we are already approaching the carrying capacity, while others maintain that we could more than double the population before the carrying capacity is reached. Furthermore, uncertainty exists about what the primary limiting factors will be and about the quality of life the inhabitants of a more populous world would have. If the human population continues to reproduce at its current rate of 1.3 percent per year, the population will double from its current 6.4 billion to over 12 billion people by 2060. Recent history shows that the rate at which the human population is increasing has been falling. Even when a declining growth rate is taken into account, however, the population is still expected to reach 9 billion people by 2050.

As with all K-strategist species, when the population increases, density-dependent limiting factors will become more forceful. Some people suggest that a lack of food, a lack of water, or increased waste will ultimately control the size of the human population. Still others suggest that, in the future, social controls will limit population growth. These social controls could be either voluntary or involuntary. In the economically developed portions of the world, families have voluntarily lowered their birthrates to fewer than two children per woman. Most of the poorer countries of the world have higher birthrates. What kinds of measures are needed to encourage them to limit their populations? Will voluntary compliance with stated national goals be enough, or will enforced sterilization and economic penalties become the norm? Others are concerned that countries will launch wars to gain control of limited resources or to simply eliminate people who compete for the use of those resources.

It is also important to consider the age structure of the world population. In most of the world, there are many reproductive and prereproductive individuals. Since most of these individuals are currently reproducing or will reproduce in the near future, even if they reduce their rate of reproduction, there will be a sharp increase in the number of people in the world in the next few years.

No one knows what the ultimate human population size will be or what the most potent limiting factors will be, but most agree that we are approaching the maximum sustainable human population. If the human population continues to increase, eventually the amount of agricultural land available will not be able to satisfy the demand for food.

Summary

A population is a group of organisms of the same species that inhabits an area. The birthrate (natality) is the number of individuals entering the population by reproduction during a certain period. The death rate (mortality) measures the number of individuals that die in a population during a certain period. Population growth is determined by the combined effects of the birthrate and death rate.

The sex ratio of a population is a way of stating the relative number of males and females. Age distribution and the sex ratio have a profound impact on population growth. Most organisms have a biotic potential much greater than that needed to replace dying organisms.

Interactions among individuals in a population, such as competition, predation, and parasitism, are also important in determining population size. Organisms may migrate into (immigrate) or migrate out of (emigrate) an area as a result of competitive pressure.

A typical population growth curve shows a lag phase followed by an exponential growth phase and a stable equilibrium phase at the carrying capacity. The carrying capacity is determined by many limiting factors that are collectively known as environmental resistance. The four major categories of environmental resistance are available raw materials, available energy, disposal of wastes, and interactions among organisms. Some populations experience a death phase following the stable equilibrium phase.

K-strategists typically are large, long-lived organisms that reach a stable population at the carrying capacity. Their population size is usually controlled by density-dependent limiting factors. Organisms that are r-strategists are generally small, short-lived organisms that reproduce very quickly. Their populations do not generally reach a carrying capacity but crash because of some density-independent limiting factor.

The human population is increasing at a rapid rate. The Earth's ultimate carrying capacity for humans is not known. The causes for human population growth are not just biological but also social, political, philosophical, and theological.

Issues – Analysis

The Lesser Snow Goose—A Problem Population

Lesser snow geese use breeding grounds in the Arctic and subarctic regions of Canada and wintering grounds on salt marshes of the Gulf of Mexico, California, and Mexico. (See map.) One population of the lesser snow goose has breeding grounds around Hudson Bay and wintering grounds on the Gulf of Mexico. Because these birds migrate south from Hudson Bay through the central United States west of the Mississippi River to the west coast of the Gulf of Mexico, they are called the midcontinent population. Populations of the midcontinent lesser snow goose have grown from about 800,000 birds in 1969 to an estimated 3.2 million in 2003, a fourfold increase. (Some experts believe the number may be closer to 6 million.) These numbers are so large that the nesting birds are destroying their breeding habitat.

What has caused this drastic population increase? A primary reason appears to be a change in feeding behavior. Originally, during the winter snow geese fed primarily on the roots and tubers of aquatic vegetation in the salt marshes of the Gulf of Mexico. Since this habitat and the food source it provided is limited, there was a natural control on the size of the population. However, many of these wetland areas have been destroyed. Simultaneously, agriculture (particularly rice farming) in the Gulf Coast region provided an alternative food source. As the geese began to use agricultural areas for food, they had an essentially unlimited source of food during the winter months, which led to very high winter survival. Furthermore, concerns about erosion have resulted in farmers using no-till and reduced tillage practices in the grain fields along the geese's flyway. The fields provided food for the fall migration that improved the survival of young snow geese. During the spring migration, geese used the same resources to store food as body fat that they took with them to their northern breeding grounds. Other factors that may have played a role are the decline in the number of people who hunt geese and a warming of the climate that leads to greater breeding success on the cold northern breeding grounds. The large goose population is causing habitat destruction on their Canadian breeding grounds.

Habitat destruction occurring in Canada is a result of several factors. Snow geese feed by ripping plants from the ground, and they nest in large colonies. Therefore, they can have a powerful local impact on the vegetation. This coupled with a huge increase in the breeding population has resulted in the destruction of large areas of the coastal vegetation around Hudson Bay. (See photo.)

What can be done to control the snow goose population and protect their breeding habitat? There are two possibilities: let nature take its course or use population management tools. If nothing is done, the geese will eventually destroy so much of their breeding habitat that they will be unable to breed successfully, and the population will crash. It could take decades for the habitat to recover, however. The alternative is to use population management tools to try to resolve the problem. Several management activities have been instituted. These include allowing hunters to kill more geese, increasing the length of the hunting season, allowing hunts during the spring as well as the fall migration seasons, and allowing more harvest by Canadian native communities. In recent years, the harvest in the United States and Canada has increased, varying between 1 million and 1.5 million geese killed. Preliminary evidence suggests that the increased harvest is starting to have an effect. It will take several more years, however, to determine if the population can be controlled.

- Do you think the killing of geese is justified to protect the coastal habitat of Hudson Bay?
- Do you think it is better to have the population crash as a result of natural forces or to reduce the size of the population by increasing the harvest?

(a) This photo shows the kind of damage done by geese. The exclosures were established by Dawn R. Bazely and Robert L. Jefferies of the Hudson Bay Project.

(b) The map shows the migration routes between the traditional nesting and wintering areas of different snow goose populations.

Key Terms

age distribution *134*

biotic potential *135*

birthrate *132*

carrying capacity *136*

death phase *137*

death rate *132*

density-dependent limiting factors *139*

density-independent limiting factors *139*

dispersal *135*

emigration *135*

environmental resistance *137*

exponential growth phase (log phase) *136*

immigration *135*

K-strategists *139*

lag phase *136*

limiting factor *137*

mortality *132*

natality *132*

population *133*

population density *135*

population growth rate *132*

r-strategists *139*

sex ratio *133*

stable equilibrium phase *136*

survivorship curve *133*

Review Questions

1. How is biotic potential related to the rate at which a population will grow?

2. List three characteristics populations might have.

3. Why do some populations grow? What factors help to determine the rate of this growth?

4. Under what conditions might a death phase occur?

5. List four factors that could determine the carrying capacity of an animal species.

6. How do the concepts of birthrate and population growth differ?

7. How does the population growth curve of humans compare with that of bacteria on a petri dish?

8. How do r-strategists and K-strategists differ?

9. As the human population continues to increase, what might happen to other species?

10. All successful organisms overproduce. What advantage does this provide for the species? What disadvantages may occur?

Critical Thinking Questions

1. Why do you suppose some organisms display high natality and others display lower natality? For example, why do cottontail rabbits show high natality and wolves relatively low natality? Why wouldn't all organisms display high natality?

2. Do you think African elephants should be managed by southern African countries like deer are managed in the United States? What values, beliefs, and perspectives lead you to your conclusion?

3. Where do you stand on the issue of birth control for elephants or culling of the herd? Why do you think the way you do? What is the position of the Humane Society in the United States? Why do you think it holds the position it does? Do you think conservationists in the United States should have a say in what is done with the elephants in Africa? Should Africans have a say in what is done in the United States? Why?

4. Consider the differences between r-strategists and K-strategists. What costs are incurred by adopting either strategy? What evolutionary benefits does each strategy enjoy?

5. Why do invading species, which can survive in a new environment, often show exponential growth rates soon after they are introduced to that environment?

Interactive Exploration

Check out the website at **www.mhhe.com/enger10e** and click on the cover of this textbook for quizzing, career information, case studies, and hot links for the following topics:

Animal Population Ecology

Field Methods for Studies of Populations

Life Histories

Field Methods for Studies of Ecosystems

Species Abundance, Diversity, and Complexity

Population Density of Animals

Population Growth

Human Population Growth

Invasive and Introduced Species

Endangered Species

8

Human Population Issues

Chapter Outline

Objectives

After reading this chapter, you should be able to:

- Apply some of the principles discussed in chapter 7 to the human population.
- Differentiate between birthrate and population growth rate.
- Describe the current population situation in the United States.
- Explain why the age distribution and the status and role of women affect population growth projections.
- Recognize that countries in the developed world are experiencing an increase in the average age of their populations.
- Recognize that most countries of the world have a rapidly growing population.

- Describe the implications of the demographic transition concept.
- Understand how an increasing world population will alter the worldwide ecosystem.
- Recognize that rapid population growth and poverty are linked.
- Explain why less-developed nations have high birthrates and why they will continue to have a low standard of living.
- Recognize that the developed nations of the world will be under greater pressure to share their abundance.

PAST PRESENT FUTURE

1983 The population of the world was 4.68 billion.

1983 The population of the United States was 234 million.

1983 The population of India was 0.730 billion.

1983 The worldwide rate of natural increase was 1.8 percent per year.

1983 The worldwide total fertility rate was 3.9 children per woman.

2004 The population of the world was 6.396 billion (a 37 percent increase).

2004 The population of the United States was 293.6 million (a 25 percent increase).

2004 The population of India was 1.087 billion (a 49 percent increase).

2004 The rate of natural increase was 1.3 percent per year (a 28 percent decrease).

2004 The worldwide total fertility rate was 2.8 children per woman (a 28 percent decrease).

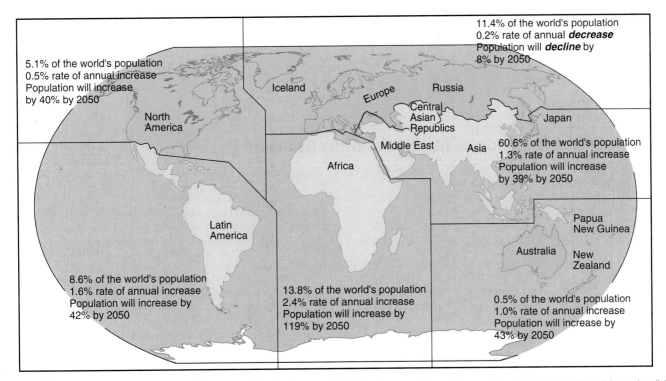

Figure 8.1 **Population Growth and Economic Development (2004)** The population of the world is not evenly distributed. It can be divided into the economically more-developed and less-developed nations. The more-developed nations are indicated in green and the less-developed in tan. Currently, about 81 percent of the world's population is in the less-developed nations of Latin America, Africa, and Asia. These areas also have the highest rates of population increase. Because of the high birthrates, they are likely to remain less developed and will constitute about 86 percent of the world's population by the year 2050.

World Population Characteristics and Implications

The human population dilemma is very complex. To appreciate it, we must understand current population characteristics and how they are related to social, political, and economic conditions.

The world can be divided into two segments based on the state of economic development of the countries. The **more-developed countries** of the world typically have a per capita income that exceeds $US10,000; they include all of Europe, Canada, the United States, Australia, New Zealand, and Japan, with a combined population of about 1.2 billion people. The remaining countries of the world are referred to as **less-developed countries** and typically have a per capita income of less than $US5000. The population of these countries totals about 5.2 billion people, nearly 3 billion of whom live on less than $US2

per day. While these definitions constitute an oversimplification and some countries are exceptions, basically this means that the majority of Asian, Latin American, and African citizens are much less well off economically than those who live in the more-developed counties. Collectively, the more-developed countries of the world have relatively stable populations and are expected to grow by about 4 percent by 2050. The less-developed regions of the world, however, have high population growth rates and are expected to grow by about 55 percent by 2050. (See figure 8.1.) If these trends continue, the total population of Africa, Asia, and Latin America will increase from the current 5.2 billion to about 8 billion by 2050, when these continents will contain over 86 percent of the world's people.

Human population growth is tied to economic development and is a contributing factor to nearly all environmental problems. Current population growth has led to famine in areas where food production cannot keep pace with increasing

numbers of people; political unrest in areas with great disparities in availability of resources (jobs, goods, food); environmental degradation (erosion, desertification) by poor agricultural practices; water pollution by human and industrial waste; air pollution caused by the human need to use energy for personal and industrial applications; extinctions caused by people converting natural ecosystems to managed agricultural ecosystems; and destructive effects of exploitation of natural resources (strip mining, oil spills, groundwater mining).

It is clear that as the population continues to increase, it will become more difficult to limit the environmental degradation that accompanies it. Since much of the growth of the human population will occur in the less-developed areas of the world that have weak economies, the money to invest in pollution control, health programs, and sustainable agricultural practices will not be present.

Several factors interact to determine the impact of a society on the resources of

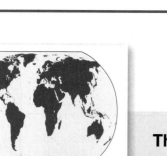

Thomas Malthus and His Essay on Population

In 1798, Thomas Robert Malthus, an Englishman, published an essay on human population. In it, he presented an idea that was contrary to popular opinion. His basic thesis was that human population increased in a geometric or exponential manner (2, 4, 8, 16, 32, 64, etc.), while the ability to produce food increased only in an arithmetic manner (1, 2, 3, 4, 5, 6, etc.). The ultimate outcome of these different rates would be that population would outgrow the ability of the land to produce food. He concluded that wars, famines, plagues, and natural disasters would be the means of controlling the size of the human population. His predictions were hotly debated by the intellectual community of his day. His assumptions and conclusions were attacked as erroneous and against the best interest of society. At the time he wrote the essay, the popular opinion was that human knowledge and "moral constraint" would be able to create a world that would supply all human needs in abundance. One of Malthus's basic postulates was that "commerce between the sexes" (sexual intercourse) would continue unchanged, while other philosophers of the day believed that sexual behavior would take less procreative forms and human population would be limited. Only within the past 50 years, however, have really effective conception-control mechanisms become widely accepted and used, and they are used primarily in developed countries.

Malthus did not foresee the use of contraception, major changes in agricultural production techniques, or the exporting of excess people to colonies in the Americas, Australia, and other parts of the world. These factors, as well as high death rates, prevented the most devastating of his predictions from coming true. However, in many parts of the world today, people are experiencing the forms of population control (famine, epidemic disease, wars, and natural disasters) predicted by Malthus in 1798. Many people feel that his original predictions were valid—only his time scale was not correct—and that we are seeing his predictions come true today.

its country. These include the natural resources available, the size of the population, the land area the people occupy, and their degree of technological development. As the population of a country increases, it puts a greater demand on its resources. Some countries have abundant natural resources, such as good agricultural land, energy resources, or mineral resources. Others are resource poor. Thus, some countries can sustain high populations while others cannot. **Population density,** the number of people per unit of land area, relates the size of the population to the resources available. A million people spread out over the huge area of the Amazon Basin have much less impact on resources than that same million people in a small island country, because the impact is distributed over a greater land surface. Countries with abundant resources can sustain higher population densities than resource-poor countries.

The degree of technological development and affluence is also important. The environmental impact of the developed world is often underestimated because the population in these countries is relatively stable and local environmental conditions are good. However, the people in highly developed countries consume huge amounts of resources. Citizens of these countries eat more food, particularly animal protein, which requires larger agricultural inputs than does a vegetarian diet. They have more material possessions and consume vast amounts of energy. Developed countries purchase goods and services from other parts of the world, often degrading environmental conditions in less-developed countries. Thus, the environmental impact of highly developed regions such as North America, Japan, Australia, New Zealand, and Europe is often felt in distant places, while the impact on resources in the developed region may be minimal.

While controlling world population growth would not eliminate all environmental problems, it could reduce the rate at which environmental degradation is occurring. It is also generally believed that the quality of life for many people in the world would improve if their populations grew less rapidly. Why, then, does the human population continue to grow at such a rapid rate?

Factors That Influence Population Growth

In chapter 7, we examined populations from a biological point of view. We looked at their characteristics, the causes of growth, and the forces that cause populations to stabilize. All of these biological factors apply to human as well as nonhuman populations. There is an ultimate carrying capacity for the human population. Eventually, limiting factors will cause human populations to stabilize. However, unlike other kinds of organisms, humans are also influenced by social, political, economic, and ethical factors. We have accumulated knowledge that allows us to predict the future. We can make conscious decisions based on the likely course of

Table 8.1 Population Characteristics of the 20 Most Populous Countries, 2004

Country	Current Population (millions)	Births per 1000 Individuals	Deaths per 1000 Individuals	Infant Mortality Rate (Deaths per 1000 Live Births)	Total Fertility Rate (Children per Woman per Lifetime)	% Married Women Using Birth Control	Rate of Natural Increase (Annual %)	Projected Population Changes (2004–50)
World	6396	21	9	56	2.8	59	1.3	45
Russia	144.1	10	17	13	1.4	67	(−0.6)	(−17)
Germany	82.6	9	10	4.1	1.3	75	(−0.2)	(−9)
Japan	127.6	9	8	3	1.3	56	0.1	(−21)
France	60.0	13	9	4.1	1.9	75	0.4	7
China	1300.1	12	6	32	1.7	83	0.6	11
United States	293.6	14	8	6.7	2.0	76	0.6	43
Thailand	63.8	14	7	20	1.7	72	0.8	15
Iran	67.4	18	6	32	2.5	74	1.2	43
Vietnam	81.5	18	6	21	2.1	75	1.2	41
Brazil	179.1	20	7	33	2.2	76	1.3	24
Turkey	71.3	21	7	39	2.5	64	1.4	37
Indonesia	218.7	22	6	46	2.6	60	1.6	41
India	1086.6	25	8	64	3.1	48	1.7	50
Egypt	73.4	26	6	38	3.2	60	2.0	74
Philippines	83.7	26	6	29	3.5	49	2.0	76
Bangladesh	141.3	30	9	66	3.3	54	2.1	98
Mexico	106.2	25	5	25	2.8	68	2.1	41
Ethiopia	72.4	41	18	105	5.9	8	2.4	139
Pakistan	159.2	34	10	85	4.8	28	2.4	85
Nigeria	137.3	42	13	100	5.7	12	2.9	124

Source: Population Reference Bureau, 2004 Population Data Sheet.

events and adjust our lives accordingly. Part of our knowledge is the certainty that as populations continue to increase, death rates and birthrates will become equal. This can happen by allowing the death rate to rise or by choosing to limit the birthrate. Controlling human population would seem to be a simple process. Once people understand that lowering the birthrate is more humane than allowing the death rate to rise, they should make the "correct" decision and control their birthrates; however, it is not quite that simple.

Biological Factors

The scientific study of human populations, their characteristics, how these characteristics affect growth, and the consequences of that growth is known as **demography.** Demographers can predict the future growth of a population by looking at several biological indicators. Currently, in almost all countries of the world, the birthrate exceeds the death rate. Therefore, the size of the population must increase. (See table 8.1.) Some countries that have high birthrates and high death rates—with birthrates greatly exceeding the death rates—will grow rapidly (Nigeria and Ethiopia). Such countries usually have an extremely high mortality rate among children because of disease and malnutrition; but because the birthrate still greatly exceeds the death rate, the populations will grow rapidly.

Some countries have high birthrates and low death rates and will grow rapidly (Mexico and Indonesia). Infant mortality rates are moderately high in these countries. Other countries have low birthrates and death rates that closely match the birthrates; they will grow slowly (Japan and France). These and other more-developed countries typically have very low infant mortality rates. The disruption caused by the political upheaval in the former Soviet Union and Eastern Europe has resulted in several countries (e.g., Russia and Germany) having death rates that are equal to or exceed birthrates, causing their populations to decline. Because of these countries and the generally low rates of growth in the rest of Europe, the European region as a whole has a declining population.

The most important determinant of the rate at which human populations grow is related to how many women in the population are having children and the number of children each woman will have. The **total fertility rate** of a population is the number of children born per woman in her lifetime. A total fertility rate of 2.1 is known as **replacement fertility,** since parents produce 2 children who will replace the parents when they die. Eventually, if the total fertility rate is maintained at 2.1, population growth will stabilize. A rate of 2.1 is used rather than 2.0 because some children do not live very long after birth and therefore will not contribute to the population for very long. When a population is not growing, and the number of births equals the number of deaths, it is said to exhibit **zero population growth.**

For several reasons, however, a total fertility rate of 2.1 will not necessarily immediately result in a stable population with zero growth. First, the death rate may fall as living conditions improve and people live longer. If the death rate falls faster than the birthrate, there will still be an increase in the population even though it is reproducing at the replacement rate.

The **age distribution,** the number of people of each age in the population, also has a great deal to do with the rate of population growth. If a population has many young people who are raising families or who will be raising families in the near future, the population will continue to increase even if the families limit themselves to two children. Depending on the number of young people in a population, it may take 20 years to a century for the population of a country to stabilize so that there is no net growth.

Social Factors

It is clear that populations in economically developed countries of the world have low fertility rates and low rates of population growth and that the less-developed countries have high fertility rates and high population growth rates. It also appears obvious that reducing fertility rates would be to everyone's advantage; however, not everyone in the world feels that way. Several factors influence the number of children a couple would like to have. Some are religious, some are traditional, some are social, and some are economic.

The major social factors that determine family size are the status and desires of women in the culture. In many male-dominated cultures, the traditional role of women is to marry and raise children. Often this role is coupled with strong religious input as well. Typically, little value

is placed on educating women, and early marriage is encouraged. In these cultures, women are totally dependent on their husbands and children in old age. Because early marriage is encouraged, fertility rates are high, since women are exposed to the probability of pregnancy for more of their fertile years. Lack of education reduces options for women in these cultures. They do not have the option to not marry or to delay marriage and thus reduce the number of children they will bear. By contrast, in much of the developed world, women are educated, delay marriage, and have fewer children. It has been said that the single most important activity needed to reduce the world population growth rate is to educate women. Whenever the educational level of women increases, fertility rates fall. Figure 8.2 compares total fertility rates and educational levels of women in the 20 most populous countries of the world. Educational level of women is strongly correlated with the total fertility rate and economic well-being of a population.

Data on the age of mothers giving birth indicate that 17 percent of births in Africa are to women in the 15- to 19-year-old range. This is true of 16 percent of births in Latin America and 9 percent of the births in Asia. The total fertility rates of these areas are 5.1, 2.6, and 2.6 children per woman per lifetime, respectively. In the developed world, the average age of first marriage is much higher, between ages 25 and 27; early marriages are rare; about 3 percent of births are to mothers who are between 15 to 19 years of age, and the total fertility rate is less than replacement fertility at about 1.6 children per woman per lifetime.

Even childrearing practices have an influence on population growth rates. In countries where breast feeding is practiced, several benefits accrue. Breast milk is an excellent source of nutrients for the infant as well as a source of antibodies against some diseases. Furthermore, since many women do not return to a normal reproductive cycle until after they have stopped nursing, during the months a woman is breast feeding her child, she is less likely to become pregnant again. Since in many cultures, breast feeding may continue for one to two years, it

serves to increase the time between successive births. Increased time between births results in a lower mortality among women of childbearing age.

As women become better educated and obtain higher paying jobs, they become financially independent and can afford to marry later and consequently have fewer children. Better-educated women are also more likely to have access to and use birth control. In economically advanced countries, a high proportion of women typically use contraception. In the less-developed countries, contraceptive use is much lower— about 28 percent in Africa, about 70 percent in Latin America, and about 62 percent in Asia (about 51 percent if China is excluded).

It is important to recognize that access to birth control alone will not solve the population problem. What is most important is the desire of women to limit the size of their families. In developed countries, use of birth control is extremely important in regulating the birthrate. This is true regardless of religion and previous historical birthrates. For example, Italy and Spain are both traditionally Catholic countries, and have a low total fertility rate of 1.3. The average for the developed countries of the world is 1.6. Obviously, women in these countries make use of birth control to help them regulate the size of their families. By contrast, Mexico, which is also a traditionally Catholic country, has a total fertility rate of 2.8, which is typical of birthrates in the less-developed world regardless of religious tradition.

Women in the less-developed world typically have more children than they think is ideal, and the number of children they have is higher than the replacement fertility rate of 2.1 children. Access to birth control will allow them to limit the number of children they have to their desired number and to space their children at more convenient intervals, but they still desire more children than the 2.1 needed for replacement. Why do they desire large families? There are several reasons. In areas where infant mortality is high, it is traditional to have large families since several of a woman's children may die before they reach adulthood. This is particularly important in the less-developed

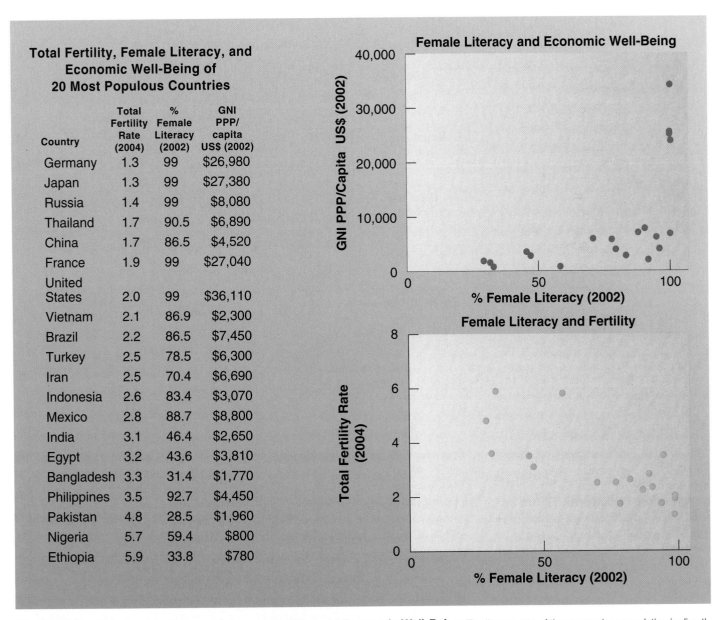

Total Fertility, Female Literacy, and Economic Well-Being of 20 Most Populous Countries			
Country	Total Fertility Rate (2004)	% Female Literacy (2002)	GNI PPP/ capita US$ (2002)
Germany	1.3	99	$26,980
Japan	1.3	99	$27,380
Russia	1.4	99	$8,080
Thailand	1.7	90.5	$6,890
China	1.7	86.5	$4,520
France	1.9	99	$27,040
United States	2.0	99	$36,110
Vietnam	2.1	86.9	$2,300
Brazil	2.2	86.5	$7,450
Turkey	2.5	78.5	$6,300
Iran	2.5	70.4	$6,690
Indonesia	2.6	83.4	$3,070
Mexico	2.8	88.7	$8,800
India	3.1	46.4	$2,650
Egypt	3.2	43.6	$3,810
Bangladesh	3.3	31.4	$1,770
Philippines	3.5	92.7	$4,450
Pakistan	4.8	28.5	$1,960
Nigeria	5.7	59.4	$800
Ethiopia	5.9	33.8	$780

Figure 8.2   Relationship Between Literacy, Fertility, and Economic Well-Being The literacy rate of the women in a population is directly correlated with the fertility rate and economic well-being.

Source: Data from Population Reference Bureau 2004 Data Sheet, and *Human Development Report 2004*, United Nations Development Program.

world, where there is no government program of social security. Parents are more secure in old age if they have several children to contribute to their needs when they can no longer work.

In less-developed countries, the economic benefits of children are extremely important. Even young children can be given jobs that contribute to the family economy. They can protect livestock from predators, gather firewood, cook, or carry water. In the developed world, large numbers of children are an economic drain. They are prevented by law from working,

they must be sent to school at great expense, and they consume large amounts of the family income. Many parents in the developed world make an economic decision about having children in the same way they buy a house or car: "We are not having children right away. We are going to wait until we are better off financially."

Political Factors

Two other factors that influence the population growth rate of a country are government policies on population growth and

immigration. Many countries in Europe have official policies that state that their population growth rates are too low. As their populations age and there are few births, they are concerned about a lack of working-age people in the future and have instituted programs that are meant to encourage people to have children. For example, Hungary, Sweden, and several other European countries provide paid maternity leave for mothers during the early months of a child's life and the guarantee of a job when the mother returns to work. Many countries provide childcare

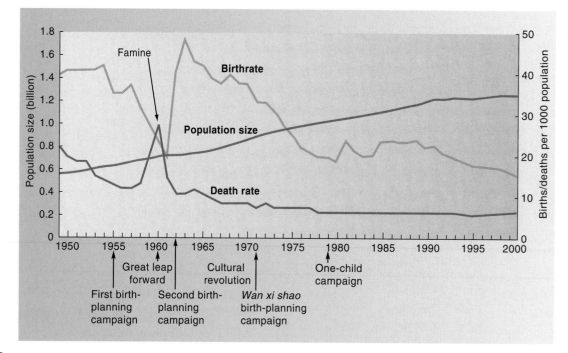

Figure 8.3 **Population Changes in China** China has a long history of actively promoting population control. This graph shows the changes that have occurred in birthrates, death rates, and total population as a result of significant policy initiatives.

Source: Data from H. Yuan Tien, "China's Demographic Dilemmas," *Population Bulletin, 1992,* Population Reference Bureau, Inc., Washington, D.C., and National Family Planning Commission of China; and more recent data from the Population Reference Bureau.

facilities and other services that make it possible for both parents to work. This removes some of the economic barriers that tend to reduce the birthrate. The tax system in many countries, such as the United States, provides an indirect payment for children by allowing a deduction for each child. Canada pays a bonus to couples on the birth of a child.

By contrast, most countries in the developing world publicly state that their population growth rates are too high. To reduce the birthrate, they have programs that provide information on maternal and child health and on birth control. The provision of free or low-cost access to contraceptives is usually a part of their population-control effort as well.

China and India are the two most populous countries in the world, each with over a billion people. China has taken steps to control its population and now has a total fertility rate of 1.7 children per woman while India has a total fertility rate of 3.1. This difference between these two countries is the result of different policy decisions over the last 50 years. The history of China's population policy is an interesting study of how government policy

affects reproductive activity among its citizens. When the People's Republic of China was established in 1949, the official policy of the government was to encourage births, because more Chinese would be able to produce more goods and services, and production was the key to economic prosperity. The population grew from 540 million to 614 million between 1949 and 1955, while economic progress was slow. Consequently, the government changed its policy and began to promote population control.

The first family-planning program in China began in 1955 as a means of improving maternal and child health. Birthrates fell. In addition, other social changes resulted in widespread famine, increased death rates, and low birthrates in the late 1950s and early 1960s. (See figure 8.3.)

The present family-planning policy began in 1971 with the launching of the *wan xi shao* campaign. Translated, the phrase means "later" (marriages), "longer" (intervals between births), and "fewer" (children). This program raised the legal ages for marriage. For women and men in rural areas, the ages were

raised to 23 and 25, respectively; for women and men in urban areas, the ages were raised to 25 and 28, respectively. These policies resulted in a reduction of birthrates by nearly 50 percent between 1970 and 1979.

An even more restrictive, one-child campaign was begun in 1978–79. The program offered incentives for couples to restrict their family size to one child. Couples enrolled in the program would receive free medical care, cash bonuses for their work, special housing treatment, and extra old-age benefits. Those who broke their pledge lost these benefits and incurred other economic penalties. By the mid-1980s, less than 20 percent of the eligible couples were signing up for the program. Rural couples particularly desired more than one child. In fact, in a country where about 70 percent of the population is rural, the rural total fertility rate was 2.5 children per woman. In 1988, a second child was sanctioned for rural couples if their first child was a girl, which legalized what had been happening anyway.

The current total fertility rate is 1.7 children per woman. Over 80 percent of couples use contraception; the most commonly used forms are male and female sterilization and the intrauterine device. Abortion is also an important aspect of this program, with a ratio of over 600 abortions per 1000 live births.

By contrast, during the same 50 years, India has had little success in controlling its population. In 2000, a new plan was unveiled that had the goal of bringing the total fertility rate from its current 3.1 children per woman to 2 (replacement rate) by 2010. In the past, the emphasis of government programs was on meeting goals of sterilization and contraceptive use, but this has not been successful. Today, about 48 percent of couples use contraceptives. This new plan emphasizes improvements in the quality of life of the people. The major thrusts are to reduce infant and maternal death, immunize children against preventable disease, and encourage girls to attend school. It is hoped that improved health will remove the perceived need for large numbers of births. Currently, less than 50 percent of the women in India can read and write. The emphasis on improving the educational status of women is related to the experiences of other developing countries. In many other countries, it has been shown that an increase in the education level of women is linked to lower fertility rates.

The immigration policies of a country also have a significant impact on the rate at which the population grows. Birthrates are currently so low in several European countries, Japan, and China that these countries will likely have a shortage of those of working age in the near future. One way to solve this problem is to encourage immigration from other parts of the world.

The developed countries are under tremendous pressure to accept immigrants. The standard of living in these countries is a tremendous magnet for refugees or people who seek a better life than is possible where they currently live. In the United States, approximately one-third of the population increase experienced each year is the result of immigration. Canada encourages immigrants and has set a goal of accepting 300,000 new immigrants each year. This is 1 percent of its current population.

Figure 8.4 **Standard of Living and Population Growth in Three Countries** Standard of living is a measure of how well one lives. It is not possible to get a precise definition, but when we compare the United States, Argentina, and Kenya, it is obvious that there are great differences in how the people in these countries live. Kenya has a high population growth rate, a low life expectancy, a high infant mortality rate, and many people without adequate food. Furthermore, their incomes are low and they are poorly educated. The United States has a low population growth rate, a high life expectancy, a low infant mortality rate, and many people who eat too much. People in the United States have high educational levels and high incomes. Argentina is intermediate in all of these characteristics.

Source: Data from The Population Reference Bureau, 2004 Population Data Sheet; UNICEF online statistics; and UNESCO Institute of Statistics online statistics.

Population Growth and Standard of Living

There appears to be an inverse relationship between the rate at which the population of a country is growing and its standard of living. The **standard of living** is an abstract concept that attempts to quantify the quality of life of people. Standard of living is a difficult concept to quantify since various cultures have different attitudes and feelings about what is desirable. However, several factors can be included in an analysis of standard of living: economic well-being, health conditions, and the ability to changes one's status in the society. Figure 8.4 lists several factors that are important in determining standard of living and contrasts three countries with very different standards of living (the United States, Argentina, and

Kenya). One important economic measure of standard of living is the average purchasing power per person. One index of purchasing power is the **gross national income (GNI).** The GNI is an index that measures the total goods and services generated within a country as well as income earned by citizens of the country who are living in other countries. Since the prices of goods and services vary from one country to another, a true comparison of purchasing power requires some adjustments. Therefore, a technique used to compare economic well-being across countries is a measure called the GNI PPP (gross national income purchasing power parity). Finally, the GNI PPP can be divided by the number of people in the country to get a per capita (per person) GNI PPP. As you can see from figure 8.4, a wide economic gap exists between economically advanced countries and those that are less developed. Yet the people of less-developed countries aspire to the same standard of living enjoyed by people in the developed world.

Health criteria reflect many aspects of standard of living. Access to such things as health care, safe drinking water, and adequate food are reflected in life expectancy, infant mortality, and growth rates of children. The United States and Argentina have similar life expectancies (over 70 years) and adequate nutrition. Kenya has a low life expectancy (51 years), many undernourished children (21 percent are underweight), and a high infant mortality rate (78 per 1000). The United States has a low infant mortality rate (6.7 per 1000). Argentina has an intermediate infant mortality rate (16.3 per 1000).

Finally, the educational status of people determines the kinds of jobs that are available and the likelihood of being able to improve one's status. In general, men are more likely to receive an education than women, but the educational status of women has a direct bearing on the number of children they will have and, therefore, on the economic well-being of the family. In Argentina and the United States over 80 percent of girls of high school age attend high school but only 24 percent do so in Kenya. Obviously, tremendous differences exist in the standard of living among these three countries. What the average U.S. citizen would consider poverty level would be considered a luxurious life for the average person in Kenya.

Population and Poverty—A Vicious Cycle?

It is clear that the areas of the world where the human population is growing most rapidly are those that have the lowest standard of living. The developed regions of the world (Europe, North America, Japan, Australia, and New Zealand) have abundant wealth and have relatively slowly growing populations. The less-developed countries of Africa, Latin America, and Asia are generally poor and have high population growth rates. Although not all cases are the same, poverty, high birthrates, poor health, and lack of education seem to be interrelated.

1. Poor people cannot afford birth control and, since they are often poorly educated, may not be able to read the directions on how to use various birth control mechanisms correctly. Therefore, they have more children than they may wish to have.

2. Poor people need to obtain income in many ways. Often this includes taking children out of school so that they are able to work on the farm or in other jobs to provide income for the family. Poorly educated people cannot get high paying jobs and so remain in poverty.

3. Poor people have little access to health care and are therefore more likely to suffer preventable illnesses that lower their ability to earn an income.

4. Women in poor countries are usually poorly educated and do not have disposable income. Therefore, they are dependent on their husbands or the family unit for their livelihood. Women who do not have independent income are more likely to have children they do not want because they cannot afford birth control.

5. High infant mortality rates result from poor health, but children (particularly sons) are desired by parents because the sons will provide for the parents when the parents are old. High infant mortality rates make it likely that parents will desire larger numbers of offspring, since some offspring will die as children.

At the United Nations International Conference on Population and Development held in Cairo, Egypt, in September 1994, attention was focused on breaking this cycle of poverty and high population growth rates. Several important conclusions were reached that have the potential to break this cycle.

1. There was recognition that economic well-being is tied to solving the population problem. However, the huge, rapidly growing population of the poor countries of the world cannot hope to consume at the rate the rich countries do. Furthermore, the rich countries of the world need to reduce their rate of consumption.

2. Improving the educational status of women was promoted. This would lead to improved financial standing for women, which could allow them to have fewer children.

3. Access to birth control and health care would reduce infant and maternal deaths.

In fact, several countries have instituted population control programs that focus on improving the health of women and children, increasing the educational levels of women, and making birth control universally available.

Hunger, Food Production, and Environmental Degradation

As the human population increases, the demand for food rises. People must either grow food themselves or purchase it. Most people in the developed world purchase

GLOBAL PERSPECTIVE

The Grameen Bank and Microcredit

Poverty and environmental degradation go together. Today, about 1.2 billion people (about 20 percent of the world's people) live on less than one U.S. dollar a day. Impoverished people do not worry about the environmental impacts of their actions when they are trying to provide food and other basic necessities for themselves and their families. Therefore, the easing of poverty can have a positive effect on environmental conditions. Typically, the poor are unable to get loans from traditional sources and must go to moneylenders who charge extremely high rates of interest. This locks them into a cycle of poverty.

The Grameen Bank is a creation of Muhammad Yunis, a professor of economics at Chittagong University in Bangladesh, that provides an alternative source of credit for poor people. It is an outgrowth of an experiment during the late 1970s that showed that it was possible to make loans to the poor with a high rate of repayment and that the loans would improve the standard of living of the borrowers. In 1983, the Bangladeshi government established the Grameen Bank (*grameen* means village or rural in the Bengali language) as an independent bank. The bank provides tiny loans without any collateral (most are for less than $US200) to the rural poor to begin a variety of businesses that can generate income that will raise the standard of living of the people. About 90 percent of the borrowers are women. The large proportion of loans to women is based on several factors. Unmarried women (widows, divorced, abandoned) are the poorest segments of society and have the greatest need. In addition, lenders feel that women are more likely to use the money to improve the conditions of their families and to repay the loan.

One of the primary reasons the bank has been successful is that the borrowers are helped to manage the repayment of their loans. Many of the conditions of the loans may be considered intrusive, but the system works. Some of the loan conditions are:

1. There is close supervision by bank personnel.
2. Borrowers must become members of a small group of their peers that advises the borrower and supervises the repayment of loans. Because most decisions about loans are made in meetings with other borrowers present, peer pressure is strong to use the loan wisely.
3. Loans are paid back on a weekly basis, which means that loans must be for activities that provide a quick return on investment and rely on skills already possessed by the borrower.
4. Compulsory saving is required of those who receive loans. This reduces the likelihood that poor decisions by the borrower will result in squandering of the new income generated by the business started with the aid of the loan.

Today, the Grameen Bank is owned by the rural poor whom it serves. Borrowers own 90 percent of the bank's shares, while the Bangladeshi government owns the remaining 10 percent. Through its loans, about 5 percent of its borrowers come out of poverty every year.

The success of the Grameen Bank has led to replication of the concept in other countries. In 1997, a Microcredit Summit was held in Washington, D.C., that attracted 2900 delegates from 137 countries. They adopted a goal of reaching 100 million of the world's poorest families with microcredit and other financial services, preferably through the women in those families, by 2005. By the end of 2002, about 2500 microcredit institutions existed in nearly 100 countries. They had served about 68 million clients, 79 percent of whom were women. It appears that the goal of reaching 100 million poor will be achieved by 2005.

what they need and have more than enough food to eat. Most people in the less-developed world must grow their own food and have very little money to purchase additional food. Typically, these farmers have very little surplus. If crops fail, people starve. Even in countries with the highest population (China and India), the majority of the people live on the land and farm. (Fifty-nine percent of Chinese and 72 percent of Indians live in rural areas.)

The human population can increase only if the populations of other kinds of plants and animals decrease. Each ecosystem has a maximum biomass that can exist within it. There can be shifts within ecosystems to allow an increase in the population of one species, but this always adversely affects certain other populations because they are competing for the same basic resources.

When humans need food, they convert natural ecosystems to artificially maintained agricultural ecosystems. The natural mix of plants and animals is destroyed and replaced with species useful to humans. If these agricultural ecosystems are mismanaged, the region's total productivity may fall below that of the original ecosystem. The dust bowl of North America, desertification in Africa, and destruction of tropical rainforests are well-known examples. In countries where

food is in short supply and the population is growing, pressure is intense to convert remaining natural ecosystems to agriculture. Typically, these areas are the least desirable for agriculture and will not be productive. However, to a starving population, the short-term gain is all that matters. The long-term health of the environment is sacrificed for the immediate needs of the population.

A consequence of the basic need for food is that people in less-developed countries generally feed at lower trophic levels than do those in the developed world. (See figure 8.5.) Converting the less-concentrated carbohydrates of plants into more nutritionally valuable animal

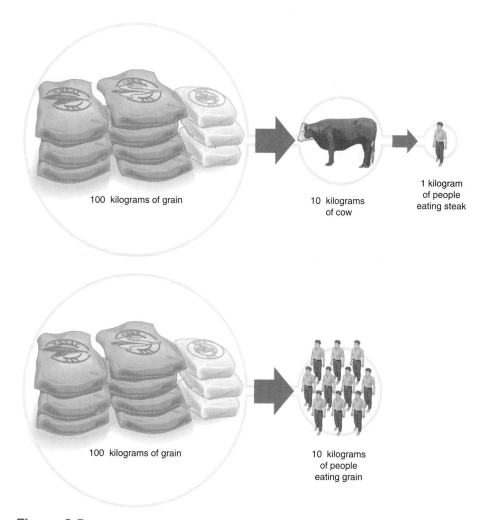

100 kilograms of grain → 10 kilograms of cow → 1 kilogram of people eating steak

100 kilograms of grain → 10 kilograms of people eating grain

Figure 8.5 **Population and Trophic Levels** The larger a population, the more energy it takes to sustain the population. Every time one organism is eaten by another organism, approximately 90 percent of the energy is lost. Therefore, when countries are densely populated, they usually feed at the herbivore trophic level because they cannot afford the 90 percent energy loss that occurs when plants are fed to animals. The same amount of grain can support 10 times more people at the herbivore level than at the carnivore level.

protein and fat is an expensive process. During the process of feeding plants to animals and harvesting animal products, approximately 90 percent of the energy in the original plants is lost. (See chapter 5.) Although many modern agricultural practices in the developed world obtain better efficiencies than this, most of the people in the developing world are not able to use such sophisticated systems. Thus, these people approach the 90 percent loss characteristic of natural ecosystems. Therefore, in terms of economics and energy, people in less-developed countries must consume the plants themselves rather than feed the plants to animals and then consume the animals. In most cases, if the plants were fed to animals, many people

would starve to death. On the other hand, a lack of protein in diets that consist primarily of plants can lead to malnutrition. It is possible to get adequate protein from a proper mixture of plant foods. In regions where food is in short supply, however, the appropriate mixture of foods is often not available. Thus, many people in the less-developed world suffer from a lack of adequate protein, which stunts their physical and mental development.

In contrast, in most of the developed world, meat and other animal protein sources are important parts of the diet. Many people suffer from overnutrition (they eat too much); they are "malnourished" in a different sense. About fifty percent of North Americans are overweight

and 25 percent are obese. The ecological impact of one person eating at the carnivore level is about 10 times that of a person eating at the herbivore level. If people in the developed world were to reduce their animal protein intake, they would significantly reduce their demands on world resources. Almost all of the corn and soybeans grown in the United States are used as animal feed. If these grains were used to feed people rather than animals, less grain would have to be grown and the impact on farmland would be reduced.

In countries where food is in short supply, agricultural land is already being exploited to its limit, and there is still a need for more food. This makes the United States, Canada, Australia, Argentina, New Zealand, and the European Union net food exporters. Many countries, such as India and China, are able to grow enough food for their people but do not have any left for export. Others, including many nations of the former Soviet Union, are not able to grow enough to meet their own needs and, therefore, must import food.

A country that is a net food importer is not necessarily destitute. Japan and some European countries are net food importers but have enough economic assets to purchase what they need. Hunger occurs when countries do not produce enough food to feed their people and cannot obtain food through purchase or humanitarian aid.

The current situation with respect to world food production and hunger is very complicated. It involves the resources needed to produce food, such as arable land, labor, and machines; appropriate crop selection; and economic incentives. It also involves the maldistribution of food within countries. This is often an economic problem, since the poorest in most countries have difficulty finding the basic necessities of life, while the rich have an excess of food and other resources. In addition, political activities often determine food availability. War, payment of foreign debt, corruption, and poor management often contribute to hunger and malnutrition.

Improved plant varieties, irrigation, and improved agricultural methods have dramatically increased food production in some parts of the world. In recent years, India, China, and much of southern Asia have moved from being food importers to being self-sufficient and, in some cases, food exporters.

The areas of greatest need are in sub-Saharan Africa. Africa is the only major region of the world where per capita grain production has decreased over the past few decades. People in these regions are trying to use marginal lands for food production, as forests, scrubland, and grasslands are converted to agriculture. Often, this land is not able to support continued agricultural production. This leads to erosion and desertification.

What should be done about countries that are unable to raise enough food for their people and are unable to buy the food they need? This is not an easy question. A simple humanitarian solution to the problem is for the developed countries to supply food. Many religious and humanitarian organizations do an excellent service by taking food to those who need it and save many lives. However, the aim should always be to provide temporary help and insist that the people of the country develop mechanisms for solving their own problem. Often, emergency food programs result in large numbers of people migrating from their rural (agricultural) areas to cities, where they are unable to support themselves. They become dependent on the food aid and stop working to raise their own food, not because they do not want to work but because they need to leave their fields to go to the food distribution centers. Many humanitarian organizations now recognize the futility of trying to feed people with gifts from the developed world. The emphasis must be on self-sufficiency.

The Demographic Transition Concept

The relationship between the standard of living and the population growth rate seems to be that countries with the highest standard of living have the lowest population growth rate, and those with the lowest

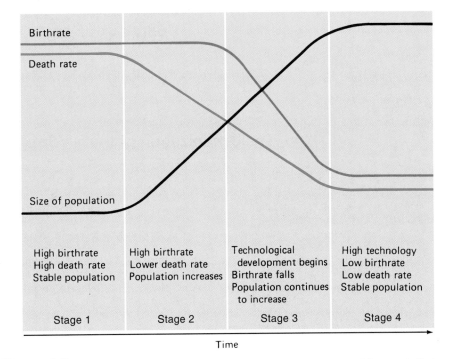

Figure 8.6 **Demographic Transition** The demographic transition model suggests that as a country develops technologically, it automatically experiences a drop in the birthrate. This certainly has been the experience of the developed countries of the world. However, the developed countries make up less than 20 percent of the world's population. It is doubtful whether the less-developed countries can achieve the kind of technological advances experienced in the developed world.

standard of living have the highest population growth rate. This has led many people to suggest that countries naturally go through a series of stages called **demographic transition.** This model is based on the historical, social, and economic development of Europe and North America. In a demographic transition, the following four stages occur (see figure 8.6):

1. Initially, countries have a stable population with a high birthrate and a high death rate. Death rates often vary because of famine and epidemic disease.
2. Improved economic and social conditions (control of disease and increased food availability) bring about a period of rapid population growth as death rates fall. Birthrates remain high.
3. As countries become industrialized, the birthrates begin to drop because people desire smaller families and use contraceptives.
4. Eventually, birthrates and death rates again become balanced, with low birthrates and low death rates.

This is a very comfortable model because it suggests that if a country can become industrialized, then social, political, and economic processes will naturally cause its population to stabilize.

However, the model leads to some serious questions. Can the historical pattern exhibited by Europe and North America be repeated in the less-developed countries of today? Europe, North America, Japan, and Australia passed through this transition period when world population was lower and when energy and natural resources were still abundant. It is doubtful whether these supplies are adequate to allow for the industrialization of the major portion of the world currently classified as less developed.

A second concern is the time element. With the world population increasing as rapidly as it is, industrialization probably cannot occur fast enough to have a significant impact on population growth. As long as people in less-developed countries are poor, there is a strong incentive to have large numbers of children. Children are a form of social security because they take care of their elderly parents.

GLOBAL PERSPECTIVE

The Urbanization of the World's Population

The United Nations estimates that currently about 48 percent of the world's population lives in cities. This is expected to increase to over 50 percent by 2005 and reach 60 percent by 2025. The economically developed world is currently about 76 percent urban, and the less-developed world is about 41 percent urban. Therefore, the increase in the urban population will occur primarily in the less-developed world, where resources to deal with urban problems are least available.

Urbanization is not necessarily bad. The economic activity of the developed world takes place primarily in cities. Cities can be planned and managed to be healthy, interesting places to live. Cities offer jobs, health care, schools, and other services that are usually lacking in rural areas. Often the rural economy of a country is considered to be of low status, and young people wish to move to cities, where higher-status jobs and desirable cultural amenities are available.

Dense populations of people have a concentrated impact on local resources. Often water must be transported long distances, wastes are difficult to get rid of, and air quality drops as each additional person places a burden on the local environment. Unfortunately, in the less-developed world, it is impossible to plan for growth and provide basic resources such as drinking water, sewers, transportation, and housing as fast as the population is growing. New migrants to the cities often lack education or training for the high wage jobs available, have lost the opportunity to raise their own food, and lack the support of their family units. Consequently, when these poor people migrate to the cities, they often live in shantytowns on the edge of the city or form squatter settlements on hillsides and other unused land. These centers of poverty lack basic services since they are often outside the bureaucratic structure of the city. The rich in such cities live a short distance from the poor and may be relatively isolated, but the negative environmental impact of the entire urban region (air and water pollution, transportation problems, etc.) affects all the residents. Even the long-established urban centers of the developed world have major problems. Many contain pockets of poverty that have led to problems of social unrest and crime.

Only people in developed countries can save money for their old age. They can choose to have children, who are expensive to raise, or to invest money in some other way.

When the countries of Europe and North America passed through the demographic transition, they had access to large expanses of unexploited lands, either within their boundaries or in their colonies. This provided a safety valve for expanding populations during the early stages of the transition. Without this safety valve, it would have been impossible to deal adequately with the population while simultaneously encouraging economic development. Today, less-developed countries may be unable to accumulate the necessary capital to develop economically, since they do not have uninhabited places to which their people can migrate and an ever-increasing population is a severe economic drain.

The U.S. Population Picture

In many ways, the U.S. population, which has a total fertility rate that is low (2.0), is similar to those of other developed countries of the world with low birthrates. One might expect the population to be stabilizing under these conditions. However, two factors are operating to cause significant change over the next 50 years. One factor has to do with the age structure of the population, and the other has to do with immigration policy. The U.S. population includes a **postwar baby boom** component, which has significantly affected population trends. These baby boomers were born during an approximately 15-year period (1947–61) following World War II, when birthrates were much higher than today, and constitute a bulge in the age distribution profile. (See figure 8.7.) As members of this group have raised families, they have had a significant influence on how the U.S. population has grown. As this population bulge ages and younger people limit their family size, the population will gradually age. By 2030, about 20 percent of the population will be 65 years of age or older.

A changing age structure will lead to social changes as well. The baby boom of the late 1940s and the 1950s encouraged growth in service industries needed by young families. Maternity wards had to be expanded, schools could not be built fast enough, baby-care companies saw unprecedented sales, and the toy industry flourished. Today, these "babies" are in their forties and fifties. They are buying homes, cars, and appliances but are raising fewer children than did their parents. However, because baby boomers are such a large segment of the population, they have contributed a large number of

Figure 8.7   Changing Age Distribution of U.S. Population (1970–2000) These graphs show the number of people in the United States at each age level. Notice that in the year 2000, a bulge begins to form at age 35 to 39 and ends at about age 50 to 54. These people represent the "baby boom" that followed World War II. As you compare the age distribution for 1970, 1980, and 1990 with 2000, you can see that this group of people moves through the population. As this portion of the population has aged, it has had a large impact on the nature of the U.S. population. In the 1970s and 1980s, baby boomers were in school. In the 1990s, they were in their middle working years. In 2000, some of them were beginning to retire, and many more will do so throughout the decade.

Source: Data from the U.S. Department of Commerce, Bureau of the Census.

children to the population. The children of the baby boomers are now having children, and in many parts of the country, schools are seeing increased numbers of children in the early grades. At the same time, there are more elderly, because people are living longer. That creates a need for additional services for the elderly. This trend toward an aging population will be accentuated as the baby boomers retire. What will the social needs be in the year 2025, when most of these baby boomers will have retired?

Both legal and illegal immigration significantly influence future population growth trends. Even with the current total fertility rate of 2.0 children per woman,

the population is still growing by about 1.1 percent per year. About 0.6 percent is the result of natural increases due to the difference between birthrates and death rates. The remainder is the result of immigration into the United States. The U.S. Census Bureau projects that immigration will increase significantly and account for 50 percent of population growth by the year 2050.

Current immigration policy in the United States is difficult to characterize. Strong measures are being taken to reduce illegal immigration across the southern border. This is in part due to pressures placed on Congress by states that receive large numbers of illegal immigrants. Ille-

gal immigrants add to the education and health care that states must fund. At the same time, some segments of the U.S. economy (agriculture, tourism) maintain that they are unable to find workers to do certain kinds of work. Consequently, special guest workers are allowed to enter the country for limited periods to serve the needs of these segments of the economy. There is also a consistent policy of allowing immigration that reunites families of U.S. residents. Obviously, the families that fall into this category are likely to include U.S. citizens who were recent immigrants themselves. Most immigration policy is the result of political decisions rather than decisions that relate to population policy

CHAPTER 8   Human Population Issues

or a concern about the rate at which the U.S. population is growing.

Projections based on the 2000 census indicate that the population will continue to grow and does not seem to be moving toward zero growth despite the fact that the total fertility rate is less than the replacement rate of 2.1. Two reasons account for this. Immigration adds about a million people per year, and new immigrants typically are young and have larger numbers of children than nonimmigrants. This is likely to result in an increase from about 293 million people in 2004 to about 420 million by 2050.

Differences in family size exist between different segments of the U.S. population. The Hispanic and Asian-American portions of the population tend to have larger families. Furthermore, many of the people in these groups are recent immigrants, so these portions of the population will grow rapidly, and the caucasian portion will decline. Thus, the U.S. population will be much more ethnically diverse in the future.

Likely Consequences of Continued Population Growth

As the world human population continues to increase, pressure for the necessities of life will become greater. Differences in the standard of living between more-developed and less-developed countries will remain large because population will increase most in less-developed countries. The supply of fuel and other resources is dwindling. Pressure for these resources will intensify as industrialized countries seek to maintain their current standard of living. People in less-developed countries will seek more land to raise crops and feed themselves unless major increases in food production per hectare occur. Since most of these people live in tropical areas, tropical forests will be cleared for farmland. The resulting erosion or alteration of the soil will make it no longer suitable for forest or crops. This conversion of natural ecosystems to agricultural ecosystems could cause profound changes in the world ecosystem.

Developed countries may have to choose between several alternatives: helping the less-developed countries, thus maintaining their friendship; allowing increased immigration from the developing world; or isolating themselves from the problems of the less-developed nations. None of these policies will be able to prevent a change in lifestyle as world population increases. The resources of the world are finite. Even if industrialized countries continue to use a disproportionate share of the world's resources, the amount available per person will decline as population rises. As world population increases, the less-developed areas will probably maintain their low standard of living. It is difficult to see how that standard of living could get much lower, since some people in these countries are already starving to death. Lifestyles in developed nations will probably change very little but may become less consumption-oriented. However, many economists and political thinkers feel that as the economies of the world become more linked and jobs flow more freely from country to country, wealth will be redistributed from the former rich countries to emerging economies in developing countries. The possibility of such a redistribution has become a major political issue in many of the developed countries of the world. If an economic redistribution takes place, it could lead to changes in the lifestyles of people in the more-developed world.

What some people currently view as necessities (meals in restaurants, vacations to remote places, and two cars per family) could become luxuries. Many people who enjoy the freedom of mobility associated with the automobile could have their travel limited as more-efficient public transportation replaces less-efficient private transportation. In many cities in Europe, public transportation already is the major form of transportation used by citizens. Recreation may cease to involve expensive, energy-demanding machines (powerboats, motorcycles, and electric-powered toys) and emphasize instead such activities as hiking, bicycling, and reading. These changes will not come quickly, unless some catastrophic political or economic force causes major worldwide adjustments.

GLOBAL PERSPECTIVE

North America—Population Comparisons

The three countries that make up North America (Canada, the United States, and Mexico) interact politically, socially, and economically. The characteristics of their populations determine how they interact. Canada and the United States are both wealthy countries with similar age structures and low total fertility rates. The United States has a total fertility rate of 2.0, and Canada has a total fertility rate of 1.5. Both countries have a relatively small number of young people (about 21 percent of the population is under 15 years of age) and a relatively large number of older people (about 12 percent are 65 or older). Without immigration, these countries would have stable or falling populations. Therefore, they must rely on immigration to supply additional people to fill the workplace. This is particularly true for jobs with low pay. The United States currently receives about 1 million immigrants per year. Canada currently receives about 230,000 immigrants per year but is planning to increase that to about 300,000 in the future.

Mexico, on the other hand, has a young, rapidly growing population. About 35 percent of the population is under 15 years of age, and the total fertility rate is 2.8. At that rate, the population will double in about 33 years. Furthermore, the average Mexican has a purchasing power about 30 percent of that in Canada and about 25 percent of that in the United States. These conditions create a strong incentive for individuals to migrate from Mexico to other parts of the world. The United States is the usual country of entry. The number is hard to evaluate, since many enter the United States illegally or as seasonal workers and eventually plan to

return to Mexico. Current estimates are that about 250,000 people enter the United States from Mexico each year.

In response to the large number of illegal immigrants from Mexico, the United States has erected fences and increased surveillance. In 2003, over a million people were apprehended attempting to cross the border from Mexico to the United States. Thus, instead of crossing at normal points of entry, illegal immigrants are likely to cross the border in remote desert areas, which has resulted in numerous deaths due to exposure.

The economic interplay between Mexico and the United States has several components. Working in the United States allows Mexican immigrants to improve their economic status. Furthermore, their presence in the United States has a significant effect on the economy of Mexico, since many immigrants send much of their income to Mexico to support family members. The North American Free Trade Agreement (NAFTA) allows for relatively free exchange of goods and services between Canada, the United States, and Mexico. Consequently, many Canadian, U.S., and European businesses have built assembly plants in Mexico to make use of the abundant inexpensive labor, particularly along the U.S.-Mexican border. Labor leaders in Canada and the United States are concerned about the effect access to low-cost labor will have on their membership and complain that many high-paying jobs have been moved to Mexico, where labor costs are lower.

Country	Population Size 2004 (millions)	Birthrate per 1000	Death Rate per 1000	Rate of Natural Increase	Total Fertility Rate (Children per woman) per lifetime)	Life Expectancy (Years)	Infant Mortality Rate (Deaths per 1000 births)	Per Capita GNI PPP 2002 US$	% under 15 Years of Age	% over 65 Years of Age	Annual Immigration/ Emigration (Estimate)*
Canada	31.9	11	7	0.3	1.5	79	5.2	28,930	18	13	+240,000
United States	293.6	14	8	0.6	2.0	77	6.7	36,110	21	12	+1,500,000
Mexico	106.2	25	5	2.1	2.8	75	25	8800	35	5	−300,000

Source: Population Reference Bureau, 2004 Population Data Sheet.

*Immigration/Emigration estimates from United States, Canadian, and Mexican sources.

The Impact of AIDS on Populations

The human immunodeficiency virus (HIV), which causes AIDS (acquired immunodeficiency syndrome), continues to spread throughout the world. Because the epidemic has affected all parts of the world, it has been called a pandemic. The disease is spread through direct transfer of body fluids containing the virus into the bloodstream of another person. Sharing of contaminated needles among intravenous drug users and sexual contact are the most likely methods of passage. In the United States, the disease was once considered a problem only for homosexuals and intravenous drug users. This perception is rapidly changing. Many of the new cases of AIDS are being found in women infected by male sex partners and in children born to infected mothers.

The World Health Organization (WHO) estimated that by 2003, about 34–46 million people were living with HIV/AIDS worldwide. WHO estimates that 3 million people died of AIDS in 2003 and that about 30 million have died since the pandemic began. The distribution of the virus is lowest in the economically developed countries and highest in the developing countries. The figure shows estimates by WHO of the numbers of HIV-infected people in various parts of the world at the end of 2003. In the less-developed world, there is little medical care to treat AIDS and a lack of resources to identify those who have HIV. Many people do not know they are infected and will continue to pass the disease to others.

Sub-Saharan Africa has been hit hardest by this disease. In Africa, AIDS has always been primarily a sexually transmitted disease spread through heterosexual contact. Many people believe that in the poor countries of central Africa, permissive sexual behavior and prostitution have created conditions for a rapid spread of the disease. This is particularly evident along major transportation routes. The World Health Organization estimates that over 70 percent of all persons infected with HIV (between 25 million and 28 million) live in this region of the world. This is about 3.5 percent of the population of the region. In some large urban hospitals in this part of Africa, over 50 percent of the hospital beds are taken up by AIDS patients, and AIDS is the leading cause of death from disease. Since those who are dying are young, there is a change in the age structure of the population of heavily infected countries. This has resulted in large numbers of AIDS orphans—about 11 million AIDS orphans in sub-Saharan Africa by the end of 2003. With the death of young infected adults, villages are composed primarily of older people and children.

The economic burden on these countries is tremendous. People with AIDS symptoms are unable to work and need medical care and medication. Because of poverty, there is often little medical care available. Also, the millions of orphaned children have resulted in an additional economic burden on relatives.

In the developed world, drugs are available that can extend life and perhaps prevent transmission between infected mothers and their infants. These drugs are expensive and therefore are unavailable to most of the people in the world who need them. In 2001, several drug companies agreed to sell some of these drugs to impoverished people at reduced prices (about $700 per person per year), but even at these prices, the drugs are still beyond the economic reach of most people, since the average per capita income is less than $2000.

Source: Data from World Health Organization of the United Nations

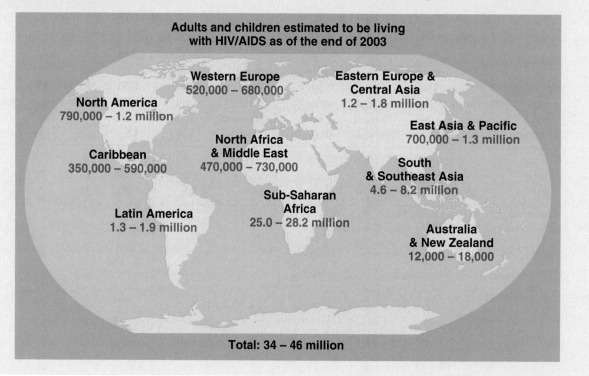

Adults and children estimated to be living with HIV/AIDS as of the end of 2003

Western Europe 520,000 – 680,000

Eastern Europe & Central Asia 1.2 – 1.8 million

North America 790,000 – 1.2 million

East Asia & Pacific 700,000 – 1.3 million

North Africa & Middle East 470,000 – 730,000

Caribbean 350,000 – 590,000

South & Southeast Asia 4.6 – 8.2 million

Sub-Saharan Africa 25.0 – 28.2 million

Latin America 1.3 – 1.9 million

Australia & New Zealand 12,000 – 18,000

Total: 34 – 46 million

Summary

Many of the problems of the world are caused or made worse by an increasing human population. Currently, the world's population is growing very rapidly. Most of the growth is occurring in the less-developed areas of the world (Africa, Asia, and Latin America), where people have a low standard of living. The more-developed regions of the world, with their high standard of living, have relatively slow population growth and, in some instances, declining populations.

Demography is the study of human populations and the things that affect them. Demographers study the sex ratio and age distribution within a population to predict future growth. Population growth rates are determined by biological factors such as birthrate, which is determined by the number of women in the population and the age of the women, and death rate. Sociological and economic conditions are also important, since they affect the number of children desired by women, which helps set the population growth rate. In more-developed countries, women usually have access to jobs. Couples marry later, and they make decisions about the number of children they will have based on the economic cost of raising children. In the less-developed world, women marry earlier, and children have economic value as additional workers, as future caregivers for the parents, and as status for either or both parents.

The U.S. population will continue to grow. The average age of the population will increase, and the racial mix of the population will change significantly.

The demographic transition model suggests that as a country becomes industrialized, its population begins to stabilize. However, there is little hope that the Earth can support the entire world in the style of the industrialized nations. It is doubtful whether there are enough energy resources and other natural resources to develop the less-developed countries or whether there is enough time to change trends of population growth. Highly developed nations should anticipate increased pressure in the future to share their wealth with less-developed countries.

Issues – Analysis

Demography, Aging Populations, and Public Policy

Europe has a very low total fertility rate of 1.4 children per woman per year, which has resulted in a declining population. The average age of the population is increasing because of the low numbers of births and the fact that people are living longer. All this adds up to a situation in which there are fewer and fewer working-age people compared to the number of retired persons. Social welfare systems that provide such services as health care and retirement income are funded by taxes on the working people of a country. With fewer workers, it will be difficult to fund the benefits currently expected by people as they retire. Policy makers are under pressure to address this problem. Several approaches can have an impact on the dilemma. One is to reduce the retirement benefits that people would receive. This is obviously not a very popular political move. Since older people tend to vote, they have a lot to say about proposed changes to their benefits.

In 2003, France experienced major work stoppages as workers protested the government's plans to increase the amount that workers would be required to pay into their pension plans and to increase the number of years they would need to contribute. France's pension system faces the prospect of bankruptcy in 2006, when it is estimated that pensioners will outnumber workers. Similar strikes took place in Germany, Italy, Austria, and other countries to protest proposed changes in pensions and benefits programs.

A second approach to alleviating this difficulty is to encourage births. Many European countries have instituted programs that give cash benefits to women who have additional children. Others provide job protection for women who leave the workforce to raise children or provide state-subsidized childcare so that women can return to work after having a child. These inducements, however, have not served to spur an increase in the number of births.

A third way to deal with the problem is to increase the number of immigrants allowed into the country in order to obtain the workers needed to maintain the system. Many potential immigrants are from the less-developed world, however, and are not highly skilled. Consequently, some European countries have specifically recruited certain kinds of people such as computer programmers or nurses for immigration and discouraged unskilled laborers. Regardless, the influx of immigrants and guest workers into many European countries has led to an increase in the number of groups that oppose immigration for a variety of political or racist reasons. Thus, it appears that national leaders know what kinds of reforms are needed but must overcome public resistance to them.

- Should governments pay women to have children?
- Would you willingly give up some of your pension benefits?
- Should countries discriminate against certain kinds of immigrants?

Key Terms

Review Questions

1. What is demography?
2. What is demographic transition? What is it based on?
3. What is a baby boom?
4. What does the age distribution of a population mean?
5. List 10 differences between your standard of living and that of someone in a less-developed country.
6. Why do people who live in overpopulated countries use plants as their main source of food?
7. Although predicting the future is difficult, describe what you think your life will be like in 10 years. Why?
8. List five changes you might anticipate if world population were to double in the next 50 years.
9. Which three areas of the world have the highest population growth rate? Which three areas of the world have the lowest standard of living?
10. How many children per woman would lead to a stable U.S. population?
11. What role does the status of women play in determining population growth rates?
12. Describe three reasons why women in the less-developed world might desire more than two children.

Critical Thinking Questions

1. Do you think it is appropriate for developed countries to persuade less-developed countries to limit their population growth? What would be appropriate and inappropriate interventions, according to your ethics? Now imagine you are a citizen of a less-developed country. What might be your reply to those who live in more-developed countries? Why?
2. China's government has been very involved in regulating that country's population growth. What do you think about this kind of government intervention in China's population problem? What values, beliefs, and perspectives do you hold that lead you to think the way you do?
3. Population growth causes many environmental problems. Identify some of these problems. What role do you think technology will play in solving these problems? Are you optimistic or pessimistic about these problems being solved through technology? Why?
4. Do you think that demographic transition will be a viable option for world development? What evidence leads you to your conclusions? What role should the developed countries play in the current demographic transition of developing countries? Why?
5. The U.S. Census Bureau projects that by 2050, immigration will account for 50 percent of the population growth in the United States. What values and perspectives should guide our immigration policy? Why?
6. Imagine a debate between an American and a Sudanese about human population and the scarcity of resources. What perspectives do you think the American might bring to the debate? What perspectives do you think the Sudanese would bring? What might be their points of common ground? On what might they differ?
7. Many people in developing countries hope to achieve the standard of living of those in the developed world. What might be the effect of this pressure on the environment in developing countries? On the political relationship between developing countries and already developed countries? What ethical perspective do you think should guide this changing relationship?
8. The demographic changes occurring in Mexico have an influence on the United States. What problems does Mexico face regarding its demographics? Should the United States be involved in Mexican population policy?

Interactive Exploration

Check out the website at www.mhhe.com/enger10e and click on the cover of this textbook for quizzing, career information, case studies, and hot links for the following topics:

Human Population Growth

Ecological Economics

Human Interactions Between Populations

Global Ecology

Urbanization and Sustainable Cities

PART
III

Energy

Xcel Corporation purchased electricity from these wind turbines to meet the directives of the Minnesota legislature.

Winds of Change

John Cronn

St. Cloud State University

In Minnesota, the use of a modern energy source—nuclear power—has coupled with the development of one of our oldest sources, wind power.

The story of this interrelationship began in 1973, when reservation land belonging to the Prairie Island Mdewakanton Dakota tribal community near Red Wing, Minnesota, became the site for a Northern States Power Company nuclear power plant. The site, on the banks of the Mississippi River, easily supplied the water the power plant needed for operation and cooling. In turn, the revenue provided by the plant was a welcomed economic stimulus for the reservation. (The late 1960s and 1970s marked a period of rapid development of nuclear power plants. More than 100 nuclear power plants are currently operating in the United States, and most of them were constructed during this time period.)

Eventually, however, spent fuel rods began filling up the storage pools at the Prairie Island nuclear power plant, as well as at plants across the country. This occurred because the federal government had promised, but not yet provided, a storage place for these nuclear wastes. Therefore, the power company (which became Xcel Energy after a merger) requested permission to store the spent fuel rods in aboveground, dry-cask storage at the Prairie Island plant. In 1994, after negotiations with the tribe and the state legislature, an agreement was reached: Xcel would be allowed to store the fuel rods aboveground at Prairie Island on the condition that it purchase 425 megawatts of wind energy capacity by the end of 2002. The company exceeded this goal with a total of 480 megawatts. In addition, in 1998, the Minnesota Public Utilities Commission required Xcel to acquire an additional 400 megawatts of wind energy, which it could require under the 1994 legislative agreement.

At the time of this agreement, development of wind power had already begun in the southwestern corner of Minnesota on a geological feature called Buffalo Ridge. The elevation of this ridge is higher than that of the surrounding land, so it has the highest sustained average wind speeds in the state. Also, because some of the land on which the wind turbines are situated is suitable only for grazing, local landowners can lease the land to power companies for wind turbines and make more money than they can by grazing cattle.

Wind turbines are well understood and provide electrical energy at a price competitive with that produced by other kinds of power plants. However, there are problems associated with them. Power generation from wind turbines is intermittent. The winds don't always blow fast enough, or they may blow at a greater speed than the blades can withstand, so the turbine must shut down. Therefore, a backup system is necessary to provide power when the turbines are not operating. There are also environmental issues related to the use of wind power. The wind turbines in Minnesota are very visible—there are hundreds of them, and they stretch to the limits of your vision. Furthermore, birds are often killed by the larger turbines, since the tips of the long blades may have top speeds of 240–320 kilometers per hour (150 to 200 miles per hour), even at modest revolutions per minute.

Although Xcel Energy has increased its reliance on wind-powered electricity, the company still uses nuclear power and fossil fuel plants to provide electricity to its customers. In all, the three nuclear reactors in Minnesota generated approximately 14 percent of the energy the state used in 2001. In 2003, the need again arose to store more spent fuel rods. A legislative agreement was drawn up allowing for the construction of more aboveground storage casks. The two Prairie Island units are presently licensed to operate until 2013 and 2014, respectively.

But times have changed. The tribe is not as economically dependent on the Prairie Island nuclear power plant as it was 30 years ago because it now runs a successful gambling casino, and tribe members have become concerned about the health issues related to living so close to a nuclear reactor. Also, the terrorist actions on September 11, 2001, have raised security questions, since aboveground storage casks are highly visible targets; an attack could lead to the release of radioactive material into the local environment. Thus, in reaching a settlement with Xcel, the Minnesota legislature imposed additional regulations on the company. The 26-page-long legislative bill contains numerous requirements that Xcel must meet for the plant to operate until its permit expires. The company will be able to request an extension to the original permit, but it would need to close down the plant if it could not store the wastes somewhere. Because Xcel's nuclear plants produce energy at a very low cost, it is advantageous for the company to keep them in operation as long as possible.

What Do You Think?

1. Compare nuclear power with wind power, listing the advantages and disadvantages of each. Do you think the original agreement, which called for increased use of wind power in exchange for continued use of nuclear power, has benefited the environment?

2. Can enough power be generated by renewable energy sources to replace the energy produced from fossil fuel and nuclear power plants?

3. Is it appropriate for the government to force utility companies to make different investments in technology, or should the companies only respond to market forces?

4. The U.S. government-promised high-level nuclear waste storage site at Yucca Mountain, Nevada (described in chapter 11) will not be ready in the near future. In the meantime, what do you see as the solution to the problem of nuclear waste storage?

9

Energy and Civilization: Patterns of Consumption

Chapter Outline

History of Energy Consumption
  Biological Energy Sources
  Increased Use of Wood
  Fossil Fuels and the Industrial Revolution
  The Role of the Automobile
  Growth in the Use of Natural Gas
How Energy Is Used
  Residential and Commercial Energy Use
  Industrial Energy Use
  Transportation Energy Use
Electrical Energy
The Economics and Politics of Energy Use
  Fuel Economy and Government Policy
  The Importance of OPEC
Energy Consumption Trends
  Growth in Energy Use
  Available Energy Sources
  Political and Economic Factors
Issues—Analysis: Are Hybrid Electric
  Vehicles the Answer?

Environmental Close-Up
Alternative-Fuel Vehicles, p. 183

Global Perspectives
Reducing Automobile Use in Cities, p. 174
Biomass Fuels and the Developing World,
  p. 176

Objectives

After reading this chapter, you should be able to:

- Explain why all organisms require a constant input of energy.
- Describe how per capita energy consumption increased as civilization developed from hunting and gathering to primitive agriculture to advanced cultures.
- Describe how advanced modern civilizations developed as new fuels were used to run machines.

- Correlate the Industrial Revolution with social and economic changes.
- Explain how cheap oil and natural gas led to a consumption-oriented society.
- Explain how the automobile changed people's lifestyles.
- Explain why energy consumption is growing more rapidly in developing countries than in the industrialized world.
- Describe the role of OPEC in determining oil prices.

PAST PRESENT FUTURE

1983 U.S. proved oil reserves were about 35.1 thousand million barrels.

1983 World proved oil reserves were about 677 thousand million barrels.

1983 The price of a barrel of oil (in 2002 US$) was about $50 per barrel.

1983 World energy consumption was about 6714 million metric tons of oil equivalent.

1983 World per capita energy consumption was about 1.43 metric tons of oil equivalent.

2003 U.S. proved oil reserves were about 30 thousand million barrels (about a 15 percent decrease).

2003 World proved oil reserves were about 1,050 thousand million barrels (about a 55 percent increase).

2003 The price of a barrel of oil (in 2002 US$) was about $26 per barrel (about a 50 percent decrease).

2003 World energy consumption was about 9500 million metric tons of oil equivalent (about a 41 percent increase).

2003 World per capita energy consumption was about 1.50 metric tons of oil equivalent (about a 5 percent increase).

History of Energy Consumption

Every form of life and all societies require a constant input of energy. If the flow of energy through organisms or societies ceases, they stop functioning and begin to disintegrate. Some organisms and societies are more energy efficient than others. In general, complex industrial civilizations use more energy than simple hunter-gatherer or primitive agricultural cultures. If modern societies are to survive, they must continue to expend energy. However, they may need to change their pattern of energy consumption as traditional sources become limited.

Biological Energy Sources

Energy is essential to maintain life. In every ecosystem, the sun provides that energy. (See chapter 5.) The first transfer of energy occurs during photosynthesis, when plants convert light energy into chemical energy in the organic molecules they produce. Herbivorous animals utilize the food energy in the plants. The herbivores, in turn, are a source of energy for carnivores. Because nearly all of their energy requirements were supplied by food, primitive humans were no different from other animals in their ecosystems. In such hunter-gatherer cultures, nearly all human energy needs were met by using plants and animals as food, tools, and fuel. (See figure 9.1.)

Early in human history, people began to use additional sources of energy to make their lives more comfortable. They domesticated plants and animals to provide a more dependable supply of food. They no longer needed to depend solely on gathering wild plants and hunting wild animals for sustenance. Domesticated animals also furnished a source of energy for transportation, farming, and other tasks. (See figure 9.2.) Wood or other plant material provided a source of fuel for heating and cooking. Eventually, this biomass energy was used in simple technologies, such as shaping tools and extracting metals.

Increased Use of Wood

Early civilizations, such as the Aztecs, Chinese, Indians, Greeks, Egyptians, and

Figure 9.1   Hunter-Gatherer Society In a hunter-gatherer society, people obtain nearly all of their energy from the collection of wild plants and the hunting of animals.

Figure 9.2   Animal Power This bas-relief panel from an Egyptian tomb (1354–1346 BCE) depicts an important stage in the development of human civilization. With the use of domesticated animals, people had a source of power other than their own muscles.

Romans were culturally advanced, but their societies used human muscle, animal muscle, and fire as sources of energy. Except for limited use of some wind-powered and water-powered devices such as ships and canoes, the controlled use of fire was the first use of energy in a form other than food. Wood was the primary fuel. (Wood was also used for building

materials and other cultural uses.) The energy provided by wood enabled people to cook their food, heat their dwellings, and develop a primitive form of metallurgy. Such advances separated humans from other animals. When dense populations of humans made heavy use of wood for fuel and building materials, they eventually used up the readily available sources and had to import wood or seek alternative forms of fuel.

Because of a long history of high population density, India and some other parts of the world experienced a wood shortage hundreds of years before Europe and North America did. In many of these areas, animal dung replaced wood as a fuel source. It is still used today in some parts of the world.

Western Europe and North America were able to use wood as a fuel for a longer period of time. The forests of Europe supplied sufficient fuel until the thirteenth century. In North America, vast expanses of virgin forests supplied adequate fuel until the late nineteenth century. Fortunately, when local supplies of wood declined in Europe and North America, coal, formed from fossilized plant remains, was available as an alternative energy source. By 1890, coal had replaced wood as the primary energy source in North America.

Fossil Fuels and the Industrial Revolution

Fossil fuels are the remains of plants, animals, and microorganisms that lived millions of years ago. (The energy in these fuels is stored sunlight, just as the biomass of wood represents stored sunlight.) During the Carboniferous period, 286 million to 362 million years ago, when the Earth's climate was warmer and wetter than it is now, conditions were conducive to the formation of large deposits of coal. (See figure 9.3.) Oil and natural gas formed primarily from one-celled marine organisms whose bodies accumulated in large quantities on the seafloor and became compressed over millions of years. Heat and pressure from overlying sediment layers eventually converted the organic matter into oil and gas by distillation. Ever since machines replaced muscle power, the major energy sources for the world have been fossil remains from the distant past.

Historically, the first fossil fuel to be used extensively was coal. In the early eighteenth century, regions of the world that had readily available coal deposits were able to switch to this new fuel and participate in a major cultural change

a.   b.

Figure 9.3 **Carboniferous Period** Approximately 300 million years ago, this kind of ecosystem was common throughout the world. Plant material accumulated in these swamps and was ultimately converted to coal.

known as the **Industrial Revolution.** The Industrial Revolution began in England and spread to much of Europe and North America. It involved the invention of machines that replaced human and animal labor in manufacturing and transporting goods. Central to this change was the invention of the steam engine, which could convert heat energy into the energy of motion. The steam engine made possible the large-scale mining of coal. Before steam engines, coal mines flooded and thus were not economical. The source of energy for steam engines was either wood or coal; wood was quickly replaced by coal in most cases. Nations without a source of coal or those possessing coal reserves that were not easily exploited did not participate in the Industrial Revolution.

Prior to the Industrial Revolution, Europe and North America were predominately rural. Goods were manufactured on a small scale in the home. However, as machines and the coal to power them became increasingly available, the factory

system of manufacturing products replaced the small home-based operation. Because expanding factories required a constantly increasing labor supply, people left the farms and congregated in areas surrounding the factories. Villages became towns, and towns became cities. Widespread use of coal in cities resulted in increased air pollution. In spite of these changes, the Industrial Revolution was viewed as progress. Energy consumption increased, economies grew, and people prospered. Within a span of 200 years, the daily per capita energy consumption of industrialized nations increased eightfold. This energy was furnished primarily by coal, but a new source of energy was about to be discovered: oil.

Although the Chinese used some oil and natural gas as early as 1000 B.C., the oil well that Edwin L. Drake, an early oil prospector, drilled in Pennsylvania in 1859 was the beginning of the modern petroleum era. For the first 60 years of production, the principal use of oil was to make kerosene, a fuel for lamps. The gasoline produced was discarded as a waste product. During this time, oil was abundant relative to its demand, and thus, it had a low price.

The Role of the Automobile

The invention of the automobile dramatically increased the demand for oil products. In 1900, the United States had only 8000 automobiles. By 1950, it had over 40 million cars, and by 2000, over 200 million. More oil was needed to make automobile fuel and lubricants. During this period, the percentage of energy provided by oil increased from about 2 percent in 1900 to about 40 percent by 1950 and has remained at about 40 percent to the present. (See figure 9.4.)

The growth of the automobile industry, first in the United States and then in other industrialized countries, led to roadway construction, which required energy. Thus, the energy costs of driving a car were greater than just the fuel consumed in travel. As roads improved, higher speeds were possible. People demanded faster cars, and automobile companies were quick to build them. Bigger and faster cars required more fuel and even better roads. So roads were continually being improved, and better cars were being produced. A cycle of *more chasing more* had begun. In North America and much of Europe, the convenience of the automobile encouraged two-car families, which created a demand for more energy. It requires energy to mine ore, process it into metals, form the metals into automobile components, and transport all the materials. As the economy grew, so did energy needs.

More cars meant more jobs in the automobile industry, the steel industry, the glass industry, and hundreds of other industries. Constructing thousands of kilometers of roads created additional jobs. From their beginnings as suppliers of lamp oil, the oil companies grew into one of the largest industries in the world. Thus, the automobile industry played a major role in the economic development of the industrialized world. All this wealth gave people more money for cars and other necessities of life. The car, orig-

Figure 9.4 **Changes in Energy Sources** This graph shows how the percentage of energy obtained from various sources in the United Sates has changed over time. Until 1900, wood and coal were the predominant sources of energy. These sources declined in importance as oil and natural gas increased. Since 1950, oil and natural gas have provided over 60 percent of the energy, while coal has fallen to about 20 percent and wood to about 3 percent. Nuclear power has grown to about 8 percent, and geothermal, hydroelectric, solar and wind collectively account for about only 3 percent.

Source: Data from *Annual Energy Review, 2002*, Energy Information Administration.

(a)   **(b)**

Figure 9.5 **Energy-Demanding Lifestyle** (a) Building private homes on large individual lots some distance from shopping areas and places of employment is directly related to the heavy use of the automobile as a mode of transportation. (b) Heating and cooling a large enclosed shopping mall, along with the gasoline consumed in driving to the shopping center, increase the demand for energy.

inally a luxury, was now considered a necessity.

The car not only created new jobs but also altered people's lifestyles. Vacationers could travel greater distances. New resorts and chains of motels, restaurants, and other businesses developed to serve the motoring public, creating thousands of new jobs. Because people could live

farther from work, they began to move to the suburbs. (See figure 9.5.) Large shopping centers in suburban areas hastened the decline of central business districts. Today, fewer than 50 percent of retail sales are made in central business districts of North American cities, which has resulted in a loss of jobs in these areas.

GLOBAL PERSPECTIVE

Reducing Automobile Use in Cities

Although people like the convenience of the personal automobile, cars are the least energy-efficient means of transportation in cities. Ironically, the desire to use cars has resulted in traffic congestion in most large cities that has further reduced efficiency and eliminated the convenience of cars. To convince people to use other more efficient public transport and relieve congestion, many cities have passed regulations that use financial or other means to discourage automobile traffic in highly congested areas. Several different mechanisms have been successful.

London

Automobiles entering central London have their license plates photographed. Drivers of cars in the central city must pay a fee equivalent to about $8 a day. Those who do not pay by 10 P.M. face a doubled fee.

Hong Kong

Electronic sensors on cars record highway travel and time of day. Drivers are issued a monthly bill (commuter hours are the most expensive).

Singapore

Automobiles and motorcycles are equipped with an in-vehicle unit that automatically deducts money from a cash card when the vehicle passes through a sensor and enters the restricted zone of downtown Singapore between 7:30 A.M. and 7 P.M.

Gothenburg, Sweden

To encourage pedestrian traffic, the central business district has been divided like a pie into zones, with cars prohibited from moving directly from one zone to another. Autos move from zone to zone by way of a peripheral ring road.

Rome and Florence

All traffic except buses, taxis, delivery vehicles, and cars belonging to area residents have been banned between 7:30 A.M. and 7:30 P.M.

Tokyo

Before closing the sale, the buyer of a standard-size vehicle must show evidence that a permanent parking space is available for the car. To comply with the law, some drivers have constructed home garages with lifts to permit parking one car above the other.

As people moved to the suburbs, they also changed their buying habits. Labor-saving, energy-consuming devices became essential in the home. The vacuum cleaner, dishwasher, garbage disposal, and automatic garage door opener are only a few of the ways human power has been replaced with electrical power. About 25 percent of the electrical energy in North America is used in homes to operate lights and appliances. Other aspects of our lifestyles illustrate our energy dependence. The small, horse-powered, family-owned farm of yesterday has grown into the huge, diesel-powered, corporate farm of today. Regardless of where we live, we expect Central American bananas, Florida oranges, California lettuce, Texas beef, Hawaiian pineapples, Ontario fruit, and Nova Scotia lobsters to be readily available at all seasons. What we often fail to consider is the amount of energy required to process, refrigerate, and transport these items. The car, the modern home, the farm, and the variety of items on our grocery shelves are only a few indications of how our lifestyles are based on a continuing supply of cheap, abundant energy.

Growth in the Use of Natural Gas

Initially, natural gas was a waste product of oil production that was burned at oil wells because it was difficult to store and transport. Before 1940, natural gas accounted for less than 10 percent of energy consumed in the United States. In 1943, in response to World War II, a federally financed pipeline was constructed to transport oil within the United States.

After the war, the federal government sold these pipelines to private corporations. The corporations converted the pipelines to transport natural gas. Thus, a direct link was established between the natural gas fields in the Southwest and the markets in the Midwest and East. By 1970, about 30 percent of energy needs were being met by natural gas. Currently, about 25 percent of the energy consumed in the United States is from natural gas, primarily for home heating and industrial purposes.

How Energy Is Used

The amount of energy consumed by countries of the world varies widely. (See table 9.1.) The highly industrialized countries consume much more energy than less-developed countries. However, even among countries with the same level of development, great differences exist in the amount of energy they use as well as in how they use it. To maintain their style of living, individuals in Canada and the United States use about twice as much energy as people in France or Japan, 10 times more energy than the people of China, and about 25 times more energy than the people of India.

Differences also exist in the purposes for which people use energy. Industrialized nations use energy about equally for three purposes: (1) residential and commercial uses, (2) industrial uses, and (3) transportation. Less-developed nations with little industry use most of their energy for residential purposes (cooking and heating). Countries that are making the transition from less-developed to industrial economies use large amounts of energy to develop their industrial base.

Residential and Commercial Energy Use

The amount of energy required for residential and commercial use varies greatly throughout the world. For example, about 16 percent of the energy used in North America is for residential purposes, while in India, 57 percent of the energy is for residential uses. The ways residential and commercial energy is used also vary widely. In North America, 75 percent is used for air conditioning, refrigeration, water heating, and space heating. However, in many parts of Africa and Asia much of the energy used in the home is for cooking.

Computer systems and the Internet are a relatively new segment of the economy that consumes energy. Although early estimates suggested that this segment of the economy would consume over 10 percent of U.S. electrical energy, more recent estimates put the energy cost at about 3 percent.

The current pattern of residential and commercial energy use in each region of the world determines what conservation methods will be effective. In Canada, which has a cold climate, 40 percent of the residential energy is used for heating. Proper conservation practices could reduce this by 50 percent. In much of the less-developed world, cooking is done over open fires. Using fuel-efficient stoves instead of open fires could reduce these energy requirements by 50 percent. Improving efficiency would protect wood resources and reduce the time or money needed to obtain firewood. (See figure 9.6.)

Table 9.1 Per Capita Energy Consumption, 2002

Country	Metric Tons of Oil Equivalent
Bangladesh	0.11
India	0.31
Egypt	0.71
China	0.78
Brazil	1.02
Mexico	1.31
Germany	3.99
Japan	4.00
France	4.33
Australia	5.73
United States	7.98
Canada	9.22

Source: Data from *BP Statistical Review of World Energy,* June 2003, and Population Reference Bureau, 2002 Population Data Sheet.

Figure 9.6 **Wood Use for Cooking** Many people in the developing world use wood as a primary fuel for cooking. Reliance on wood is a major cause of deforestation. (a) In many parts of the world such as Africa cooking is often done over open fires. This is a very inefficient use of fuel. (b) The use of simple mud stoves as in this home in Nepal greatly increases efficiency.

GLOBAL PERSPECTIVE

Biomass Fuels and the Developing World

Although most of the world uses fossil fuels as energy sources, much of the developing world relies on *biomass* as its source of energy. The biomass can be wood, grass, agricultural waste, or dung. According to the United Nations, 2 billion people (30 percent of the world's population) use biomass as fuel for cooking and heating dwellings. In developing countries, nearly 40 percent of energy used comes from biomass. In some regions, however, the percentage is much higher. For example, in sub-Saharan Africa, fuelwood provides about 80 percent of energy consumed. Worldwide, about 60 percent of wood removed from the world's forests is used for fuel.

This dependence on biomass has several major impacts:

- Often women and children must walk long distances and spend long hours collecting firewood and transporting it to their homes.

- Because the fuel is burned in open fires or inefficient stoves, smoke contaminates homes and affects the health of the people.

The World Health Organization estimates that in the developing world, 40 percent of acute respiratory infections are associated with poor indoor air quality related to burning biomass. A majority of those who become ill are women and children because the children are in homes with their mothers who spend time cooking food for their families.

- Often the fuel is harvested unsustainably. Thus, the need for an inexpensive source of energy is a cause of deforestation. Furthermore, deforested areas are prone to soil erosion.

- When dung or agricultural waste is used for fuel, they cannot be used as an additive to improve the fertility or organic content of the soil. Thus, use of these materials for fuel negatively affects agricultural productivity.

Industrial Energy Use

The amount of energy countries use for industrial processes varies considerably. Nonindustrial countries use little energy for industry. Countries that are developing new industries dedicate a high percentage of their energy use to them. They divert energy to the developing industries at the expense of other sectors of their economy. Highly industrialized countries use a significant amount of their energy in industry, but their energy use is high in other sectors as well. In the United States, industry claims about 26 percent of the energy used.

The amount of energy required in a country's industrial sector depends on the types of industrial processes used. Many countries use inefficient processes and could reduce their energy consumption by converting to more energy-efficient ones. However, they need capital investment to upgrade their industries and reduce energy consumption. Some countries cannot afford the upgrade. For example, many countries of the former Soviet Union still use outdated, inefficient open-hearth fur-

naces to produce steel. These furnaces require nearly double the worldwide energy average to produce a metric ton of steel compared to more efficient processes. By contrast, China, the world's largest steel producer, has closed all its open-hearth steelmaking operations in favor of more efficient processes.

Transportation Energy Use

As with residential, commercial, and industrial uses, the amount of energy used for transportation varies widely throughout the world. In some of the less-developed nations, transportation uses are very small. Per capita energy use for transportation is larger in developing countries and highest in highly developed countries. (See table 9.2.) Bangladesh, India, and China use about 10 percent of their energy for transportation, while most developing and industrialized countries use about 30 percent or more for the same purpose.

Once a country's state of development has been taken into account, the specific combination of bus, rail, waterways,

and private automobiles is the main factor in determining a country's energy use for transportation. In Europe, Latin America, and many other parts of the world, rail and bus transport are widely used because they are more efficient than private automobile travel, governments support these transportation methods, or a large part of the populace is unable to afford an automobile. In countries with high population densities, rail and bus transport is particularly efficient. (See figure 9.7.) In these countries, automobiles require about four times more energy per passenger kilometer than does bus or rail transport. In addition, most of these countries have high taxes on fuel, which raise the cost to the consumer and encourage the use of public transport.

In North America, the situation is different. Government policy has kept the cost of energy artificially low and supported the automobile industry while removing support for bus and rail transport. Consequently, the automobile plays a dominant role, and public transport is primarily used only in metropolitan areas. Rail and bus transport are about twice as

Table 9.2 Per Capita Energy Use for Transportation 2002

Country	Metric Tons of Oil Equivalent	Percent of Energy Used for Transportation
Bangladesh	0.009	8.6
India	0.036	11.7
China	0.076	9.7
Egypt	0.183	25.8
Brazil	0.321	31.5
Mexico	0.531	40.5
Germany	1.069	26.8
Japan	1.124	28.1
France	1.325	30.6
Australia	2.223	38.8
Canada	2.628	28.5
United States	3.072	39.5

Sources: Data from BP Statistical Review of World Energy, June 2003; Population Reference Bureau, 2002 Population Data Sheet; and International Energy Agency online statistics.

Figure 9.7 **Public Transportation** In regions of the world where energy is expensive, people make maximum use of cheaper public transportation.

energy efficient as private automobiles. Private automobiles in North America, with about 5 percent of the world's population, consume over 40 percent of the gasoline produced in the world. (See table 9.3.) Air travel is relatively expensive in terms of energy, although it is slightly more efficient than private automobiles. Passengers, however, are paying for the convenience of rapid travel over long distances.

Electrical Energy

Electrical energy is such a large proportion of energy consumed in most countries that it deserves special comment. Electricity is both a way that energy is consumed and a way that it is supplied. Almost all electric energy is produced as a result of burning fossil fuels. Thus, we can look at electrical energy as a use to which fossil fuel energy is put. In the same way we use natural gas to heat homes, we can use natural gas to produce electricity. Because the transportation of electrical energy is so simple and the uses to which it can be put are so varied, electricity is a major energy source for many people of the world.

As with other forms of energy use, electrical consumption in different regions of the world varies widely. The industrialized countries of the world, with about 20 percent of the world's population, consume 60 percent of the world's electricity. Less-developed nations of the world, which have about 80 percent of the world's population, use 40 percent of the world's electricity. The per capita use of electricity in North America is 25 times greater than average per capita use in the less-developed countries. In Bangladesh, the annual per capita use of electricity is about 130 kilowatt hours, which is enough to light a 100-watt lightbulb for less than two months. The per capita consumption of electricity in North America is about 100 times greater than in Bangladesh. Electrical consumption among developed countries also varies widely. For example, the per capita use in Japan and the 15 countries of the European Union is about 60 percent of that in North America.

The production and distribution of electricity is a major step in the economic development of a country. In developed nations, about a quarter of the electricity is used by industry. The remainder is used primarily for residential and commercial purposes. In nations that are developing their industrial base, over half of the electricity is used by industry. For example, industries consume about 50 percent of the electricity used in South Korea.

The Economics and Politics of Energy Use

A direct link exists between economic growth and the availability of inexpensive energy. The replacement of human and animal energy with fossil fuels began with the Industrial Revolution and was greatly accelerated by the supply of cheap, easy-to-handle, and highly efficient fuels.

Table 9.3 U.S. Driving Habits

Average travel time to work in the United States in minutes:	25.5
In New York state	31.7
In North Dakota	15.8
Percent of Americans who get to work by:	
Driving alone	75.7
Carpooling	12.2
Taking public transportation	4.7
Staying home	3.3
Walking	2.9
Other	1.2

Source: U.S. Census Bureau, 2000.

Figure 9.8 **Gasoline Taxes and Fuel Efficiency** The price paid for fuel is greatly influenced by the amount of tax paid. High fuel prices cause consumers to choose automobiles with greater fuel efficiency.

Source: International Energy Agency.

Because the use of inexpensive fossil fuels allows each worker to produce more goods and services, productivity increased. The result was unprecedented economic growth in Europe, North America, and the rest of the industrialized world.

Because of this link between energy and productivity, most industrial societies want to ensure a continuous supply of affordable energy. The higher the price of energy, the more expensive goods and services become. To keep costs down, many countries have subsidized their energy industries and maintained energy prices at artificially low levels. International trade in fossil fuels has a major influence on the world economy and politics. The emphasis on low-priced fuels has encouraged high rates of consumption.

Fuel Economy and Government Policy

Governments fashion policies that influence how people use energy. Automobile fuel efficiency is one area in which government policy has had significant impact. For example, the price of a liter of gasoline is determined by two major factors: (1) the cost of purchasing and processing crude oil into gasoline and (2) various taxes. Most of the differences in gasoline prices among countries are a result of taxes and reflect differences in govern-

ment policy toward motor vehicle transportation. The cost of taxes to the U.S. consumer is less than 30 percent of the retail gasoline price, and in Canada, about 40 percent of the price of gasoline is taxes, while in Japan and many European countries, taxes account for 50 percent to 75 percent of the cost of gasoline. (See figure 9.8.) When we compare the kinds of automobiles driven, we find a direct relationship between the cost of fuel and fuel efficiency. In the United States and Canada, the average fleet fuel economy is about 11.2 liters per 100 kilometers (20.8 miles per gallon) and about 9.2 liters per 100 kilometers (25 miles per gallon), respectively. This compares with a European average of about 7.7 liters per 100 kilometers (30.4 miles per gallon). The average European car driver pays more than twice as much as U.S. and Canadian drivers and uses about 30 percent less fuel to drive the same distance as a U.S. driver. Since taxes make up the majority of the price of gasoline in Europe, government tax policy has provided an incentive for people to purchase fuel-efficient automobiles.

Another objective of governments is to have a mechanism for generating the money needed to build and repair roads. Many European countries raise more money from fuel taxes than they spend on building and repairing roads. The United States, on the other hand, raises approximately 60 percent of the monies needed for roads from fuel taxes. The relatively low cost of fuel in the United States encourages more travel, which increases road repair costs.

The European Union has also taken steps to comply with the Kyoto treaty that mandates reductions in the amount of carbon dioxide released into the atmosphere. To achieve carbon dioxide reduction targets, European automobile manufacturers have agreed to lower fuel consumption to about 6 liters per 100 kilometers (39.2 miles per U.S. gallon) by 2008. By contrast, the United States has not signed the Kyoto treaty and has seen falling fuel efficiency as more people buy and drive SUVs.

Since electricity is such an important source of energy for many uses, it is interesting to look at how different countries structure electricity prices. Because of the

Table 9.4 The Relationship Between Price and Consumption of Electricity

Country	Price for Industry (US Cents/Kilowatt-Hour) 2001	Price for Households (US Cents/Kilowatt-Hour) 2001	Annual per Capita Consumption (Kilowatt-Hours) 2001
Canada	3.9	6	16,746
Sweden	3.4	10.3	16,013
Finland	4.6	9.1	15,687
United States	4.7	8.3	12,896
Japan	14.3	21.4	7,907
Austria	9.2	13.6	7,500
Germany	7.9	16.7	6,806
United Kingdom	5.4	11	6,192
South Korea	6.6	8.5	5,607
Spain	4.1	10.9	5,482

Source: Data from International Energy Agency, Key World Energy Statistics 2003.

Table 9.5 Source of U.S. Crude Oil Imports

Country	Thousand Barrels per Day (2003)	Percent of Total Imported
Saudi Arabia (OPEC)	1,754	19.2%
Mexico	1,572	17.2%
Canada	1,537	16.7%
Venezuela (OPEC)	1,181	12.9%
Nigeria (OPEC)	828	9.0%
Iraq (OPEC)	450	4.9%
Angola (OPEC)	370	4.0%
United Kingdom	355	3.9%
Kuwait (OPEC)	205	2.2%
Norway	169	1.8%
Other nations	756	8.2%
Total imports	9,177	100%
Total from OPEC countries	4,788	52.2%
U.S. production	5,710	

Source: U.S. Energy Information Administration.

nature of the electrical industry, most countries regulate the industry and influence the price utilities are able to charge. Furthermore, the cost to industrial users is typically about half that charged to residential customers. Table 9.4 shows industrial and residential prices for electricity in several countries. Obviously, higher prices discourage use. The United States, Canada, Sweden, and Finland, which have some of the lowest prices for electricity, have the highest rates of consumption. Canada, Sweden, and Finland have high proportions of their electricity generated by hydroelectric plants, which contributes to the low cost of electricity in those countries. Countries such as Japan, which must import fossil fuels to generate electricity, have a higher cost to generate the electricity, which is reflected in the price.

The Importance of OPEC

OPEC began in September 1960, when the governments of five of the world's leading oil-exporting countries agreed to form a cartel. Three of the original members—Saudi Arabia, Iraq, and Kuwait—were Arab countries, while Venezuela and Iran were not. Today, 12 countries belong to OPEC. These include seven Arab states—Saudi Arabia, Kuwait, Libya, Algeria, Iraq, Qatar, and United Arab Emirates—and five non-Arab members—Iran, Indonesia, Nigeria, Gabon, and Venezuela. OPEC nations control over 78 percent of the world's estimated oil reserves of 1000 billion barrels of oil. Middle Eastern OPEC countries control over 60 percent of this total, which makes OPEC and the Middle East important world influences.

During the Arab-Israeli War of 1973, OPEC was supplying 55 percent of the world's oil. To protest the war, seven Arab members of OPEC reduced their production of oil. This resulted in a worldwide oil shortage, which caused all oil companies, in OPEC and non-OPEC countries, to increase their prices. In the United States, the price of oil skyrocketed from $3.39 to $13.93 per barrel, resulting in oil shortages, long lines at gas stations, and increased domestic exploration.

Continued action by OPEC led to continued price increases until the price peaked in 1980 at over $35 a barrel. During the 1980s, OPEC members had important differences over oil pricing and production rules. These differences weakened the cartel, and prices fell until they reached about $15 a barrel in 1988. Then, the 1990 invasion of Kuwait by Iraq deeply divided the OPEC countries. Owing in part to friction caused by the Kuwait conflict, the power of OPEC declined, and world oil prices fluctuated between $15 and $20 a barrel.

The price of oil fell to about $13 per barrel in 1998, began to rise in 1999, and by 2002 was about $25 per barrel. Gasoline prices in the United States began to exceed $2 per gallon. The United States exerted diplomatic pressure on OPEC to increase production levels to lower the cost of gasoline. OPEC stated that gasoline production in the United States, rather than a shortage of OPEC crude oil output, was to blame for high prices. Today, OPEC countries still control about 40 percent of the world's oil production and are a major force in determining price. (See table 9.5.)

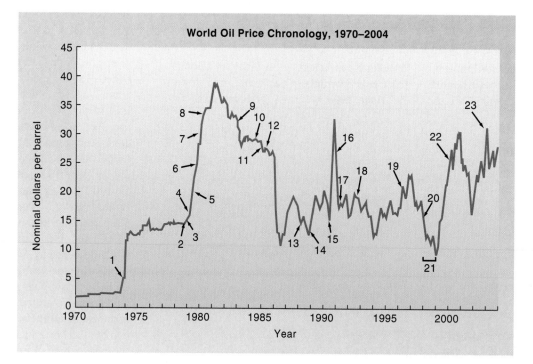

World Oil Price Chronology, 1970–2004

1. Organization of Petroleum Exporting Countries (OPEC) oil embargo begins (October 19–20, 1973).
2. OPEC decides on 14.5 percent price increase for 1979.
3. Iranian revolution; Shah deposed.
4. OPEC raises prices 14.5 percent on April 1, 1979.
5. OPEC raises prices 15 percent.
6. Iran takes hostages; President Carter halts imports from Iran; Iran cancels U.S. contracts; non-OPEC output hits 17.0 million barrels per day.
7. Saudis raise market crude price from $19 to $26 per barrel.
8. Kuwait, Iran, and Libya production cuts drop OPEC oil production to 27 million barrels per day.
9. OPEC cuts prices by $5 per barrel and agrees to 17.5 million barrels per day output.
10. Norway, United Kingdom, and Nigeria cut prices.
11. OPEC accord cuts Saudi Light price to $28 per barrel.
12. OPEC output falls to 13.7 million barrels per day.
13. OPEC/non-OPEC meeting fails.
14. Exxon's *Valdez* tanker spills 11 million gallons of crude oil.
15. Iraq invades Kuwait.
16. Operation Desert Storm begins; 17.3 million barrels of Strategic Petroleum Reserves crude oil sales are awarded.
17. Persian Gulf war ends.
18. OPEC production reaches 25.3 million barrels per day, the highest in over a decade.
19. Extremely cold weather in the United States and Europe.
20. OPEC raises its production ceiling by 2.5 million barrels per day to 27.5 million barrels per day. This is the first increase in four years.
21. Oil prices continue to plummet as increased production from Iraq coincides with no growth in Asian oil demand due to the Asian economic crisis and increases in world oil inventories following two unusually warm winters.
22. Oil prices triple between January 1999 and September 2000 due to strong world oil demand, OPEC oil production cutbacks, and other factors, including weather and low oil stock levels.
23. United States invades Iraq in March 2003.

Figure 9.9 **The Impact of Events on Oil Prices** This chronology shows how various events can affect the price of crude oil. Most of these events involve actions by OPEC or political changes in the Middle East.

Source: U.S. Energy Information Administration.

Since 2002, OPEC countries have agreed to regulate production to maintain a price greater than $22 per barrel. With increased solidarity among OPEC countries, oil prices have continued to rise and reached over $40 per barrel in 2004.

Figure 9.9 summarizes the fluctuations in oil prices and the events with which they are correlated.

Energy Consumption Trends

From a historical point of view, it is possible to plot changes in energy consumption. Economics, politics, public attitudes, and many other factors must be incorporated into an analysis of energy use trends.

Growth in Energy Use

In 2002, world energy consumption was around 9405 million metric tons of oil equivalent, an increase of 30 percent since 1985. Of this total, conventional fossil fuels—oil, natural gas, and coal—accounted for nearly 90 percent.

Over half of world energy is consumed by the 25 countries that are members of the Organization for Economic Cooperation and Development (OECD). These countries (Australia, New Zealand, Japan, Canada, Mexico, the United States, and the countries of Europe) are the developed nations of the world. In the last decade, energy consumption in OECD countries has risen moderately (slightly over 1 percent per year) while economic growth has continued. There has also been a shift toward service-based economies, with energy-intensive industries moving to non-OECD countries. In contrast, in countries that are becoming more economically advanced, energy consumption is increasing at a faster rate (about 4 percent per year). We should expect to see this pattern continue and countries with emerging economies increasingly demanding more energy. One segment of the world economy has actually had a decline in energy usage. The countries of the former Soviet Union have had a decline of about 25 percent in the last 10 years. (See figure 9.10.)

Available Energy Sources

Oil remains the world's major source of energy, accounting for about 38 percent of primary energy demand. Coal accounts for 26 percent and natural gas for 24 percent; the remainder is supplied mainly by nuclear energy and hydropower.

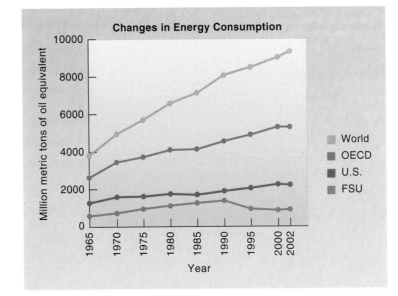

Figure 9.10 **Changes in World Energy Consumption by Region** Energy consumption has increased steadily. The growth in energy consumption of the economically developed nations (OECD) accounts for about half of world consumption. The former Soviet Union (FSU) has had declining energy use in recent years. In general, the economically developed countries have a slower rate of increase in energy use than the world as a whole.

Source: Data from *BP Statistical Review of World Energy*, 2003.

The share of total world energy consumption attributed to oil is projected to remain unchanged over the 2002–20 period at about 40 percent. Oil's market share does not increase because countries in many parts of the world are expected to switch to natural gas and other fuel, particularly for electricity generation. World oil consumption is projected to increase by 2.3 percent annually over the 20-year projection period, from 75 million barrels per day in 1999 to 120 million barrels per day in 2020.

Political and Economic Factors

Since 1950, North American energy consumption has steadily increased. However, there were two periods of decline, one beginning in 1973 and the other in 1979. Both of these episodes were the result of political turmoil in the Middle East and the increasing influence of the Organization of Petroleum Exporting Countries (OPEC). Figure 9.11 shows these trends in the United States, which mirror the consumption patterns in all of North America.

The same trend occurred in Western Europe, Japan, and Australia, resulting in a slowing of the world's growth in energy consumption. Several factors contributed to these declines in energy consumption. Increased prices for oil and all other forms of energy forced businesses and individuals to become more energy conscious and to expand efforts to conserve energy. From 1970 to 1983, the amount of energy used for heat per dwelling in the United States declined by 20 percent. Comparable reductions were made in Denmark, West Germany, and Sweden.

Since 1970, however, energy consumption has increased in North America and Europe. Like the stabilization of energy use in 1973 and 1979, the increase since 1980 is due to the price of oil. In 1979, oil was selling for about $40 a barrel. Beginning in 1980, the price began to drop, and in 1998, oil was selling for less than $15 a barrel. When Iraq invaded Kuwait in August 1990, prices rose dramatically but decreased at the end of hostilities.

During the 1980s, with energy costs declining, people in North America and Europe became less concerned about their

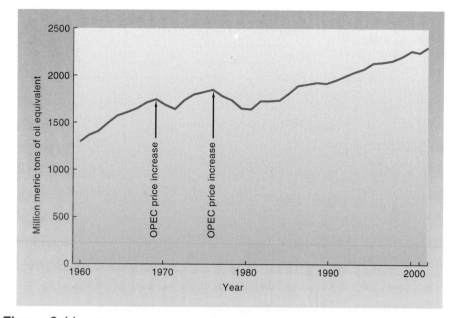

Figure 9.11 **Changes in U.S. Energy Consumption** Energy consumption in the United States rose gradually until 1973, when the OPEC countries increased the price of oil. The rise in price reduced consumption. A similar action in 1979 resulted in another decrease. Since 1983, however, consumption has steadily increased.

Source: Data from *BP Statistical Review of World Energy*, 2003.

energy consumption. They used more energy to heat and cool their homes and buildings, used more home appliances, and bought bigger cars. Obviously, the two primary factors that determine energy use are political stability in parts of the world that supply oil and the price of that oil. Insecurity leads to a price rise, but governments can also manipulate prices by changing taxes, granting subsidies, and using other means. The energy consumption behavior of most people is motivated by economics rather than by a desire to wisely use energy resources.

Over the past several years, world oil prices have been extremely volatile. In 1999, consumers benefited from oil prices that fell to $10 per barrel—a result of oversupply caused by lower demand for oil in both southeast Asia, which was suffering from an economic recession, and North America and Western Europe, which had warmer than expected winters. In 2000, however, world oil prices rebounded strongly, reaching a daily peak of $37 per barrel, rates not seen since the Persian Gulf War or 1990–91. The high prices can be traced to a tightening of production by OPEC and several key non-OPEC countries, such as Russia, Mexico, and Norway. In addition, oil companies did not want to commit capital to major development efforts for fear of a return to low prices. Unrest in the Middle East has also affected the price volatility.

The industrialized world was impacted by the high world oil prices of 2000. Concerns in the United States about a recurrence of the previous winter's shortage of home heating fuel oil for the Northeast led the Clinton administration to allow industry access to as much as 30 million barrels of crude oil from the nation's Strategic Petroleum Reserve. By 2001, however, the price of oil was again declining. The slowing of the world's economy and mild winter temperatures led to the price of oil approaching five-year lows.

The war in Iraq began in March 2003, leading to an immediate increase in the price of oil that continued through 2004.

Alternative-Fuel Vehicles

Electricity is the only zero-emission option now available for vehicles, but many experts believe that other fuels will prove viable in *reduced*-emission vehicles, although emissions are still produced in the generation of electricity at the power plant. Here are some of the options.

Compressed Natural Gas (CNG)

Status:

Fuels thousands of vehicles already on U.S. highways. Mass production is still a couple of years away, at the earliest.

Advantages:

- Substantial U.S. supply of natural gas
- Easy—though expensive—to convert conventional gasoline cars to CNG
- Cheaper: the equivalent of buying gasoline at 70 cents a gallon

Disadvantages:

- Requires heavy, bulky fuel tanks
- Public refueling stations expensive to build
- Lower performance and shorter range than gasoline vehicles

Emissions Benefits:

- Burns 80 percent cleaner than gasoline

Methanol/Flex-Fuel

Status:

Some heavy vehicles—including 354 buses operated by Southern California's Metropolitan Transportation Authority—run on pure methanol. More than 10,000 flex-fuel vehicles—running on methanol-gasoline combinations—are now on U.S. roads.

Advantages:

- Can be made from natural gas (98 percent of current production) and a range of renewable sources
- Requires only modest changes in gasoline engines and fueling infrastructure
- Higher octane than gasoline, giving 5 percent better performance

Disadvantages:

- Highly corrosive
- Lower energy content than gasoline, so requires bigger fuel tanks
- Fuel cost slightly higher than gasoline

Emissions Benefits:

- An 85 percent methanol mix burns 30 to 50 percent cleaner than pure gasoline; pure methanol vehicles potentially could be much cleaner

Hydrogen

Status:

Still largely experimental. Vehicles could be powered by hydrogen fuel cells or by engines that burn the gas. Practical, wide-scale use is at least a decade away.

Advantages:

- A high-energy fuel, boosting vehicle range
- Potentially unlimited fuel supply if produced from water; currently made from natural gas
- Produces no carbon dioxide, the gas blamed for global warming

Disadvantages:

- Highly explosive, though ultimately could be less dangerous than gasoline
- Costly and difficult to produce
- Costly and difficult to store

Emissions Benefits:

- If produced from water using solar energy and then used in a fuel cell, it would be a zero-emissions technology.

Propane

Status:

Broadly used in transportation worldwide

Advantages:

- A proven, low-cost fuel
- Public refueling infrastructure in place
- Nontoxic, so storage tanks are exempt from environmental regulations

Disadvantages:

- Supplies are limited, compared with other alternatives. At most, it could replace 10 percent of gasoline
- Highly flammable

Emissions Benefits:

- At least 50 percent cleaner than conventional gasoline

Sources: Data from California Air Resources Board, California Energy Commission, South Coast Air Quality Management District, Ward's Communications, Ford Motor Co., General Motors Corp., Chrysler Corp., and LP Gas Coalition.

A constant supply of energy is required by all living things. Energy has a major influence on society. A direct correlation exists between the amount of energy used and the complexity of civilizations.

Wood furnished most of the energy and construction materials for early civilizations. Heavy use of wood in densely populated areas eventually resulted in shortages, so fossil fuels replaced wood as a prime source of energy. Fossil fuels were formed from the remains of plants, animals, and microorganisms that lived about 300 million years ago. Fossil-fuel consumption in conjunction with the invention of labor-saving machines resulted in the Industrial Revolution, which led to the development of technology-oriented societies today in the developed world.

The invention of the automobile caused major changes in the lifestyles of people that led to greater consumption of energy. Because of the high dependence of modern societies on oil as a source of energy, OPEC countries, which control a majority of the world's oil, can set the price of oil through collective action.

Throughout the world, residential and commercial uses, industries, transportation, and electrical utilities require energy. Because of financial, political, and other factors, nations vary in the amount of energy they use as well as in how they use it. In general, rich countries use large amounts of energy and poor countries use much less. Analysts expect the worldwide demand for energy to increase steadily and the growth in energy usage by those countries that are becoming industrialized to be greater than that of the countries that are already industrialized.

Issues – Analysis

Are Hybrid Electric Vehicles the Answer?

In many metropolitan areas, emissions from automobiles are a major contributor to air-quality problems. An increase in the efficiency with which the chemical energy of fuel is converted to the motion of automobiles would greatly reduce air pollution. Because sources of fossil fuels are limited, they will become less available and greater efficiency will extend the limited supplies. Several kinds of modifications to automobiles have been suggested. Reducing vehicle weight by using lighter materials and streamlining the shape are techniques that have been used for many years. To further reduce energy consumption (increase energy efficiency), more innovative techniques are required.

A great deal of energy is needed to get a vehicle to begin moving from a stop (accelerate), and it takes an equal amount of energy to stop a vehicle once it is moving (decelerate). Internal combustion engines operate most efficiently when they are running at a specific speed (rpm) and work at less than peak efficiency when the vehicle is accelerating or decelerating. Thus, using an internal combustion engine to accelerate contributes significantly to air pollution. When the brakes are applied to stop the vehicle, the kinetic energy possessed by the moving vehicle is converted to heat in the braking system. So the energy that has just been used to accelerate the vehicle is lost as heat when the vehicle is brought to a stop. Furthermore, in metropolitan areas, an automobile sits in traffic with its engine running a significant amount of the time. Thus, the stop-and-go traffic common in city driving provides conditions that significantly reduce the efficient transfer of the chemical energy of fuel to the kinetic energy of turning wheels.

Today, the automobile industry is convinced that future generations of automobiles will be propelled by hydrogen fuel cells, which generate electricity directly by combining hydrogen with oxygen. However, the fuel cell vehicle is many years from becoming the standard vehicle in use. Because the technology is still being developed, its cost is very high. Before technology and developing infrastructure bring down the costs of fuel cell vehicles, the next generation of the automobile appears to be the hybrid electric vehicle (HEV). Because fuel cell vehicles would be electric vehicles, hybrid electric vehicles are a logical transition, since much of the technology needed for hybrids could be used by fuel cell vehicles.

Current hybrid electric vehicles use electric motors to drive the car when it is traveling at low speed and an internal combustion engine to provide the extra power needed to reach higher speeds on the highway. While hybrids still use petroleum, they increase gas mileage by about 25 percent. The gas-saving technology of the hybrid is in its combined use of the internal combustion engine with an electric motor and battery. A computer within the vehicle decides when to use the engine, the electric motor, or a combination of both.

In conventional vehicles, about 17 percent of the fuel efficiency is lost while idling in traffic or at stoplights. The hybrid's gas engine does not need to keep running at these times so its electric motor takes control. The battery is recharged when the car switches back to the gas engine, and in a process called regenerative braking, the kinetic energy of the car is captured during braking and stored as electrochemical energy in the battery. Since the internal combustion engine is only used to assist at high speeds, it can be smaller, weigh less, and consume less fuel than engines of conventional vehicles.

Although Toyota and Honda have sold hybrid electric vehicles since 1999, U.S. automakers are just beginning to think about producing them. In 1993, they received 50 percent funding from the Department of Energy to develop hybrid electric vehicle technology and produce vehicles for the market by 2003. None produced a vehicle by the 2003 deadline, but Ford offered a hybrid SUV in 2004. DaimlerChrysler is scheduled to produce a pickup in 2005 or 2006, and General Motors a pickup in 2007.

- If the hybrid electric vehicle has so many advantages, why doesn't everyone drive one?
- Why do you think American car companies have been slow to market a hybrid electric vehicle?
- Which is most important to you, fuel economy or power?
- Currently, the federal government provides a tax incentive if you purchase a hybrid electric vehicle. Should subsidies be provided for hybrid electric vehicles?

Key Terms

fossil fuels *172*
Industrial Revolution *172*

Review Questions

1. Why was the sun able to provide all energy requirements for human needs before the Industrial Revolution?
2. In addition to food, what energy requirements does a civilization have?
3. Why were some countries unable to use the technologies developed during the Industrial Revolution?
4. What factors caused a shift from wood to coal as a source of energy?
5. How were energy needs in World War II responsible for the subsequent increased consumption of natural gas?
6. What part does government regulation play in changing the consumption of natural gas and oil?
7. Why was much of the natural gas that was first produced wasted?
8. What was the initial use of oil? What single factor was responsible for a rapid increase in oil consumption?
9. List the three purposes for which a civilization uses energy.
10. Why is OPEC important in the world's economy?
11. Give examples of how political and economic events affect energy prices and usage.

Critical Thinking Questions

1. Imagine you are a historian writing about the Industrial Revolution. Imagine that you also have your new knowledge of environmental science and its perspective. What kind of a story would you tell about the development of industry in Europe and the United States? Would it be a story of triumph or tragedy, or some other story? Why?
2. What might be some of the effects of raising gasoline taxes in the United States to the rate that most Europeans pay for gasoline? Why? What do you think about this possibility?
3. Some argue that the price of gasoline in the United States is artificially low because it does not take into account all of the costs of producing and using gasoline. If you were to figure out the "true" cost of gasoline, what kinds of factors would you want to take into account?
4. How has the ubiquitous nature of automobiles changed the United States? Do you feel these changes are, in balance, positive or negative? What should the future look like regarding automobile use in the United States? How can this be accomplished?
5. The Organization of Petroleum Exporting Countries (OPEC) controls nearly 80 percent of the known oil reserves. What political and economic effects do you think this has? Does this have any effect on energy use?
6. How do you think projected energy consumption will affect world politics and economics, given current concerns about global warming?

Welcome to McGraw-Hill's Online Learning Center

Location: www.mhhe.com/enger10e

McGraw Hill

Interactive Exploration

Check out the website at www.mhhe.com/enger10e and click on the cover of this textbook for quizzing, career information, case studies, and hot links for the following topics:

Conventional Energy Sources

Oil, Coal, and Gas as Energy Sources

Nuclear Energy

Sustainable Energy

Air Pollution

10 Energy Sources

Chapter Outline

Objectives

After reading this chapter, you should be able to:

- Differentiate between resources and reserves.
- Identify peat, lignite, bituminous coal, and anthracite coal as steps in the process of coal formation.
- Recognize that natural gas and oil are formed from ancient marine deposits.
- Explain how various methods of coal mining can have negative environmental impacts.
- Explain why surface mining of coal is used in some areas and underground mining in other areas.
- Explain why it is more expensive to find and produce oil today than it was in the past.
- Recognize that secondary recovery methods have been developed to increase the proportion of oil and natural gas obtained from deposits.
- Recognize that transport of natural gas is still a problem in some areas of the world.
- Explain why the amount of energy supplied by hydroelectric power is limited.
- Describe how wind, geothermal, and tidal energy are used to produce electricity.
- Recognize that wind, geothermal, and tidal energy can be developed only in areas with the proper geologic or geographical features.
- Describe the use of solar energy in passive heating systems, active heating systems, and the generation of electricity.
- Recognize that fuelwood is a major source of energy in many parts of the less-developed world and that fuelwood shortages are common.
- Describe the potential and limitations of biomass conversion and waste incineration as sources of energy.
- Recognize that energy conservation can significantly reduce our need for additional energy sources.

PAST PRESENT FUTURE

1983 The world use of oil and gas was 2993 million tons of oil equivalent (MTOE) for oil and 1683 MTOE for gas.

1983 The average cost of wind-generated electricity was about 44 cents (in 2003 $US) per kilowatt-hour.

1983 Few automobile companies or utilities were openly pursuing the fuel cell as a production power plant.

1984 The global photovoltaic capacity was less than 200 megawatts.

2004 The world use of oil and gas was 4297 MTOE for oil and 3100 MTOE for gas.

2004 The average cost of wind-generated electricity was about 4 to 6 cents (in 2003 $US) per kilowatt-hour (at good wind sites).

2004 Every major automobile company in the United States, Europe, and Japan had an active fuel cell program.

2004 The global photovoltaic capacity exceeded 2200 megawatts.

Energy Sources

Chapter 9 outlined the historical development of energy consumption and how advances in civilizations were closely linked to the availability and exploitation of energy. New manufacturing processes relied on dependable sources of energy. Technology accelerated in the twentieth century. Between 1900 and 2003, world energy consumption increased by a factor of 16, the quantity of products manufactured increased by nearly fortyfold, but population increased only threefold. (See table 10.1.)

The energy sources most commonly used by industrialized nations are the fossil fuels: oil, coal, and natural gas, which supply about 90 percent of the world's commercially traded energy. Fossil fuels were formed hundreds of millions of years ago. They are the accumulation of energy-rich organic molecules produced by organisms as a result of photosynthesis over millions of years. We can think of fossil fuels as concentrated, stored solar energy. The rate of formation of fossil fuels is so slow that no significant amount of fossil fuels will be formed over the course of human history. Since we are using these resources much faster than they can be produced and the amount of these materials is finite, they are known as **nonrenewable energy sources.** Eventually, human demands will exhaust the supplies of coal, oil, and natural gas.

In addition to nonrenewable fossil fuels, there are several renewable energy sources. **Renewable energy sources** replenish themselves or are continuously present as a feature of the solar system. Some forms of renewable energy can also be referred to as *perpetual energy*. For example, in plants, photosynthesis converts light energy into chemical energy.

$$\begin{matrix} \text{Carbon} \\ \text{dioxide} \end{matrix} + \text{Water} + \begin{matrix} \text{Sunlight} \\ \text{energy} \end{matrix} \rightarrow \begin{matrix} \text{Biomass} \\ \text{(Chemical} + \text{Oxygen} \\ \text{energy)} \end{matrix}$$

This energy is stored in the organic molecules of the plant as wood, starch, oils, or other compounds. Any form of biomass—plant, animal, alga, or fungus—can be traced back to the energy of the sun. Since biomass is constantly being produced, it is a form of renewable energy. Solar, geothermal, and tidal energy are renewable energy sources because they are continu-

	Population (Billions)	Gross World Product (Trillion 2000 $US)	Fossil-Fuel Consumption (Billion Tons Coal Equivalent)
1900	1.6	0.6	1
1950	2.5	2.9	3
2003	6.2	36.0	16

Table 10.1 World Population, Economic Output, and Fossil-Fuel Consumption

Source: Data from *CIA World Fact Book*.

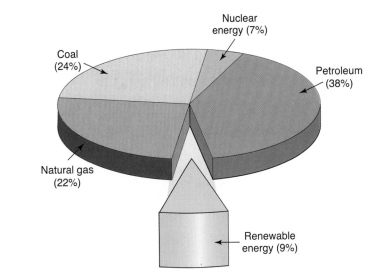

Figure 10.1 **All Energy Sources** The use of nonrenewable energy sources, including fossil fuels and nuclear energy, far outweighs the use of renewable sources in the industrialized world today.

ously available. Anyone who has ever lain in the sun, seen a geyser or hot springs, or been swimming in the surf has experienced these forms of energy. Progress has been made in the use of renewable sources of energy, as in heating and cooling homes and businesses with solar energy. However, technical, economic, and cultural challenges must be addressed before renewable energy will meet a significant percentage of humans' energy demands. Figure 10.1 shows the proportions of renewable and nonrenewable energy sources used in the world today.

Resources and Reserves

When discussing deposits of nonrenewable resources, such as fossil fuels, we must differentiate between deposits that

can be extracted and those that cannot. From a technical point of view, a **resource** is a naturally occurring substance of use to humans that can *potentially* be extracted using current technology. **Reserves** are known deposits from which materials *can* be extracted profitably with existing technology *under certain economic conditions*. It is important to recognize that *reserves* is an economic concept and is only loosely tied to the total quantity of a material present in the world. Therefore, reserves are smaller than resources. (See figure 10.2.) Both terms are used when discussing the amount of mineral or fossil-fuel deposits a country has at its disposal. This can cause considerable confusion if the difference between these concepts is not understood. The total amount of a resource such as coal or oil changes only by the amount used each year. The amount of a reserve changes as technology advances, new deposits are discovered, and economic

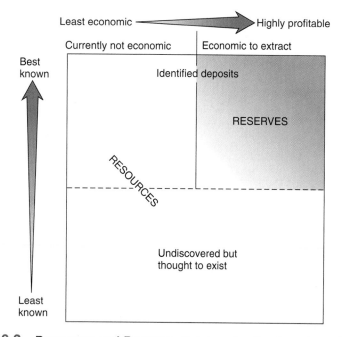

Figure 10.2 **Resources and Reserves** Each term describes the amount of a natural resource present. Reserves are those known deposits that can be profitably obtained using current technology under current economic conditions. Reserves are shown in the box in the upper right-hand corner in this diagram. The darker the color, the more valuable the reserve. Resources are much larger quantities that include undiscovered deposits and deposits that currently cannot be profitably used, although it might be feasible to do so if technology or market conditions change.

Source: Adapted from the U.S. Bureau of Mines.

conditions vary. There can be large changes in the amount of reserves, while the resource remains almost constant.

When we read about the availability of fossil fuels, we must remember that energy is needed to extract the wanted material from the Earth. If the material is dispersed or hard to reach, it is expensive to extract. If the cost of removing and processing a fuel is greater than the fuel's market value, no one is going to produce it. Also, if the amount of energy used to produce, refine, and transport a fuel is greater than its potential energy, the fuel will not be produced. A net useful energy yield is necessary to exploit the resource. However, in the future, new technology or changing prices may permit the profitable removal of some fossil fuels that currently are not profitable. If so, those resources will be reclassified as reserves.

To further illustrate the concept of reserves and how technology and economics influence their magnitude, let us look at the history of oil. The ancient Chinese are said to have been the first to use oil as a fuel. The only oil available to them was the small amount that naturally seeped out

of the ground. These seepages represented the known oil reserves as well as the known oil resources at that time.

Nearly 2000 years passed before oil reserves increased significantly. When the first oil well in North America was drilled in Pennsylvania in 1859, it greatly expanded the estimate of the amount of oil in the Earth. There was a sudden increase in the known oil reserves. In the years that followed, new deposits were discovered. Better drilling techniques led to the discovery of deeper oil deposits, and offshore drilling established the location of oil under the ocean floor. At the time of their discovery, these deep deposits and the offshore deposits added to the estimated size of the world's oil resources. But they did not necessarily add to the reserves because it was not always profitable to extract the oil. With advances in drilling and pumping methods and increases in oil prices, it eventually became profitable to obtain oil from many of these deposits. As it became economical to extract them, they were reclassified as reserves.

The amount of fossil fuel reserves is in a constant state of flux. For example,

prior to 1973, many oil wells were capped because it was not profitable to remove the oil. The oil still in the ground at these wells was not economical to produce and was not included in the reserves category. During the oil embargoes of 1973–74 and 1979–80, when some members of the Organization of Oil Exporting Countries (OPEC) reduced oil production, the price of oil increased. With the increase in oil prices, these wells became profitable, and the oil was reclassified as reserves. Then, in the 1980s, an oil surplus developed because the high oil prices resulting from the embargo caused increased production worldwide. Prices fell. Many of these same wells were capped again, and the oil was no longer classified as part of the reserve. (See figure 10.3.) With the exception of 1990 and 1991, when Iraq invaded Kuwait, the price of oil in the 1990s was relatively stable.

By the end of the decade, however, oil prices were once again rising. Beginning in 1999, the price of oil began to climb, and, by the first half of 2001, oil prices had risen to over $40 per barrel, up from $12 a barrel only two years earlier. The United States met frequently with OPEC leaders in the hope of having OPEC increase its production of oil and thus reduce the price at the pump in the United States. By the summer of 2000, the price of a gallon of gasoline exceeded $2 in many parts of the United States—double what it was a year earlier.

Many reasons were given for this sharp increase. Gasoline and petroleum stocks were low. With little crude oil or gasoline in inventory, refiners began to purchase crude oil in a market short on supplies, and that translated to higher prices. At the same time, refineries, squeezed by the high crude prices, were not operating at maximum capacity. By spring of 2002, refineries on average were running at about 85 percent capacity, compared with nearly 92 percent at the same time a year earlier. In addition to the supply issues, the oil companies claimed that new environmental rules governing the manufacture of gasoline contributed to the increased cost of gasoline at the pump. By the summer of 2004, the price of a gallon of gasoline in the United States had declined to the 1999 range. This decline was

GLOBAL PERSPECTIVE

Potential World Petroleum Resources

Oil is a finite, nonrenewable resource. It will eventually be exhausted. The primary questions about when we will run out of oil revolve around price, rates of use, and rates at which new discoveries are made. In 2003, the U.S. Geological Survey (USGS) issued an assessment of potential world petroleum resources. The report stated that undiscovered oil and gas resources of the world had increased, with a 20 percent increase in undiscovered oil and a slight decrease in undiscovered natural gas. This assessment estimates the volume of oil and gas, exclusive of the United States, that may be added to the world's reserves in the next 30 years.

The assessment indicates that there is more oil and gas in the Middle East and in the offshore areas of western Africa and eastern South American than previously reported, less oil and gas in Canada and Mexico, and significantly lower volumes of natural gas in the former Soviet Union.

With the evolution of technology and new understandings of petroleum systems, the USGS *World Petroleum Assessment 2003* is the first of its kind to provide a rigorous geologic foundation for estimating undiscovered energy resources for the world. The results have important implications for energy prices, policy, security, and the global resource balance.

The assessment provides a snapshot of current information about the location and abundance of undiscovered oil and gas resources. Such

an overview provides exploration geologists, economists, and investors with a general picture of where oil and gas resources are likely to be developed in the future. The USGS periodically estimates the amount of oil and gas remaining to be found. Since 1981, the last three of these studies have shown a slight increase in the combined volume of identified reserves and undiscovered resources.

In the USGS 2003 report, the world was divided into approximately 1000 petroleum provinces, based primarily on geologic factors, and grouped into eight regions. Estimates of reserve growth were based on the following:

- As drilling and production within discovered fields progresses, new pool or reservoirs are found that were not previously known.
- Advances in exploration technology make it possible to identify new targets within existing fields.
- Advances in drilling technology make it possible to recover oil and gas not previously considered recoverable in the initial reserve estimates.

To view the entire report, go to the USGS home page at www.usgs.gov.

Figure 10.3 **Changes in Proved Oil Reserves** The figure shows the changes in oil reserves over a 20-year period. The changes that occurred in 1987 and 1989 are the result of changes in reporting, not the result of new discoveries. Since 1989, new discoveries and revisions have matched consumption. Thus, reserves have remained nearly constant.

Source: Data from *BP Statistical Review of World Energy*, June 2004.

a result of a global economic recession, OPEC disunity, and the September 11, 2001, terrorist attacks. By the end of 2004, however, the price of a barrel of oil was up again to over $50 per barrel. Perhaps the only certainty of gasoline pricing is its continuing uncertainty.

Fossil-Fuel Formation

Fossil fuels are the remains of once living organisms that were preserved and altered as a result of geologic forces. Significant differences exist in the formation of coal from that of oil and natural gas.

Coal

Tropical freshwater swamps covered many regions of the Earth 300 million years ago. Conditions in these swamps

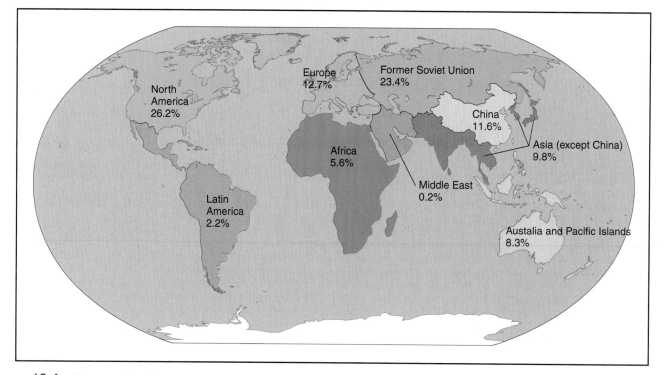

Figure 10.4 **Recoverable Coal Reserves of the World 2003** The percentage indicates the coal reserves in different parts of the world. This coal can be recovered under present local economic conditions using available technology.

Source: Data from *BP Statistical Review of the World Energy*, 2004.

favored extremely rapid plant growth, resulting in large accumulations of plant material. Because this plant material collected under water, decay was inhibited, and a spongy mass of organic material formed, called peat. Peat moss (peat) deposits are 90 percent water, 5 percent carbon, and 5 percent volatile materials. In some parts of the world such as Ireland, Latvia, and parts of Russia, peat is cut, dried, and used as fuel. However, because of its high water content, it is regarded as a low-grade fuel.

Due to geological changes in the Earth, some of the swamps containing peat were submerged by seas. The plant material that had collected in the swamps was then covered by sediment. The weight of the plant material plus the weight of the sediment on top of it compressed it into a harder form of low-grade coal known as lignite, which contains less water and a higher proportion of burnable materials.

Given sufficient thickness of overlying sediment, heat from the Earth, and the passing of geologic time, the lignite would have changed into bituminous (soft) coal. The major change from lignite to bituminous coal is a reduction in the

water content from over 40 percent to about 3 percent. If the heat and pressure continued over time, some of the bituminous coal could have changed to a hard-grade coal known as anthracite, which is about 96 percent carbon. Through this combination of events, which occurred over hundreds of millions of years, present-day coal deposits were created. Most parts of the world have coal deposits. (See figure 10.4.)

Oil and Natural Gas

Oil and natural gas, like coal, are products from the past. They probably originated from microscopic marine organisms. When these organisms died and accumulated on the ocean bottom and were buried by sediments, their breakdown released oil droplets. Gradually, the muddy sediment formed rock called shale, which contained dispersed oil droplets. Although shale is common and contains a great deal of oil, extraction from shale is difficult because the oil is not concentrated. If a layer of porous rock called sandstone formed on the top of the oil-containing shale and an impermeable layer of rock formed on top of

the sandstone, conditions might be suitable for oil pools to form. Usually, the trapped oil does not exist as a liquid mass but rather as a concentration of oil within sandstone pores, where it accumulates because water and gas pressure force it out of the shale. (See figure 10.5.) These accumulations of oil are more likely to occur if the rock layers were folded by geological forces.

Natural gas, like coal and oil, forms from fossil remains. If the heat generated within the Earth reached high enough temperatures, natural gas could have formed along with or instead of oil. This would have happened as the organic material changed to lighter, more volatile (easily evaporated) hydrocarbons than those found in oil. The most common hydrocarbon in natural gas is the gas methane (CH_4). Water, liquid hydrocarbons, and other gases may be present in natural gas as it is pumped from a well.

The conditions that led to the formation of oil and gas deposits were not evenly distributed throughout the world. Figure 10.6 illustrates the geographic distribution of oil and natural gas reserves. Some of these deposits are easy to extract, while others are not.

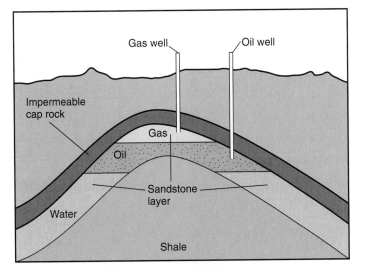

Figure 10.5 **Crude Oil and Natural Gas Pool** Water and gas pressure force oil and gas out of the shale and into sandstone capped by impermeable rock.

Source: Adapted with permission from Arthur N. Strahler, *Planet Earth*. Copyright ©1972 by Arthur N. Strahler.

Issues Related to the Use of Fossil Fuels

As previously mentioned, of the world's commercial energy, over 90 percent is furnished by the three nonrenewable fossil-fuel resources: coal, oil, and natural gas. Coal supplies about 24 percent, oil supplies about 40 percent, and natural gas supplies about 25 percent. Each fuel has advantages and disadvantages and requires special techniques for its production and use.

Coal Use

Coal is the world's most abundant fossil fuel, but it supplies only about 24 percent of the energy used in the world. It varies in quality and is generally classified in three categories: lignite, bituminous, and anthracite. Lignite coal has a high moisture content and is crumbly in nature, which

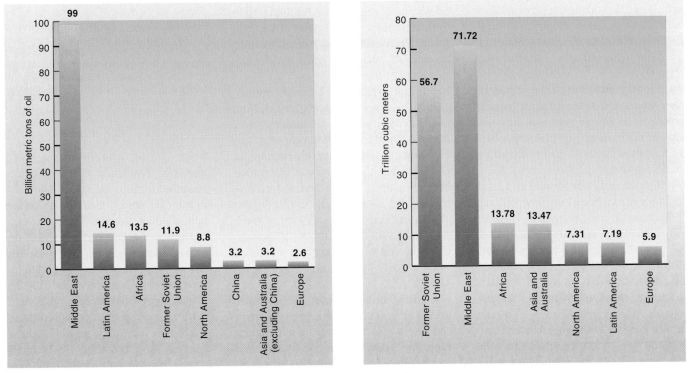

(a) Oil Reserves

(b) Natural Gas Reserves

Figure 10.6 **World Oil Reserves and Natural Gas Reserves, 2003** (a) The world's supply of oil is not distributed equally. Certain areas of the world enjoy an economic advantage because they control vast amounts of oil. Reserves are given in billions of metric tons of oil.
(b) Natural gas reserves, like oil and coal, are concentrated in certain regions of the world. Figures are in trillion cubic meters.

Source: Data from *BP Statistical Review of World Energy*, 2003.

(a) Vertical shaft

(b) Drift mine

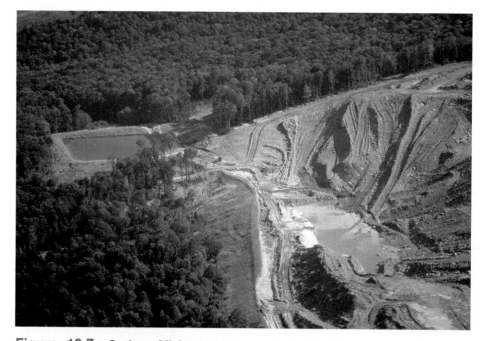

Figure 10.7 **Surface Mining** Aerial view F&M Coal strip mining site, Preston County, West Virginia.

Figure 10.8 **Underground Mining** If the overburden is too thick to allow surface mining, underground mining must be used. (*a*) If the coal vein is not exposed, a vertical shaft is sunk to reach the coal. (*b*) In hilly areas, if the vein is exposed, a drift mine is used in which miners enter from the side of the hill.

makes it the least desirable form. Bituminous coal is the most widely used because it is the easiest to mine and the most abundant. It supplies about 20 percent of the world's energy requirements. Coal is primarily used for electric power generation and other industrial uses. For most uses, anthracite coal is the most desirable because it furnishes more energy than the other grades and is the cleanest burning. Anthracite is not as common, however, and is usually more expensive because it is found at great depths and is difficult to obtain.

Because coal was formed as a result of plant material being buried under layers of sediment, it must be mined. There are two methods of extracting coal: surface mining and underground mining. **Surface mining** (strip mining) involves removing the material located on top of a vein of coal, called **overburden,** to get at the coal beneath. (See figure 10.7.) Coal is usually surface mined when the overburden is less than 100 meters (328 feet) thick. This type of mining operation is efficient because it removes most of the coal in a vein and can be profitably used for a seam of coal as thin as half a meter. For these reasons, surface mining results in the best utilization of coal reserves. Advances in the methods

of surface mining and the development of better equipment have increased surface mining activity in the United States from 30 percent of the coal production in 1970 to more than 60 percent today. This trend toward increased surface mining has also occurred in Canada, Australia, and the former Soviet Union.

If the overburden is thick, surface mining becomes too expensive, and the coal is extracted through **underground mining.** The deeply buried coal seam can be reached in two ways. In the first, in flat country where the vein of coal lies buried beneath a thick overburden, the coal is reached by a vertical shaft. (See figure 10.8a.) In the second, in hilly areas where the coal seam often comes to the surface along the side of a hill, the coal is reached from a drift-mine opening. (See figure 10.8b.)

The mining, transportation, and use of coal as an energy source present several significant problems. Surface mining disrupts the landscape, as the topsoil and overburden are moved to access the coal. It is possible to minimize this disturbance by reclaiming the area after mining operations are completed. (See figure 10.9.) However, reclamation rarely, if ever, returns the land to its previous level of pro-

ductivity. The cost of reclamation is passed on to the consumer in the form of higher coal prices. Underground mining methods do not disrupt the surface environment as much as surface mining does, but subsidence (sinking of the land) occurs if the mine collapses. In addition, large waste heaps are produced around the mine entrance from the debris that must be removed and separated from the coal.

Health and safety are important concerns related to coal mining, which is one of the most dangerous jobs in the world. This is particularly true with underground mining. Many miners suffer from **black lung disease,** a respiratory condition that results from the accumulation of fine coal-dust particles in the miners' lungs. The

a.   b.

Figure 10.9 **Surface-Mine Reclamation** (a) This photograph shows a large area that has been surface mined with little effort to reclaim the land. The windrows created by past mining activity are clearly evident, and little effort has been made to reforest the land. By contrast, (b) is an example of effective surface-mining reclamation. The sides of the cut have been graded and planted with trees. The topsoil has been returned, and the level land is now productive pasture.

coal particles inhibit the exchange of gases between the lungs and the blood. The health care costs and death benefits related to black lung disease are an indirect cost of coal mining. Since these costs are partially paid by the federal government, their full price is not reflected in the price of coal but is paid by taxpayers in the form of federal taxes and higher health premiums.

Because coal is bulky, shipping presents a problem. Generally, the coal can be used most economically near where it is produced. Rail shipment is the most economic way of transporting coal from the mine. Rail shipment costs include the expense of constructing and maintaining the tracks, as well as the cost of the energy required to move the long strings of railroad cars. In some areas, the coal is transferred from trains to ships.

Coal mining and transport generate a great deal of dust. The large amounts of coal dust released into the atmosphere at the loading and unloading sites can cause local air-pollution problems. If a boat or railroad car is used to transport coal, there is the expense of cleaning it before other types of goods can be shipped. In some cases, the coal can be ground and mixed with water to form a slurry that can be pumped through pipelines. This helps to alleviate some of the air-pollution problems without causing significant water-pollution problems.

Air pollution from coal burning releases millions of metric tons of material into the atmosphere and is responsible for millions of dollars of damage to the environment. The burning of coal for electric generation is the prime source of this type of pollution.

Since coal is a fossil fuel formed from plant remains, it contains sulfur, which was present in the proteins of the original plants. Sulfur is associated with **acid mine drainage** and air pollution. Acid mine drainage occurs when the combined action of oxygen, water, and certain bacteria causes the sulfur in coal to form sulfuric acid. Sulfuric acid can seep out of a vein of coal even before the coal is mined. However, the problem becomes worse when the coal is mined and the overburden is disturbed, allowing rains to wash the sulfuric acid into streams. Streams may become so acidic that they can support only certain species of bacteria and algae. Today, many countries regulate the amount of runoff allowed from mines, but underground and surface mines abandoned before these regulations were enacted continue to contaminate the water.

Currently, a form of acid pollution called acid deposition is becoming a serious problem. Acid deposition occurs when coal is burned and sulfur oxides are released into the atmosphere, causing acid-forming particles to accumulate. Each year, over 150 million metric tons of sulfur dioxide are released into the atmosphere worldwide. This problem is discussed in greater detail in chapter 17.

The release of carbon dioxide from the burning of coal has become a major issue in recent years. Increasing amounts of carbon dioxide in the atmosphere are strongly implicated in global warming. Environmentalists have suggested that the use of coal be decreased, since the other fossil fuels (oil and natural gas) produce less carbon dioxide for an equivalent amount of energy.

Because coal is difficult to transport and often has a high sulfur content resulting in air pollution, people seek alternative sources of fuel. The most common alternatives for coal are oil and natural gas.

Oil Use

Oil has several characteristics that make it superior to coal as a source of energy. Its extraction causes less environmental damage than does coal mining. It is a more concentrated source of energy than coal, it burns with less pollution, and it can be moved easily through pipes. These characteristics make it an ideal fuel for automobiles. However, it is often difficult to find. Today, geologists use a series of tests to locate underground formations that may contain oil. When a likely area is identified, a test well is drilled to determine if oil is actually present. Since the many easy-to-reach oil fields have already been tapped, drilling now focuses on smaller amounts of oil in less accessible sites, which means that the cost of oil from most recent discoveries is higher than that from

Drilling derrick

Crew quarters

Ocean

Seabed

Platform legs imbedded in seabed

Directionally drilled wells

Impermeable cap rock

Oil and gas reservoir

Reservoir rock

Impermeable cap rock

Oil and gas reservoir

Reservoir rock

a.

b.

Figure 10.10 **Offshore Drilling** Once the drilling platform is secured to the ocean floor, a number of wells can be sunk to obtain the gas or oil.

Source: (b) American Petroleum Institute.

the large, easy-to-locate sources of the past. As oil deposits on land have become more difficult to find, geologists have widened the search to include the ocean floor. Building an offshore drilling platform can cost millions of dollars. To reduce the cost, as many as 70 wells may be sunk from a single platform. (See figure 10.10.) Total proven world reserves of oil in 2003 were estimated at about 1100 billion barrels. Of this, more than three-quarters are in OPEC countries, and more than half lie in four Middle Eastern countries—Saudi Arabia, Iraq, Kuwait, and Iran.

One of the problems of extracting oil is removing it from the ground. If the water or gas pressure associated with a pool of oil is great enough, the oil is forced to the surface when a well is drilled. When the natural pressure is not great enough, the oil must be pumped to the surface. Present technology allows only about one-third of the oil in the ground to be removed. This means that two barrels of oil are left in the ground for every barrel produced. In most oil fields, **secondary recovery** is used to recover more of the oil. Secondary recovery methods include pumping water or gas into the well to drive the oil out or even starting a fire in the oil-soaked rock to liquefy thick oil. As oil prices increase, more expensive and aggressive secondary recovery methods will need to be used.

Processing crude oil to provide useful products generates a variety of problems.

Oil, as it comes from the ground, is not in a form suitable for use. It must be refined. The various components of crude oil can be separated and collected by heating the oil in a distillation tower. (See figure 10.11.) After distillation, the products may be further refined by "cracking." In this process, heat, pressure, and catalysts are used to produce a higher percentage of volatile chemicals, such as gasoline, from less volatile liquids, such as diesel fuel and furnace oils. It is possible, within limits, to obtain many products from one barrel of oil. In addition, petrochemicals from oil serve as raw materials for a variety of synthetic compounds. (See figure 10.12.) All of these processing activities are opportunities for accidental or routine

PART THREE Energy

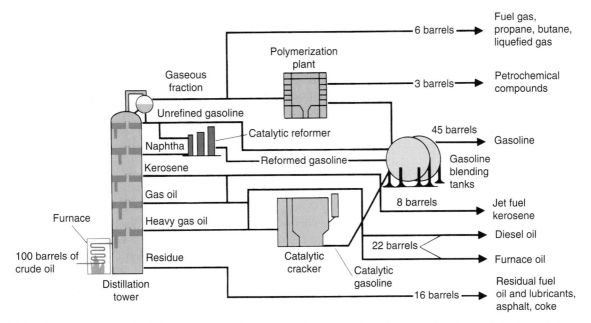

Figure 10.11 **Processing Crude Oil** A great variety of products can be obtained from distilling and refining crude oil. A barrel of crude oil produces slightly less than half a barrel of gasoline. This figure shows the many steps in the refining process and the variety of products that can be obtained from crude oil.

Source: From *Man, Energy, and Society* by Earl Cook. ® 1976 by W. H. Freeman and Company. Used with permission.

releases that may cause air or water pollution. The petrochemical industry is a major contributor to air pollution.

The environmental impacts of producing, transporting, and using oil are somewhat different from those of coal. Oil spills in the oceans have been widely reported by the news media. However, these accidental spills are responsible for only about one-third of the oil pollution resulting from shipping. The number of oil spills has dropped over the past two decades. In 1985, a reported 15,500 metric tons of oil were spilled. In 2003, that figure was 5680 metric tons. Nearly 60 percent of the oil pollution in the oceans is the result of routine shipping operations. Oil spills on land can contaminate soil and underground water. The evaporation of oil products and the incomplete burning of oil fuels contribute to air pollution. These problems are discussed in chapter 17.

Figure 10.12 **Oil-Based Synthetic Materials** These common household items are produced from chemicals derived from oil. Although petrochemicals represent only about 3 percent of each barrel of oil, they are extremely profitable for the oil companies.

The Arctic National Wildlife Refuge and Oil

The Arctic National Wildlife Refuge (ANWR) has been a source of controversy for many years. The major players are environmentalists who seek to preserve this region as wilderness; the state of Alaska, which funds a major portion of its activities with dividends from oil production; Alaska residents, who receive a dividend payment from oil revenues; oil companies that want to drill in the refuge; and members of Congress who see the oil reserves in the region as important economic and political issues.

In 1960, 3.6 million hectares (8.9 million acres) were set aside as the Arctic National Wildlife Range. Passage of the Alaskan National Interest Lands Conservation Act in 1980 expanded the range to

8 million hectares (19.8 million acres) and established 3.5 million hectares (8.6 million acres) as wilderness. The act also renamed the area the Arctic National Wildlife Refuge. There are international implications to this act. The refuge borders Canada's Northern Yukon National Park. Many animals, particularly members of the Porcupine caribou herd, travel across the border on a regular yearly migration. The United States is obligated by treaty to protect these migration routes.

The act requires specific authorization from Congress before oil drilling or other development activities can take place on the coastal plain in the refuge. The coastal plain has the greatest concentration of wildlife, is the calving ground for the Porcupine caribou, and has the greatest potential for oil production. In each case, the potential authorization caused a collision of three forces: environmental protection, economic development, and political benefit. Furthermore, great differences of opinion exist within each of the competing interest groups. Some Alaskan citizens support drilling; others oppose it. The Inupiat Eskimos who live along the north Alaskan coast mostly are in favor of drilling in ANWR. The Inupiat believe oil revenues and land-rental fees from oil companies will raise their living standards. The other Native American tribe in the region, the Gwich'in, who live on the southern fringe of the refuge, oppose drilling. They argue that the drilling will impact the caribou migration through the area every fall and thus affect their ability to provide food for their families. Members of Congress are similarly split. Even members of the Department of the Interior have provided conflicting testimony about the risks and benefits of drilling for oil in the refuge.

In 1998, the secretary of the Interior, acting under the recommendation of President Bill Clinton, cleared the way for oil development on Alaska's North Slope. Under the plan, about a third of a 4.6-million-acre study area in the northeastern corner of the federal reserve would be off limits to drilling. Oil leases would be sold on 4 million acres of the government's National Petroleum Reserve west of the Prudhoe Bay oil fields. Some of the leases would allow only slant drilling because the surface is to be protected. In 2000, the Energy Information Administration (EIA) released a report on the potential oil production from the coastal plain of ANWR. The coastal plain region, which comprises approximately 8 percent of the 7.7 million hectares (19 million acres) ANWR, is the largest unexplored, potentially productive geologic onshore basin in the United States. A decision on permitting the exploration and development is up to the U.S. Congress. In 2002, President George W. Bush reconfirmed his support for drilling. The EIA report estimated a 95 percent probability that at least 5.7 billion barrels of technically recoverable undiscovered oil are in the ANWR coastal plain. There is a 5 percent probability that at least 16 billion barrels of oil are recoverable. The report states that once oil has been discovered, more than 80 percent of the technically recoverable oil is commercially developable at a price of $25 per barrel (oil was $25 per barrel in July 2002). The value of the oil in 2002 dollars could be between $125 billion and $350 billion. Oil companies have repeatedly stated that the oil can be recovered without endangering wildlife or the fragile Arctic ecosystem. Conservationists have argued that none of the reserve should be developed when improvements in energy could reduce demand. They argue that drilling in the reserve will harm the habitat of millions of migratory birds, caribou, and polar bears. Only time will tell which side, if either, is correct.

Natural Gas Use

Natural gas, the third major source of fossil-fuel energy, supplies 25 percent of the world's energy. The drilling operations to obtain natural gas are similar to those used for oil. In fact, a well may yield both oil and natural gas. As with oil, secondary recovery methods that pump air or water into a well are used to obtain the maximum amount of natural gas from a deposit. After processing, the gas is piped to the consumer for use.

Transport of natural gas still presents a problem in some parts of the world. In the Middle East, Mexico, Venezuela, and Nigeria, wells are too far from consumers to make pipelines practical, so much of the natural gas is burned as a waste product at the wells. However, new methods of transporting natural gas and converting it into other products are being explored. At $-162°C$ ($-126°F$), natural gas becomes a liquid and has only 1/600 of the volume of its gaseous form. Tankers have been designed to transport **liquefied natural gas** from the area of production to an area of demand. In 2003, over 170 billion cubic meters (6000 billion cubic feet) of natural gas were shipped between countries as liquefied natural gas. This is nearly 6.5 percent of the natural gas consumed in the world. Of that amount, Japan alone imported 80 billion cubic meters (2800 billion cubic feet). As the demand for natural gas increases, the amount of it wasted will decrease and new methods of transportation will be employed. Higher prices will make it profitable to transport natural gas greater distances between the wells and the consumers.

One concern about transporting liquefied natural gas is accidents that might cause tankers to explode. Another, safer process converts natural gas to methanol, a liquid alcohol, and transports it in that form.

Of the three fossil fuels, natural gas is the least disruptive to the environment. A natural gas well does not produce any unsightly waste, although there may be local odor problems. Except for the danger of an explosion or fire, natural gas poses no harm to the environment during transport. Since it is clean burning, it causes almost no air pollution. The products of its combustion are carbon dioxide and water. However, the burning of natural gas does produce carbon dioxide, which contributes to global warming. Global warming is discussed in chapter 17.

Although natural gas is used primarily for heat energy, it does have other uses, such as the manufacture of petrochemicals and fertilizer. Methane contains hydrogen atoms that are combined with nitrogen from the air to form ammonia, which can be used as fertilizer.

More than two-thirds of the world's natural gas reserves are located in Russia and the Middle East. In fact, more than a third of the total reserves are located in 10 giant fields; six are located in Russia and the remainder are in Qatar, Iran, Algeria, and the Netherlands.

From 1985 to 2003, world production of natural gas rose by more than 56 percent.

Because oil reserves have been located beneath the North Sea, several European countries (the United Kingdom, Norway, and the Netherlands) have sizable reserves of natural gas. During oil shortages in the 1970s, they increased their use of natural gas to compensate for the increased cost of oil. This trend has continued; natural gas consumption increased by 55 percent from 1985 to 2003.

Renewable Sources of Energy

At the dawn of the twenty-first century, civilization still runs on the compost of the distant past—fossil fuels. The burning of oil, natural gas, and coal provides 85 percent of the world's commercial energy and 80 percent of all human-caused carbon dioxide emissions. Energy demand has nearly doubled in the past three decades and is expected to increase another 60 percent by 2020. But cleaner renewable energy use is growing fast. Currently, alternative sources–mostly hydropower, but also wind turbines, geothermal, solar cells, biomass fuels, and hydrogen fuel—supply not quite 10 percent of the world's total energy. (See figure 10.13.) According to many studies, however, these sources could provide half of the world's energy needs by 2050. These data do not include the use of wood, animal dung, and other locally produced energy sources typically used in the less-developed world.

Hydroelectric Power

People have long used water to power a variety of machines. Some early uses of water power were to mill grain, saw wood, and run machinery for the textile industry. Flowing water creates energy that can be captured and turned into electricity. This is called hydroelectric power, or *hydropower.* The most common type of hydroelectric power plant uses a dam on a river to store water in a reservoir. Water released from the reservoir flows through a turbine, spinning it, which in turn activates a generator to produce electricity. But hydroelectric power does not necessarily require a large dam. Some hydroelectric power plants just use a small canal to channel the river water through a turbine.

Another type of hydroelectric power plant—called a pumped storage plant—can even store power. A pumped storage plant uses two reservoirs, one located at a much higher elevation than the other. During periods of low demand for electricity, such as nights and weekends, energy is stored by reversing the turbines and pumping water from the lower to the upper reservoir. The stored water can later be released to turn the turbines and generate electricity as it flows back into the lower reservoir.

A small or microhydroelectric power system can produce enough electricity for a home, farm, or ranch.

Today, water power is used almost exclusively to generate electricity. (See figure 10.14.) Hydroelectric power plants are commonly located on human-made reservoirs. (See figure 10.15.) The impounded water represents a potential energy source. In some areas of the world where the streams have steep gradients and a constant flow of water, hydroelectricity may be generated without a reservoir. Such sites are usually found in mountainous regions and can support only small power-generating stations. At present, hydroelectricity produces about 6 percent of the world's energy supply, or about 15 percent of the world's electricity.

Hydroelectric power potential is distributed among the continents in rough proportion to land area; China alone possesses one-tenth of the world's potential. Mountainous regions and large river valleys are the most promising. Besides the United States, the eastern area of the former Soviet Union, and southern Canada, the regions that have done the most to harness hydroelectric energy are Europe and Japan. Europe has exploited almost

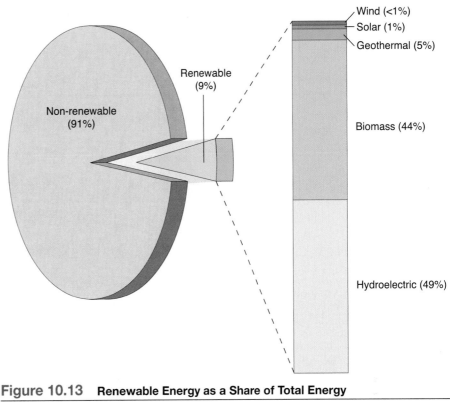

Figure 10.13 **Renewable Energy as a Share of Total Energy**
Source: U.S. Department of Energy.

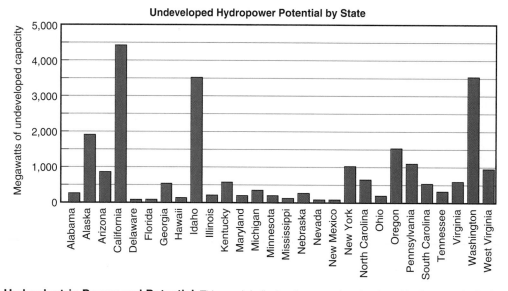

Undeveloped Hydropower Potential by State

Figure 10.14 **Hydroelectric Power and Potential** This graph indicates the current undeveloped hydropower in the United States. (For comparison, many large coal-fired power plants have about 1000 megawatts of generating capacity.)

Source: U.S. Geological Survey and the Foundation for Water and Energy Education, 2000.

60 percent of its potential. Although it has only one-fourth of Asia's resources, it generates nearly twice as much hydroelectric power. In contrast, Africa has developed only 5 percent of its potential, half of which comes from only three dams: Kariba in East Africa, Aswan on the Nile, and Akosombo in Ghana.

In some areas of the world, hydroelectric power is the main source of electricity. More than 35 nations already obtain more than two-thirds of their electricity from falling water. In South America, 73 percent of the electricity used comes from hydroelectric power, compared to 44 percent in the developing world as a whole. Norway gets 99 percent of its electricity and 50 percent of all its energy from falling water.

It is important to recognize that the construction of a reservoir for a hydroelectric plant causes environmental and social problems. The building of the Tellico Dam in Tennessee was delayed for several years because it might have caused the extinction of the snail darter, a fish that lived only in streams that would be flooded. The construction of the Aswan Dam in Egypt resulted in the displacement of 80,000 people. It created an environment that increased schistosomiasis, a waterborne disease caused by flatworm parasites that spend part of their life cycle in snails that live in slowly moving

water. The irrigation canals built to distribute the water from the Aswan Dam provide ideal conditions for the snails. Now many of the people who use the canal water for cooking, drinking and bathing are infected with the flatworm parasites.

From 1985 to 2002, the energy furnished by hydroelectricity for world use increased by over 30 percent. Approximately 17 percent of the potential hydroelectric sites of the world have been developed. The World Energy Conference estimates that the electricity produced by hydropower will increase six times by the year 2020. The less-developed countries, which have developed about 10 percent of their hydropower, will experience most of this growth.

The projected increase will come mainly from the development of plants on large reservoirs. However, construction of "minihydro" (less than 10 megawatts) and "microhydro" (less than 1 megawatt) plants is also increasing. Such plants can be built in remote places and supply electricity to small areas. China has over 80,000 such small stations, and the United States has nearly 1500.

About 50 percent of the U.S. hydroelectric capacity has been developed. However, this statistic is somewhat misleading. The Wild and Scenic Rivers Act (1968) prevents the construction of dams

on designated streams. Presently, 37 potential hydroelectric sites are on streams protected by this act. The hydroelectric generating potential often quoted for the United States includes these areas, even though, at present, construction of plants at these sites is not possible. In fact, some dams on rivers in states such as Maine and Washington are scheduled to be destroyed in order to improve native fisheries. The U.S. political climate, which now favors the protection of certain rivers, might change if the demand for more energy becomes acute.

China is constructing a huge hydroelectric dam, known as the Three Gorges Dam, on the Yangtze River. When completed, it will be the largest hydroelectric dam in the world. (See Global Perspectives: The Three Gorges Dam.)

Today's large dams rank among humanity's greatest engineering feats. The Itaipu hydroelectric power plant is the largest development of its kind in operation in the world (until the completion of the Three Gorges Dam in China). Built from 1975 to 1991 in a development on the Parana River, Itaipu is a joint project of Brazil and Paraguay. The power plant's 18 generating units add up to a total production capacity of 12,600 megawatts. The magnitude of the project also can be demonstrated by the fact that in 1997, Itaipu provided 25 percent of the energy

(a)

(b)

Figure 10.15 Hydroelectric Power Plant (a) The water impounded in this reservoir is used to produce electricity. In addition, this reservoir serves as a means of flood control and provides an area for recreation. (b) This figure shows how a hydroelectric dam operates.

Source: Tennessee Valley Authority.

hydropower development in these countries has occurred primarily because energy-intensive industries need cheap electricity and because global lending institutions have been willing to advance multibillion-dollar loans for these projects. According to the World Bank, developing countries will need to raise an estimated $US 100 billion by the year 2002 for hydroelectric plants currently in the planning stage. Given the environmental concerns of large dam construction, the trend for the future will probably be to build small-scale hydro plants that can generate electricity for a single community.

Environmental Effects of Hydroelectric Power

Hydroelectric power plants have many environmental impacts, some of which are just beginning to be understood. These impacts, however, must be weighted against the environmental impacts of alternative sources of electricity. Until recently, there was an almost universal belief that hydropower was a clean and environmentally safe method of producing electricity. Hydroelectric power plants do not emit any of the standard atmospheric pollutants, such as carbon dioxide or sulfur dioxide given off by fossil fuel-fired power plants. In this respect, hydropower is better than burning coal, oil, or natural gas to produce electricity, because it does not contribute to global warming or acid rain. Similarly, hydroelectric power plants do not carry the risk of radioactive contamination associated with nuclear power plants.

A few recent studies of large reservoirs created behind hydro dams have suggested that decaying vegetation, submerged by flooding, may give off quantities of greenhouse gases equivalent to those from other sources of electricity. If this turns out to be true, hydroelectric facilities such as the James Bay project in Quebec, Canada, that flood large areas of land might be significant contributors to global warming. Run-of-the-river hydro plants without dams and reservoirs would not be a source of these greenhouse gases.

The most obvious impact of hydroelectric dams is the flooding of vast areas

supply in Brazil and 80 percent in Paraguay. By means of comparison, Itaipu produces enough power to meet most of California's annual need. The lakes created by major dams also number among the planet's largest freshwater bodies. Ghana's Lake Volta covers 8500 square kilometers (3280 square miles), an area the size of Lebanon. Hydroelectric dam projects figure prominently in the economic and investment plans of many developing countries. Egypt electrified virtually all of its villages with power from Aswan. Since the 1950s, large-scale

Energy Development in China

China is the second-largest energy consumer in the world, following the United States (about 41 quadrillion British thermal units [Btu] in 2002 versus 92 quadrillion Btu in the United States). China's rapidly growing economy will drive energy demand growth of about 4 percent to 5 percent annually through 2015 (compared with growth of about 1 percent in the industrialized countries). China consumes about 10 percent of the world's energy and accounts for about 10 percent of the world's energy production.

China's relentless industrial development has outstripped its supply of power. The Chinese government forecasts shortages of 10 percent to 15 percent in key manufacturing areas, estimating that China needs about $108 billion worth of new generating capacity to close the gap. Factories are now cutting production, while hotels and restaurants dim lights and switch off heat. Many governmental leaders worry that the scarcity could hurt enterprises that have become critical employers at a time when millions of government-supported jobs are being eliminated by the country's transition from Communism to a free market economy.

China is now in the midst of a building boom in power plants, though most will not be completed until 2006 or 2008. The Three Gorges Dam, the largest hydroelectric project in history, is beginning to provide power, though it will not be completed before 2009. Even as these facilities come on-line, China confronts a fundamental energy deficit. Despite the vastness of its landmass, it has nowhere near enough reserves to meet its needs, necessitating a widening stream of imports.

Only a decade ago, China was a net exporter of oil. Now, it is the third-largest importer, behind only the United States and Japan. Imports leaped nearly a third during the first 10 months of 2003. By the government's figures, China will need as much as 600 million tons of oil annually by 2020, more than three times its expected output. More than half of China's oil imports now come from the Middle East, stoking fears in Beijing that the country's economic stability is overly dependent on access to supplies from the world's most volatile region.

In 2003, coal accounted for 70 percent of China's primary energy production. In the same year, petroleum accounted for 19 percent, hydroelectricity 11 percent, natural gas 2 percent, and nuclear power 0.5 percent. China's electricity is generated overwhelmingly by coal (about 68 percent), followed by hydroelectricity at about 20 percent. Because of China's extensive domestic coal resources and its wish to minimize dependence on foreign energy sources, it is expected that coal will remain the main energy source for electricity generation in China for the foreseeable future. Since 1979, the government has pursued a policy of replacing oil consumption with coal in the generation of electric power. The proportion of coal-fired electric-power generation has grown from 60 percent in 1980 to 70 percent in 2003.

High coal use has serious environmental repercussions. China plans to diversify by doubling its 1990 level of hydropower production by 2003 and increasing nuclear capacity. The Three Gorges Dam, when completed, will contribute significantly to the country's hydropower production.

One of the biggest concerns regarding energy consumption in China focuses on carbon emissions and the threat of increasing global warming. China's carbon emissions are expected to increase 4 percent annually through 2015, driven by fast economic growth and a rapid increase in coal use. The country's total carbon emissions should exceed 2 billion metric tons (2.2 tons) shortly after 2015, over three times the 1986 emissions and more than double the 1995 emissions.

The growing demand for power has driven reforms in China's energy-sector process and structure. The traditional, centralized regulatory regime is moving toward an increasingly decentralized and market-oriented regime.

Until recently, the Ministry of Electric Power (MOEP) regulated the electric power industry of China. The Ninth People's Congress held in 1998 introduced reform measures to streamline the central government. This resulted in the abolition of a number of industrial ministries, including MOEP. Much of the former MOEP's function has now been taken over by

of land, much of it previously forested or used for agriculture. The size of reservoirs created can be extremely large. The La Grande project in the James Bay region of Quebec has already submerged over 10,000 square kilometers (3861 square miles) of land; and if future plans are carried out, the eventual area of flooding in northern Quebec will be larger than the country of Switzerland. Reservoirs can be used for ensuring adequate water supplies, for providing irrigation, and for recreation. But in several cases, they have flooded the homelands of native peoples,

whose way of life has then been destroyed. Many rare ecosystems are also threatened by hydroelectric development.

Large dams and reservoirs can have other impacts on a watershed. Damming a river can alter the amount and quality of water in the river downstream of the dam, as well as prevent fish from migrating upstream to spawn. These impacts can be reduced by requiring minimum flows downstream of a dam and by creating fish ladders that allow fish to move upstream past the dam. Silt, normally carried downstream to the lower reaches of a river, is

trapped by a dam and deposited on the bed of the reservoir. This silt can slowly fill up a reservoir, decreasing the amount of water that can be stored and used for electrical generation. The river downstream of the dam is also deprived of silt, which fertilizes the river's flood plain during high-water periods.

Bacteria present in decaying vegetation can also change mercury, present in rocks underlying a reservoir, into a form that is soluble in water. The mercury accumulates in the bodies of fish and poses a health hazard to those who depend on the State Power Corporation of China (SPCC). SPCC's main duties in this respect are:

- Formulating China's electric-power development strategies, legislation, and policies, including investment policy, technical policy, and major energy production and consumption policies.
- Formulating unified energy-industry planning in collaboration with the State Development Planning Commission and other governmental agencies.
- Supervising the implementation of related national policies decrees and plans.

- Providing services to regional and provincial electric power enterprises.

Furthermore, SPCC owns all of China's state-owned power-generating assets, other than those owned by companies not directly managed by SPCC and a few smaller units directly owned by local governments.

Source: Data from Energy Information Administration, U.S. Department of Energy, *International Energy Outlook,* 2001. Source: History data from Energy Information Administration, *International Petroleum Statistics Report,* DOE/EIA–520 (92/08), August 1992, Washington, D.C. Projection from EIA, National Energy Modeling System, International Energy Module, 2000.

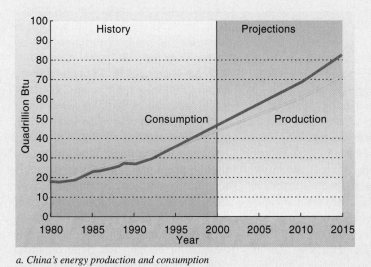

a. China's energy production and consumption

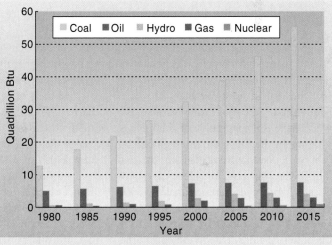

b. Energy production by fuel type, 1980–2015

these fish for food. The water quality of many reservoirs also poses a health hazard due to new forms of bacteria that grow in many of the hydro rivers. Therefore, run-of-the-river hydro plants generally have a smaller impact on the environment.

Tidal Power

In coastal areas with large tides, flowing tidal waters contain large amounts of potential energy. The principle of harnessing the energy of the tides dates back to eleventh-century England, when tides were used to turn waterwheels, producing mechanical power. More recently, rising and falling tides have been used to generate electricity in much the same manner as hydroelectric power plants.

Tides, the daily rise and fall of ocean levels relative to coastlines, are a result of the gravitational force of the moon and sun as well as the revolution of the Earth. The moon and the sun both exert a gravitational force of attraction on the Earth. The magnitude of the gravitational attraction of an object depends on the mass of an object and its distance. The moon exerts a larger gravitational force on the Earth because, although it is much smaller in mass, it is a great deal closer than the sun. This force of attraction causes the oceans, which make up 71 percent of the Earth's surface, to bulge along an axis pointing toward the moon. Tides are produced by the rotation of the Earth beneath this bulge in its watery coating, resulting in the rhythmic rise and fall of coastal ocean levels.

Certain coastal regions experience higher tides than others. This is a result of the amplification of tides caused by local

Figure 10.16 Tidal Generating Station La Rance River Estuary Power Plant in France is the world's largest tidal electrical generating station.

geographical features such as bays and inlets. To produce practical amounts of power, a difference between high and low tides of at least 5 meters (16 feet) is required. About 40 sites around the world have this magnitude of tidal range. The higher the tides, the more electricity can be generated from a given site and the lower the cost of electricity produced. Worldwide, approximately 3000 gigawatts (1 gigawatt = 1 billion watts) of energy are continuously available from the action of tides. Due to the constraints outlined above, it has been estimated that only 2 percent, or 60 gigawatts, can potentially be recovered for electricity generation.

The technology required to convert tidal energy into electricity is very similar to that used in traditional hydroelectric power plants. The first requirement is a dam or "barrage" across a tidal bay or estuary. Building such dams is expensive. Therefore, the best tidal sites are those where a bay has a narrow opening, thus reducing the length of dam required. At certain points along the dam, gates and turbines are installed. When the difference in the elevation of the water on the different sides of the barrage is adequate, the gates are opened. This "hydrostatic head"

that is created causes water to flow through the turbines, turning an electric generator to produce electricity.

Although the technology required to harness tidal energy is well established, tidal power is expensive, and only one major tidal generating station is in operation. This is a 240-megawatt (1 megawatt = 1 million watts) at the mouth of the La Rance river estuary on the northern coast of France (a large coal or nuclear power plant generates about 1000 megawatts of electricity). La Rance generating station has been in operation since 1966 and has been a very reliable source of electricity for France. (See figure 10.16.) La Rance was supposed to be one of many tidal power plants in France, until the nation's nuclear program was greatly expanded in the late 1960s. Elsewhere, there is a 20-megawatt experimental facility at Annapolis Royal in Nova Scotia, Canada, and a 0.4 megawatt tidal power plant near Murmansk, Russia.

Studies have been undertaken to examine the potential of several other tidal power sites worldwide. It has been estimated that a barrage across the Severn River in western England could supply as much as 10 percent of the country's electricity needs (12 megawatts). Similarly,

several sites in the Bay of Fundy, Cook Inlet in Alaska, and the White Sea in Russia have been found to have the potential to generate large amounts of electricity.

Environmental Impacts of Tidal Power

Tidal energy is a renewable source of electricity that does not result in the emission of gases responsible for global warming as does fossil-fuel–generated electricity. However, changing tidal flows by damming a bay or estuary could result in negative impacts on aquatic and shoreline ecosystems, as well as navigation and recreation.

Studies undertaken to identify the environmental impacts of tidal power have determined that each specific site is different and the impacts depend greatly on local geography. Local tides changed only slightly due to the La Rance barrage, and the environmental impact has been negligible, but this may not be the case for all other sites. It has been estimated that in the Bay of Fundy, tidal power plants could decrease local tides by 15 centimeters (6 inches). This does not seem like much when you consider that natural variations such as winds can change the level of tides by several meters.

Very little is understood about how altering the tides can affect incredibly complex aquatic and shoreline ecosystems. One fear is that enhanced mixing of water could be caused by tidal barrages in the Bay of Fundy, potentially stimulating the growth of the "red tide" organism *Gonyalaux excavata,* which causes paralysis in shellfish. Unfortunately, one of the only methods of increasing knowledge about how tidal barrages affect ecosystems may be to study the effects after such facilities have been built.

A new style of tidal power generator is being developed in the Dalupiri Ocean Power Plant in the Philippines. This new system is a submerged turbine, referred to as a hydro turbine. Similar to an underwater windmill, the hydro turbine will not obstruct fish migration, silt transport, or water flow, problems often associated with traditional tidal power generation facilities.

Changing tides in the area generate ocean currents that can turn hydro turbines. The hydro turbines will have slow-moving blades, pose little danger to marine life, and not impede the flow of silt and water. A protective fence prevents large marine mammals from approaching the turbines. It is estimated that a 1-kilometer (0.62-mile) "tidal bridge" made up of a series of these turbines could generate more electricity than a large nuclear plant. The completed project in the Philippines will span 4 kilometers (2.5 miles) and will include 274 turbines with a generating capacity of 2.3 gigawatts during peak tidal flow.

Geothermal Power

Geothermal energy flow from the Earth is not uniformly distributed throughout the Earth's surface area but is linked to geologically active areas where heat from the Earth can reach the surface through thinner crust. In particular, countries around the Pacific plate (in a region called the circum-Pacific "Ring of Fire"), mid-ocean ridges (such as Iceland), continental rift zones, and other hot spots experience significantly higher geothermal energy fluxes from the Earth's interior.

In areas where steam is trapped underground, **geothermal energy** is tapped by drilling wells to obtain the steam. The steam is then used to power electrical generators. At present, geothermal energy is practical only in areas where this hot mass is near the surface. (See figure 10.17.) Geothermal energy is not considered a true renewable energy source but rather an alternative one.

⑤ Steam from the hot water generates electricity.

① Recirculated water is pumped at high pressure 6.5 to 10 kilometers (4 to 6 miles) below the Earth's surface, where the temperature reaches up to 220°C (400°F) hotter than that at the Earth's surface.

④ Pipes recapture the hot water and carry it back to the Earth's surface.

③ "Hot rocks" heat the water as it percolates through the rock fissures.

② As the water is forced into the rock, the extreme pressure causes the rock to break.

Figure 10.17 Geothermal Power Plant Steam obtained from geothermal wells is used in the production of electricity. Figure b. shows how a geothermal plant operates.

a.

b.

The United States has about half of the world's geothermal electrical generating capacity. More than 130 generating plants are operating in 12 other countries. The Philippines, Italy, Mexico, Japan, New Zealand, and Iceland produce sizable amounts of electricity by geothermal methods.

California alone produces 40 percent of the world's geothermal electricity, about 1900 megawatts. One megawatt provides energy for about 1000 households. The Pacific Gas and Electric Company (PG&E) has been producing electricity from geothermal energy since 1960. PG&E's complex of generating units located north of San Francisco is the largest in the world and provides 700 megawatts of power, enough for 700,000 households, or 2.9 million people.

In addition to producing electricity, geothermal energy is used directly for heating. In Iceland, half of the geothermal energy is used to produce electricity and half is used for heating. In the capital, Reykjavik, all of the buildings are heated with geothermal energy at a cost that is less than 25 percent of what it would be if oil were used.

Geothermal energy creates some environmental problems. The steam contains hydrogen sulfide gas, which has the odor of rotten eggs and is an unpleasant form of air pollution. (The sulfides from geothermal sources can, however, be removed.) The minerals in the steam corrode pipes and equipment, causing maintenance problems. The minerals are also toxic to fish.

All objects contain heat energy, which can be extracted from and transferred to other locations. A refrigerator extracts heat from the interior and exports it to the coils on the back of the unit. Similarly, pipes placed within the Earth can extract heat and transfer it to a home. It is important to recognize that such an application of geothermal energy requires the expenditure of energy through an electric-powered device. In some instances, obtaining heat in this manner is less expensive than other traditional sources, such as oil or natural gas, but it is not a "free" source of heat. Geothermal energy use, however, does conserve fossil fuels and does not contribute to global warming.

Electricity from the Ground Up

The naturally occurring heat beneath the planet's surface powers volcanoes, hot springs, and geysers such as Yellowstone National Park's Old Faithful. In a few places in the world, such as Iceland and parts of California, natural heat and natural water come together in sufficient quantities to provide energy. In most places where there is abundant subsurface water, useful heat is 5 kilometers (3 miles) or more below the surface, beyond the reach of economical drilling technology. In places where heat is within reach, there is often no subsurface water. (See figure 10.17b.)

Searching for cheap, nonpolluting energy, scientists at the Los Alamos National Laboratory in New Mexico have begun a process to "mine heat." Mining heat is a new effort to use geothermal energy. In this process, water is pumped at high pressure 3 kilometers (1.9 miles) down into the Earth through a well. When the water comes back up through a parallel well, it has been heated beyond the boiling point by the Earth's natural heat. The Hot Dry Rock Project is an attempt to bring heat and water together as a geothermal energy source.

On average, temperatures below the Earth's surface increase about 27°C per kilometer (80°F per mile). In much of the western United States, however, residual heat from ancient volcanoes increases temperatures by more than 69°C per kilometer (200°F per mile). At the Los Alamos site, the temperature at the base of the well is about 240°C (430°F).

According to researchers at Los Alamos, a fairly conservative estimate is that there are at least 500,000 quads (quadrillion Btus) of useful heat in hot dry rock at accessible drilling depths beneath the United States. This is about 6000 times the total amount of energy used in the country in one year.

The fundamental question now being asked is if the energy can be developed economically for commercial use. At this point, the answer is no; however, if new technology is developed, the answer could change. Other questions also must be addressed: Will the heat in a particular well last long enough to justify the cost? Can the system be sealed to minimize water leakage? Can a constant flow of water be sustained?

Even if it proves commercially competitive, hot dry rock mining might encounter obstacles involving site access, hookups to distant power grids, and legal disputes over water rights. If such problems can be overcome, maybe one day the energy used to run your computer will come from water heated kilometers beneath the surface of the Earth.

Wind Power

As the sun's radiant energy strikes the Earth, that energy is converted into heat, which warms the atmosphere. The Earth is unequally heated because various portions receive different amounts of sunlight. Since warm air is less dense and rises, cooler, denser air flows in to take its place. This flow of air is wind. For centuries, wind has been used to move ships, grind grains, pump water, and do other forms of work. In more recent times, wind has been used to generate electricity. (See figure 10.18.)

In 2003, California was using 15,800 wind turbines to produce 390 megawatts of electricity at a cost competitive with newly constructed coal or nuclear plants. By 2010, according to the American Wind Energy Association, wind could power 10 million U.S. households. The U.S. Department of Energy has stated that the Great Plains could supply 48 states with 75 percent of their electricity. Wind power is now considered competitive with new coal and natural gas plants and cheaper than nuclear plants.

Globally, wind power use has multiplied nearly fourfold over the last five years, a growth rate matched only by the computer industry. In the United States, the American Wind Energy Association projects a 60 percent growth in wind-generating capacity by 2005.

Wind power was once confined to California, but since 2000, wind farms coming on-line in Minnesota, Iowa, Texas, Colorado, Wyoming, Oregon, and Pennsylvania have boosted U.S. capacity

Figure 10.18 **Wind Energy** Fields of wind-powered generators such as these in California can produce large amounts of electricity.

by half, from 1680 megawatts to 2550 megawatts. A 300-megawatt wind farm on the Oregon-Washington border is currently the world's largest.

The Bonneville Power Administration (BPA) indicated in 2003 that it wanted to buy 1000 megawatts of wind-generating capacity and requested proposals. Much to its surprise, it received enough to build 2600 megawatts of capacity in five states, with the potential of expanding these sites to over 4000 megawatts. BPA, which may accept most of these proposals, expects to have at least one site on-line by the end of 2005.

A 3000-megawatt wind farm in the early planning stages in South Dakota, near the Iowa border, is 10 times the size of the Oregon-Washington wind farm. Named Rolling Thunder, this project is designed to feed power to the midwestern region around Chicago. This proposed project is not only large by wind power standards, it is one of the largest energy projects of any kind in the world today.

Advances in wind turbine technology, drawing heavily from the aerospace industry, have lowered the cost of wind power from 38 cents per kilowatt hour in the early 1980s to 3 cents to 6 cents in 2004, depending on the wind site. Wind, now competitive with fossil fuels, is already cheaper in some locations than oil- or gas-fired power. With major corporations such as ABB and Shell International, putting

resources into this field, further cost cuts are likely.

Today, Denmark, a world leader in wind turbine technology and manufacture, is getting 15 percent of its electricity from wind power. For Schleswig-Holstein, the northernmost state of Germany, it is 19 percent, and for some parts of the state, 75 percent. Spain's industrial state of Navarra, starting from scratch in 1998, now gets 24 percent of its electricity from wind.

As wind-generating costs fall and as concern about climate change escalates, more and more countries are developing wind energy programs. In 2003, France announced it will develop 5000 megawatts of wind power by 2010. Also in 2003, Argentina announced a plan to develop 3000 megawatts of wind power in Patagonia by 2010. In April 2003, the United Kingdom accepted bids for 1500 megawatts of wind power. In May of the same year, a report from Beijing indicated that China plans to develop some 2500 megawatts of wind power by 2005.

The growth in wind power is consistently outrunning earlier estimates. The European Wind Energy Association, which in 1966 had set a target of 40,000 megawatts for Europe in 2010, recently upped the target to 60,000 megawatts.

A push for energy deregulation and concerns about smog, acid rain, and global warming are driving policy makers to

require utilities to sell electricity from renewable sources. Eight states—Texas, Wisconsin, Massachusetts, Connecticut, New Jersey, Minnesota, Nevada, and Pennsylvania—require utilities to provide some "green" electricity. At least 36 utilities include wind energy as a component of their green power programs.

Most U.S. wind sources remain untapped. Texas and the Dakotas alone have enough wind to power the nation, but that's not likely to happen. Variable wind speeds make it unreliable as a primary energy source; energy companies and regulators view it as supplementary to fossil fuels. While places such as the Dakotas have the strongest winds, they are remote from energy-using population centers and lack suitable transmission grids.

As U.S. companies begin building 1- or 2-megawatt wind turbines, European firms are exploring 5-megawatt machines. In inner Mongolia, nomadic herders carry small, portable, wind-driven generators that provide electricity for light, television, and movies in their tents as well as for electric fences to contain their animals.

A steady and dependable source of wind makes the use of wind power more productive in some regions than others. Wide open areas, such as the Great Plains in North America, are better suited for wind power than are heavily wooded areas. Wind-generated electricity is usually used in conjunction with other sources of electricity that take over when the wind does not blow.

Wind generators do have some negative effects. The moving blades are a hazard to birds and produce a noise that some find annoying. Newer windmills, however, have slower moving rotors that many birds such as the golden eagle find easier to avoid. Vibrations from the generators can also cause structural problems. In addition, some people consider the sight of a large number of wind generators, such as are found in California, to be visual pollution.

Since 1998, wind power has been the fastest-growing new source of electricity in the world, expanding an average of 30 percent a year. The value of the world's electrical power generated from renewable sources such as wind and solar is about $9 billion—up from less than $1 billion a decade ago but still a small fraction of the total electricity market. The market for renewable energy is expected to exceed $100 billion by 2010 as technological advances lower the price and make renewables easier to use.

Solar Energy

The sun is often mentioned as the ultimate answer to the world's energy problems. It provides a continuous supply of energy that far exceeds the world's demands. In fact, the amount of energy received from the sun each day is 600 times greater than the amount of energy produced each day by all other energy sources combined. The major problem with solar energy is its intermittent nature. It is available only during the day and when it is sunny. All systems that use solar energy must store energy or use supplementary sources of energy when sunlight is not available. Because of differences in the availability of sunlight, some parts of the world are more suited to the use of solar energy than others. Solar energy is also very diffuse, which is another disadvantage of its use.

Solar energy is utilized in three ways:

1. In a passive heating system, the sun's energy is converted directly into heat for use at the site where it is collected.
2. In an active heating system, the sun's energy is converted into heat, but the heat must be transferred from the collection area to the place of use.
3. The sun's energy also can be used to generate electricity, which may be used to operate solar batteries or may be transmitted along normal transmission lines.

Passive Solar Systems

Anyone who has walked barefoot on a sidewalk or blacktopped surface on a sunny day has experienced the effects of passive solar heating. In a **passive solar system,** light energy is transformed to heat energy when it is absorbed by a surface. Some of the earliest uses of passive solar energy were to dry food and clothes and to evaporate seawater to produce salt. Homes and buildings may now be designed to use passive solar energy for heating. (See figure 10.19.)

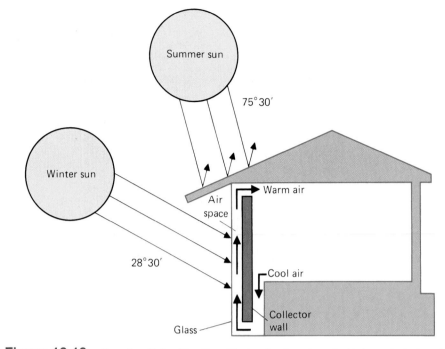

Figure 10.19 **Passive Solar Heating** The length of overhang in this home is designed for solar heating at the latitude of St. Louis, Missouri (38°N). In this design, a wall 30 to 40 centimeters (12 to 16 inches) thick is used to collect and store heat. The collector wall is located behind a glass wall and faces south. During a midwinter day, when the sun's angle is 28 degrees, light energy is collected by the wall and stored as heat. At night, the heat stored in the wall is used to warm the house. Natural convection causes the air to circulate past the wall, and the house is heated. During a midsummer day, when the sun's angle is 75 degrees, the overhang shades the collector wall from the sun.

The south side of a building always receives the most sunlight. Therefore, buildings designed for passive solar heating usually have large, south-facing windows. Materials that absorb and store the sun's heat can be built into the sunlit floors and walls. The floors and walls heat up during the day and slowly release heat at night, when the heat is needed most. This passive solar design feature is called *direct gain.*

Other passive solar heating features include sunspaces and trombe walls. A *sunspace* (which is much like a greenhouse) is built on the south side of a building. As sunlight passes through glass or other glazing, it warms the sunspace. Proper ventilation allows the heat to circulate into the building. On the other hand, a *trombe wall* is a very thick, south-facing wall painted black and made of a material that absorbs a lot of heat. A pane of glass or plastic glazing, installed a few centimeters in front of the wall, helps hold in the heat. The wall heats up slowly during the day; then, as it cools gradually during the night, it gives off its heat inside the building.

Many of the passive solar heating design features also provide daylighting. *Daylighting* is simply the use of natural sunlight to brighten a building's interior. To lighten north-facing rooms and upper levels, a clerestory—a row of windows near the peak of the roof—is often used along with an open floor plan inside that allows the light to bounce throughout the building.

Of course, too much solar heating and daylighting can be a problem during hot summer months. There are design features that can help keep passive solar buildings cool in the summer. For instance, overhangs can be designed to shade windows when the sun is high in the summer. Sunspaces can be closed off from the rest of the building. And a building can be designed to use fresh-air ventilation in the summer.

Active Solar Systems

An **active solar system** requires a solar collector, a pump, and a system of pipes to transfer the heat from the site of

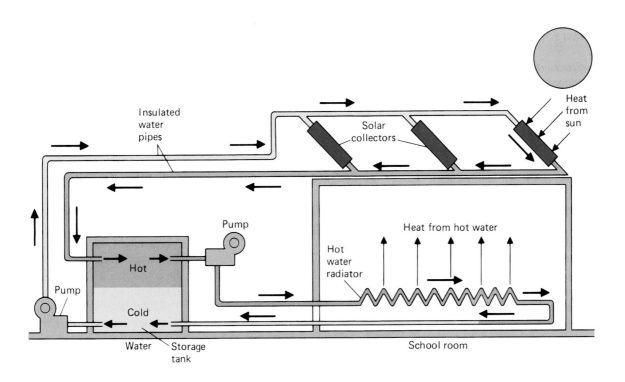

a.

production to the area to be heated. (See figure 10.20a.)

Active solar collector systems take advantage of the sun to provide energy for domestic water heating, pool heating, ventilation air preheating, and space heating. Water heating for domestic use is generally the most economical application of active solar systems. The demand for hot water is fairly constant throughout the year, so the solar system provides energy savings year-round. Successful use of solar water heating systems requires careful selection of components and proper sizing.

An active solar water heating system can be designed with components sized large enough to provide heating for pools or a combined function of heating both domestic water and space. Space heating requires a heat-storage system and additional hardware to connect with a heat distribution system. An active solar space heating system makes economic sense if it can offset considerable amounts of heating energy from conventional systems over the life of the building or the system.

Active solar systems are most easily installed in new buildings, but in some cases can be installed in existing structures. A major consideration in the use of

b.

Figure 10.20 **Solar Heating Designs** (a) An active solar system requires a solar collector, a pump, and a system of pipes. (b) In some remote areas, solar energy is an economical method of electricity production.

an active solar system is the initial cost of installation. An active system requires a specially designed collector, consisting of a series of liquid-filled tubes; a pump; and pipes to transfer warm liquid from the collector to the space to be heated. Because an active system has moving parts, it also

has operation and maintenance costs like any other heating or cooling system.

Rock, water, or specially produced products are used to store heat. The hot liquid in the pipes heats the storage medium, which releases its heat when the sun is not shining.

generate power for as little as 5 cents per kilowatt-hour. In co-firing, biomass fuel can cost less than coal when low-cost biomass fuels are used; modifications to the coal plant can have payback periods of two to three years.

With an estimated 14,000 megawatts of annual worldwide installed generation capacity, biomass power is the largest source of nonhydropower renewable electricity in the world. The United States is the largest biomass power generator, representing a $15 billion investment and 66,000 jobs. In the future, the need for on-site industrial power, waste reduction, stricter environmental regulations, and rising consumer demand for renewable energy will provide the main impetus for the industry's growth.

Worldwide biomass power generation is expected to grow more than 30,000 megawatts by 2020. Developing countries are the top markets because of rapid economic growth, burgeoning demand for electricity, mounting environmental problems, and the need for rural electrification and reliable electricity. Most of them also have significant quantities of agricultural or forestry residues.

Recent studies indicate that additional (presently unused) quantities of economically available biomass may exceed 39 million tons per year in the United States—enough to supply about 7500 megawatts of new biomass power, a doubling of the existing U.S. capacity.

Biomass Conversion Technologies

There exist several technologies capable of converting biomass into energy. These include anaerobic digestion, pelletising, direct combustion and co-generation, pyrolysis, gasification, and ethanol production.

Anaerobic Digestion

Anaerobic digestion is the decomposition of wet and green biomass, through bacterial action in the absence of oxygen, to produce a mixed gas output of methane and carbon dioxide known as biogas. The anaerobic digestion of municipal solid waste buried in landfill sites produces a gas known as landfill gas. This process occurs naturally as the bacterial decompo-

sition of the organic matter continues over time. The methane gas produced in landfill sites eventually escapes into the atmosphere. However, the landfill gas can be extracted by inserting perforated pipes into the landfill. In this way, the gas will travel through the pipes, under natural pressure, to be used as an energy source, rather than simply escaping into the atmosphere to contribute to greenhouse gas emissions.

Pelletising

Pelletising involves the compaction of biomass at high temperatures and very high pressures. The biomass particles are compressed in a die to produce briquettes or pellets. These products have significantly smaller volume than the original biomass and thus have a higher volumetric energy density, making them a more compact source of energy. They are also easier to transport and store than natural biomass. The pellets can be used directly on a large scale as direct combustion feed or on a small scale in domestic stoves or wood heaters.

Direct Combustion and Cogeneration

Direct combustion is the main process adopted for utilizing biomass energy. The energy produced can be used to provide heat or steam for cooking, space heating, and industrial processes, or for electricity generation. Large biomass power-generation systems can have comparable efficiencies to fossil-fuel systems, but this comes at a higher cost due to the design of the burner to handle the higher moisture content of biomass. However, by using the biomass in a combined heat- and electricity-production system (or cogeneration system), the economics are significantly improved.

Pyrolysis

Pyrolysis is the basic thermochemical process for converting solid biomass to a more useful liquid fuel. Biomass is heated in the absence of oxygen, or partially combusted in a limited oxygen supply, to produce a hydrocarbon-rich gas mixture, an oil-like liquid, and a carbon-rich solid

residue. Traditionally, in developing countries, the solid residue produced is charcoal, which has a higher energy density than the original fuel. The traditional charcoal kilns are simply mounds of wood covered with earth or pits in the ground. However, the process of carbonization is very slow and inefficient in these kilns, and more sophisticated kilns are replacing the traditional ones. The pyrolitic residue or "bio-oil" produced can be easily transported and refined into a series of products. The process is similar to refining crude oil.

Gasification

Gasification is a form of pyrolysis, carried out with more air and at high temperatures, to optimize the gas production. The resulting gas, known as producer gas, is a mixture of carbon monoxide, hydrogen and methane, together with carbon dioxide and nitrogen. The gas is more versatile than the original solid biomass, and it can be used as a source of heat or used in internal combustion engines or gas turbines to produce electricity. During the Second World War, countries such as Australia and Germany even used it to power vehicles.

Ethanol Production

Ethanol can be produced from certain biomass materials that contain sugars, starch, or cellulose. The best-known feedstock for ethanol production is sugarcane, but other materials can be used, including wheat, corn, other cereals, and sugar beets. Starch-based biomass is usually cheaper than sugar-based materials but requires additional processing. Similarly, cellulose materials, such as wood and straw, are readily available but require expensive preparation.

Ethanol is produced by a process known as fermentation. Typically, sugar is extracted from the biomass crop by crushing and mixing with water and yeast, and then keeping the mixture warm in large tanks called fermenters. The yeast breaks down the sugar and converts it to ethanol. A distillation process is required to remove the water and other impurities from the dilute alcohol product. Brazil has a successful industrial-scale ethanol project, which produces ethanol from sugarcane for blending with gasoline. In the United States, corn is used for ethanol production and then blended with gasoline to produce "gasohol."(See figure 10.23.)

Environmental Impacts of Biomass Conversion

A consensus exists among scientists that biomass fuels used in a sustainable manner result in no net increase in atmospheric carbon dioxide. Some scientists would even go as far as to declare that sustainable use of biomass would result in a net decrease in atmospheric carbon dioxide. This is based on the assumption that all the carbon dioxide given off by the use of biomass fuels was recently taken in from the atmosphere by photosynthesis. Increased substitution of fossil fuels with biomass fuels would therefore help reduce the potential for global warming, caused by increased atmospheric concentrations of carbon dioxide.

Things may not be as simple as has been assumed. The number of people globally who face hunger and starvation continues to grow. It has been argued that the use of land to grow fuel crops will increase this problem. The cause of hunger in developing countries, however, is more complex than just a lack of agricultural land. Many countries in the world today, such as the United States, have food surpluses. Much fertile agricultural land is also used to grow tobacco, flowers, food for domestic pets, and other "luxury" items, rather than staple foods. Similarly, a significant proportion of agricultural land is used to grow feed for animals to support the meat-centered diet of the industrialized world. By feeding grain to livestock, we end up with only about 10 percent of the caloric content of the grain. When looked at in this light, it does not seem to be so unreasonable to use some fertile land to grow fuel. Marginal land and underutilized agricultural land can also be used to grow biomass for fuel.

Another issue surrounding biomass energy may be the loss of biodiversity. Transforming natural ecosystems into energy plantations with a very small number of crops, as few as one, can reduce the biodiversity of a region. Such "monocul-

Figure 10.23 **Gasohol** Gasohol is available for purchase in many parts of North America.

Figure 10.24 **Desertification** The demand for fuelwood in many regions has resulted in the destruction of forests. This is a major cause of desertification.

tures" lack the balance achieved by a diverse ecosystem and are susceptible to widespread damage by pests or disease.

Fuelwood

In less-developed countries, wood has been the major source of fuel for centuries. In fact, wood is the primary source of energy for nearly half of the world's population. In these regions, the primary use of wood is for cooking.

The use of wood as a prime energy source, a rapid population increase, and the high cost of other types of fuel have combined to create some serious environmental problems in many areas of the world. It is estimated that 1.3 billion people are not able to obtain enough wood or must harvest wood at a rate that exceeds its growth. This has resulted in the destruction of much forest land in Asia and Africa and has hastened the rate of desertification in these regions. (See figure 10.24.)

Because of its bulk and low level of energy compared to equal amounts of coal or oil, wood is not practical to transport over a long distance, so most of it is used locally. In the United States, Norway, and Sweden, wood furnishes 10 percent of the energy for home heating. Canada obtains 3 percent of its total energy, not just home-heating energy, from wood. Most of this energy is used in forest product industries, such as lumbering and paper mills.

Burning wood is also a source of air pollution. Studies indicate that more than 75 organic compounds are released when wood is burned, 22 of which are hydrocarbons known or suspected to be carcinogens. Often, woodstoves are not operated in the most efficient manner, and high amounts of particulate matter and other products of incomplete combustion, such as carbon monoxide, are released, contributing to ill health and death.

In areas with a high population density, the heavy use of wood releases large amounts of fly ash into the air. In Missoula, Montana, in recent years, 55 percent of the particles in the air during the summer were from burning wood. In the winter, wood was responsible for 75 percent of the particles. A number of steps have been taken to reduce air pollution resulting from burning wood. Some cities, such as London, England, have a total ban on burning wood. Vail, Colorado, permits only one wood-burning stove per dwelling. Many areas require woodstoves to have special pollution controls that reduce the amount of particulates and other pollutants released.

Solid Waste

Residents of New York City discard in excess of 25,000 metric tons of waste each day, which is 1.8 kilograms (4 pounds) per person. In fact, New Yorkers lead the world in the production of municipal trash. High-income cities such as New York usually produce more waste than low-income cities. (See table 10.2.) About 80 percent of this waste is combustible and, therefore, represents a potential energy source. (See figure 10.25.)

"Trash power," the use of municipal waste as a source of energy, requires several steps. First, the waste must be sorted

Table 10.2 Amount of Solid Waste Produced per Capita	
High-Income Cities	**Kilograms per Day**
New York	1.8
Singapore	1.6
Tokyo	0.94
Rome	0.72
Low-Income Cities	
Tunis, Tunisia	0.56
Medellin, Colombia	0.54
Calcutta, India	0.51
Kano, Nigeria	0.46

Figure 10.25 **Waste to Energy** Municipal trash can be burned to produce heat and electricity. This refuse pit is used to feed hoppers of high-temperature furnaces.

so that the burnable organic material is separated from the inorganic material. The sorting is accomplished most economically by the person who produces the waste. That means the producer must separate trash before putting it out for pickup. It must be separated into garbage, burnable materials, glass, and metals, and picked up by compartmentalized collection trucks. Second, the efficient use of waste as a source of fuel requires a large volume and a dependable supply. If a community constructs a facility to burn 200 metric tons of waste a day, the community must generate and collect 200 metric tons of waste each day.

Communities have been burning trash as a means of reducing the volume of waste for a number of years. The first consolidated incineration of waste was done in Nottingham, England, in 1874. Burning the municipal waste of Munich, Germany, not only reduces the volume of the waste but supplies the energy for 12 percent of the city's electricity. Rotterdam, the Netherlands, operates a 55-megawatt power plant from its garbage. In the United States, only 3 percent of household waste is burned. This is low compared to the 26 percent burned in Japan, 51 percent in Sweden, and 75 percent in Switzerland.

Burning trash is not profitable. It is a way to decrease the cost of trash disposal because it reduces the need for landfill sites. Baltimore produces gas from 1000 metric tons of waste per day. This gas generates steam for the Baltimore Gas and Electric Company. The daily waste also yields 80 metric tons of a charcoal-like material (char), 70 metric tons of ferrous metals, and 170 metric tons of glass. These products all have potential uses. The char can be burned as an additional source of energy. The ferrous metals can be sold, reducing the cost of operating the plant, as well as conserving mineral resources. The glass can be sold for recycling, further reducing the costs of the plant.

Although the burning of trash reduces the trash volume and furnishes energy, it poses environmental concerns, one of which is air pollution. Many of the older incinerating plants do not comply with today's air-quality standards. Also, much of the waste material, such as bleached paper and plastics, have chlorine-containing organic compounds. When burned, these compounds can form dioxins, which are highly toxic and suspected carcinogens. Another problem associated with waste-to-energy systems is the popularity of recycling. Many of the items that are now recycled, such as plastics and wood, have high heat content. The reduction in the amount of these items in the waste stream reduces its value as an energy source.

Energy Conservation

Many observers have pointed out that demanding more energy while failing to conserve is like demanding more water to fill a bathtub while leaving the drain open. To be sure, conservation and efficiency strategies by themselves will not eliminate demands for energy, but they can make the demands much easier to meet, regardless of what options are chosen to provide the primary energy.

Energy efficiency and conservation is a quiet but effective energy resource, contributing substantially to the United States' economic growth and increased standard of living over the past 30 years. Energy efficiency improvements since 1973 accounted for approximately 25 quadrillion Btus in 2003, which is about 26 percent of U.S. energy use and more energy than we now get annually from coal, natural gas, or domestic oil sources. Consider these facts, which are based primarily on data published by the federal Energy Information Administration:

- Total primary energy use per capita in the United States in 2003 was almost identical to that in 1973. Over the same 30-year period, economic output (gross domestic product, or GDP) per capita increased 74 percent.
- National energy intensity (energy use per unit of GDP) fell 43 percent between 1973 and 2002. About 60 percent of this decline is attributable to real energy efficiency improvements, and about 40 percent is due to structural changes in the economy and fuel switching.
- If the United States had not dramatically reduced its energy intensity over the past 30 years, consumers and businesses would have spent at least $430 billion more on energy purchases in 2003.
- Between 1996 and 2002, GDP increased 21 percent while primary energy use increased just 2 percent. Imagine how much worse our energy problems would be today if energy use had increased 10 percent or 20 percent during 1996–2002.

Even though the United States is much more energy-efficient today than it was 25 years ago, the potential is still enormous for additional cost-effective energy savings. Some newer energy efficiency measures have barely begun to be adopted. Other efficiency measures could be developed and commercialized in coming years.

The typical family in the United States spends close to $1300 a year on their home's utility bills. Unfortunately, a large portion of that energy is wasted. The amount of energy wasted through poorly insulated windows and doors alone is about as much energy as the United States receives from the Alaskan pipeline each year. The electricity generated by fossil fuels for a single home puts more carbon dioxide into the air than two mid-sized cars. It is estimated that by using inexpensive, energy-efficient measures, the average energy bills of a single home could be reduced by 10 percent to 50 percent and help reduce the emissions of carbon dioxide into the atmosphere.

Much of the energy we consume is wasted. This statement is not meant as a reminder to simply turn off lights and lower furnace thermostats; it is a technological challenge. Our use of energy is so inefficient that most potential energy in fuel is lost as waste heat, becoming a form of environmental pollution.

Many conservation techniques are relatively simple and highly cost effective. More efficient and less energy-intensive industry and domestic practices could save large amounts of energy. Improved automobile efficiency, better mass transit, and increased railroad use for passenger and freight traffic are simple and readily available means of conserving transportation energy. In response to the 1970s' oil price shocks, automobile gas-mileage averages in the United States more than doubled, from 5.55 kilometers per liter (13 miles per gallon) in 1975 to 12.3 kilometers per liter (28.8 miles per gallon) in 1988. Unfortunately, the oil glut and falling fuel prices of the late 1980s discouraged further conservation. Between 1990 and 1997, the average slipped to only 11.8 kilometers per liter (27.6 miles per gallon). It remains to be seen if the sharp increase of gasoline in the early years of the twenty-first century will translate into increased miles per gallon in new car design.

Conservation is not a way of generating energy, but it is a way of reducing the need for additional energy consumption and saves money for the consumer. Some conservation technologies are sophisticated, while others are quite simple. For example, if a small, inexpensive wood-burning stove were developed and used to replace open fires in the less-developed world, energy consumption in these regions could be reduced by 50 percent.

Several technologies that reduce energy consumption are now available. (See figure 10.26.) Highly efficient fluorescent light bulbs that can be used in regular

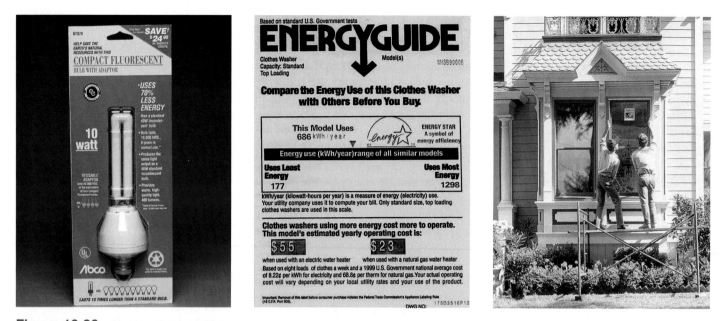

Figure 10.26 Energy Conservation The use of fluorescent light bulbs, energy-efficient appliances, and low-emissive glass could reduce energy consumption significantly.

incandescent fixtures give the same amount of light for 25 percent of the energy, and they produce less heat. Since lighting and air conditioning (which removes the heat from inefficient incandescent lighting) account for 25 percent of U.S. electricity consumption, widespread use of these lights could significantly reduce energy consumption. Low-emissive glass for windows can reduce the amount of heat entering a building while allowing light to enter. The use of this glass in new construction and replacement windows could have a major impact on the energy picture. Many other technologies, such as automatic dimming devices or automatic light-shutoff devices, are being used in new construction.

The shift to more efficient use of energy needs encouragement. Often, poorly designed, energy-inefficient buildings and machines can be produced inexpensively. The short-term cost is low, but the long-term cost is high. The public needs to be educated to look at the long-term economic and energy costs of purchasing poorly designed buildings and appliances.

Electric utilities have recently become part of the energy conservation picture. In some states, they have been allowed to make money on conservation efforts; previously, they could make money only by building more power plants. This encourages them to become involved in energy conservation education, because teaching their customers how to use energy more efficiently allows them to serve more people without building new power plants.

The Hydrogen Economy

Hydrogen economy is the term coined to mark the shift from fossil fuels such as coal, oil, and gas to hydrogen. All fuel cells are fueled by some form of hydrogen. Hydrogen is abundant and generates heat and pure water when it reacts with air. The vision of a hydrogen economy is one of an unlimited source of fuel that generates energy without releasing carbon and other pollutants into the atmosphere. As fossil fuels become rarer and more expensive to extract and burn, and as new technological breakthroughs bring down the cost of using hydrogen in fuel cells, we move closer to a day when the hydrogen economy will become a reality (See table 10.3).

Table 10.3 Key Drivers Affecting Hydrogen Energy Development

Supports	Challenges	Both supports and challenges
National security and the need to reduce oil imports	Inability to build and sustain national consensus on energy policy priorities	Rapid pace of technological change in hydrogen and competing energy sources and technologies
Global climate change and the need to reduce greenhouse gas emissions and pollution	Lack of hydrogen infrastructure and substantial costs of building one	Current availability of relatively low-cost fossil fuels, but the inevitable depletion of these resources
Global population and economic growth and the need for clean energy supplies at affordable prices	Lack of commercially available, low-cost hydrogen production, storage, and conversion devices, such as fuel cells	Simultaneous consumer preference for a clean environment and affordable energy supplies
Air quality and the need to reduce emissions from vehicles and power plants	Hydrogen safety issues	

Source: Courtesy of U.S. Department of Energy (2002).

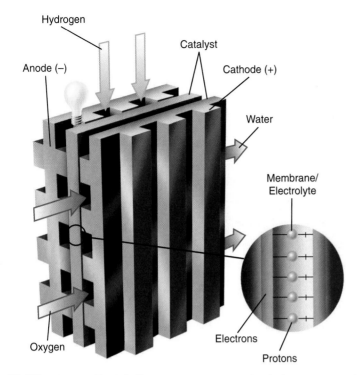

Figure 10.27 **Simple Fuel Cell** A fuel cell is an electrochemical energy conversion device.

At the forefront of this development are hydrogen fuel cells, which, although first conceived over a century ago, are now beginning to rise in popularity due to their high efficiency and low emissions. (See figure 10.27.)

Fuel cells are often likened to batteries. The primary difference between fuel cells and batteries, however, is that fuel cells do not need to be recharged; while batteries require a long recharging period and release toxic heavy metals when disposed of, fuel cells run continuously if provided with adequate fuel input. This fuel may take the form of natural gas, methanol, gasoline, or pure hydrogen—almost anything from which hydrogen may be extracted.

Although numerous types of fuel cells are available, the most common is the proton exchange membrane (PEM) fuel cell. Although the specific reactions within the fuel cell involve several steps, they can be summarized as follows. Pressurized hydrogen gas enters the fuel cell and comes in contact with a platinum catalyst that causes the hydrogen molecules (H_2) to split into hydrogen ions (H^+), also known as protons, and electrons. A proton

exchange membrane allows protons (H^+)—but not the electrons—to flow through it. The electrons instead flow through an electric circuit to do work such as generating light or running a motor. The hydrogen ions flow through the proton exchange membrane and recombine with the electrons that have passed through the circuit and with oxygen to form water.

Numerous obstacles are delaying the process of fuel cell development. Most prominently, hydrogen used to power the cell is difficult to obtain in a pure state. While pure hydrogen gas produces the highest efficiency and lowest emissions, it is highly flammable, difficult to store, and not available to consumers as readily as oil-derived fuels. The hydrogen may be extracted from hydrocarbons through a process known as fuel reformation, but this results in slightly impure hydrogen, decreasing the efficiency of the fuel cell. Fuel reformation also releases carbon dioxide, nitrous oxide, and particulate matter, although in vastly lesser quantities than are released by traditional electricity production techniques. Manufacturing costs are currently high because

the economies of scale experienced with mass production have not yet developed.

Fuel cells have been developed for specialized use in numerous fields. The National Aeronautics and Space Administration developed cells for use aboard spacecrafts in the 1960s due to their size, weight, and nontoxic emissions, which, as an added bonus, provided pure drinking water for the crew. Some schools, banks, and offices have installed fuel cells to provide electricity. While large-scale power generation and industrial applications have not yet been developed, fuel cells may soon be adapted for these purposes, and residential applications may soon follow.

The advantages of fuel cells over other power sources are numerous. With a low operating temperature and minimal noise generation, fuel cells are safe to install in semiexposed areas. Cells are also self-sustaining, making them ideal for locations far removed from traditional electricity production facilities. Because they can operate separately from the power lines associated with most electricity distribution systems, the risk of weather-related power outages is also diminished. Their most important benefit, however, is that they do not pollute. When fueled by pure hydrogen, water, heat, and electricity are the only by-products. Although currently powered by modified fossil fuels, fuel cells are capable of extracting far greater quantities of energy from each unit of fuel than other methods of power generation, thereby reducing the need for fossil fuel extraction.

One of the most significant prospects for future development of fuel cells is in the automotive industry. Although there are some technical difficulties arising from adapting fuel cells for vehicular rather than stationary use, many leading automotive companies agree that fuel cells will be the successor to the internal combustion engine. Experimental buses are already on the road in selected European and American cities. (See figure 10.28.) Similar cars are expected to follow.

Although the widespread introduction of hydrogen energy systems in any sector will clearly take a long time, specific plans are developing in several countries. The

Figure 10.28 **Experimental Fuel Cell Bus** Citaro Fuel Cell bus is the latest model with fuel cells from DaimlerChrysler AG in Germany. This picture shows one of 30 buses which were delivered to 10 European cities in 2003.

Japanese have stated that they plan on having 50,000 hydrogen fuel cell vehicles on the road by 2010. The U.S. Department of Energy is planning on 5000 vehicles in the same time frame. Whether these and other targets can be achieved depends on several factors, including government policy (local, national, and international) on issues such as carbon dioxide emissions, local pollutants, energy diversity, and energy security. On the technological side, the automotive companies need to achieve desired performance and cost targets for hydrogen vehicles, as well as address the issue of developing an infrastructure for the refueling of the gas. In 2003, the world's first commercial hydrogen filling station for cars and buses opened in Iceland under a European Union–sponsored environmental project. The station will initially serve three DaimlerChrysler hydrogen-powered buses that will run on regular routes in Iceland's capital, Reykjavik.

GLOBAL PERSPECTIVE

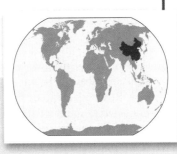

The Three Gorges Dam

A project that began in late 1997 is already being compared to the building of the pyramids and the Great Wall of China. China's Three Gorges Dam across the world's third longest river, the Yangtze, is a project that when completed will be visible from the moon and will be the largest dam in the world. The Three Gorges Dam will stretch 2 kilometers (1.3 miles) across the Yangtze River, tower 185 meters (610 feet) into the air, and create a 600-kilometer (385-mile) reservoir behind it. The dam is expected to cost in excess of $40 billion before the estimated date of completion in 2009. The government of China contends that the dam is necessary for several reasons. The primary reason is the 18,000 megawatts of electric power that the dam will generate. China's rapidly developing economy needs the energy for its expanding industrial centers. It is also argued that the project will transform the Yangtze River, especially the upper regions, into a more navigable and, hence, economic waterway. The final major argument for the dam relates to flood control in the middle and lower reaches of the river. These parts of the river have been prone to frequent and disastrous floods. The economic arguments, however, are not the only issues being debated in relation to the Three Gorges Dam.

Scientists in China and from other nations have warned that the project could threaten migratory fish, concentrate water pollution, and endanger to the point of extinction the Chinese alligator, river dolphins, the Siberian white crane, and the Chinese sturgeon. Environmental concerns relating to the project were instrumental in the 1993 decision of the United States to withdraw technical assistance. The impact on the regional Chinese population is also a critical issue. Upon completion, the lake formed by the dam will inundate 153 towns, 4500 villages, numerous archeological sites, and the scenic canyons of the Three Gorges that have inspired poets and painters for centuries.

Source: Magellan Geographix/ABCNEWS.com. Data from International Rivers Network.

The Three Gorges Dam, currently under construction, will have a major impact on a large area of China.

A resource is a naturally occurring substance of use to humans, a substance that can potentially be extracted using current technology. Reserves are known deposits from which materials can be extracted profitably with existing technology under present economic conditions.

Coal is the world's most abundant fossil fuel. Coal is obtained by either surface mining or underground mining. Problems associated with coal extractions are disruption of the landscape due to surface mining and subsidence due to underground mining. Black lung disease, waste heaps, water and air pollution, and acid mine drainage are additional problems. Oil was originally chosen as an alternative to coal because it was more convenient and less expensive. However, the supply of oil is limited. As oil becomes less readily available, multiple offshore wells, secondary recovery methods, and increased oil exploration will become more common. Natural gas is another major source of fossil-fuel energy. The primary problem associated with natural gas is transport of the gas to consumers.

Fossil fuels are nonrenewable: The amounts of these fuels are finite. When the fossil fuels are exhausted, they will have to be replaced with other forms of energy, probably renewable forms. Hydroelectric power can be increased significantly, but its development must flood areas and in so doing may require the displacement of people. The use of geothermal and tidal energy is limited by geographic locations. Wind power may be used to generate electricity but may require wide, open areas and a large number of wind generators. Solar energy can be collected and used in either passive or active systems and can also be used to generate electricity. Lack of a constant supply of sunlight is solar energy's primary limitation. Fuelwood is a minor source of energy in industrialized countries but is the major source of fuel in many less-developed nations. Biomass can be burned to provide heat for cooking or to produce electricity, or it can be converted to alcohol or used to generate methane. In some communities, solid waste is burned to reduce the volume of the waste and also to supply energy.

Energy conservation can reduce energy demands without noticeably changing standards of living.

Issues – Analysis

Public Perceptions of Energy

Rising energy costs and political concerns in the oil-rich nations of the Middle East have made energy a topic of discussion over the past few years. The Gallup Organization has conducted numerous polls regarding energy to gauge public opinion on energy issues. The results of several of the polls are of particular interest.

A 2004 poll showed that while relatively few Americans view the current energy situation as a "crisis," seven in 10 believe the availability and high cost of energy is at least a major problem for the country. The poll also indicated that Americans are most likely to blame oil and electric companies for the country's energy problems, rather than politicians or consumers. This poll also indicated that two-thirds of the public believes that Americans must make real changes to their lifestyle to help resolve the energy problems.

Another poll indicated that Americans express widespread support for several measures to deal with the current energy situation, including both new production and conservation initiatives. In a trade-off between the approaches, however, more Americans choose conservation over new production—although the margin in favor of conservation has declined over the past few years. The poll also shows that while much support exists for steps to mandate more energy-efficient products such as appliances, cars, and buildings, there is less support for additional investments in some of the infrastructures that would be needed to produce more power. Also, Americans are divided over whether to increase the use of nuclear power, and they remain opposed to opening up the Arctic National Wildlife Refuge for oil exploration.

- How do you feel about the Gallup poll results?
- Do you feel the poll results are accurate in terms of national feelings?
- What, if anything, do you feel would significantly change these poll findings?

Key Terms

acid mine drainage *193*
active solar system *206*
biomass *208*
black lung disease *192*
geothermal energy *203*

liquefied natural gas *196*
nonrenewable energy sources *187*
overburden *192*
passive solar system *206*
renewable energy sources *187*

reserves *187*
resources *187*
secondary recovery *194*
surface mining *192*
underground mining *192*

Review Questions

1. Why are fossil fuels important?
2. Distinguish between reserves and resources.
3. What are the advantages of surface mining of coal compared to underground mining? What are the disadvantages of surface mining?
4. Compare the environmental impacts of the use of coal and the use of oil.
5. What are some limiting factors in the development of new hydro-electric generating sites?
6. What factors limit the development of tidal power as a source of electricity?
7. In what parts of the world and why is geothermal energy available?
8. Why can wind be considered a form of solar energy?
9. Compare a passive solar-heating system with an active solar-heating system.
10. What problems are associated with the use of solid waste as a source of energy?
11. List three energy conservation techniques.

Critical Thinking Questions

1. Given what you know about the economic and environmental costs of different energy sources, would you recommend that your local utility company use hydroelectricity or coal to supplement electric production? What criteria would you use to make your recommendation?
2. Coal-burning electric power plants in the Midwest have contributed to acid rain in the eastern United States. Other energy sources would most likely be costlier than coal, thereby raising electricity rates. Should citizens of another state be able to pressure these utility companies to change the method of generating electricity? What mechanisms might be available to make these changes? How effective are these mechanisms?
3. Imagine you are an official with the Department of Energy and are in the budgeting process for alternative energy research. Where would you put the money? Why?
4. Given your choices from question 3, what do you think the political repercussions of your decision would be? Why?
5. Do you believe that large dam projects like the Three Gorges Dam project in China are, on the whole, beneficial or not? What alternatives would you recommend? Why?
6. Energy conservation is one way to decrease dependence on fossil fuels. What are some things you can do at home, work, or school that would reduce fossil-fuel use and save money?
7. What alternative energy resources that the text has outlined are most useful in your area? How might these be implemented?

Welcome to McGraw-Hill's Online Learning Center

Location: www.mhhe.com/enger10e

Interactive Exploration

Check out the website at **www.mhhe.com/enger10e** and click on the cover of this textbook for quizzing, career information, case studies, and hot links for the following topics:

Conventional Energy Sources

Oil, Coal, and Gas as Energy Sources

Nuclear Energy

Sustainable Energy

Solar Energy

Hydropower

Wind Energy

11 Nuclear Energy:
Benefits and Risks

Chapter Outline

Objectives

After reading this chapter, you should be able to:

- Explain how nuclear fission has the potential to provide large amounts of energy.
- Describe how a nuclear reactor produces electricity.
- Describe the basic types of nuclear reactors.
- Explain the steps involved in the nuclear fuel cycle.
- List concerns regarding the use of nuclear power.
- Explain the problem of decommissioning a nuclear plant.
- Describe how high-level radiation waste is stored.
- Describe the accident at Chernobyl.
- Explain how a breeder reactor differs from other nuclear reactors.
- List the technical problems associated with the design and operation of a liquid metal fast-breeder reactor.
- Explain the process of fusion.

PAST PRESENT FUTURE

1983 Globally, the installed electrical generating capacity of nuclear power plants was 189 gigawatts.

1983 There were 18 new nuclear reactor construction starts.

1980s The global nuclear capacity growth rate was 240 percent.

2003 Globally, the installed electrical generating capacity of nuclear power plants was 361 gigawatts.

2004 There were no new nuclear reactor construction starts.

2004 Since 1990, global nuclear capacity has risen just 7 percent.

The Nature of Nuclear Energy

Energy from disintegrating atomic nuclei has tremendous potential. (See figure 11.1.) In terms of benefiting the people of the world, we routinely use X rays to examine bones for fractures, treat cancer with radiation, and diagnose disease with the use of radioactive isotopes. About 17 percent of the electrical energy generated in the world comes from nuclear power plants. Engineers in the former Soviet Union used nuclear explosions to move large amounts of earth and rock to construct dams, canals, and underground storage facilities. The U.S. government briefly considered the possibility of using nuclear explosions to build a new Panama Canal.

On the other hand, nuclear energy has the potential to do great harm. The Japanese cities of Hiroshima and Nagasaki were destroyed by nuclear bombs. The manufacture and storage of nuclear weapons have left a legacy of radioactive wastes. In many cases, these wastes have been mismanaged or carelessly disposed of.

To understand where nuclear energy comes from, it is necessary to review some of the aspects of atomic structure presented in chapter 4. All atoms are composed of a central region called the nucleus, which contains positively charged protons and neutrons that have no charge. Moving around the nucleus are smaller, negatively charged electrons. Since the positively charged particles in the nucleus repel one another, energy is needed to hold the protons and neutrons together. However, some isotopes of atoms are **radioactive;** that is, the nuclei of these atoms are unstable and spontaneously decompose. Neutrons, electrons, protons, and other larger particles are released during nuclear disintegration, along with a great deal of energy. The rate of decomposition is consistent for any given isotope. It is measured and expressed as **radioactive half-life,** which is the time it takes for one-half of the radioactive material to spontaneously decompose. Table 11.1 lists the half-lives of several radioactive isotopes.

Nuclear disintegration releases energy from the nucleus as **radiation,** of which there are three major types:

Alpha radiation consists of a moving particle composed of two neutrons and two protons. Alpha radiation usually travels through air for less than a meter and can be stopped by a sheet of paper or the outer layer of the skin.

Beta radiation consists of electrons released from nuclei. Beta particles travel more rapidly than alpha particles and will travel through air for a couple of meters. They are stopped by a layer of clothing, glass, or aluminum.

Gamma radiation is a type of electromagnetic radiation that includes X rays, light, and radio waves. It can pass through several centimeters of concrete.

If the radiation reaches living tissue, equivalent doses of beta and gamma radiation cause equal amounts of biological damage. Alpha particles are more massive than beta particles (electrons) or gamma rays (photons); therefore, they cause more damage to biological tissues. The ingestion or inhalation of radioactive materials places the source of the radiation in direct

Figure 11.1 Uses of Nuclear Energy
The energy available from the splitting of atoms has been put to many uses: (a) nuclear test detonation, Bikini Atoll, May 1956; (b) a nuclear power plant in San Onofre, California; (c) a nuclear medicine camera used to diagnose medical conditions.

Table 11.1 Half-Lives of Some Radioactive Isotopes

Radioactive Isotope	Half-Life
Iodine 132	2.4 hours
Technetium 99	6.0 hours
Rhodium 105	36.0 hours
Xenon 133	5.3 days
Barium 140	12.8 days
Cerium 144	284.0 days
Cesium 137	30 years
Carbon 14	5730 years
Uranium 234	250,000 years
Chlorine 36	300,000 years
Beryllium 10	4.5 million years
Potassium 40	1.3 billion years
Helium 4	12.5 billion years

contact with cells; thus, this kind of exposure is much more dangerous than exposure from distant sources.

The release of neutrons is particularly important in obtaining energy from nuclear disintegration. In addition to releasing alpha, beta, and gamma radiation when they disintegrate, the nuclei of a few kinds of atoms release neutrons. When moving neutrons hit the nuclei of certain other atoms, they can cause those nuclei to split as well. This process is known as **nuclear fission.** If these splitting nuclei also release neutrons, they can strike the nuclei of other atoms, which also disintegrate, resulting in a continuous process called a **nuclear chain reaction.** Only certain kinds of atoms are suitable for the development of a nuclear chain reaction. The two materials commonly used in nuclear reactions are uranium-235 and plutonium-239. In addition, there must be a certain quantity of nuclear fuel (a critical mass) in order for a nuclear chain reaction to occur. It is this process that results in the large amounts of energy released from bombs or nuclear reactors.

The History of Nuclear Energy Development

The first controlled fission of an atom occurred in Germany in 1938, but the United States was the first country to develop an atomic bomb. In 1945, the U.S. military dropped atomic bombs on the Japanese cities of Hiroshima and Nagasaki. The incredible devastation of these two cities demonstrated the potential of nuclear energy for destruction. For many years, most atomic research involved military applications of nuclear energy as bombs and as power sources for ships. During the 50 years following World War II, the two major military powers of the world—the United States and the former Soviet Union—conducted secret nuclear research projects related to the building and testing of bombs. This continued to be a primary focus of nuclear research until the recent changes in the former Soviet Union, which led to a world in which nuclear war

is much less of a concern—although the explosion of nuclear devices in June of 1998 by India and Pakistan has heightened concern about nuclear war somewhat. A legacy of this military research is a great deal of soil, water, and air contaminated with radioactive material. Many of these contaminated sites have come to light recently and require major cleanup efforts.

After World War II, people began to see the potential for using nuclear energy for peaceful purposes rather than as weapons. The world's first electricity-generating reactor was constructed in the United States in 1951, and the Soviet Union built its first reactor in 1954. In December 1953, President Dwight D. Eisenhower, in his "Atoms for Peace" speech, made the following prediction:

> Nuclear reactors will produce electricity so cheaply that it will not be necessary to meter it. The users will pay an annual fee and use as much electricity as they want. Atoms will provide a safe, clean, and dependable source of electricity.

More than fifty years have passed since Eisenhower's predictions. Although nuclear power is being used throughout the world as a reliable source of electricity, it has not fulfilled such overly optimistic promises. Several serious accidents have caused worldwide concern about safety, and construction of most new nuclear power projects has stopped, except in Asia, as we will examine later in this chapter. At the same time, many energy experts predict a rebirth of the nuclear power industry as energy demands increase and a new generation of safer nuclear power plants is designed. These experts believe the public will favor nuclear power plants because they do not generate carbon dioxide, which contributes to global warming.

Nuclear Fission Reactors

A **nuclear reactor** is a device that permits a controlled fission chain reaction. In the reactor, neutrons are used to cause a con-

trolled fission of heavy atoms, such as uranium. **Uranium-235 (U-235)** is an isotope used to fuel nuclear fission reactors. When the nucleus of a U-235 atom is struck by a slowly moving neutron from another atom, the nucleus of the atom of U-235 is split into several smaller particles. Because the nucleus will split, it is said to be **fissionable.** When the nucleus is split, two to three rapidly moving neutrons are released, along with large amounts of energy. This energy is an important product of nuclear fission reactions. The neutrons released strike the nuclei of other atoms of U-235 and cause them to undergo fission, which in turn releases more energy and more neutrons, thus resulting in a chain reaction. (See figure 11.2.) Once begun, this chain reaction continues to release energy until the fuel is spent or the neutrons are prevented from striking other nuclei.

In addition to fuel rods containing uranium, reactors contain control rods of cadmium, boron, graphite, or other nonfissionable materials used to control the rate of fission by absorbing neutrons. When control rods are lowered into a reactor, they absorb the neutrons produced by fissioning uranium. There are fewer neutrons to continue the chain reaction, and the rate of fission decreases. If the control rods are withdrawn, more fission occurs, and more particles, radiation, and heat are produced.

The fuel rods housed in a reactor are surrounded by water or some other type of moderator. A **moderator** absorbs energy, which slows neutrons, enabling them to split the nuclei of other atoms more effectively. Fast-moving neutrons are less effective at splitting atoms than slow-moving neutrons. As U-235 undergoes fission, the energy of the fast-moving neutrons is transferred to water; the neutrons slow down, and the water is heated.

In the production of electricity, a nuclear-powered reactor serves the same function as any fossil-fueled boiler. It produces heat, which converts water to steam to operate a turbine that generates electricity. After passing through the turbine, the steam must be cooled, and the water is returned to the reactor to be heated again. Various types of reactors have been constructed to furnish heat for the production of steam. They differ in the moderator

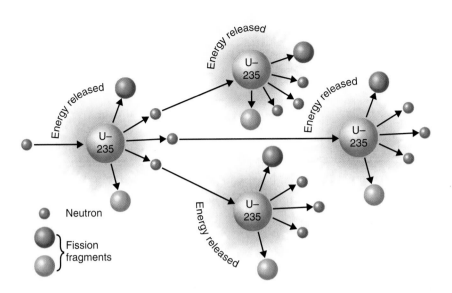

Figure 11.2 **Nuclear Fission Chain Reaction** When a neutron strikes a nucleus of U-235, energy is released, and several fission fragments and neutrons are produced. These new neutrons may strike other atoms of U-235, causing their nuclei to split. This series of events is called a chain reaction.

Figure 11.3 **Boiling-Water Reactor** A boiling-water reactor is a type of light-water reactor that produces steam to directly power the turbine and produce electricity. Water is used as a moderator and as a reactor-core coolant.

used, in how the reactor core is cooled, and in how the heat from the core is used to generate steam. Water is the most commonly used reactor-core coolant and also serves as a neutron moderator. **Light-water reactors (LWR),** which make up 90 percent of reactors operating today, use ordinary water, which contains the lightest, most common isotope of hydrogen, having an atomic mass of one. The two types of LWRs are **boiling-water reactors (BWR)** and **pressurized-water reactors (PWR).**

In a BWR (the type of construction used in about 20 percent of the nuclear reactors in the world), the water functions as both a moderator and a reactor-core coolant. (See figure 11.3.) Steam is formed within the reactor and transferred directly to the turbine, which generates electricity. A disadvantage of the BWR is that the steam passing to the turbine must be treated to remove any radiation. Even then, some radioactive material is left in the steam; therefore, the generating building must be shielded.

In a PWR (the type of construction used in over 70 percent of the nuclear reactors in the world), the water is kept under high pressure so that steam is not allowed to form in the reactor. (See figure 11.4.) A secondary loop transfers the heat from the pressurized water in the reactor to a steam generator. The steam is used to turn the turbine and generate electricity. Such an arrangement reduces the risk of radiation in the steam but adds to the cost of construction by requiring a secondary loop for the steam generator.

A third type of reactor that uses water as a coolant is a **heavy-water reactor (HWR),** developed by Canadians. It uses water that contains the hydrogen isotope deuterium in its molecular structure as the reactor-core coolant and moderator. Since the deuterium atom is twice as heavy as the more common hydrogen isotope, the water that contains deuterium weighs slightly more than ordinary water. An HWR also uses a steam generator to convert regular water to steam in a secondary loop. Thus, it is similar in structure to a PWR. The major advantage of an HWR is that naturally occurring uranium isotopic mixtures serve as a suitable fuel. This is possible because heavy water is a better neutron moderator than is regular water, while other reactors require that the amount of U-235 be enriched to get a suitable fuel. Since it does not require enriched fuel, the cost of operating an HWR is less than that of an LWR.

The **gas-cooled reactor (GCR)** was developed by atomic scientists in the United Kingdom. Carbon dioxide serves as a coolant for a graphite-moderated core. As in the HWR, natural isotopic mixtures of uranium are used as a fuel. (See figure 11.5.)

Figure 11.4 **Pressurized-Water Reactor** A pressurized-water reactor is a type of light-water reactor that uses a steam generator to form steam and a secondary loop to transfer this steam to the turbine.

Figure 11.5 **Advanced Gas-Cooled Reactor** This type of reactor uses graphite as a moderator and the gas carbon dioxide as the reactor-core coolant. A steam generator forms steam, and a secondary loop is used to transfer steam to the turbine.

The various types of reactors represent differing approaches to building safe, economical plants. Each method has its advantages and disadvantages, as well as its supporters and critics. Modifications will continue to be made to these basic types. Figure 11.6 shows the distribution of nuclear power plants in North America, and table 11.2 shows the numbers of reactors present in each country.

Plans for New Reactors Worldwide

Currently, there are 439 nuclear power reactors in 31 countries, with a combined capacity of 354 gigawatts. In 2003, these provided 2447 billion kilowatt hours, over 16 percent of the world's electricity. Although some countries, notably Japan, China, Taiwan, India, Russia, Slovak Republic, and the Republic of Korea, intend to continue major nuclear power construction programs, the rate of growth of installed nuclear-generating capacity over the next 10 years is expected to be low.

Some 32 power reactors are currently being constructed in 10 countries. Construction is well advanced on many of them and, based on reported progress and allowing for delays in countries, 13 with a total net capacity of over 9000 megawatts are expected to be in operation before 2005. This excludes two Russian and two Ukrainian reactors for which funding is uncertain.

After about 2005, forecasts of installed nuclear capacity become much less certain. When the present construction programs are completed, most significant nuclear power growth is expected to continue only in Asia and parts of the former Soviet Union. The International Atomic Energy Agency forecasts that the total installed nuclear capacity in 2015 will be little more than that in 2000, with the nuclear share of world electricity output decreased from 17 percent in 1997 to 13 percent in 2015.

Eight countries have plans to build new power reactors beyond those now under construction. In all, 32 power reactors with a total net capacity of over 34,460 megawatts are planned. Only one of these is in Western Europe or North or South America; it is in Argentina.

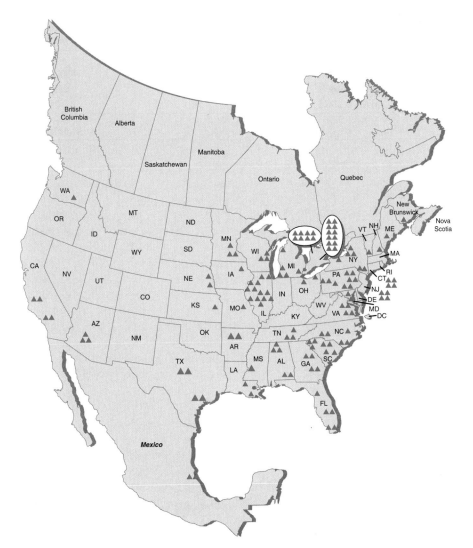

Figure 11.6 **Distribution of Nuclear Power Plants in North America** The United States has the largest number of nuclear power plants of any country in the world—over 100. Canada has about 20 nuclear power plants, and Mexico has two.

economies and rapidly rising electricity demand. Nuclear power will continue to play a major role in the future electricity supply mix in both South Korea and Japan. South Korea plans to bring a further 12 reactors, with a total capacity of 27,607 megawatts, into operation by the year 2025. Japan has plans and, in most cases, designated sites and announced timetables for a further 15 power reactors, totaling about 20,000 megawatts, and some of these are negotiating the governmental approval process. Fulfilling the necessary conditions for approval can take over a decade.

China, with seven operating reactors, has begun the next phase of its nuclear power program. Construction has started on seven reactors. These are expected to start up from 2004 to 2007 and to add some 5770 megawatts to China's existing 2767 megawatts nuclear capacity.

Plant Life Extension

Most nuclear power plants originally had a nominal design lifetime of up to 40 years, but engineering assessments of many plants over the last decade have established that many can operate longer. In the United States, most reactors now have confirmed life spans of 40 to 60 years, and in Japan, 40 to 70 years. In the United States, the first two reactors have been granted license renewals, which extends their operating lives to 60 years.

When the oldest commercial nuclear power stations in the world, Calder Hall and Chapelcross in the United Kingdom, were built in the 1950s, it was assumed that they would have a useful lifetime of 20 years. As of 2001, they were still authorized to operate.

Sweden's oldest reactor, which started up in 1971, has been fully rebuilt at a cost equivalent to 8 percent of a replacement unit, and all Sweden's reactors are maintained to operate until 2010.

It should be noted, however, that economic, regulatory, and political considerations have led to the premature closure of some power reactors. In the United States, reactor numbers have fallen from 110 to 104. Germany, as previously mentioned, plans to close all of its plants by the mid-2020s.

In France, the only Western European country that has had an active nuclear power construction program, the national utility announced in 1994 that it would order no new generating capacity, nuclear or conventional, in the near future. Sites have, however, been designated for new power reactors, and new reactor construction is expected to resume by 2006.

In 2000, Germany announced that it would close all of its 19 nuclear power plants by 2021, making it the world's largest industrialized nation to willingly forgo using the technology.

Germany, Europe's largest economy, currently receives one-third of its electrical power from nuclear sources. The decision reached in Germany limits nuclear plants to an average 32 years of operation.

That means that the most modern plants likely will close around 2021, and Germany will join nations such as Italy and Austria in abandoning nuclear power.

In Eastern Europe, the Russian government in 1997 approved a nuclear power construction program. In addition to the three reactors under construction, another three, taking total capacity to about 26,400 megawatts, are planned to be operating by 2010. Several of the oldest Russian reactors are expected to be retired by 2010, and Russia has announced its intention to replace retired nuclear capacity with new construction at the same site, to optimize the use of established infrastructure and personnel.

Most planned reactors are in the Asian region, which has fast-growing

Table 11.2 World Nuclear Power Reactors, 2001–2004, and Uranium Requirements

Country	Reactors Operating March 2002		Reactors Building March 2002		On Order or Planned July 2004		Uranium Required 2004
	No.	MWe	No.	MWe	No.	MWe	Metric tons U
Argentina	2	935	0	0	1	692	133
Armenia	1	376	0	0	0	0	68
Belgium	7	5728	0	0	0	0	1163
Brazil	2	1855	0	0	1	1245	303
Bulgaria	6	3538	0	0	0	0	618
Canada	14	9998	6	3598	2	1030	1692
China	4	2767	7	5770	4	3800	2127
Taiwan	6	4884	2	2600	0	0	966
Czech Republic	5	2560	1	912	0	0	519
Finland	4	2656	0	0	1	1600	553
France	59	63,203	0	0	0	0	10,159
Germany	19	21,141	0	0	0	0	3712
Hungary	4	1755	0	0	0	0	271
India	14	2548	3	1916	3	1340	312
Iran	0	0	1	950	1	950	0
Japan	54	444,301	3	3696	12	15,858	7661
Korea DPR (North)	0	0	0	0	2	1900	0
Korea RO (South)	16	12,970	4	3800	8	9200	2819
Lithuania	2	2370	0	0	0	0	359
Mexico	2	1310	0	0	0	0	232
Netherlands	1	452	0	0	0	0	115
Pakistan	2	425	0	0	1	300	56
Romania	1	655	0	0	1	620	90
Russia	30	20,793	3	2625	3	2950	3013
Slovak Republic	6	2472	2	840	0	0	370
Slovenia	1	679	0	0	0	0	131
South Africa	2	1842	0	0	0	0	363
Spain	9	7345	0	0	0	0	1613
Sweden	11	9460	0	0	0	0	1533
Switzerland	5	3170	0	0	0	0	599
Ukraine	13	11,195	0	0	2	1900	1512
United Kingdom	33	12,528	0	0	0	0	2488
USA	104	98,248	0	0	0	0	22,353
WORLD	**439**	**354,159**	**32**	**26,707**	**39**	**43,385**	**67,903**

Source: Reprinted by permission of Uranium Information Centre Ltd., Melbourne, Australia.

Investigating Nuclear Alternatives

Breeder Reactors

During the early stages of the development of nuclear power plants, breeder reactor construction was seen as the logical step after nuclear fission development. Although a regular fission reactor produces some additional radioactive material that can serve as nuclear fuel, it is smaller than the quantity of fuel consumed. A **nuclear breeder reactor** is a nuclear fission reactor that produces heat to be converted to steam to generate electricity and also forms a new supply of radioactive isotopes. If a fast-moving neutron hits a uranium-238 (U-238) nucleus and is absorbed, an atom of fissionable **plutonium-239 (Pu-239)** is produced. (See figure 11.7.) In a breeder reactor, water is not used as a moderator because water slows the neutrons too much and Pu-239 is not produced. Breeder reactors need a moderator that allows the neutrons to move more rapidly and has good heat transfer properties.

Figure 11.7 **Formation of Pu-239 in a Breeder Reactor** When a fast-moving neutron (N) is absorbed by the nucleus of a U-238 atom, a series of reactions results in the formation of Pu-239 from U-238. Two intermediate atoms are U-239 and Np-239, which release beta particles (electrons) from their nuclei. (A neutron in the nucleus can release a beta particle and become a proton.) This is an important reaction because, while the U-238 does not disintegrate readily and therefore is not a nuclear fuel, the Pu-239 is fissionable and can serve as a nuclear fuel.

The **liquid metal fast-breeder reactor (LMFBR)** appears to be the most promising model. In this type of reactor, the fuel rods in the core are surrounded by rods of U-238 and liquid sodium. The energy of the neutrons released from U-235 in the fuel rods heats the sodium to a temperature of 620°C. In addition to furnishing heat, these fast-moving neutrons are absorbed by the rods containing U-238, and some of these atoms are converted into Pu-239. After approximately 10 years of operation, during which electricity is produced, the LMFBR will have also produced enough radioactive material to operate a second reactor.

There are some serious drawbacks to the LMFBR. Sodium reacts violently if it comes into contact with water or air. Therefore, costly, highly specialized equipment is required to contain and pump the sodium. If these systems fail, the sodium boils, which allows the chain reaction to proceed at a faster rate and could damage the reactor, leading to a nuclear accident.

There are also problems in the startup of an LMFBR. When a breeder reactor is starting up or going on-line after a shutdown, the solid sodium cannot be moved by the pumps until it becomes a liquid. This presents a technical problem in developing LMFBRs.

Another problem is that reaction rates are extremely rapid and very difficult to regulate. The instrumentation needed to monitor an LMFBR must be of extremely high quality because control of the reaction involves precise adjustments over very short periods of time.

Finally, the product of a breeder reactor, plutonium-239, is extremely haz-ardous to humans who come in contact with it. And, because plutonium-239 can be made into nuclear weapons, it must be transported, processed, or produced under very close security. The more breeder reactors in use, the more difficult the security problems and the more likely the chance that the small amount of plutonium needed to manufacture a bomb could be stolen.

Because of these problems, no breeder reactors are scheduled for commercial use in the United States. In fact, the only experimental U.S. breeder reactor is scheduled to be shut down. The Clinch River Fast-Breeder Reactor Plant near Knoxville, Tennessee, which was to be operational in the 1970s, was never completed.

In Europe, as in the United States, the development of breeder reactors has slowed. In 1981, after running smoothly for eight years, a 250-megawatt French prototype plant developed a leak in the cooling system that resulted in a sodium fire. This raised some serious questions about the safety of breeder reactors. In 1982, the French government announced that it was scaling down the planned construction of breeder reactors from five to one. Today, it is doubtful that even one will be built. A total of five LMFBRs are in operation in the world today, in France, Japan, Kazakhstan, and Russia.

Nuclear Fusion

Another aspect of nuclear power involves the even more advanced technology of nuclear fusion. When two lightweight atomic nuclei combine to form a heavier nucleus, a large amount of energy is released. This process is known as **nuclear fusion.** The energy produced by the sun is the result of fusion. Most studies of fusion have involved small atoms such as hydrogen. Most hydrogen atoms have one proton and no neutrons in the nucleus. The hydrogen isotope deuterium (H^2) has a neutron and a proton. Tritium (H^3) has a proton and two neutrons. When deuterium and tritium isotopes combine to form heavier atoms, large amounts of energy are released. (See figure 11.8.) The energy that would be released by combining the deuterium in 1 cubic kilometer of ocean water would be greater than that contained in the world's entire supply of fossil fuels.

Although fusion could solve the world's energy problems, technology must answer several questions before we can actually use fusion power. Three conditions must be met simultaneously for fusion to occur: high temperature, adequate density, and confinement. If heat is used to provide the energy necessary for fusion, the temperature must approach that of the center of the sun. At the same time, the walls of the vessel confining the atoms must be protected from the heat, or they will vaporize. However, the main problem is containment of the nuclei. Because they have a positive electric charge, the nuclei repel one another.

Even though, in theory, fusion promises to furnish a large amount of energy, technical difficulties appear to prevent its commercial use in the near future. Even the governments of nuclear nations are budgeting only modest amounts of money for fusion research. And, as with nuclear fission and the breeder reactor, economic costs and fear of accidents may continue to delay the development of fusion reactors.

The Nuclear Fuel Cycle

To appreciate the consequences of using nuclear fuels to generate energy, it is important to understand how the fuel is processed. The nuclear fuel cycle begins

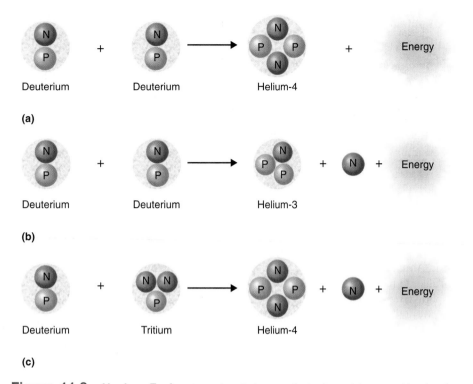

Figure 11.8 Nuclear Fusion In nuclear fusion, small atomic nuclei are combined to form heavier nuclei. Large amounts of energy are released when this occurs. Different isotopes of hydrogen can be used in the process of fusion. There are three possible types of fusion: (*a*) two deuterium isotopes can combine to form helium-4 and energy; (*b*) two deuterium isotopes can combine to form helium-3, a free neutron, and energy; and (*c*) a deuterium and a tritium isotope can combine to form helium-4, a free neutron, and energy.

with the mining operation. (See figure 11.9.) Low-grade uranium ore is obtained by underground or surface mining. The ore contains about 0.2 percent uranium by weight. After it is mined, the ore goes through a milling process. It is crushed and treated with a solvent to concentrate the uranium. Milling produces yellowcake, a material containing 70 to 90 percent uranium oxide.

Naturally occurring uranium contains about 99.3 percent nonfissionable U-238 and only 0.7 percent fissionable U-235. This concentration of U-235 is not high enough for most types of reactors, so the amount of U-235 must be increased by enrichment. Since the masses of the isotopes U-235 and U-238 vary only slightly, and the chemical differences are very slight, enrichment is a difficult and expensive process. However, it increases the U-235 content from 0.7 percent to 3 percent.

Fuel fabrication converts the enriched material into a powder, which is then compacted into pellets about the size of a pencil eraser. These pellets are sealed in metal fuel rods about 4 meters (13 feet) in length, which are then loaded into the reactor.

As fission occurs, the concentration of U-235 atoms decreases. After about three years, a fuel rod does not have enough radioactive material to sustain a chain reaction, and the spent fuel rods must be replaced by new ones. The spent rods are still very radioactive, containing about 1 percent U-235 and 1 percent plutonium. These rods are the major source of radioactive waste material produced by a nuclear reactor and are classified as *high-level* nuclear waste. Other high-level wastes result from obsolete nuclear weapons and the wastes generated by their manufacture. *Low-level* radioactive waste is generated by nuclear power plants, military facilities, hospitals, and research institutions.

When nuclear reactors were first being built, scientists proposed that spent fuel rods could be reprocessed. The remaining U-235 could be enriched and used to manufacture new fuel rods. Since

plutonium is fissionable, it could also be fabricated into fuel rods. Besides providing new fuel, reprocessing would reduce the amount of nuclear waste. At present, India, Japan, Russia, France, and the United Kingdom operate reprocessing plants that reprocess spent fuel rods as an alternative to storing them as a nuclear waste.

Each step in the nuclear fuel cycle involves the transport of radioactive materials. The uranium mines are some distance from the processing plants. The fuel rods must be transported to the power plants, and the spent rods must be moved to a reprocessing plant or storage area. Each of these links in the fuel cycle presents the possibility of an accident or mishandling that could release radioactive material. Therefore, the methods of transport are extremely carefully designed and tested before they are used. Many people are convinced that the transport of radioactive materials is hazardous, while others are satisfied that the utmost care is being taken and that the risks are extremely small.

Ultimately, both high-level and low-level radioactive wastes must be stored. Thus, each step in the nuclear fuel cycle, from the mining of uranium to the storage of nuclear waste, poses health and environmental concerns.

Nuclear Material and Weapons Production

Producing nuclear materials for weapons and other military uses involves many of the same steps used to produce nuclear fuel for power reactors. In fact, the nuclear power industry is an outgrowth of the weapons industry. In the United States, the Department of Energy is responsible for nuclear research for both weapons and peaceful uses and stewardship of the facilities used for research and weapons production. Some facilities are used for both processing nuclear fuel and providing materials for weapons. In both cases, uranium must be mined, concentrated, and transported to sites of use. Furthermore,

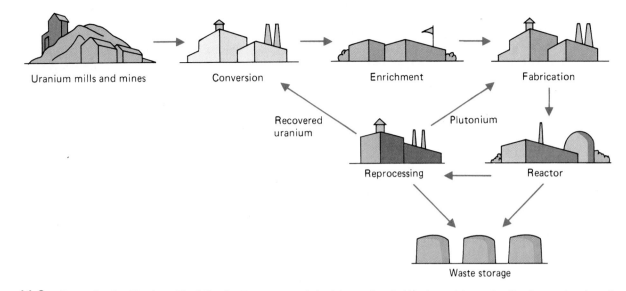

Figure 11.9   Steps in the Nuclear Fuel Cycle The process of obtaining nuclear fuel involves mining, extracting the uranium from the ore, concentrating the U-235, fabricating the fuel rods, installing and using the fuel in a reactor, and disposing of the waste. Some countries reprocess the spent fuel as a way of reducing the amount of waste they must deal with.

Source: U.S. Department of Energy.

military uses involve the production and concentration of plutonium. The production and storage facilities invariably become contaminated, as does the surrounding land.

Research and production facilities have typically dealt with hazardous chemicals and *low-level* radioactive wastes by burying them, pumping them into the ground, storing them in ponds, or releasing them into rivers. Despite environmental regulations that prevented such activities, the Department of Energy (DOE) (formerly the Atomic Energy Commission) maintained that it was exempt from such federal environmental legislation. As a result, the DOE has become the steward of a large number of sites that are contaminated with both hazardous chemicals and radioactive materials. The magnitude of the problem is huge. There are 8600 square kilometers (3365 square miles) of DOE properties that are or have been involved in weapons development or production. These include:

- 3700 contaminated sites,
- 330 underground storage tanks with high-level radioactive waste,
- more than a million 55-gallon drums of radioactive, hazardous, or mixed waste in storage,

- 5700 sites where wastes are moving through the soil, and
- millions of cubic meters of low-level and high-level radioactive wastes.

The Department of Energy has pledged to clean up these sites; however, the process could take decades and cost billions of dollars. Several U.S. sites are being cleaned up, but things are not going smoothly. Local residents and the states that are hosts to these facilities distrust the DOE and are insisting that the sites be cleaned completely. This may not be technologically or economically possible. Furthermore, environmental cleanup is a new mission for the DOE, and the department is having difficulty adjusting.

An additional problem has arisen as a result of the reduced importance of nuclear weapons. The political disintegration of the Soviet Union and Eastern Europe has made large numbers of nuclear weapons, in both the East and West, unnecessary. But how are nations of the world to dispose of their nuclear weapons? Some nuclear material can be diverted to fuel use in nuclear reactors, and some reactors can be modified to accept enriched uranium or plutonium. The security of these materials, particu-

larly in the former Soviet Union, is a concern.

Nuclear Power Concerns

Nuclear power has provided a significant amount of electricity for the people of the world. Currently, over 7 percent of the energy consumed worldwide and 17 percent of the electricity consumed worldwide comes from nuclear power. However, several accidents have raised questions about safety, radiation has been released into the air and water, the production and transportation of fuel have caused contamination, and the disposal of wastes is a continuing problem.

After September 11, 2001, fear arose regarding nuclear plants as potential targets for terrorist attacks. The image of a hijacked airliner crashing into the core of a nuclear reactor brought new concerns into the minds of many, and security measures were heightened at power plants nationwide. But just how vulnerable are nuclear plants to terrorist attack?

The consensus is that there is not a great cause for concern. An airliner flying at high speed would not be able to do significant damage to a reactor or containment building. The reactor itself is very small and is encased in steel and concrete in order to survive any natural disaster; this enclosure could also withstand the impact of a small aircraft. The segments of the plant that may be vulnerable to attack, such as cooling and auxiliary equipment, are supplied with emergency cooling systems and power supplies. If all of these should fail, the reactor core could overheat and melt, but its materials would remain contained.

The areas of most concern within a nuclear plant are the storage facilities for spent fuel. These depositories contain more total radioactivity than the reactor itself, but many of the more dangerous radioactive isotopes have already decayed. Also, while an attack could drain the water out of cooling ponds for spent fuel rods, causing the fuel to overheat and melt, little radioactive material would be dispersed. Similarly, the crash of an airliner into a dry storage facility would release little radioactive particulate matter. If any of these facilities were the target of an attack, residents would have at least eight hours to evacuate the region before unsafe levels of radioactivity were reached; the worst contamination would occur through dispersal of airborne radioactive material into the natural environment.

Reactor Safety: The Effects of Three Mile Island and Chernobyl

Although accidents have occurred at nuclear power plants since they were first used, two relatively recent accidents have had the greatest effect on people's attitudes toward nuclear power plant safety: Three Mile Island in 1979 and Chernobyl in 1986.

On March 14, 1979, the valves closed to three auxiliary pumps at one of the reactors at the Three Mile Island nuclear plant in Pennsylvania. On March 28, 1979, the main pump to the reactor broke down. The auxiliary pumps failed to operate because of the closed valves, and the electrical-generating turbine stopped. At this point,

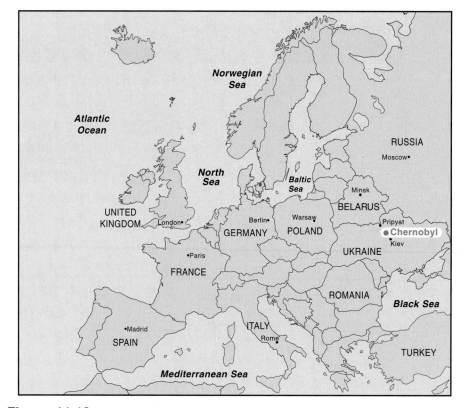

Figure 11.10 **Chernobyl** The town of Chernobyl became infamous as the location of the world's worst nuclear power plant accident.

an emergency coolant should have flooded the reactor and stabilized the temperature. The coolant did start to flow, but a faulty gauge indicated that the reactor was already flooded. Relying on this faulty reading, an operator overrode the automatic emergency cooling system and stopped it. Without the emergency coolant, the reactor temperature rose rapidly. The control rods eventually stopped fission, but a partial core meltdown had occurred. Later that day, radioactive steam was vented into the atmosphere.

The crippled reactor was eventually defueled in 1990 at a cost of about $1 billion. It has been placed in monitored storage until the companion reactor, which is still operating, reaches the end of its useful life. At that time, both reactors will be decommissioned.

Chernobyl is a small city in Ukraine near the border with Belarus, north of Kiev. (See figure 11.10.) As is true of many small cities in the world, most people had never heard of it until the spring of 1986, when the world's largest nuclear accident catapulted Chernobyl into the news.

At 1 A.M. on April 25, 1986, at Chernobyl Nuclear Power Station-4, a test was begun to measure the amount of electricity that the still-spinning turbine would produce if the steam were shut off. This was important information since the emergency core cooling system required energy for its operation and the coasting turbine could provide some of that energy until another source became available. The amount of steam being produced was reduced by lowering the control rods into the reactor. But the test was delayed because of a demand for electricity, and a new shift of workers came on duty. The operators failed to program the computer to maintain power at 700 megawatts, and output dropped to 30 megawatts. This presented an immediate need to rapidly increase the power, and many of the control rods were withdrawn. Meanwhile, an inert gas (xenon) had accumulated on the fuel rods. The gas absorbed the neutrons and slowed the rate of power increase. In an attempt to obtain more power, operators withdrew all the control rods. This was a second serious safety violation.

GLOBAL PERSPECTIVE

The Nuclear Legacy of the Soviet Union

The tremendous political changes that have occurred in the former Soviet Union and Eastern Europe have brought to light the magnitude of nuclear contamination caused by those nations' unwise use of nuclear energy. The Soviet Union used nuclear energy for several purposes: generating power for electricity and nuclear-powered ships, testing and producing weapons, and blasting to move earth. The accident at Chernobyl was only the most recent and most public of many problems associated with the use of nuclear energy in the Soviet Union. About 13 nuclear reactors of the Chernobyl type are still operating in the former Soviet Union. These are considered unsafe by most experts, including scientists from the former Soviet Union.

The production of nuclear fuels and weapons and the reprocessing of nuclear fuels, all of which produced nuclear wastes, occurred at three major sites: Chelyabinsk, Tomsk, and Krasnoyarsk. These facilities were secret until the recent breakup of the Soviet Union. At Chelyabinsk, radioactive waste was dumped into the Techa River and Lake Karachai. Other nuclear wastes were secretly dumped into the ocean or were buried in shallow dumps. It is estimated that there are 600 secret nuclear dump sites in and around Moscow alone.

More than 100 nuclear bombs were exploded to move earth for mining purposes, to create underground storage caverns, or to "dig" canals. In addition, 467 nuclear blasts were conducted to test weapons at a site near Semipalatinsk in Kazakhstan. As a result, nuclear fallout contaminated farmland to the northeast.

The Barents Sea and the Kara Sea have been polluted by nuclear tests and the dumping of nuclear waste. At least 15 nuclear reactors from obsolete ships were dumped into the Kara Sea. Seals that live in the area have high rates of cancer.

None of these problems is unique to the former Soviet Union. Similar secrecy surrounded the early development of nuclear power in the United States and the rest of the world. What makes the situation in the former Soviet Union different is its magnitude, caused by decades of secrecy and a commitment to nuclear weapons as the primary deterrent against enemies. It can be argued that the threat posed to the Soviet Union by its former enemies contributed to the reckless development of nuclear power.

Known locations of nuclear waste in the former Soviet Union are shown in red.

At 1 A.M. on April 26, the operators shut off most emergency warning signals and turned on all eight pumps to provide adequate cooling for the reactor following the completion of the test. Just as final stages of the test were beginning, a signal indicated excessive reaction in the reactor. In spite of the warning, the operators blocked the automatic reactor shutdown and began the test.

As the test continued, the power output of the reactor rose beyond its normal level and continued to rise. The operators activated the emergency system designed to put the control rods back into the reactor and stop the fission. But it was too late. The core had already been deformed, and the rods would not fit properly; the reaction could not be stopped. In 4.5 seconds, the energy level of the reactor increased 2000 times. The fuel rods ruptured, the cooling water turned into steam, and a steam explosion occurred. The lack of cooling water allowed the reactor to explode. The explosion blew the 1000-metric ton (1102-ton) concrete roof from the reactor, and the reactor caught fire.

In less than 10 seconds, Chernobyl became the scene of the world's worst nuclear accident. A core meltdown had occurred. (See figure 11.11.) It took 10 days to bring the runaway reaction under control. By November, the damaged reactor was entombed in a concrete covering, but the hastily built structure, known as the sarcophagus, may have structural flaws. The Ukrainian government is planning to have a second containment structure built around the current structure. The immediate consequences were 31 fatalities; 500 persons hospitalized, including

Figure 11.11   **The Accident at Chernobyl** An uncontrolled chain of reactions in the reactor of unit four resulted in a series of explosions and fires. See the circled area in the photograph.

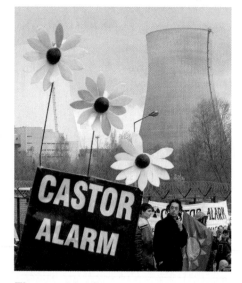

Figure 11.12   **Anti-nuclear Demonstration** This anti-nuclear demonstration was held in 2000 in Germany as members of the public protested the transport of nuclear waste across the country to a storage site.

237 with acute radiation sickness; and 116,000 people evacuated. Of the evacuees, 24,000 received high doses of radiation. The delayed effects are more difficult to assess. Many people suffer from illnesses they feel are related to their exposure to the fallout from Chernobyl. In 1996, it became clear that at least one delayed effect was strongly correlated with exposure. Children or fetuses exposed to fallout are showing increased frequency of thyroid cancer. The thyroid gland accumulates iodine, and radioactive iodine 131 was released from Chernobyl. It is still too early to tell if there are other delayed health effects.

More than a year after the disaster at Chernobyl, the decontamination of 27 cities and villages within 40 kilometers of Chernobyl was considered finished. This does not mean that the area was cleaned up, only that all practical measures were completed. Some areas were simply abandoned. The largest city to be affected was Pripyat, which had a population of 50,000 and was only 4 kilometers from the reactor. A new town was built to accommodate those displaced by the accident, and Pripyat remains a ghost town. Sixteen other communities within the 40-kilometer radius have been cleaned up, and the original inhabitants have returned, although there is still controversy about the safety

of these towns. Seventeen years after the Chernobyl explosion, some scientists fear that the worst is yet to come. Compared to the general population, rates of some non-cancer diseases—endocrine disorders and stroke, for instance—appear to be rising disproportionately among the roughly 600,000 "liquidators" who cleaned up the heaviest contamination in the plant's vicinity and entombed unit four's lethal remnants in a concrete sarcophagus. Whether people living in the shadow of Chernobyl remain at risk for health problems is a question now under intense scrutiny.

One important impact of Chernobyl is that it deepened public concern about the safety of nuclear reactors. Even before Chernobyl, between 1980 and 1986, the governments of Australia, Denmark, Greece, Luxembourg, and New Zealand had officially adopted a "no nuclear" policy. Since 1980, 10 countries have canceled nuclear plant orders or mothballed plants under construction. Argentina canceled 4 plants; Brazil, 8; Mexico, 18; and the United States, 54. There have been no orders for new plants in the United States since 1974. Sweden, Austria, Germany, and the Philippines have decided to phase out and dismantle their nuclear power plants.

Before Chernobyl, 65 percent of the people in the United Kingdom were opposed to nuclear power plants; after Chernobyl, 83 percent were against them. In Germany, opposition increased from 46 percent to 83 percent. (See figure 11.12.) Opposition in the United States rose from 67 to 78 percent. Even the French, who have the greatest commitment to nuclear power, were against it by 52 percent. The number of new nuclear plants being constructed has been reduced. In addition, many plants have been shut down prematurely, resulting in the prediction that the amount of energy furnished by nuclear fission reactors will actually decline in the future.

At the same time, some people are predicting the death of nuclear power as a viable energy source, others are examining ways to make nuclear power facilities safer. They expect nuclear power to be an important energy source in the future. At both Three Mile Island and Chernobyl, operator error caused or contributed to the accidents. Operators manually stopped normal safety actions from taking place. However, a contributing factor was the design: active mechanical processes had to work properly to shut the reactors down. Many of the new designs for reactors include passive mechanisms that will shut down the reactor, special catching basins for the reactor core should it melt down, and better containment buildings.

Nuclear power will continue to be a part of the energy mix, particularly in countries that lack fossil-fuel reserves. Furthermore, as fossil fuels are used up, the pressure for energy will probably create a market for nuclear power technology.

Exposure to Radiation

Although nuclear accidents are spectacular and frightening, and can cause immediate death because of the incredible amount of energy involved, the long-term problems resulting from exposure to radiation are even more worrisome. Radiation is converted to other forms of energy when it is absorbed by matter. When organisms are irradiated, this energy conversion causes damage at a cellular, tissue, organ, or organism level. The degree and kind of damage vary with the kind of radiation, the amount of radiation, the duration of the exposure, and the types of cells irradiated.

Radiation can also cause mutations, which are changes in the genetic messages within cells. Mutations can cause two quite different kinds of problems. Mutations that occur in the ovaries or testes can form mutated eggs or sperm, which can lead to abnormal offspring. Care is usually taken to shield these organs from unnecessary radiation. Mutations that occur in other tissues of the body may manifest themselves as abnormal tissue growths known as cancer. Two common cancers that are strongly linked to increased radiation exposure are leukemia and breast cancer. Because mutations are essentially permanent, they may accumulate over time. Therefore, the accumulated effects of radiation over many years may result in the development of cancer later in life.

Human exposure to radiation is usually measured in **rems** (*r*oentgen *e*quivalent *m*an), a measure of the biological damage to tissue. The effects of large doses (1000 to 1 million rems) are easily seen and can be quantified, because there is a high incidence of death at these levels, but demonstrating known harmful biological effects from smaller doses is much more

difficult. (See table 11.3.) Moderate doses (10 to 100 rems) are known to increase the likelihood of cancer and birth defects. The higher the dose, the higher the incidence of abnormality. Lower doses may cause temporary cellular changes, but it is difficult to demonstrate long-term effects.

It should be pointed out, that no human is subject to zero exposure and that the average person is exposed to 0.2 to 0.3 rems per year from natural and medical sources. Current research is trying to assess the risks associated with repeated exposure to low-level radiation. Thus, the effects of low-level, chronic radiation generate much controversy. Some people feel that all radiation is harmful, that there is no safe level, and that special care must be taken to prevent exposure. (See figure 11.13.)

Each step in the nuclear fuel cycle poses a radiation exposure problem, beginning with the mining of uranium. Although the radiation level in the ore is low, miners' prolonged exposure to low-level radiation increases their rates of certain cancers, such as lung cancer. Uranium miners who smoke have an even higher lung cancer rate. After mining, the ore must be crushed in a milling process. This releases radioactive dust into the atmosphere, so workers are chronically exposed to low levels of radioactivity. The crushed rock is left on the surface of the ground as mine tailings. There are over

150 million metric tons of low-level radioactive mine tailings in the United States and at least as much in the rest of the world. These tailings constitute a hazard because they are dispersed into the environment. (See figure 11.14.)

In the enrichment and fabrication processes, the main dangers are exposure to radiation and accidental release of radioactive material into the environment. Transport also involves exposure risks. Most radioactive material is transported by highways or railroads. If a transporting vehicle were involved in an accident, radioactive material could be released into the environment. Even after fuel rods have been loaded into the reactor, people who work in the area of the reactor risk radiation exposure. Countries that reprocess spent fuel rods must be concerned about the exposure of workers and the possible theft of rods or reprocessed material by terrorist organizations that would use the radioactive material to construct an atomic or "dirty" atomic bomb in which a conventional explosive is used to spread radioactive material.

Thermal Pollution

Thermal pollution is the addition of waste heat to the environment. This is a problem particularly in aquatic environments, since many aquatic organisms are

Table 11.3 Radiation Effects

Source	Dose	Biological Effect
Nuclear bomb blast or exposure in a nuclear facility	100,000 rems/incident	Immediate death
	10,000 rems/incident	Coma, death within one to two days
X rays for cancer patients (It is important to distinguish between a whole-body dose of 1000 rems, which is lethal, and a similar dose targeted at a cancerous organ, which often saves lives.)	1000 rem/incident	Nausea, lining of intestine damaged, death in one to two weeks
	100 rems/incident	Increased probability of leukemia
	10 rems/incident	Early embryos may show abnormalities
Upper limit for occupationally exposed people	5 rems/year	Effects difficult to demonstrate
X ray of the intestine	1 rem/procedure	Effects difficult to demonstrate
Upper limit for release from nuclear installations (except nuclear power plants)	0.5 rem/year	Effects difficult to demonstrate
Natural background radiation	0.2–0.3 rem/year	Effects difficult to demonstrate
Upper limit for release by nuclear power plants	0.005 rem/year	Effects difficult to demonstrate

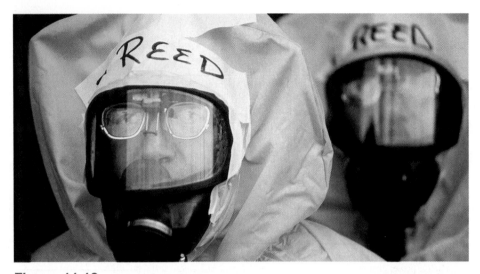

Figure 11.13 **Protective Equipment** Persons working in an area subjected to radiation must take steps to protect themselves. These workers are wearing protective clothing and filtering the air they breathe.

③ Radon diffusion

④ Windblown particles

② Water erosion

⑤ Windblown deposition

⑥ Direct gamma

① Leaching

Potable Aquifer

Figure 11.14 **Uranium Mine Tailings** Even though the amount of radiation in the tailings is low, the radiation still represents a threat to human health. Radioactivity may be dispersed throughout the environment and come in contact with humans in several ways: (1) radioactive materials may leach into groundwater; (2) radioactive materials may enter surface water through erosion; (3) radon gas may diffuse from the tailings and enter the air; or (4) persons living near the tailings may receive particles in the air, (5) come in contact with objects that are coated with particulate, or (6) receive direct gamma radiation.

Source: U.S. Environmental Protection Agency.

very sensitive to changes in temperature. When temperatures increase, oxygen solubility decreases while metabolic rates of aquatic organisms increase, raising the de-

mand for oxygen. All industrial processes release waste heat, so the problem of thermal pollution is not unique to nuclear power plants. In both fossil-fuel and nu-

clear plants, generating steam to produce electricity results in a great deal of waste heat. In a fossil-fuel plant, half of the heat energy produces electricity, and half is lost as waste heat. In a nuclear power plant, only one-third of the heat generates electricity, and two-thirds is waste heat. Therefore, the less efficient nuclear power plant increases the amount of thermal pollution more than does an equivalent fossil-fuel power plant. To reduce the effects of this waste heat, utilities build costly cooling facilities. Cooling processes involve water, so nuclear power plants often are constructed next to a water source. In some cases, water is drawn directly from lakes, rivers, or oceans and returned. In other cases, it is supplied by giant cooling towers. (See figure 11.15.)

Decommissioning Costs

All industrial facilities have a life expectancy—that is, the number of years they can be profitably operated. The life expectancy for an electrical generating plant, whether fossil-fuel or nuclear, is about 30 to 40 years, after which time the plant is demolished. With a fossil-fuel plant, the demolition is relatively simple and quick. A wrecking ball and bulldozers reduce the plant to rubble, which is trucked off to a landfill. The only harm to the environment is usually the dust raised by the demolition.

Demolition of a nuclear plant is not so simple. In fact, nuclear plants are not demolished, they are decommissioned. **Decommissioning** involves removing the fuel, cleaning surfaces, and permanently preventing people from coming into contact with the contaminated buildings or equipment. Today, over 70 nuclear plants in the world are shut down and waiting to be decommissioned, and the number of plants scheduled for decommissioning will grow. By 2005, 68 of the 104 nuclear plants in the United States will be 20 years old or older. The Nuclear Regulatory Commission authorizes their operation for 40 years but is considering extending authorization for an additional 20 years. This would put off the need for decommissioning for a few years, but eventually, it will be necessary. In the United States, many small experimental facilities have been

Figure 11.15 **Cooling Towers** Cooling towers draw air over wet surfaces. The evaporation of the water cools the surface, removing heat from the process and releasing the heat into the air.

Table 11.4 Low-Level Radioactive Contaminants from Decommissioning an 1100-Megawatt Pressurized-Water Reactor

Material	Volume (Cubic Meters)
Radioactive	618
Activated	
Metal	484
Concrete	707
Contaminated	
Metal	5465
Concrete	10,613
Total	17,887

Source: Data from *Worldwatch Paper 69*, "Decommissioning: Nuclear Power's Missing Link," and U.S. Nuclear Regulatory Commission.

decommissioned, and 14 plants have been removed from service and are already awaiting decommissioning. Canada has decommissioned three plants and has one scheduled for decommissioning. Dozens of nuclear reactors have been shut down or are expected to close soon in Europe, Canada, the former Soviet republics, and Japan. The great majority of near-term decommissioning work will take place in Europe: France, Germany, and Britain each are decommissioning 20 or more reactors, while nations using less nuclear power, such as Italy, Belgium, Spain, and Sweden, are decommissioning one or more reactors.

The European Commission (EC) is working to standardize national decommissioning policies. Currently, policies vary widely on key questions such as how long sites will be put in safe storage to allow radiation to decay (from a maximum of 30 years in Finland to 135 years in Britain). Waste management also poses a serious challenge. The EC estimates that by 2060, decommissioning will produce more than 2 million metric tons of metallic and concrete waste, but many member states currently have, at best, limited storage and disposal options to handle such amounts of waste.

The ongoing international effort to close unsafe Soviet-designed reactors in central and Eastern Europe has important decommissioning implications. The European Commission has made closure of these reactors a condition for membership in the European Union and has pledged financial support for decommissioning eight

reactors in Lithuania, Bulgaria, and the Slovak Republic that are expected to be closed by 2008. The European Bank for Reconstruction and Development is collecting international contributions for these projects and supports predecommissioning work at Ukraine's Chernobyl plant.

Because most Asian reactors are not as far into their licensed operating lives, decommissioning there is a less urgent issue—although it will become a concern as reactors age in Japan, South Korea, and Taiwan, all of which derive major shares of their electricity from nuclear power. However, none of these countries currently has a viable plan for long-term management of radioactive waste, so decommissioning may loom as a larger problem in coming decades. In September 2000, the Taiwanese economics minister raised concerns by arguing that Taiwan should scrap its partially constructed fourth nuclear power plant and phase out nuclear power by 2025 because it lacked a way to safely dispose of nuclear waste.

The decommissioning of a plant is a two-step process. In stage 1, the plant is shut down and all the fuel rods are removed; all of the water used as a reactor-core coolant, as a moderator, or to produce steam is drained; and the reactor and generator pipes are cleaned and flushed. The spent fuel rods, the drained water, and the material used to clean the pipes are all radioactive and must be safely stored or disposed of. This removes 99 percent of the radioactivity. Stage 2 consists of disman-

tling all the parts of the plant except the reactor and safely containing the reactor.

Utilities have three decommissioning options: (1) decontaminate and dismantle the plant as soon as it is shut down; (2) shut the plant down for 20 to 100 years to allow radioactive materials that have a short half-life to disintegrate and then dismantle the plant; (3) entomb the plant by covering the reactor with reinforced concrete and placing a barrier around the plant.

Originally, entombment was thought to be the best method. But because of the long half-life of some of the radioactive material and the danger of groundwater contamination, this method now appears to be the least favorable. Most U.S. utilities plan to allow about 60 years to elapse between stage 1 and stage 2. Japan, which has greater needs for land, plans to mothball its plants for 5 to 10 years before dismantling them. The costs associated with decommissioning include dismantling the plant and then packaging, transporting, and burying the wastes. Because of contamination and activated material, ordinary dismantling methods cannot be used to demolish the buildings. Special remote-control machinery must be developed. Techniques that do not release dust into the atmosphere must be used, and workers must wear protective clothing. Waste disposal is estimated to be 40 percent of the decommissioning costs. Table 11.4 shows the volume of material that would be handled from an 1100-megawatt power plant.

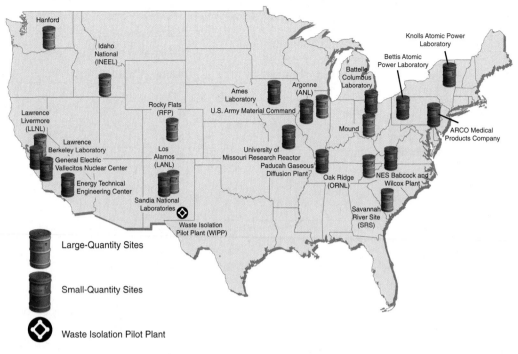

Figure 11.16 **Department of Energy High-Level Radioactive Transuranic Waste Sites** The high-level transuranic radioactive waste generated by the U.S. Department of Energy will be shipped to the Waste Isolation Pilot Plant near Carlsbad, New Mexico, for storage.

Source: Nuclear Regulatory Commission.

Recent experience indicates that the cost for decommissioning a large plant will be between $200 million and $400 million, about 5 percent of the cost of generating electricity. Although the mechanisms vary among countries, the money for decommissioning is generally collected over the useful life of the plant.

Radioactive Waste Disposal

When the world entered the atomic age, the problem of the disposal of nuclear waste was not fully appreciated. Today, in the United States, 380,000 cubic meters of highly radioactive military waste are temporarily stored at several sites. (See figure 11.16.) Near Carlsbad, New Mexico, the Waste Isolation Pilot Plant was built to house high-level radioactive waste. The wastes, commonly referred to as **transuranic wastes,** consist primarily of various isotopes of plutonium. This facility began accepting waste in March 1999.

In addition to the high-level waste from weapons programs, 2 million cubic meters (525 million gallons) of low-level radioactive military and commercial waste

are buried at various sites. In addition, about 30,000 metric tons of highly radioactive spent fuel rods are stored in special storage ponds at nuclear reactor sites. Many plants are running out of storage space and have been authorized by the Nuclear Regulatory Commission to store radioactive waste in aboveground casks, because the storage ponds will not accommodate additional waste.

In other countries, spent fuel rods are either stored in special ponds or sent to reprocessing plants. As previously mentioned, even though reprocessing is more expensive than manufacturing fuel rods from ore, some countries reprocess as an alternative to waste storage.

High-Level Radioactive Waste

At this time, no country has a permanent storage solution for the disposal of high-level radioactive waste. Most experts feel the best solution is to bury it in a stable geologic formation. Several countries are investigating storage in salt deposits. The waste is mixed with silica, borax, and other chemicals, melted, and poured into

metallic (stainless steel) canisters to form borosilicate glass. Each container would then be sealed in a thick-walled, stainless steel canister. Due to the high temperature of radioactive waste, the containers would be stored aboveground for 10 years. At the end of this time, the temperature of the container would have decreased, and the material could be buried in a salt deposit 600 meters (1968 feet) below the surface.

Sweden intends to store its waste 500 meters (1640 feet) underground in granite. Site construction is not planned until 2010, when Sweden's 12 reactors are scheduled to be decommissioned (see The Politics of Nuclear Power later in this chapter).

High-Level Waste Storage in the United States

In December 1982, the U.S. Congress passed legislation calling for an additional high-level radioactive disposal site for the storage of spent fuel rods from nuclear power plants to be selected by March 1987 and to be completed by 1998. The project quickly fell behind schedule. In 2002, the Secretary of Energy indicated that the choice of a site at Yucca Mountain in Nevada was based on "scientifically sound and suitable science." The site is in an unpopulated area near the Nevada Test Site, where several nuclear devices were exploded. It is a very dry area, and the water table is about 600 meters (1968 feet) below the mountain. It was also considered geologically stable; however, the site has witnessed seismic activity since being chosen as a disposal site.

Work has begun on the series of tunnels that would serve as storage places for the waste. (See figure 11.17.) However, current work is primarily exploratory and is seeking to characterize the likelihood of earthquake damage and the movement of water through sediments. If completed, the facility would hold about 70,000 metric tons of spent fuel rods and other highly radioactive material. It will not be completed before 2015, and by that time, the total amount of waste produced by nuclear power plants will exceed the storage capacity of the site. The volume of reprocessing waste from nuclear weapons production is larger than the volume of spent fuel from nuclear power plants, but the total activity of the spent fuel is much greater than that of the defense wastes.

A new milestone in the history of Yucca Mountain was passed in February 2002, when President George W. Bush notified Congress that he considered the Yucca Mountain site qualified to receive a construction permit. In response, the governor of Nevada, Kenny Guinn, filed an official notice of disapproval. Subsequently, both the House of Representatives and the Senate voted to override Governor Guinn's disapproval, and in July 2002, President Bush signed a joint resolution allowing construction to proceed. However, this is not the final step, since some members of Congress have vowed to continue to fight this designation, and the state of Nevada has several lawsuits pending in federal courts.

Low-Level Radioactive Waste

Low-level radioactive waste includes the cooling water from nuclear reactors, material from decommissioned reactors,

(a)

(b)

Figure 11.17 **High-Level Nuclear Waste Disposal** (a) Current plans for the disposal of high-level nuclear waste involve placing materials in tunnels underground at Yucca Mountain, Nevada. The material could be stored about 300 meters (984 feet) below the surface and about 300 meters above the water table. Because of the dry climate, it is considered unlikely that water infiltrating the soil would move nuclear material downward to the water table. (b) Yucca Mountain, United States: This image obtained 22 February, 2004 shows the entrance to the Yucca Mountain nuclear waste repository located in Nye County, Nevada, about 100 miles northwest of Las Vegas. Yucca Mountain is the U.S. Department of Energy's potential geologic repository designed to store and dispose of spent nuclear fuel and high-level radioactive waste. By the year 2010, if all licenses are granted, the facility will begin receiving casks of spent atomic fuel from aircraft carriers, submarines and power plants and will eventually become the largest repository of nuclear waste in the world.

radioactive materials used in the medical field, protective clothing worn by persons working with radioactive materials, and materials from many other modern uses of radioactive isotopes. Disposal of this type of radioactive waste is very difficult to control. Estimates indicate that much of it is not disposed of properly.

In 1970, a U.S. moratorium halted the dumping of radioactive waste in the oceans. Before this, the United States placed some 90,000 barrels of radioactive waste on the ocean floor. European countries also have dumped both high-level and low-level radioactive material into the Atlantic Ocean. Before 1983, when ocean dumping was halted, these countries disposed of 90,000 metric tons of radioactive waste in the ocean.

The United States produces about 800,000 cubic meters of low-level radioactive waste per year. This is presently being buried in disposal sites in South Carolina and Washington. Temporary storage sites have also been opened in Utah and Texas. However, these states balked at accepting all the country's low-level waste. In 1980, Congress set a deadline of 1986 (later extended to 1993) for each state to provide for its own low-level radioactive waste storage site. Later, a change allowed several states to cooperate and form regional coalitions, called compacts. Under this arrangement, one state would provide a disposal site for the entire compact. Several states did not join compacts, and several others have since withdrawn from the compacts they were originally a part of. Figure 11.18 shows the current compacts and the status of their disposal sites. Several of the proposed sites are under severe pressure from activists who do not want such sites in their locality. As a result, most states still do not have a permanent place to deposit low-level radioactive wastes and are relying on temporary storage facilities in Richland, Washington, or Barnwell, South Carolina. (See Environmental Close-Up: The Hanford Facility: A Storehouse of Nuclear Remains.) As a further example of the creative politics engendered by the need to have low-level radioactive waste disposal sites and the unwillingness to host such a site, the states of Texas, Maine, and Vermont have applied to be a

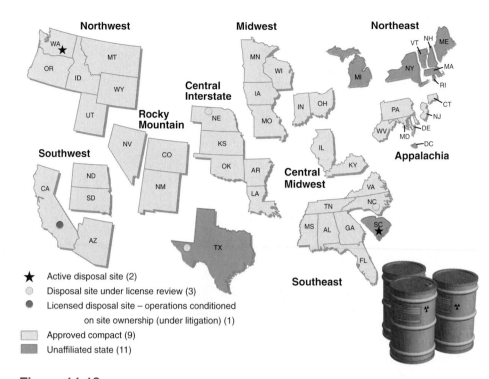

Figure 11.18 Low-Level Radioactive Waste Sites Each state is responsible for the disposal of its low-level radioactive waste. Many states have formed compacts and selected a state to host the disposal site. None of the proposed sites is currently accepting waste. Many other states do not have an acceptable site and are temporarily relying on a site in South Carolina to accept their waste. Eventually, they will need to find other acceptable sites. The map shows the current compacts and proposed disposal sites.

Source: Nuclear Regulatory Commission.

compact, with Texas serving as the host for the site.

The Politics of Nuclear Power

Nuclear power is projected to represent a shrinking share of the world's electricity consumption from 2004 through 2025, despite a net increase in world nuclear capacity as a result of new construction and plant life extensions.

In 2004, 19 countries depended on nuclear power for at least 20 percent of their electricity generation. In absolute terms, the world's total nuclear power usage is projected to increase from 353 gigawatts in 2004 to 366 gigawatts in 2025. Most nuclear additions are expected to be in Asia, where China, India, Japan, and South Korea are projected to add a combined total of approximately 45 gigawatts of nuclear capacity between

2004 and 2025, while the rest of the world sheds some 32 gigawatts of existing capacity. In addition, life extensions, higher capacity factors, and improved outputs are expected to offset some of the capacity lost through plant retirements in other parts of the world. Life extension and higher capacity factors will play a major role in sustaining the U.S. nuclear industry throughout the forecast period. Russia also has an ambitious life extension program. Thus, despite a declining share of global electricity production, nuclear power is projected to continue in its role as an important source of electric power.

Nuclear power projections are subject to considerable uncertainty, both economic and political. On the optimistic side, for example, emerging technologies could change the economics and perceived safety of nuclear power plants, as well as public sentiment about radioactive waste disposal and nuclear weapons proliferation. In the high end, world nuclear capacity is projected to grow from 353 gigawatts in 2004 to 567 gigawatts in 2025.

On the pessimistic side, whatever public support for nuclear power is currently in evidence could be eroded quickly if a serious nuclear mishap occurred anywhere in the world; expected technology breakthroughs might not materialize; and future delays or cost overruns on nuclear power construction projects could adversely affect economics. In fact, no new nuclear power plants have been ordered since 1978 in the United States and none since 1993 in the European member countries of the Organization for Economic Cooperation and Development. Nuclear power development generally depends on government support or sanction, and political developments can bring into power political parties that are opposed to the nuclear option, as has happened in Western Europe in recent years. In the low nuclear growth case, world nuclear capacity is projected to shrink from 353 gigawatts in 2004 to 253 gigawatts in 2025. The low nuclear growth case does include new capacity additions in other regions, specifically Asia.

In large part, government support for nuclear power has waxed and waned with the changing of governing regimes, depending on whether the nation's ruling party is liberal or conservative. In recent years, public officials and industry representatives from various nations have called for a reevaluation of nuclear power. For example, France, the Netherlands, Italy, and the United States have governments more favorably inclined to nuclear options. In 2002, the interim head of the Italian environmental protection agency stated that the country should review its nuclear energy options and consider the potential national benefits of new-generation technologies. In the Netherlands, representatives of the ruling coalition have proposed construction of a new plant. In the United States in 2003 the Bush administration's energy plan calls for the expansion of nuclear energy "as a major component of our national energy policy." Current U.S. energy goals include an intended new plant by the end of the current decade. Further, the Bush administration's budget for 2003 included a provision to increase spending on nuclear technology research to $46.5 million from $12 million in 2002.

In contrast, liberal governments in Sweden and Germany have committed both nations to the early retirement of their nuclear power sectors, and their successes at the ballot box (2002) lowered the odds of reviving nuclear power programs in both countries. Since June 2000, Germany has been committed to the shutdown of its nuclear power industry by the mid-2020s, or after German reactors have been operational for an average of 32 years. Germany's Social Democratic chancellor, Gerhard Schröder, with the strong backing of political allies in the environmentalist Green Party, negotiated the terms of the nuclear phaseout with Germany's electricity industry. It remains unclear, however, whether the goals will be met. Shortly after the 2001 election, the German nuclear supply industry showed some hesitancy about meeting the agreed target date. In 2003, Energie Badfen Warttemberg AG applied for government permission to delay the scheduled closure of its Obrigheim nuclear power station for five years; and the chief executive officer of E.ON, Germany's second-largest electricity company, has called for the retention of nuclear power.

Political and economic considerations clearly can affect national plans for moving away from nuclear power. For instance, Sweden is committed to closing down its nuclear power industry entirely by the time the youngest of its nuclear power reactors reaches the end of its expected life span—which was generally assumed to be around the year 2010—but the first two plant closures in the nuclear phaseout plan were repeatedly delayed. Barsebäck 1, originally scheduled for shutdown in July 1998, continued operating until November 1999; and Barsebäck 2, originally scheduled for closure in 2001, remains in operation. Only two months after the Swedish elections in November 2002, two reports commissioned by the government pointed to the difficulties that might arise from closing Barsebäck 2 on schedule. In March 2003, the Swedish government admitted that the necessary conditions for closing Barsebäck 2 (i.e., finding an alternative source of power) could not be met.

Sweden's goal of phasing out its nuclear generation and simultaneously attempting to meet its commitment to greenhouse gas reductions following its ratification of the Kyoto Protocol in March 2002 poses a particular dilemma for this resource-intensive nation. Energy-intensive industries, such as forest products, iron, and steel, contribute a sizable sum to Sweden's gross domestic product and exports, and it has been estimated that 5 percent of the nation's GDP could be lost when nuclear power is phased out entirely. Combining a nuclear phaseout with climate change commitments could cost Sweden roughly one-third of its annual GDP.

Another factor being weighed by European nations in deciding whether to abandon, continue, or expand their nuclear power programs is the influence of the multilateral European Union (EU). Although the EU does not set the energy policies of its members, its voice can influence the debate. European Commission Vice President (and also Transport and Energy Commissioner) Loyola de Palacio has stated that it would be "irresponsible" for countries to ignore nuclear power, and in mid-2002, the commission published a report that called for keeping the nuclear option open.

Developments in Asia

In Asia, nuclear power plants are currently under construction in China, South Korea, India, Taiwan, and Japan. In contrast to most of the rest of the world, developing Asia, in particular, still supports a buoyant nuclear power plant construction industry. The developing countries of Asia (excluding Japan, which is part of the industrialized Asia country grouping) are projected to have a 17-percent share of the world's total nuclear power capacity in 2025, up from 7 percent in 2003.

China

In 2001, China had only three nuclear power units in operation: Guangdong 1 and 2 (944 megawatts each), and Qinshan 2 (279 megawatts). Four new units were opened in 2002, adding a total of 3151 megawatts of nuclear capacity:

Environmental Close-Up

The Hanford Facility: A Storehouse of Nuclear Remains

In March 1943, the U.S. government obtained 1650 square kilometers (640 square miles) in southeastern Washington to house a large-scale plutonium production facility. The Hanford site, located on the Columbia River, offered a mild climate, isolation, and abundant water, all of which were necessary for the operation of reactors and reprocessing plants.

Plutonium production on this scale had never before been attempted. However, due to the urgent pressures to develop nuclear weapons during World War II, the normal process of using small-scale research facilities before initiating large production facilities was sidestepped. General Leslie Groves, the Army's head of Hanford's development, expressed concern regarding the safety of industrial-scale plutonium production but accepted the risks because of the urgent need of the war effort. Residents were given three weeks' notice to vacate the area, and construction began immediately.

The first nuclear reactor and chemical processing plants were completed in late 1944. Over the course of the following two decades, eight additional reactors and several processing plants were constructed.

During the height of its production, the Hanford facility was the primary source of plutonium for the United States: the world's first nuclear explosion in 1945 and the atomic bomb dropped on Nagasaki at the end of World War II each contained plutonium isolated at Hanford.

Hanford's plutonium and uranium production facility is no longer in operation. However, it has left a legacy of contamination and waste that has aroused the concern of scientists, citizens, and policy makers alike; this reservation is one of the largest and most complex environmental cleanup sites in the United States.

Nuclear waste at the Hanford reservation exists in three forms: contained liquid and sludge, buried and stored solid waste, and released liquids. Of the 2 million cubic meters (525 million gallons) of high-level waste produced at Hanford over the four decades of its operation, 202,000 cubic meters (53 million gallons) remain onsite, contained within underground tanks, buildings, or concrete basins. This comprises approximately 60 percent of all high-level waste existing in the United States today.

For the first 20 years of operation, Hanford's liquid waste was poured into single-shell carbon-steel tanks for storage. Of the 149 single-shell tanks on the reservation, 67 have cumulatively leaked over 3811 cubic meters (1 million gallons) of waste into the surrounding soil. Following the initial discoveries of these leaks, 28 double-shell tanks were constructed to house the waste. Most of the liquids present in the single-shell tanks have been pumped into the newer tanks, and no leaks have been discovered. However, the design life of the oldest double-shell tanks has been reached.

The vast majority of the remaining irradiated uranium fuel, which never underwent reprocessing, is stored in two water-filled concrete basins. Some of this fuel has corroded, and radionuclides have been detected in the basin water, walls, and sludge, as well as surrounding soil and groundwater. Approximately 11 metric tons of plutonium are also present onsite and are contained within three separate facilities.

Buried and stored solid waste is composed of boxes, crates, and drums holding materials such as clothing, rags, and tools that were contaminated with chemicals or radioactivity. About 97 percent of this waste is buried in one of 75 solid waste burial grounds on the Hanford reservation, containing nearly 600 metric tons of uranium and 360 kilograms (800 pounds) of plutonium. In 2000, groundwater near one of the burial sites was found to contain 400 times more tritium than drinking water standards allow.

Release of waste has contributed nearly 1 million curies of radioactivity and 90,000 to 270,000 metric tons of chemicals to the soil and groundwater beneath Hanford. In the first months of Hanford's operation, low-level liquid waste was discharged directly onto the ground, from whence contamination leached into the soil, sediment, and groundwater. Evaporated waste also left surface residues that were dispersed by wind to surrounding wetlands.

When this contamination became evident, liquid wastes were instead pumped into wells. However, this only accelerated the process of contamination by injecting the waste closer to underground aquifers. Cribs (underground boxlike structures), backfilled trenches, vertical buried pipes, and gravel-filled fields were then constructed to receive the liquid waste. Non- or slightly contaminated liquids were released into open ponds and ditches.

Today, plumes of groundwater contaminated with metals, chemicals, and radionuclides cover 400 square kilometers (150 square miles). Intentionally discharged tank waste has contributed additional radioactive material, and sections of the underlying aquifer have been contaminated with uranium, plutonium, and other metallic isotopes.

Radionuclides were also released into the atmosphere from Hanford's reactors and reprocessing plants throughout the course of their production, exposing residents downwind of Hanford to radioactive argon and iodine. Radioactive material was also released into the Columbia River, where it now remains buried in sediment.

Today, Hanford contains 440 million curies of radioactivity, or 40 percent of all human-made radioactivity in the nuclear weapons complex. The Department of Energy, the Environmental Protection Agency, and the Washington Department of Energy signed an agreement, known as the Tri-Party Agreement, in 1989, calling for cleanup and better management of hazardous materials. However, the continued dispersal of radionuclides and sheer immensity of the project present a formidable challenge to the effective removal of radioactive material.

Lingao 1 and 2 (938 megawatts each), Qinshan 2 unit 1 (610 megawatts), and Qinshan 3 unit 1 (665 megawatts). China's nuclear capacity is projected to grow from 2167 megawatts in 2001 to 19,593 megawatts in 2025—the largest increase projected for any country in the world.

China has been attempting to develop an indigenous nuclear technology base for some time. Thus far, China's nuclear power program has used a variety of nuclear technologies, some imported and some domestic. A goal of the program, as stated by the chairman of China's Atomic Energy Authority, is to "attain independence in the design, manufacture, and operation of large nuclear power units on the basis of learning [from the] advanced experience of other countries." China's first reactors, Guangdong 1 and 2, were designed by French Framatome ANP and came on-line in 1993 and 1994. Qinshan 1, which came on-line in 1991, was China's first domestically designed unit, and its design was scaled up for Qinshan 2 units 1 and 2. Qinshan 3 unit 1 is China's first reactor based on Canadian Candu technology. The two Lingao reactors that came on-line in 2002 use French technology supplied by Framatome ANP.

South Korea

South Korea's nuclear power capacity is projected to grow from 12,990 megawatts in 2003 to 27,607 megawatts in 2025. Two 960-megawatt units are currently under construction. The country has pursued an aggressive nuclear power program since the late 1970s and has announced plans to build 10 new nuclear power reactors by 2025.

Japan

Japan is one of the few advanced industrialized nations projected to build additional reactors over the 2004–2025 time frame. Japan—the world's third-largest producer of nuclear power, after the United States and France—completed its fifty-third nuclear reactor in 2001, the 798-megawatt Onagawa 3. Japan's nuclear power capacity is projected to grow from 43,245 megawatts in 2004 to 51,899 megawatts in 2025.

Recent events could stall Japan's effort to expand its nuclear power industry. A scandal of major proportions emerged in August 2002, when it was disclosed that Japan's largest nuclear power company, Tokyo Electric Power Company, had filed falsified inspection documents for 13 reactors. The documents concealed from government regulators knowledge about cracks in structures holding nuclear fuel in place in reactor cores at several power plants. As a result of the disclosures, several senior executives, including the company's president, were forced to resign.

Japan's Nuclear and Industrial Safety Agency ordered the shutdown of the Fukushima plant for up to one year and by early 2003 had suspended operations at all of its 17 nuclear reactors. Several of the other nine nuclear utilities in Japan also reported similar wrongdoings. In September 2002, two of Japan's producers of nuclear power, Chubu Electric Power Company and Tohoku Electric Power Company, reported "questionable handling of nuclear reactor inspections," and by the end of 2002, a reported 13 nuclear reactors had been shut down. In reaction to the falsification of inspections and repairs, the Japanese minister of Economy, Trade and Industry stated, "It is absolutely abominable that this incident caused the people's confidence to be largely lost in nuclear power."

These industrial improprieties have heightened public concern over the reporting practices at Japan's nuclear power plants and the integrity of its nuclear industry. Whether they will result in a major reevaluation of the country's nuclear power future by Japanese policy makers and industry is uncertain.

Summary

Nuclear fission is the splitting of the nucleus of the atom. The resulting energy can be used for a variety of purposes. The splitting of U-235 in a nuclear reactor can be used to heat water to produce steam that generates electricity. Various kinds of nuclear reactors have been constructed, including boiling-water reactors, pressurized-water reactors, heavy-water reactors, gas-cooled reactors, and experimental breeder reactors. Scientists are also conducting research on the possibilities of using fusion to generate electricity. All reactors contain a core with fuel, a moderator to control the rate of the reaction, and a cooling mechanism to prevent the reactor from overheating.

The nuclear fuel cycle involves mining and enriching the original uranium ore, fabricating it into fuel rods, using the fuel in reactors, and reprocessing or storing the spent fuel rods. The fuel and wastes must also be transported. At each step in the cycle, there is danger of exposure. During the entire cycle, great care must be taken to prevent accidental releases of nuclear material.

The use of nuclear materials for military purposes has resulted in the same kinds of problems caused by nuclear power generation. The reduced risk of nuclear warfare has led to a need to destroy nuclear weapons and clean up the sites contaminated by their construction.

Public acceptance of nuclear power plants has been declining. Initial promises of cheap electricity have not been fulfilled because of expensive construction, cleanup, and decommissioning costs. The accidents at Three Mile Island in the United States and Chernobyl in Ukraine have accelerated concerns about the safety of nuclear power plants. In Europe and Asia the future of nuclear power is very different. Europe is phasing out nuclear power while parts of Asia are expanding nuclear power capacity.

Other concerns are the unknown dangers associated with exposure to low-level radiation, the difficulty of agreeing to proper long-term storage of high-level and low-level radioactive waste, and thermal pollution.

Issues – Analysis

Predicting the Future of Nuclear Power

History is characterized by both slow forces that bring change over long periods and rapid leaps that either through visionary leadership or by necessity bring about completely new paradigms.

The events that have helped to shape the United States during the past half-century are the very same events that have shaped the development of nuclear technology. World War II and the events since that time saw the United States, a nation that was content to focus on matters within its borders, assume the leadership of the Western world.

Nuclear technology grew up with America internationalism and industrial preeminence. It is forever shaped by its role at the center of the Manhattan Project, the Cold War that followed, and the Atoms for Peace programs of the 1950s and 1960s.

It was shaped by visionaries such as Admiral Hyman Rickover and chemist Glenn Seaborg, and by evolutionary forces such as economics and changing public perceptions of science, government, and environment.

We do not pretend to know the future; predictions and projections over the long term are famously unreliable—particularly in the energy arena. Nevertheless, we can see the formation of the forces that will shape the next few decades:

1. **Globalization:** It is perhaps one of the most overused terms today, but nonetheless a major force that will prompt change in many sectors of economics and politics. For nuclear energy, globalization has already had huge impacts. The suppliers of nuclear technology had rapidly consolidated, and the utilities that implement energy technologies are not far behind. There now appear to be three main blocks of suppliers in the world; further consolidation is likely. On the utility side, international alliances and ownership relationships are already occurring. Thus far, the market for nuclear technology is still a country-by-country market. If the market is to mature and thrive, it must become global as well.

2. **The rise of the developing world:** For the first time in the history of the world, it is possible to envision a realistic course by which, in a relatively short period, most of the world's people will enjoy modern conveniences such as electricity, instead of merely a fortunate few. The rapid modernization in Asia and the maturing democracies of Latin America will place demands on the world's energy resources that will far outstrip past experience. Efficiency of energy use will be critical for these countries, as will extensive use of renewable energy resources. But their industrial needs will require the deployment of power plants. This represents a major new opportunity for nuclear power, one that already accounts for much of the near-term growth in its use.

3. **The expansion of free markets:** The expanded dominance of free-market economics and decision making is reshaping both the developing world and the industrialized world in a number of ways. In many parts of the world, centralized planning is giving ground to free-market economics, even in the business of U.S.

electric utilities. The cost of providing electricity will determine what technologies are successful. Government policies, preferences, and goals that are not directly prescriptive or do not have a direct impact on cost will be overwhelmed by the desire of energy providers to be reliable, cost-effective providers of electricity.

4. **National and international environmental issues:** The market forces, as always, will be tempered by government regulation. Environmental rules are likely to evolve over the coming decades as public concern over air quality and climate change intensifies. It is impossible to predict what shape these government policies will take in the future. In many countries, they are already acting to discourage construction of new power plants (of any kind). As power costs rise and reliability suffers, these policies will come under increased scrutiny. As countries adapt their rules to allow for growth in supply while minimizing harm to the environment, the window of opportunity for nuclear power will open wider. At the same time, however, how nations deal with nuclear waste will be a key environmental issue in the future. It is popularly understood that dealing successfully with spent fuel disposition is required if a new generation of nuclear plants is to become a reality. For many countries, this will become an international issue as the subject of regional high-level waste repositories is explored.

5. **The post–Cold War world:** Overlying many of the preceding issues, the aftermath of the Cold War will continue to have an impact on nuclear energy. With the end of the Cold War, government uranium and plutonium inventories are being released for commercial use. The implications are proving to be significant. The fragile nuclear fuel market is faltering, and concerns about the potential that materials could fall into the wrong hands are growing. Further, the Cold War discipline of a bipolar world is being replaced by a much more complicated structure that finds some nations in turmoil and other nations going their own way on issues of concern to the international community. In this challenging environment, the construction of new nuclear power plants throughout the world presents a complicating factor for many analysts concerned about proliferation of nuclear materials. Ignoring these concerns could be fatal to any effort to expand the use of nuclear energy in industrializing nations.

The future of nuclear energy will be shaped by these forces. Recognizing them will help us define a vision for the next generation of nuclear power systems and enable us to lay the groundwork for activities that can ensure, or not, the ability of nuclear technology to meet the challenges of the future. While we cannot predict the future, we can see that an opportunity exists to shape it.

- Do you agree with the above five points?
- Are there other points that need to be raised?
- What do you feel is the role of nuclear power in the future?

Key Terms

alpha radiation *221*
beta radiation *221*
boiling-water reactor (BWR) *223*
decommissioning *234*
fissionable *222*
gamma radiation *221*
gas-cooled reactor (GCR) *223*
heavy-water reactor (HWR) *223*
light-water reactor (LWR) *223*

liquid metal fast-breeder reactor
    (LMFBR) *227*
moderator *222*
nuclear breeder reactor *226*
nuclear chain reaction *222*
nuclear fission *222*
nuclear fusion *227*
nuclear reactor *222*
plutonium-239 (Pu-239) *226*

pressurized-water reactor (PWR) *223*
radiation *221*
radioactive *221*
radioactive half-life *221*
rems *233*
thermal pollution *233*
transuranic waste *236*
uranium-235 (U-235) *222*

Review Questions

1. How does a nuclear power plant generate electricity?
2. Name the steps in the nuclear fuel cycle.
3. What is a rem?
4. What is a nuclear chain reaction?
5. How will nuclear fuel supplies, the cost of decommissioning facilities, and the storage and ultimate disposal of nuclear wastes influence the nuclear power industry?
6. What happened at Chernobyl, and why did it happen?
7. Describe a boiling-water reactor, and explain how it works.
8. How is plutonium-239 produced in a breeder reactor?
9. Why is plutonium-239 considered dangerous?
10. Why is fusion not being used as a source of energy?
11. What are the major environmental problems associated with the use of nuclear power?
12. List three environmental problems associated with the construction and subsequent de-emphasis of nuclear weapons.

Critical Thinking Questions

1. Recent concerns about global warming have begun to revive the nuclear industry in the United States. Do you think nuclear power should be used instead of coal for generating electricity? Why?
2. Why has nuclear power not become the power source that is "too cheap to meter," as foreseen by President Eisenhower?
3. Disposal of radioactive wastes is a big problem for the nuclear energy industry. What are some of the things that need to be evaluated when considering nuclear waste disposal? What criteria would you use to judge whether a storage proposal is adequate or not?
4. Nuclear weapons testing has released nuclear radiation into the environment. These tests have always been justified as necessary for national security. Do you agree or not? What are the risks? What are the benefits?
5. Some states allow consumers to choose an electric supplier. Would you choose an alternative to nuclear or coal even if it cost more?

Welcome to McGraw-Hill's Online Learning Center

Location: www.mhhe.com/enger10e

Interactive Exploration

Check out the website at **www.mhhe.com/enger10e** and click on the cover of this textbook for quizzing, career information, case studies, and hot links for the following topics:

Nuclear Energy

PART IV

Resource Management

The coastal sage scrub is one of the most diverse and endangered of southern California's ecosystems, that could be protected through the natural community conservation planning process.

Natural Community Conservation Planning

Morgan E. Barrows

Saddleback College

According to the 2000 U.S. census, 24.5 million people live in California, and the population continues to grow by approximately 670,000 people per year. This population growth brings about urban sprawl, which is the leading cause of species imperilment in California. This trend is particularly significant because California is home to eight of the 21 most endangered ecosystems in the United States. Over the past decade, it has become quite clear that the state and federal endangered species acts have not been effective in creating the broad solutions necessary to resolve conflicts between land use and conservation by California's growing population.

To help resolve the conflict, Governor Pete Wilson proposed on April 22, 1991 (Earth Day), the Natural Community Conservation Planning Act, a new policy intended to change the way California's natural resources would be managed and protected. As enacted by the California legislature, the stated goal of the act is to provide long-term, regional protection of biodiversity while still allowing development to occur. This goal is to be accomplished by preparing and implementing natural community conservation plans. These plans would establish permanent, multispecies reserves, with corridors and linkages to other natural lands while allowing compatible and appropriate land development and economic growth.

The Natural Community Conservation Planning Act represents a paradigm shift from endangered species legislation that is strictly a reactive, single-species, project-based permitting approach to a proactive, multispecies, ecosystem approach. The act focuses on this new approach to preserving biodiversity with the explicit goal of avoiding species listings, economic displacement, and legal battles. It is important to note that the act does not supercede the California or federal endangered species acts or their protections; rather, it provides a way to anticipate the consequences of such laws at a regional scale and plan for them. Thus, the California Department of Fish and Game and the U.S. Fish and Wildlife Service work closely together to ensure that natural community conservation plans can be approved under the act as well as satisfy the habitat conservation plan requirement of the federal Endangered Species Act.

In September 1991, with a memorandum of understanding between the California Department of Fish and Game and the U.S. Fish and Wildlife Service, the natural community conservation plan pilot program was established. This program focused on a portion of the coastal sage scrub ecosystem that consists of approximately 15,000 square kilometers (6000 square miles) that transect five counties: San Diego, Orange, Riverside, Los Angeles, and San Bernardino. Coastal sage scrub, among the most diverse and endangered of southern California's ecosystems, occurs in coastal terraces and foothills from sea level to 900 meters (3000 feet). The desire to have beachfront and ocean-view property or to convert land to agricultural uses has led to the destruction of about 90 percent of coastal sage scrub habitat, leaving only 1600 square kilometers (600 square miles), primarily on private land. This land is some of the most valuable undeveloped land in the United States since parcels close to the coast may be valued at more than $2.5 million per hectare ($1 million per acre). But at the same time, the coastal sage scrub habitat is home to approximately 100 plant and animal species that are considered rare, sensitive, threatened, or endangered by the California Department of Fish and Game or the U.S. Fish and Wildlife Service. Because private landowners want to develop their land and environmentalists hope to preserve the land, it is the perfect test site for a natural community conservation plan.

Typically, one natural community conservation plan would be coordinated across the five counties. However, since the region is so large, it has been divided into 11 subregions. Some subregions are further divided into subareas. Each subregion prepares a plan that will ultimately be integrated with the others to form one overall plan. So far, plans have been approved in San Diego and Orange counties, while plans in the other three counties are still being developed.

On March 25, 1993, the California gnatcatcher (*Polioptila californica californica*) was listed as a threatened species under the Endangered Species Act. Therefore, to protect the coastal sage scrub while subregional plans were being completed, U.S. Interior Secretary Bruce Babbitt invoked Section 4(d), a special rule under the Endangered Species Act. This rule formally recognizes the natural community conservation plan and allowed for no more than 5 percent of the existing coastal sage scrub habitat in the test site to be developed during the process of completing the plan.

Although it is too soon to determine the success of the pilot program in southern California, natural community conservation plans have now been established in Contra Costa, Merced, and Placer counties. Plans are also being considered in other northern California areas. Additionally, there have been discussions about implementing this program throughout the United States.

What Do You Think?

1. Should natural community conservation plans be implemented throughout the United States? Why or why not?

2. If you were designing a conservation management strategy, would you focus on individual species or on ecosystems? Explain your answer.

3. What are the benefits and limitations to adopting an ecosystem approach?

4. Do you believe that biodiversity can be successfully protected while still allowing appropriate development to occur?

Biodiversity Issues

12

Chapter Outline

Objectives

After reading this chapter, you should be able to:

- Recognize that humans significantly modify natural ecosystems.
- State the major causes of biodiversity loss.
- Give examples of genetic diversity, species diversity, and ecosystem diversity.
- Describe the values of biodiversity.
- Appreciate the ways humans modify forests.
- Identify causes of desertification.
- Describe the role of endangered species legislation and the biodiversity treaty.
- Describe techniques that foster sustainable use of wildlife and fisheries resources.

PAST PRESENT FUTURE

1973 The U.S. Congress passed the Endangered Species Act. In 1980, the U.S. Fish and Wildlife Service listed 224 species as endangered.

1992 The Convention on Biological Diversity was an outgrowth of the Earth Summit in Rio de Janeiro, Brazil.

1975 The Convention on International Trade in Endangered Species of Wild Fauna and Flora (CITES) entered into force, with 80 participating countries.

2002 The U.S. Fish and Wildlife Service listed 988 species as endangered.

2003 A total of 187 countries had ratified the convention. The United States is one of eight countries that have not ratified the Convention on Biological Diversity.

2003 By this year, 164 countries were participating in CITES.

Biodiversity Loss and Extinction

Biodiversity is a broad term used to describe the diversity of genes, species, and ecosystems in a region. **Extinction** is the death of a species, the elimination of all the individuals of a particular kind. Biodiversity is lost when all the individuals of a particular kind die off—that is, when a species becomes extinct. Extinction is a natural and common event in the long history of biological evolution. However, as we will see later in this section, extinction—and the resulting loss of biodiversity—is also a major consequence of human domination of the Earth.

Causes of Extinction

Complete extinction occurs when all the individuals of a species are eliminated. Besides complete extinction, we commonly observe *local extinctions* of populations. Although not as final, a local extinction indicates that the future of the species is not encouraging. Furthermore, as a population is reduced in size, some of the genetic diversity in the population is likely to be lost.

Studies of modern local extinctions suggest that certain kinds of species are more likely than others to become extinct. Table 12.1 lists the major factors that affect extinction. First, species that have small populations of dispersed individuals (low population density) are more prone to extinction because successful breeding is more difficult for them than it is for species that have large populations of relatively high density. Second, organisms in small, restricted areas, such as islands, are also prone to extinction because an environmental change in their locale can eliminate the entire species at once. Organisms scattered over large areas are much less likely to be negatively affected by one event. Third, specialized organisms are more likely to become extinct than are generalized ones. Since specialized organisms rely on a few key factors in the environment, anything that negatively affects these factors could result in their extinction, whereas generalists can use alternate resources. Finally, some kinds of organisms,

Table 12.1 Probability of Becoming Extinct

Most Likely to Become Extinct	Least Likely to Become Extinct
Low population density	High population density
Found in small area	Found over large area
Specialized niche	Generalized niche
Low reproductive rates	High reproductive rates

Most Likely to Become Extinct

Least Likely to Become Extinct

such as carnivores at higher trophic levels in food chains, typically have low populations but also have low rates of reproduction compared to their prey species.

Rabbits, raccoons, and rats are good examples of animals that are not likely to become extinct soon. They have high population densities and a wide geographic distribution. In addition, they have high reproductive rates and are generalists that can live under a variety of conditions and use a variety of items as food. The cheetah is much more likely to become extinct because it has a low population density, is restricted to certain parts of Africa, has low reproductive rates, and has very specialized food habits. It must run down small antelope, in the open, during daylight, by itself. Similarly, the entire wild whooping crane species consists of about 300 individuals restricted to small winter and summer ranges that must have isolated marshes. (Captive and experimental populations bring the total number to about 420 individuals.) In addition, their rate of reproduction is low.

Extinction as a Result of Human Activity

At one time, a human was just another consumer somewhere in the food chain. Humans fell prey to predators and died as a result of disease and accident just like other animals. The simple tools they used would not allow major changes in their surroundings, so these people did not have a long-term effect on their surroundings. They only minimally exploited mineral and energy resources. They mined chert, obsidian, and raw copper for the manufacture of tools and used certain other mineral materials, as well. Salt was used as a nutrient, clay for pottery, and ocher as a pigment. Aside from occasional use of surface seams of coal or surface oil seeps, early humans met most of their energy needs by muscle power or the burning of biomass.

As human populations grew, and as their tools and methods of using them became more advanced, the impact that a single human could have on his or her surroundings increased tremendously. The purposeful use of fire was one of the first events that marked the capability of humans to change ecosystems. Although in many ecosystems fires were natural events, the use of fire by humans to capture game and to clear land for gardens could destroy climax communities and return them to earlier successional stages more frequently than normal.

As technology advanced, wood was harvested for fuel and building materials,

Figure 12.1 **Changes in the Ability of Humans to Modify Their World** As technology has advanced, the ability of people to modify their surroundings has increased significantly. (*a*) When humans lacked technology, they had only minor impacts on the natural world. (*b*) The agricultural revolution resulted in many of the suitable parts of the Earth being converted to agriculture. (*c*) Modern agricultural technology allows major portions of the Earth to be transformed into agricultural land.

land was cleared for farming, streams were dammed to provide water power, and various mineral resources were exploited to provide energy and build machines. These modifications allowed larger human populations to survive, but always at the expense of previously existing ecosystems. (See figure 12.1.)

Humans are among the most successful organisms on Earth. We are adaptable, intelligent animals with few enemies. As our population increased, we displaced other kinds of organisms. This has resulted in an accelerated rate of extinction. Wherever humans become the dominant organism, extinctions occur. Sometimes, we use other animals directly as food. In doing so, we reduce the population of our prey species. Since our population is so large and because we have an advanced technology, catching or killing other animals for food is relatively easy. In some cases, this has led to extinctions. The passenger pigeon in North America, the moas (giant birds) of New Zealand, and the bison and wild cattle of Europe were certainly helped on their way to extinction by people who hunted them for food. Today, with over 6 billion people on Earth, nearly all of the Earth's surface has been affected in some way by human activity. One of the major impacts of human activity has been to reduce biodiversity.

Describing Biodiversity

Biodiversity is a broad term used to describe the diversity of genes, species, and

ecosystems in a region. Loss of biodiversity is recognized by many as a major consequence of human domination of the Earth. Furthermore, a loss of biodiversity is a threat to the human occupants of the planet. To understand the implications of the threat, we must have a clear understanding of what the concept of biodiversity includes. Biodiversity can be examined at the genetic, species, and ecosystem levels.

Genetic Diversity

Genetic diversity is a term used to describe the number of different kinds of genes present in a population or a species. High genetic diversity indicates that there are many different kinds of genes present and that individuals within the population will have different structures and abilities. Low genetic diversity indicates that nearly all the individuals in the population have the same characteristics. Several things influence the genetic diversity of a population.

1. **Mutations** are changes in the genetic information of an organism. Mutations introduce new genetic information into a population by modifying genes that are already present. Most of the mutations we observe are harmful, but occasionally a mutation results in a new valuable characteristic. For example, at some time in the past, mutations occurred in the DNA of certain insect species that made some individuals tolerant to the insecticide DDT, even though the chemical had not yet been invented. These mutant

characteristics remained very rare in these insect populations until DDT was used. Then, this characteristic became very valuable to the insects that carried it. Because insects that lacked the ability to tolerate DDT died when they came in contact with DDT, more of the DDT-tolerant individuals were left to reproduce the species, and therefore, the DDT-tolerance became much more common in these populations. If we think about the evolution of organisms, it is clear that incredible numbers of mutations were required for the evolution of complex plants, animals, and fungi from their single-celled ancestors. In the final analysis, all the characteristics shown by a species are the result of mutations.

2. *Migration* of individuals of a species from one population to another is also an important way to alter the genetic diversity of a population. Most species consist of many separate populations that are adapted to local environmental conditions. Whenever an organism leaves one population and enters another, it subtracts its genetic information from the population it left and adds it to the population it joins. If the migrating individual contains rare characteristics, it may significantly affect the genetic diversity of both populations. The extent of migration need not be great. As long as genes are entering or leaving a population, genetic diversity will change.

3. *Sexual reproduction* is another process that influences genetic diver-

sity. Although the process of sexual reproduction does not create new genetic information, it tends to generate new *genetic combinations* when the genetic information from two individuals mixes during fertilization, generating a unique individual. This doesn't directly change the genetic diversity of the population, but the new member may have a unique combination of characteristics so superior to those of other members of the population that the new member will be much more successful in producing offspring and will influence the genetic diversity of future generations.

4. *Population size* is a very important factor related to genetic diversity. The smaller the population, the less genetic diversity it can contain and the fewer the variations in the genes for specific characteristics. In addition, random events often can significantly alter the genetic diversity in small populations, with rare characteristics being lost from the population. For example, consider a population of a species of rose that consists of 100 individuals in which 95 of them have red blossoms and five have white blossoms. The death of five plants with red blossoms would not change genetic diversity very much. However, the death of the five plants with white blossoms could eliminate the characteristic from the population.

5. *Selective breeding* also can affect the genetic diversity of a species. Domesticated plants and animals have been modified over many generations by our choosing certain desired characteristics. Undesirable characteristics were eliminated, and desirable ones selected for. One of the consequences of this process is a loss of genetic diversity. Many modern domesticated organisms are very different from their wild ancestors and could not survive without our help. Often when genetic diversity is reduced, whether by reduced population size or selective breeding, deleterious genes for a characteristic may be present in both parents. When this occurs and the deleterious genes from both par-

ents are passed to the offspring, the survival of the offspring may be jeopardized.

Species Diversity

Species diversity is a measure of the number of different species present in an area. Some localities naturally have high species diversity, while others have low species diversity. For example, it is well known that tropical rainforests have very high species diversity, while Arctic regions have low species diversity. Some people find it useful to distinguish two kinds of species diversity. One way is to simply count the number of different kinds of species in an area. This is often referred to as *species richness*. Another way to look at species diversity is to take into account the number of different taxonomic categories of the species present. This can be called *taxonomic richness*. A region with many different taxonomic categories of organisms (frogs, birds, mammals, insects, pine trees, mosses) would have a higher species diversity compared to one that had fewer taxonomic categories (grasses, insects, birds, mammals).

Several things influence the species diversity of a particular location. Over the long term, the evolutionary and geologic history of a region impacts its species diversity. As mentioned earlier, some regions such as tropical rainforests naturally have greater species diversity compared to polar regions. Perhaps this is due to relatively recent climatic events. Twenty thousand years ago, glaciers covered much of northern Europe, the more northerly parts of North America, and parts of Asia. The glaciers would have drastically reduced the number of species present, and species would have migrated into the region as the glaciers receded. This same kind of drastic change did not take place in tropical regions; therefore, they retain more species. In the shorter term, migration can introduce new species to an area where they were not present previously. While such introductions would seem to increase species diversity, some invading species may have the effect of eliminating species that had been present. For example, the introduction of the zebra mussel to North America has had the effect of reduc-

ing the population size of many native mussels. Many of these native mussels are endangered.

The size of the area being considered also affects species diversity. In general, the larger the area being considered, the larger the species diversity. This is a natural consequence of including more kinds of spaces and the organisms that are adapted to them. For example, if species diversity were being measured in a desert, the diversity would increase greatly if a stream or water hole were included in the area under investigation.

Human activities have a great effect on the species diversity of a region. When humans exploit an area, they influence species diversity. They convert natural ecosystems to human-managed agricultural, forest, aquaculture, and urban ecosystems. They harvest certain species for their use. They specifically eliminate species that compete with more desirable species. They introduce species that are not native to the area. These topics will be discussed in greater depth later in the chapter.

Measuring species diversity is not easy. For one thing, we do not know how many species there are. Estimates of the actual number of species range from a few million to 100 million. About 1.4 million species have been described, but each year, new species are discovered, and in some groups, such as the bacteria, there may be millions of species yet unnamed. Another problem associated with measuring species diversity is the difficulty of discovering all the species in an area. Many species are naturally rare, and others live in difficult to reach places or are active only at certain times. For example, an examination of the kinds of organisms in the tops of trees in tropical forests yielded many species previously unknown. Table 12.2 gives current data about the number of different kinds of species in specific taxonomic groups.

Ecosystem Diversity

Ecosystem diversity is a measure of the number of kinds of ecosystems present in an area. Many regions of the world appear to be quite uniform in terms of the kinds of ecosystems present. For example, large

Table 12.2 Numbers of Described Species by Taxonomic Group

Taxonomic Group	Approximate Number of Described Species
Insects	751,000
Plants	248,428
Noninsect arthropods (mites, spiders, crustaceans, etc.)	123,151
Mollusks	50,000
Fungi	46,983
Protozoa	30,800
Algae	26,900
Fish	19,056
Flatworms	12,200
Roundworms	12,000
Earthworms and related organisms	12,000
Birds	9040
Jellyfish, corals, comb jellies	9000
Reptiles	6300
Starfish, sea urchins, sand dollars, etc.	6100
Sponges	5000
Bacteria, blue-green algae	4760
Amphibians	4184
Mammals	4000
Total	**1,380,902**

Source: World Resources Institute.

parts of central Australia, North Africa, and southwestern United States and adjacent Mexico are deserts. While there are general similarities (low rainfall, thorny woody plants, and animals that can survive on little water), each of these deserts is different and has specific organisms typical to the region. Furthermore, within each of these deserts are local regions where water is available. These areas may include rivers, springs, or rocky outcrops that collect and hold water. These locations support species of organisms not found elsewhere in the desert. In many other cases, natural events such as hurricanes, fires, floods, or volcanic eruptions may have destroyed the original vegetation, resulting in patches of early successional stages that contribute much to the diversity of organisms present.

In addition, local topographic conditions may create patches of the landscape that differ from the prevailing type.

Ridges typically have different vegetation on north- and south-facing slopes. Differences in soil type will support unique vegetation. Rolling terrain may provide pockets that contain temporary ponds that support different vegetation types. Each of these unique localities also can have animals that are specialized to the locality. Since many microhabitats exist within a region, the larger the area studied, the more the kinds of ecosystems present.

The Value of Biodiversity

Although most would agree that maintaining biodiversity is a good thing, a need exists to place a value on preserving biodiversity so that governments and individuals can apply logic to the daily decisions that affect biodiversity. There are many different ways to assign value to biodiversity. Some involve understanding the ecological roles played by organisms, others involve cold economic analysis, and still others stem from ethical considerations.

Biological and Ecosystem Services Values

Our species is totally dependent on the diversity of organisms on Earth. As consumer organisms in the food chain, we rely on the organic molecules produced by other organisms for food. Photosynthesis is directly responsible for agricultural and forest products production. Furthermore, since energy flows from plants to animals, photosynthesis is indirectly responsible for protein production in the form of animal products. Agriculture makes use of many kinds of plants to provide us with food. If we cause the extinction of a potentially useful organism, we have lost the opportunity to use it for our own ends. Most of the wild ancestors of our most important food grains, such as maize (corn), wheat, and rice, are thought to be extinct. What would our world be like if these plants had not been domesticated before they went extinct as wild populations? Because of this concern, many agricultural institutes and universities maintain "gene banks" of wild and primitive stocks of crop plants.

However, agriculture is more than just growing plants. It involves understanding fungal and insect pests and the organisms that either spread or inhibit the pests. There are many examples of innocuous organisms that turned out to have great value once their life history was understood. For example, many different kinds of insects are pollinators that are extremely important to the successful fruiting of plants. The careless elimination of these beneficial insects by broad use of insecticides can negatively affect agricultural production.

It is important to recognize that each organism is involved in a vast network of relationships with other organisms. Symbiotic nitrogen-fixing bacteria live in the roots of certain plants. Soil-building organisms live on the dead organic matter provided by plants and animals. Animals

GLOBAL PERSPECTIVE

Biodiversity "Hot Spots"

In an attempt to prioritize efforts to protect the biodiversity, scientists have identified the most endangered habitats around the world. Conservation International has identified 25 biodiversity "hot spots." Several criteria were used to help identify these critical habitats. The primary criteria are the amount of the original habitat remaining and the number of unique species present in the region. The primary cause of habitat loss is human activity. Therefore, protection of these areas requires actions that discourage the continued modification of the habitat. Species that are only found in specific locations are said to be endemic to the region. Some regions have large numbers of such unique organisms and are said to have high rates of endemism. Since it is impossible to study all the organisms in an area, scientists have looked at two groups of rea-

sonably well-known organisms, plants and vertebrate animals, and determined rates of endemism for a region.

High rates of endemism are found in areas that have been isolated from other areas for a long time. If you look at the list of 25 biodiversity hot spots in the table, you will notice that eight of them consist of islands or groups of islands. Other areas may be isolated because they are surrounded by other kinds of systems that have very different types of organisms. For example, the tropical rainforest areas of western Africa are isolated by dryer savanna to the north, east, and south. Whenever ecosystems have existed in relative isolation for millions of years, the likelihood is high that unique species will have evolved in the region that are not found elsewhere in the world.

Source: *National Geographic* 2002.

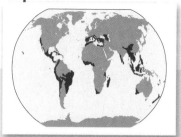

Biodiversity Hot Spots

Hot Spot	Characteristics	% Original Area Remaining	Number of Endemic Plants (%)	Number of Endemic Vertebrate Animals (%)
Tropical Andes	This region is quite varied. At lower elevations (500-1500 meters; 1640-4920 feet) tropical wet and moist forests occur. Various types of cloud forests occur at intermediate elevations (800 to 3,500 meters; 2625-11,480 feet). At higher elevations (3000-4800 meters; 9840-15,745 feet), grassland and scrubland ecosystems are found up to the snow line.	25%	20,000 (44.4%)	1567 (46.2%)
Chocó-Darién-Western Ecuador	This region is quite varied and includes mangroves, beaches, rocky shorelines, and coastal rainforests. The only remaining coastal dry forests in South America occur in this region.	24.2%	2250 (25%)	418 (25.7%)
Atlantic Forest of Brazil, Paraguay, and Argentina	Coastal and inland tropical and subtropical rainforests of many types exist in this area.	7.5%	8000 (40%)	563 (33.7%)
Brazilian Cerrado	The Cerrado is the most extensive woodland/savanna region in South America. There are many different vegetation types, including tree and scrub savanna, grassland with scattered trees, and occasional patches of a dry, closed canopy forest.	20%	4400 (44%)	117 (9.2%)
Central Chile	The Central Chile region contains two dry habitat types. A Mediterranean-type climate in the south is characterized by a high number of woody plant species, woodlands, and closed forest vegetation. The arid desert occurs in the north. Both areas have winter rainfall.	30%	1605 (46.8%)	61 (18.2%)
Caribbean	Since the climate in the Caribbean is highly variable, the vegetation differs from island to island and often is quite variable on a single island. Several kinds of dry habitats are found: dry evergreen thickets, savanna, cactus scrub, and spiny shrub vegetation in a few of the driest spots, dominate the dry and flat islands. In wetter areas, marsh forests, forests, and woodlands occur.	11.3%	7000 (58.3%)	779 (51.3%)
California Floristic Province	The climate of the California Floristic Province consists of hot, dry summers and cool, wet winters. The habitats types are quite variable. Sagebrush, prickly pear, coastal sage scrub, chaparral, and juniper-pine woodland are found at lower elevations. At higher elevations and where there is more moisture, there are evergreen forests of cypress, Douglas fir, sequoia, or redwood. There are also dunes and salt marshes along the coast.	24.7%	2125 (48%)	71 (12.2%)
Mesoamerica	There is great variety in the ecosystems of this region. They include dry forests, lowland moist-forests, subtropical wet forests, rainforests, coastal swamps, and mangrove forests with broad-leaved and coniferous forests at higher altitudes.	20%	5000 (20.8%)	1159 (40.5%)
Caucasus	The vegetation of the Caucasus is quite diverse, but the primary vegetation types are dry-grasslands, semidesert, desert, and arid woodlands with scattered broadleaf forests, coniferous forests, and shrublands.	10%	1600 (25.4%)	59 (9.3%)
Mediterranean Basin	The climate of the Mediterranean Basin consists of cool, wet winters and hot, dry summers. Much of the region was once covered by deciduous forests, which have been modified by human activity. Today, the most widespread vegetation consists of shrubs with dry hard leaves.	4.7%	13,000 (52%)	235 (30.5%)
Madagascar and Indian Ocean Islands	Madagascar has been separate from Africa for about 160 million years. Although it is only about 370 kilometers (230 miles) from Africa, Africa and Madagascar share few species of organisms. The vegetation types are quite varied and include tropical rainforest, dry deciduous forests, a spiny desert, and several high mountain ecosystems. The islands have high marine biodiversity with large areas of coral reefs.	9.9%	9704 (80.9%)	771 (78.1%)

specific monetary value to. For example, much firewood, fish, game meat, and other wild foods are consumed directly by the people who harvest them. Medicinal plants, ivory, skins, and many other natural products are all used and sold. There are many historical examples of valued species that were or are being harvested beyond their capacity to reproduce. The harvesting of ivory was steadily reducing the number of elephants, and the harvesting of alligators for their hides was cutting alligator numbers. In both cases, regulation of the harvest has preserved the resource. Alligators are again being harvested in the United States, and many countries in Africa maintain that harvesting of elephants for meat and ivory should be allowed as well.

The potential value of biodiversity must also be considered. About 25 percent of drug prescriptions filled in the United States contain active substances that were discovered in plants. Other drugs are derived from microorganisms or animals. How many potential drugs are yet undiscovered? Protecting biodiversity can be thought of as protecting future economically important resources.

Ethical Values

In our pragmatic world, it is often difficult to make arguments based on purely ethical precepts. However, a case can be made that all species have an intrinsic value and a fundamental right to exist without being needlessly eliminated by the unthinking activity of the human species. This is an ethical position that is unrelated to social or economic considerations. According to those who support this philosophical position, extinction by itself is not bad, but human-initiated extinction is. This contrasts with the philosophical position that humans are simply organisms that have achieved a preeminent position on Earth, and therefore, extinctions we cause are no different from extinctions caused by other forces. Others would argue that experiencing natural landscapes and processes is an important human right. The beauty of a forested hillside in full autumn color, the graceful movements of deer, the raw power of a grizzly bear, or the wonder shown by a child when it discovers a garter snake or a butterfly are all emotional events. There is a value in their just being, so that they can be observed. Many people have difficulty relating to organisms at this level because they have little experience with the natural world.

The values held by a person are typically shaped by experience. In 1800, 94 percent of North Americans lived in rural settings and had daily encounters with plants and animals in a natural setting. Furthermore, there was a basic understanding of how natural ecosystems operated. They milked their own cows, slaughtered animals for meat, and observed the effect of weather, seasons, and other cycles of change on the natural systems of which they were a part. Although they often changed ecosystems to provide for agriculture, they had a basic understanding of how ecosystems function. Today, about 80 percent of North Americans live in urban settings and have limited contact and experience with these natural processes. While the developed world shows a greater degree of urbanization, the move toward urbanization is a worldwide phenomenon. In 1950, about 70 percent of the world's people were rural. Today, about 50 percent are rural. Urban residents' nature experiences are likely to be in city parks or nature centers, or brief visits to the countryside. Because they have little personal experience with naturally functioning ecosystems, they have a different set of ethical values about what natural ecosystems provide. As the shift from rural to urban continues, there is a continual erosion of the nature experiences that can shape the values of people. These values are important in determining how society will approach threats to biodiversity.

Threats to Biodiversity

Efforts to preserve biodiversity involve a tension between the desire to use biotic resources and a wish to maintain biodiversity. The value of exploiting a resource can be measured in economic terms. Farmland, lumber, or animal products can be given a measurable monetary value by the economic marketplace. On the other hand, it is often difficult to put an economic value on the preservation of biodiversity and the environmental services provided by organisms. Therefore, environmentalists often must rely on ethical or biological arguments to make their points. Often the decisions that are made involve a compromise that allows some utilization of a resource while preserving some of the biodiversity.

Four major human activities threaten to reduce biodiversity. *Habitat loss* occurs when human activities result in the conversion of natural ecosystems to human-dominated systems. The resulting changes eliminate or reduce the numbers of species that were a part of the original ecosystem. *Overexploitation* occurs when humans harvest organisms faster than the organisms are able to reproduce. Overexploitation has driven some organisms to extinction and threatens many others. *Introduction of exotic species* can also have a significant effect on biodiversity. Often exotic species compete with native species and drive them to extinction. Another major threat to biodiversity is *persecution of pest organisms*. Many large carnivores were hunted to extinction because of their threat to humans and their livestock. The sections that follow will examine these threats in more detail.

Habitat Loss

Because several factors may contribute to the decline of a species, often it is difficult to assign an exact cause for a species's endangerment. However, the World Conservation Union (IUCN[1]) estimates that about 80 percent to 90 percent of threatened species are under threat because of habitat loss or fragmentation. Similarly, habitat loss and fragmentation are thought to be a major cause of past extinctions. (See figure 12.2.) The primary activities that result in habitat loss are farming, forestry, grazing by livestock, modification of aquatic habitats, and conversion to urban and industrial landscapes.

---

[1]The International Union for the Conservation of Nature changed its name to the World Conservation Union but kept IUCN as its acronym.

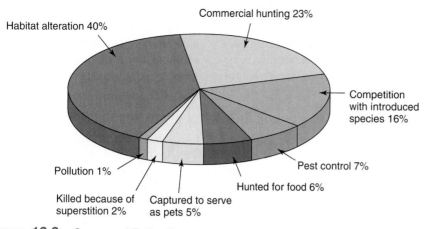

Commercial hunting 23%

Habitat alteration 40%

Competition
with introduced
species 16%

Pest control 7%

Hunted for food 6%

Captured to serve
as pets 5%

Killed because of
superstition 2%

Pollution 1%

Figure 12.2 **Causes of Extinction** Many organisms have become extinct as a result of human activities. On a worldwide basis, the most important of these activities is the destruction of habitat by humans modifying the environment for their own purposes.

Often the destruction of natural habitats is not total but leaves patches of relatively unaltered habitat interspersed with human-modified landscapes. Scientists studying the effects of this activity on forest birds have noticed that as the forest is reduced to small patches, many species of birds disappear from the area. The same kinds of activities certainly happened in Europe, where little of the original forest is left. In North America, the eastern deciduous forests were reduced, which probably resulted in the extinctions of some animals and plants. Almost all the original prairie in the United States has been replaced with agricultural land, resulting in the loss of some species.

Conversion to Agriculture

About 40 percent of the world's land surface has been converted to cropland and permanent pasture. Typically, the most productive natural ecosystems (forests and grasslands) are the first to be modified by human use and the most intensely managed. Since nearly all agricultural practices involve the removal of the original vegetation and substitution of exotic domesticated crops and animals, the loss of biodiversity is significant. As the human population grows, it needs more space to grow food. The pressures to modify the environment are greatest in areas that have high population density. Often the changes brought about by intense agricultural use can degrade the ecosystem and permanently alter the biotic nature of the area. For example, much of the Mediterranean and Middle East, once supported extensive forests that were converted to agriculture but now consist of dry scrubland. Today, agricultural land is being pushed to feed more people, and its wise use is essential to the health and welfare of the people of the world. Chapters 14 and 15 take a close look at patterns of use of agricultural land.

Forestry Practices

The history and current status of the world's forests are well known. The economic worth of the standing timber can be assessed, and the importance of forests for wildlife and watershed protection can be given a value. Originally, almost half of the United States, three-fourths of Canada, almost all of Europe, and significant portions of the rest of the world were forested. The forests were removed for fuel, for building materials, to clear land for farming, and just because they were in the way. These forest-destroying activities are often called **deforestation.** This activity destroyed much of the forested land and returned other forests to an earlier successional stage, which resulted in the loss of certain animal and plant species that required mature forests for their habitat. In general, all continents still contain significant amounts of forested land, although most of these forest ecosystems have been extensively modified by human activity. Figure 12.3 shows that temperate parts of the world (particularly North America and Europe, which have large forested areas) are maintaining their forests, while tropical forests are being lost.

Modern forest-management practices involve a compromise that allows economic exploitation while maintaining some of the biodiversity of the forest. Logging activities are often selective and preferentially remove certain species of trees. This causes a shift in species diversity of plants. Logging also destroys the habitat for many kinds of animals that require mature stands of timber. The pine martin, grizzly bear, and cougar all require forested habitat that is relatively untouched by human activity. The removal of mature forests reduces their population size and consequently the genetic diversity of their populations.

For example, today in Australia, a squirrel-sized marsupial known as a numbat is totally dependent on old forests that have termite- and ant-infested trees. Termites are numbats' major source of food, and the hollow trees and limbs caused by the insects' activities provide numbats with places to hide. Clearly, tree harvesting will have a negative impact on this species, which is already in danger of extinction. (See figure 12.4.)

In addition to serving as habitats for many species of plants and animals, forests provide many other ecosystem services. Forested areas modify the climate, reduce the rate of water runoff, protect soil from erosion, and provide recreational opportunities. Because trees transpire large quantities of water and shade the soil, their removal often leads to a hotter, drier climate. Trees and other plants hold water on their surfaces, thus reducing the rate of runoff. Slowing runoff also allows more water to sink into the soil and recharge groundwater resources. Therefore, removal of trees results in more rapid runoff so that flooding and soil erosion are more common. Soil particles wash into streams, where they cause siltation. The loss of soil particles reduces the soil's fertility. The particles that enter streams may cover spawning sites and reduce the biodiversity of fish populations. If the trees along the stream are

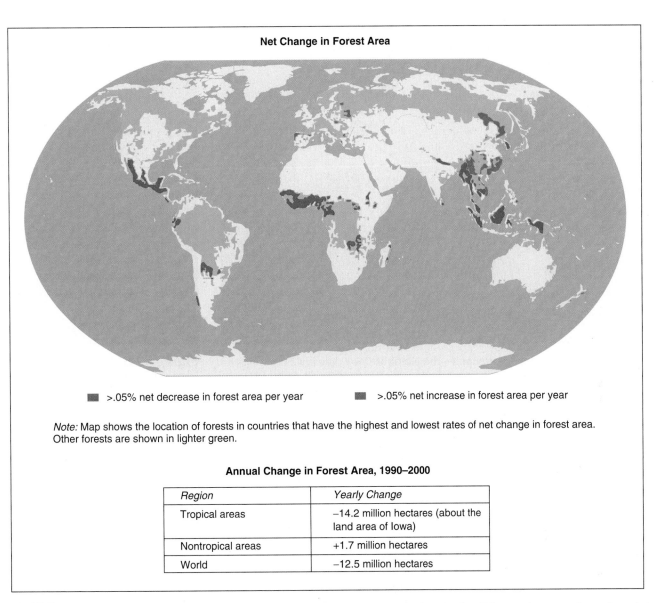

Net Change in Forest Area

■ >.05% net decrease in forest area per year   ■ >.05% net increase in forest area per year

Note: Map shows the location of forests in countries that have the highest and lowest rates of net change in forest area. Other forests are shown in lighter green.

Annual Change in Forest Area, 1990–2000

Region	Yearly Change
Tropical areas	−14.2 million hectares (about the land area of Iowa)
Nontropical areas	+1.7 million hectares
World	−12.5 million hectares

Figure 12.3 **Changes in Forest Area** Areas shown in red are losing forests at a rate of greater than 0.5 percent per year. Areas shown in brown are increasing at a rate of greater than 0.5 percent per year. It is obvious that most tropical forests in Central America, Africa, and much of Asia are being lost, while nontropical areas are holding their own or growing. The table shows actual land area affected.

Source: Food and Agriculture Organization of the United Nations.

removed, the water in the stream will be warmed by increased exposure to sunlight. This may also negatively affect fish populations.

The process of cutting the trees and transporting the logs also disturbs the wildlife in the area. The roads necessary for moving equipment and removing logs are a special problem. Constant travel over these roads removes vegetation and exposes the bare soil to more rapid erosion. When the roads are not properly located and constructed, they eventually become gullies and serve as channels

for the flow of water. Most of these environmental problems can be minimized by using properly engineered roads and appropriate harvesting methods.

In many parts of the world, the construction of logging roads increases access to the forest and results in colonization by peasant "squatters" who seek to clear the forest for agriculture. The roads also permit poachers to have greater access to wildlife in the forest. Finally, the "wilderness" nature of the area is destroyed, which results in a loss of value for many who like to visit mature forests for recre-

ation. An area that has recently been logged is not very scenic, and the roads and other changes often irreversibly alter the area's wilderness character.

Environmental Implications of Various Harvesting Methods

One of the most controversial logging practices is **clear-cutting.** (See figure 12.5.) As the name implies, all of the trees in a large area are removed. This is a very economical method of harvesting, but it exposes the soil to significant erosive

Figure 12.4 **A Specialized Marsupial—the Numbat** This small marsupial mammal requires termite- and ant-infested trees for its survival. Termites serve as food, and the hollow limbs and logs provide hiding places. Loss of old-growth forests with diseased trees will lead to the numbat's extinction.

Figure 12.5 **Extensive Clear-cutting** Large, clear-cut plots of land can lead to a loss of species and accelerated soil erosion.

forces. If large blocks of land are cut at one time, it may slow reestablishment of forest and have significant effects on wildlife. On some sites with gentle slopes, clear-cutting is a reasonable method of harvesting trees, and environmental damage is limited. This is especially true if a border of undisturbed forest is left along the banks of any streams in the area. The roots of the trees help to stabilize the stream banks and retard siltation. Also, the shade provided by the trees helps to prevent warming of the water, which might be detrimental to some fish species.

Clear-cutting can be very destructive on sites with steep slopes or where regrowth is slow. Under these circumstances, it may be possible to use **patchwork clear-cutting.** With this method, smaller areas are clear-cut among patches of untouched forest. This reduces many of the problems associated with clear-cutting and can also improve conditions for species of game animals that flourish in successional forests but not in mature forests. For example, deer, grouse, and rabbits benefit from a mixture of mature forest and early-stage successional forest.

Clear-cut sites where natural reseeding or regrowth is slow may need to be replanted with trees, a process called **reforestation.** Reforestation is especially important for many of the conifer species, which often require bare soil to become established. Many of the deciduous trees will resprout from stumps or grow quickly from the seeds that litter the forest floor, so reforestation is not as important in deciduous forests.

Selective harvesting of some species of trees is also possible but is not as efficient or as economical as other methods, from the point of view of the harvesters. It allows them, however, to take individual, mature, high-value trees without completely disrupting the forest ecosystem. In many tropical forests, high-value trees, such as mahogany, are often harvested selectively. However, there may still be extensive damage to the forest by the construction of roads and to noncommercial trees by the felling of the selected species.

Special Concerns About Tropical Deforestation

Tropical forests have greater species diversity than any other terrestrial ecosystem. The diverse mixture of tree species requires harvesting techniques different from those traditionally used in northern temperate forests. In addition, because tropical soils have low fertility and are highly erodible, tropical forests are not as likely to regenerate after logging as are temperate forests. Currently, few tropical forests are being managed for long-term productivity; they are being harvested on a short-term economic basis only, as if they were nonrenewable resources. Worldwide tropical forests are being lost at a rate of about 0.6 percent per year.

Several concerns are raised by tropical deforestation. First, the deforestation of large tracts of tropical forest is significantly reducing the species diversity of the world. Second, because tropical forests very effectively trap rainfall and prevent rapid runoff and the large amount of water transpired from the leaves of trees tends to increase the humidity of the air, the destruction of these forests also can significantly alter climate, generally resulting in a hotter, more arid climate. Third, high rainfall coupled with the nature of the soils results in the deforested lands being easily eroded. Finally, people have become concerned about preserving the potential of forests to trap carbon dioxide. As they carry on photosynthesis, trees trap large amounts of carbon dioxide. This may help to prevent increased carbon dioxide levels that contribute to global warming. (See chapter 17 for a discussion of global warming and climate change.)

Another complicating factor is that the human population is growing rapidly in tropical regions of the world. More people need more food, which means that forestland will be converted to agriculture and the value of forest for timber, fuel, watershed protection, wildlife habitat, biodiversity, and carbon dioxide storage will be lost.

Plantation Forestry

Many forest products companies manage forest plantations in the same way farmers manage crops. They plant single-species, even-aged forests of fast-growing hybrid trees that have been developed in the same way as high-yielding agricultural crops. Fertilizer is applied if needed, and weeds and pests are controlled. Often controlled fires and aerial application of pesticides are used to control competing species and pests. Such forests have low species diversity and are not as valuable for wildlife and other uses as are more natural, mixed species forests. Furthermore, the trees planted in many managed forests may be exotic species. *Eucalyptus* trees from Australia have been planted in South America, Africa, and other parts of the world; and most of the forests in northern England and Scotland have a mixture of native pines and imported species from the European mainland and North America. In these intensively managed forests, some single-species plantations mature to harvestable size in 20 years, rather than in the approximately 100 years typical for naturally reproducing mixed forests. However, the quality of the lumber products is reduced. In many of these forests, clear-cutting is the typical method of harvest, and the cut-over area is immediately replanted.

A single-species, even-aged forest is ideal for producing wood but does not support as wide a variety of wildlife as does a mixed-age, mixed-species forest. However, with proper planning, trees can be harvested in small enough patches to encourage some wildlife but in large

enough sections to be economical. Since some species require old trees as part of their niche, some older trees or patches of old-stand timber can be left as refuges for such species. For example, in the southern pine forests of the United States, the endangered red-cockaded woodpecker requires old, diseased pine trees in which to build its nests. (See figure 12.6.) A well-managed forest plantation does not provide these sites, but forest products companies in the southeastern United States have agreed to modify forestry practices on their lands to allow for the retention of some old, diseased trees needed by the woodpeckers. This is important because not enough suitable woodpecker habitat exists on state and federally owned lands to protect the woodpecker from extinction.

Rangeland and Grazing Practices

Rangelands consist of the many arid and semiarid lands of the world that support grasses or a mixture of grasses and drought-resistant shrubs. These lands are too dry to support crops but are often used to raise low-density populations of domesticated or semidomesticated animals. In some cases, the animals are maintained on permanent open ranges, while in others, nomadic herds are moved from place to place in search of suitable grazing. Usually, the animals are introduced species not native to the region. Sheep, cattle, and goats that are native to Europe and Asia have been introduced into the Americas, Australia, New Zealand, and many areas of Africa.

The conversion of rangelands to grazing by domesticated animals has major impacts on biodiversity. In an effort to increase the productivity of rangelands, management techniques may specifically eliminate certain species of plants that are poisonous or not useful as food for the grazing animals, or specific grasses may be planted that are not native to the area. In some cases, native animals are reduced if they are a threat to livestock because they are predators or because they may spread disease to the livestock. In addition, the selective eating habits of livestock tend to reduce certain species of native plants and encourage others.

Figure 12.6 **Habitat Needs of the Red-Cockaded Woodpecker** The red-cockaded woodpecker excavates its nest in old living pines that have a disease known as red heart. Old diseased trees are rare in forests that are intensely managed for wood products. Modification of its habitat by removing old, diseased trees has led to this bird being endangered.

Because rainfall is low and often unpredictable, it is important to regulate the number of livestock on the range. In areas where the animals are on permanent pastures, the numbers of animals can be adjusted to meet the capacity of the range to provide forage. In many parts of the world, nomadic herders simply move their animals from areas where forage is poor to areas that have better forage. This is often a seasonal activity that involves movement of animals to higher elevations in the summer or to areas where rain has fallen recently. (See figure 12.7.)

In many parts of the world where human population pressures are great, overgrazing is a severe problem. As populations increase, desperate people attempt to graze too many animals on the land. They also cut down the trees for firewood. If overgrazed, many plants die, and the loss of plant cover subjects the soil to wind erosion, resulting in a loss of fertility, which further reduces the land's ability to support vegetation. The cutting of trees for firewood has a similar effect, but it is especially damaging because many of

Nomadic herding   Stock raising on ranges

Figure 12.7 **Use of Rangelands** The arid and semiarid regions of the world will not support farming without irrigation. In many of these areas, livestock can be raised. Permanent ranges occur where rainfall is low but regular. Nomadic herders can utilize areas that have irregular, sparse rainfall.

Freshwater lakes, streams, and rivers are modified for navigation, irrigation, flood control, or power production purposes, all of which may alter the natural ecosystem and change the numbers or kinds of aquatic organisms present. These topics are discussed in greater detail in chapter 16. In the Pacific Northwest, the extensive development of dams to provide power and aid navigation has made it nearly impossible for adult salmon to migrate upstream to spawn and difficult for young fish to migrate downstream to the ocean. Fish ladders and other techniques have not been successful in allowing the fish to pass the dams. As a result, many populations of Pacific salmon are nearly extinct. The only solution to the problem is to remove or greatly modify several of the dams.

Conversion to Urban and Industrial Uses

About 4.3 percent of U.S. land is developed as urban centers, industrial sites, and the transportation infrastructure that allows for the movement of people and products about the country. Although this is a relatively small percentage of the total land area of the United States, urban areas are the most heavily impacted by human activity. A large proportion of such areas are covered with impermeable surfaces that prevent plant growth and divert rainfall to local streams and rivers. In addition, streams and other natural features are altered to serve the needs of the people. Biodiversity is drastically reduced, and only the most adaptable organisms can survive in such settings.

Many industrial sites are associated with urban centers, although some industries such as mining and oil and natural gas production may be located far from urban centers. Their impact, however, is similar to that of an urban center. The land is altered in such a way that the natural ecosystems are destroyed.

On a worldwide basis, it is difficult to generalize about the impact of urban centers. Some countries are sparsely populated and urban centers are not a major part of the landscape, while other countries are heavily urbanized. Regardless, the trend is toward greater urbanization.

these trees are legumes, which are important in nitrogen fixation. Their removal further reduces soil fertility. This severe overuse of the land results in conversion of the land to a more desertlike ecosystem. This process of converting arid and semiarid land to desert because of improper use by humans is called **desertification.** Desertification can be found throughout the world but is particularly prevalent in northern Africa and parts of Asia, where rainfall is irregular and unpredictable, and where many people are subsistence farmers or nomadic herders who are under considerable pressure to provide food for their families. (See figure 12.8.)

Habitat Loss in Aquatic Ecosystems

Habitat loss is also a problem in aquatic systems. In marine ecosystems, much of

the harvest is restricted to shallow parts of the ocean where bottom-dwelling fish can be easily harvested. The typical method used to harvest bottom-dwelling fish and shellfish involves the use of trawls which are nets that are dragged along the bottom. These nets capture various species, many of which are not commercially valuable. The trawls disturb the seafloor and create conditions that make it more difficult for the fish populations to recover. In addition, 25 percent of the catch typically consists of species that have no commercial value. These are discarded by being thrown overboard. However, they are usually dead, and their removal further alters the ecological nature of the seafloor. Some people have even advocated that the trawl should be banned as a fishing technique because of the damage done to the ocean bottom.

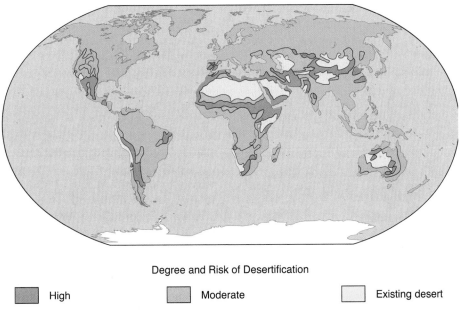

Degree and Risk of Desertification

High   Moderate   Existing desert

a.

b.

Figure 12.8 **Desertification** (a) Areas of the world where desertification is prevalent. (b) Arid and semiarid areas can be converted to deserts by overgrazing or unsuccessful farming practices. The loss of vegetation increases erosion by wind and water, increases the evaporation rate, and reduces the amount of water that infiltrates the soil. All of these conditions encourage the development of desertlike areas.

Sources: *United Nations Map of World Desertification,* United Nations Food and Agriculture Organization, United Nations Educational, Scientific, and Cultural Organization, and the World Meteorological Organization for the United Nations Conference on Desertification, 1977, Nairobi, Kenya.

Therefore, in the future, a greater proportion of all countries will be modified by the increase in size of urban centers and the transportation corridors that support them. Much of the impact of urban centers can be reduced with proper land-use planning. Chapter 13 on land use planning explores this issue in greater depth.

Overexploitation

According to the World Conservation Union, overexploitation is responsible for over 30 percent of endangered species of animals and about 8 percent of plants. Organisms are harvested for a wide variety of purposes. In much of the less-developed world, animals of all kinds are killed and eaten as a source of protein. We use organisms for a variety of purposes in addition to food. Many plants and animals are used as ornaments. Flowers are picked, animals skins are worn, and animal parts are used for their purported aphrodisiac qualities. In the United States, many species of cactus are being severely reduced because people like to have them in their front yards. In other parts of the world, rhinoceros horn is used to make dagger handles or is powdered and sold as an aphrodisiac. Because some people are willing to pay huge amounts of money for these products, unscrupulous people are willing to break the law and poach these animals for the quick profit they can realize. Table 12.4 shows some of the organisms, or their parts, that are highly prized by buyers. The World Wildlife Fund estimates that illegal trade in wild animals globally produces $2 billion to $30.5 billion per year. These activities have already resulted in local extinctions of some plants and animals and may be a contributing factor to the future extinction of additional species.

Overexploitation is a serious problem in several situations. Among the most severe are overfishing of fisheries, unsustainable harvesting of wildlife, and capture for the pet and aquarium trade.

Overfishing of Marine Fisheries

The United Nations estimates that 70 percent of the world's marine fisheries are being overexploited or are being fully exploited and are in danger of overexploitation as the number of fishers increases. Figure 12.9 shows both the amount of fish captured and the amount produced by aquaculture. While the amount of fish captured by fishers has remained essentially constant since 1989, the amount produced by fish farming has increased. It is important to note that more effective capture methods and increased fishing effort were required to maintain the production levels for the capture of marine fish.

Another indication that marine fishery resources are being overexploited is the change in the kinds of fish being caught. Figure 12.10 shows the change in species composition for both bottom-dwelling species and surface-dwelling open-ocean fish. The commercial fishing industry has been attempting to market fish species that previously were regarded as unacceptable to the consumer. These activities are the result of reduced catches of desired species. Examples of "newly discovered" fish in this category are monkfish and orange roughy.

Open-ocean species of fish are also subject to overexploitation. During the 1960s, anchovy fishing off the coast of Peru was a major industry. From 1971 to 1972, the catch dropped dramatically. Overfishing was believed to be one of the major contributing factors. This was aggravated by an increase in the area's water temperature, which prevented nutrient-rich layers from rising to the euphotic zone. Thus, productivity at all trophic levels decreased. Currently, the Peruvian anchovy fishery has rebounded but is considered to be overexploited and not sustainable. (See figure 12.10b.)

Aquaculture

Fish farming (aquaculture) is becoming increasingly important as a source of fish production. Salmon farming has been

Table 12.4 Typical Species Prices (1990 Rates)

International Species	Price ($US)	North American Species	Price ($US)
Olive python	1500	Bald eagle	2500
Rhinoceros horn	12,500/pound	Golden eagle	200
Tiger skin (Siberian)	3500	Gila monster	200
Tiger meat	130/pound	Peregrine falcon	10,000
Cockatoo	2000	Grizzly bear	5000
Leopard	8500	Grizzly bear claw necklace	2500
Snow leopard	14,000	Polar bear	6000
Elephant tusk	250/pound	Black bear paw pad	150
Walrus tusk	50/pound	Reindeer antlers	35/pound
Mountain gorilla	150,000	Mountain lion	500
Giant panda	3700	Mountain goat	3500
Ocelot	40,000/coat	Saguaro cactus	15,000
Imperial Amazon macaw	30,000		

Source: Data from U.S. Fish and Wildlife Service.

Figure 12.9 **Trends in World Fish Production** The amount of fish captured increased steadily until about 1989. Since then, the amount has remained essentially constant. This indicates that the world fisheries are being exploited to their capacity. Aquaculture has continued to increase so the total production continues to increase. These data have been called into question by many because it is thought that China has grossly overreported its fish production statistics. This would mean that the capture fisheries may actually have been declining in recent years.

Source: Food and Agriculture Organization of the United Nations.

PART FOUR Resource Management

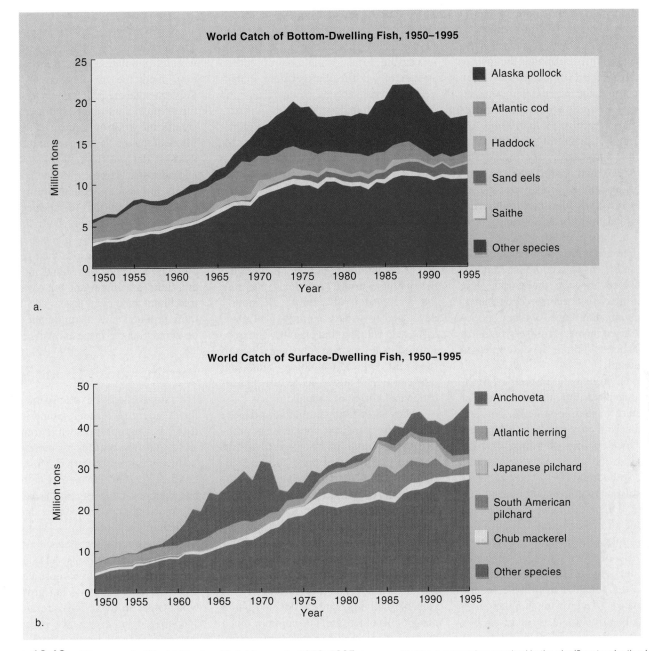

World Catch of Bottom-Dwelling Fish, 1950–1995

Alaska pollock
Atlantic cod
Haddock
Sand eels
Saithe
Other species

a.

World Catch of Surface-Dwelling Fish, 1950–1995

Anchoveta
Atlantic herring
Japanese pilchard
South American pilchard
Chub mackerel
Other species

b.

Figure 12.10 **Changes in World Marine Fish Harvests 1950–1995** Increased fishing pressure has resulted in the significant reduction in many of the traditional fishing stocks. As traditional species have been overfished, other, less desirable species have been replacing them. This is true for both the bottom-dwelling species (a) and those that live in the open ocean (b).

Source: Food and Agriculture Organization of the United Nations.

particularly successful. This involves raising various species of salmon in "pens" in the ocean, which allows the introduction of food and other management techniques to achieve rapid growth of the fish. Norway, Chile, Canada, and Scotland are the leading countries in salmon production. The production of salmon from fish farms has increased rapidly. In 1988, less than 20 percent of the salmon sold were from fish farms, compared to over 60 percent in 2002. During the same period of time, the production of wild-caught salmon has been relatively constant. However, the farming method of producing fish is not without its environmental effects.

Raising fish in such concentrated settings results in increased nutrients in the surrounding water from uneaten food and the wastes the fish release. This can cause local algal blooms that negatively affect other fisheries, such as shellfish. Many of the salmon species raised in farms are not native to the waters in which they are raised. Inevitably, some of these fish escape. The introduction of exotic species can have a negative impact on native species. Some people even worry about the genetic stocks of fish that are raised.

Many wild salmon populations are in danger of extinction. The introduction of new genetic stocks that escape and interbreed with wild fish can alter the genetic makeup of the wild populations. On an economic front, the increase in the amount of farmed salmon is reducing the prices salmon fishers receive for their catch.

Currently, about 60 percent of all aquaculture production is from freshwater systems, and production is growing rapidly. Aquaculture of freshwater species typically involves the construction of ponds, which allow the close management of the fish, shrimp, or other species. This can be done with a very low level of technology and is easily accomplished in underdeveloped areas of the world. Very little rearing of fish occurs in open waters, although immature stages of many species may be raised in special nursery facilities (fish hatcheries) before they are released into the wild.

The environmental impacts of freshwater aquaculture are similar to those of aquaculture in marine systems. Nutrient overloads from concentrations of fish can pollute local bodies of water, and the escape of exotic species may harm native species. In addition, freshwater aquaculture involves the conversion of land to a new use. Often the lands involved are mangrove swamps or other wetlands that many people feel should be protected. Regardless of environmental concerns, the productivity of freshwater aquaculture has the potential to provide for the protein needs of a growing population, so it is likely to continue to increase at a rapid rate.

Unsustainable Harvest of Wildlife and Plants

Wildlife may be harvested for a variety of reasons. Primary among them is the need for food in much of the world. Any animal that can be captured or killed is used for food. Meat from wild animals is often referred to as **bush meat,** since it is gathered from the "bush," a common term for the wild. Estimates about the extent of the problem vary, but there is wide agreement that the harvest is unsustainable. One estimate by the Wildlife Conservation Society is that about 70 percent of wildlife species

in Asia and Africa and about 40 percent of species in Latin America are being hunted unsustainably. It is clear that this practice is causing local extinctions of certain species of wildlife. Furthermore, endangered species such as chimpanzees and gorillas are often harvested.

Trade in bush meat and other animal products has become a major problem for several reasons. Hunting of wildlife is a part of all subsistence cultures, so it is considered a normal activity. Poor people can earn money by harvesting and selling wild animals for meat or other purposes. Many kinds of wildlife are considered delicacies and are highly prized for the home and restaurant trade. In many parts of the developing world, regulations regarding the preservation of wildlife are poorly developed or widely ignored. The roads associated with logging operations in tropical forests give access to a greater part of the forest. Modern technology (guns and artificial lights) allows more efficient location and killing of animals.

Harvest of living animals for the pet and aquarium trade is a significant problem. Many kinds of birds, reptiles, and fish are threatened because they are desired as pets. Generally, it is not the isolated taking of an individual organism that is the problem but rather, the method of capture, which may have far-reaching effects. For example, many nests are destroyed when nestling birds are taken. In the case of tropical marine fishes, toxins are often used to stun the fish, and the few valuable individuals are recovered while many of the others die. Furthermore, since much of the market in exotic pets involves transporting animals long distances, the mortality rate among those captured is high.

Wildlife species also are hunted because parts of the animal may have particular value. For example, ivory and animal skins are highly valued in many cultures as art objects or clothing. Parts of some animals are thought to have particular medicinal properties, and others are hunted because of their use in traditional cultural practices. For example, rhinoceros horn is thought to be an aphrodisiac, and some cultures prize rhinoceros horn as handles for knives. These beliefs lead to a brisk trade in the desired species.

Traditional medicine makes use of a great number of different plants and animals. Since about 80 percent of the world's population uses traditional cultural medicine as opposed to modern Western-style medical treatments, harvesting of medicinal plants and animals is a significant problem in much of the world. Many of the medicines used in these practices are derived from endangered plants and animals. Harvest of some species of plants has led to reductions in populations to the point that the species are endangered. For example, both the Asian and American ginseng plants have been greatly reduced because of their use in traditional medicine. Tiger bone and rhinoceros horn are important for a variety of traditional medicinal uses. The trade in these products puts a significant pressure on populations of these animals.

Introduction of Exotic Species

The introduction of exotic species occurs for a variety of reasons. Some introductions are purposeful, while others are accidental. The reasons for purposeful introductions are quite varied. Most agriculturally important plants and animals are introductions. Other organisms were introduced because people had a fondness for a particular plant or animal. Many plants were introduced for horticultural purposes. Some introduced species are easily controlled. For example, most agriculturally important plants do not grow in the wild. Other introductions are accidental and arrived as stowaways on imported materials. Not all exotic species become problems; however, some organisms disperse broadly and displace native species.

The World Conservation Union estimates that about 30 percent of birds and 15 percent of plants are threatened because they are unable to successfully compete against invasive exotic species. On many islands, the introduction of rats has had a major impact on nesting birds because the rats eat the eggs and kill the nestlings. The introduction of cats and foxes into Australia has resulted in the reduction of populations of many species of native marsupials. Cats and foxes are efficient predators of the native wildlife and caused the extinction of many kinds of native mammals on the Australian mainland. Some species of native marsupials are confined to isolated islands off the coast where cats and foxes are not found.

Introduction of diseases has also had considerable impact on American forests. Two fungal diseases of trees have significantly changed the nature of American forests. Chestnut blight essentially eliminated the American chestnut, and Dutch elm disease has greatly reduced the numbers of American elms. Various kinds of insects have also had an effect on the structure of ecosystems. The gypsy moth has spread throughout North America. Its larvae eat the leaves of forest trees and have significantly altered forest ecosystems because mature oaks are more likely to die as a result of defoliation than some other forest trees. The Asian longhorned beetle is a recent arrival to North America. It was discovered in New York in 1996. Its larvae feed on the wood of deciduous trees and often cause their death. Asian longhorned beetles probably enter North America inside solid-wood packing material from China. Although currently confined to a few locations in the eastern United States and Canada, the beetle has the potential to do great damage to the deciduous forest of the region, and both the U.S. and Canadian governments are taking action to control the problem. The only method of controlling the infestations involves the destruction of infected trees.

Freshwater ecosystems also have been greatly affected by accidental and purposeful introductions. The zebra mussel and Eurasian milfoil are two accidental introductions that have caused great problems. Zebra mussels were first discovered in Lake St. Clair near Detroit about 1985. They were most likely released from ballast water from European ships. The zebra mussel has spread through much of the eastern United States and adjacent Canada in the Great Lakes region and the Mississippi River watershed. It has three major impacts. It clogs intake pipes for water treatment plants and other industrial users. It establishes colonies on the surface of native mussels, often resulting in their death. And it has altered freshwater ecosystems by filtering much plankton from the water and allowing more aquatic plants to grow.

a.

c.

d.

b.

e.

Figure 12.11 **Invasive Species** The routes by which each of these species entered the United States is different. (*a, b*) The gypsy moth was purposely brought to the United States and accidentally escaped. (*c*) The Asian longhorned beetle appears to have arrived in packing crates. (*d*) The zebra mussel arrived in ballast water from Europe, and (*e*) the Eurasian milfoil arrived as an aquarium plant.

Eurasian milfoil was first discovered in Washington, D.C., in the 1940s but has since spread to much of the United States and southern Canada. Although its introduction may have been on purpose, its spread has occurred by the accidental transfer by boats, wildlife, and drifting. It forms thick mats in many bodies of water that can interfere with boat traffic. Figure 12.11 shows examples of these exotic organisms.

The introduction of exotic fish species also has greatly affected naturally occurring freshwater ecosystems. The Great Lakes, for example, have been altered considerably by the accidental and purposeful introduction of fish species. The sea lamprey, smelt, carp, alewife, brown trout, and several species of salmon are all new to this ecosystem. (See figure 12.12.) The sea lamprey is parasitic on lake trout and other species and nearly eliminated the native lake trout population. Controlling the lamprey problem requires use of a very specific larvicide that kills the immature lamprey in the streams.

Lake trout

Yellow perch

Native	Introduced
Brook lamprey	Sea lamprey
Lake trout	Brown trout
Brook trout	Rainbow trout
Whitefish	Pink salmon
Herring	Coho salmon
Ciscoes	Chinook salmon
	Atlantic salmon
Suckers	
Chubs	Carp
Shiners	
Catfish	
Bullheads	
	Smelt
	Alewife
White bass	White perch
Smallmouth bass	Sunfish
Rock bass	
Yellow perch	Ruffe
Walleye	

Brown trout

Smelt

Figure 12.12 **Native and Introduced Fish Species in the Great Lakes** The Great Lakes have been altered considerably by the introduction of many non-native fish species. Some were introduced accidentally (lamprey, alewife, and carp), and others were introduced on purpose (salmon, brown trout, rainbow trout).

This technique works because mature lamprey migrate upstream to spawn, and the larvae spend several years in the stream before migrating downstream into the lake. With the partial control of lamprey, lake trout populations have been increasing. (See figure 12.13.) Recovery of the lake trout is particularly desirable, since at one time it was an important commercial species.

Another accidental introduction to the Great Lakes, the alewife, a small fish of little commercial or sport value, became a problem during the 1960s, when alewife populations were so great that they died in large numbers and littered beaches. Various species of salmon were introduced about this time in an attempt to control the alewife and replace the lake trout population, which had been depressed by the lamprey. While this salmon introduction has been an economic success and has generated millions of dollars for the sport fishing industry, it may have had a negative impact on some native fish such as the lake trout, with which the salmon probably compete. Salmon also migrate upstream, where they disrupt the spawning of native fish. Most

species of salmon die after spawning, which causes a local odor problem.

Other more recent arrivals in the Great Lakes include a clam, the zebra mussel, and two fishes, the round goby and river ruffe. All of these probably arrived in ballast water from fresh or brackish water bodies in Europe. These exotic introductions are increasing rapidly and changing the mix of species present in Great Lakes waters. (See Environmental Close-Up: Population Growth of Invading Species in chapter 7.)

Control of Pest Organisms

The systematic killing of certain organisms because they interfere with human activities also results in reduced biodiversity. Many large predators have been locally exterminated because they preyed on the domestic animals that humans use for food. Mountain lions and grizzly bears in North America have been reduced to small, isolated populations, in part because they were hunted to reduce livestock loss. Tigers in Asia and the lion and wolf in Europe were reduced or elimi-

nated for similar reasons. Even though commercial hunters killed thousands of passenger pigeons, their ultimate extinction was caused primarily by the increased conversion of forest to agriculture. Passenger pigeons ate the acorns of oaks and the beechnuts of beech, and relied on the forests for communal nesting sites. When the forests were cleared, the pigeons became pests to farmers, who shot them to protect their crops from being eaten by the birds. Some extinctions of pest species are considered desirable. Most people would not be disturbed by the extinction of black widow spiders, mosquitoes, rats, or fleas. In fact, people work hard to drive some species to extinction. For example, in the *Morbidity and Mortality Weekly Report* (October 26, 1979), the U.S. Centers for Disease Control triumphantly announced that the virus that causes smallpox was extinct in the human population after many years of continuous effort to eliminate it.

At one time, it was thought that populations of game species could be increased substantially if predators were controlled. In Alaska, for example, the salmon-canning industry claimed that bald eagles

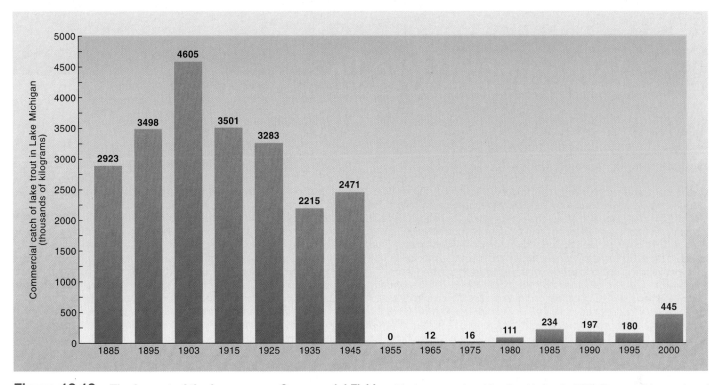

Figure 12.13   The Impact of the Lamprey on Commercial Fishing The lamprey entered the Great Lakes in 1932. Because it is an external parasite on lake trout, it had a drastic effect on the population of lake trout in the Great Lakes. As a result of programs to prevent the lamprey from reproducing, the number of lamprey has been reduced somewhat, and the lake trout population is recovering with the aid of stocking programs.

were reducing the salmon population, and a bounty was placed on eagles. From 1917 to 1952 in Alaska, 128,000 eagles were killed for bounty money. This theory of predator control to increase populations of game species has not proven to be valid in most cases, however, since the predators do not normally take the prime animals anyway. They are more likely to capture sick or injured individuals not suitable for game hunting.

Although predator and competitor control activities have become less common recently, they are still used in some special situations. In intensely managed European systems, gamekeepers are often charged with the responsibility of killing predators or unwanted competitors. For some species, such as ground-nesting birds, it may make some sense to control predators that eat eggs or kill the young, but, in most cases, humans have a greater impact through habitat modification and hunting than do the natural predators. Although at one time they were thought to be helpful, bounties and other forms of predator management have been largely eliminated in North America. In fact, the

pendulum has swung the other way, and predator control is not considered to be cost-effective in most cases. One exception to this general trend is the hunting and trapping of wolves in Alaska and Canada. One rationale for the taking of wolves is that they kill moose and caribou. Since many of the people who live in these areas rely on wild game as a significant part of their food source, controlling wolf populations is politically popular. However, the number of wolves taken is regulated. Since, in many parts of the world, the major predators of game species are humans, regulation of hunting is a form of predator control.

The control of cowbird populations has been used to enhance the breeding success of the Kirtland's warbler. Cowbirds lay their eggs in the nests of other birds, including Kirtland's warblers. When the cowbird egg hatches, the hatchling pushes the warbler chicks out of the nest. Thus, trapping and killing of cowbirds in the vicinity of nesting Kirtland's warblers has been used to enhance the reproduction of this endangered species.

Figure   12.14   Habitat   Protection The area on the left side of the fence has been protected from cattle grazing. This area provides a haven for many native species of plants and animals that cannot survive in heavily grazed areas.

Many species may require refuges where they are protected from competing introduced species or human interference. Many native rangeland species of wildlife benefit from the exclusion of introduced grazing animals such as cattle and sheep because the absence of grazing livestock allows a more natural grassland community to be reestablished. (See figure 12.14.)

Environmental Close-Up

The Northern Spotted Owl

In June 1990, the northern spotted owl (*Strix occidentalis caurina*) was listed as a threatened species under the Endangered Species Act. Threatened species are those that are likely to become endangered if current conditions do not change. The northern spotted owl lives in old-growth coniferous forests in the Pacific Northwest. A primary reason for the listing is that the owl requires mature forests as its habitat. Several characteristics of the owl's biology make old-growth forests important. It prefers a relatively closed canopy with rather open spaces under the canopy. These conditions allow it to fly freely as it searches for food. One of its primary food sources is the northern flying squirrel, which feeds primarily on fungi that are common in old-growth forests that have many dead trees and much down timber on the forest floor. Dead and diseased trees are also important because they provide cavities and platforms for the owls to use as nesting sites. As with most carnivores at the top of the food chain, the northern spotted owl requires large areas of forest for hunting. Therefore, relatively large areas of forest are needed for nesting pairs to be successful.

Logging in the Pacific Northwest has already used up most of the trees on private land. Most of the remaining old-growth forest, which is about 10 percent of its original area, is on land managed by the U.S. Forest Service. Many of the trees in these forests are several centuries old, and it takes 150 to 200 years for regenerating forests to become suitable habitat for the northern spotted owl. This timescale would not allow for the continued existence of the northern spotted owl if major sections of the remaining old-growth forest were logged. The reduction in logging has had significant economic impact on companies that relied on harvesting timber from federally owned lands, and as the companies are hurt economically, so too are the communities where these companies function.

Listing the owl as threatened required that the U.S. Forest Service develop and implement a plan to protect it. Since the owl is found only in mature old-growth forests and since it requires large areas for hunting, the plan set aside large tracts of forested land and protected them from logging. The plan to protect the owl resulted in several actions. Most federally owned lands that were suitable habitat for the owl have indeed been protected from logging, although some salvage logging has been permitted when forests are damaged by fire or other events. Furthermore, private land that is known to have nesting owls was protected as well. As a result of this, individuals and companies have filed lawsuits claiming that since they cannot cut the trees on land they own, they have been deprived of the use of their land. Therefore, they have asked for compensation from the government. One lumber company was awarded $1.8 million in compensation for 22.7 hectares (56 acres) of forest that was "taken" for use as owl habitat. Court cases continue in the area. In 2000, a federal district judge ruled that cutting down trees does not necessarily harm northern spotted owls and that a logging company could proceed to cut a stand of timber. He ruled that to prevent logging of privately held land, the federal government must prove that cutting the trees would harm the owl. In 2001, a coalition of environmental groups filed a lawsuit charging that the U.S. Fish and Wildlife Service had not enforced the Endangered Species Act and had not adequately protected the habitat of the northern spotted owl. It is clear that, at this point, the current arguments about the northern spotted owl are legal maneuvers rather than serious discussions about the protection of a threatened species.

What Is Being Done to Preserve Biodiversity?

Efforts to preserve biodiversity involve a variety of approaches. Foremost among them is the need to understand the life history of organisms so that effective measures can be taken to protect species from extinction. Since the primary causes of endangerment are habitat loss, overexploitation, introduction of exotic species, and persecution of pest species, efforts to prevent further extinctions need to focus on reducing the impact of these human activities. Several international organizations

work to prevent the extinction of organisms. The World Conservation Union (IUCN) estimates that by the year 2000, at least 500,000 species of plants and animals may have been exterminated. The IUCN classifies species in danger of extinction into four categories: endangered, vulnerable, rare, and indeterminate. Endangered species are those whose survival is unlikely if the conditions threatening their extinction continue. These organisms need action by people to preserve them, or they will become extinct. Vulnerable species are those that have decreasing populations and will become endangered unless causal factors, such as habitat destruction, are eliminated. Rare species are primarily those that have

small worldwide populations and that could be at risk in the future. Indeterminate species are those that are thought to be extinct, vulnerable, or rare, but so little is known about them that they are impossible to classify.

Although the IUCN is a highly visible international conservation organization, it has very little power to effect change. It generally seeks to protect species in danger by encouraging countries to complete inventories of plants and animals within their borders. It also encourages the training of plant and animal biologists within the countries involved. (There is currently a critical shortage of plant and animal biologists who are familiar with the organisms of the tropics.) The IUCN also

GLOBAL PERSPECTIVE

The History of the Bison

The original ecosystem in the central portions of North America was a prairie dominated by a few species of grasses. The eastern prairie, where moisture was greater, had grasses up to 2 meters (6 feet) tall, while the dryer western grasslands were populated with shorter grasses. Many kinds of animals lived in this area, including prairie dogs, grasshoppers, many kinds of birds, and bison. The bison was the dominant organism. Millions of these animals roamed the prairies of North America with few predators other than the Native Americans, who used the bison for food, their hides for shelter, and their horns for tools and ornaments. The relationship between the bison and Native Americans was a predator-prey relationship in which the humans did not significantly reduce bison numbers.

When Europeans came to North America, they changed this relationship drastically. European-born Americans sought to convert the prairie to agriculture and ranching. However, two things stood in their way: the Native Americans, who resented the intrusion of the "white man" into their territory, and the bison. Since many of the Native American tribes had horses and a history of warlike encounters with other tribes, they attempted to protect their land from this intrusion. It was impossible to use the land for agriculture or ranching with the millions of bison occupying so much area because the groups of migrating bison would damage farmers' fields and compete with livestock for grasses. To facilitate the settling of the West, the U.S. government established a policy of controlling the bison and the Native Americans: Since bison were the primary food source of Native Americans in many areas, eliminating them would result in the starvation of many Native Americans, which would eliminate them as a problem for the frontier settler. In 1874, the secretary of the Interior stated that "the civilization of the Indian was impossible while the buffalo remained on the plains." Another example of this kind of thinking was expressed by Colonel Dodge, who was quoted as saying, "Kill every buffalo you can; every buffalo dead is an Indian gone." Bison were killed by the millions. Often, only their hides and tongues were taken; the rest of the animal was left to rot. Years later, the bones from these animals were collected

and ground up to provide fertilizer. By 1888, the bison was virtually eliminated.

A few bison were left in the Canadian wild and in remote mountain areas of the United States, while others survived in small captive herds. Eventually, in the early 1900s, the U.S. government established the national Bison Range near Missoula, Montana. The Canadian government established a bison reserve in Alberta.

The current status of the bison is fairly secure. There are over 370,000 bison in the United States and Canada. The majority of these animals (about 344,000) are in private herds. One place where bison have existed continuously in a wild state is Yellowstone National Park and the area immediately surrounding it. But even these bison have been managed in several ways over the years. Bison from captive herds were introduced into Yellowstone National Park, and the size of the bison herd was managed until 1966, when management activities were stopped and natural forces were allowed to determine the herd's size. Consequently, the size of the herd increased substantially. Since the Yellowstone bison are not fenced in, they occasionally leave the park area and enter surrounding ranch land. This has caused considerable concern because they carry a disease (brucellosis) that is also found in cattle. Fearing the disease could be transmitted to their cattle, ranchers in the area want the wandering bison controlled. Several management activities have been instituted to help alleviate the problem. The reintroduction of wolves to the Yellowstone ecosystem has had some effect, since wolves kill a few bison, particularly in the winter. But the major management activity involves reducing the number of bison that leave the park by capturing them and shipping them to slaughter or by allowing authorized persons to kill them. The carcasses are sold at auction or given to Native American groups.

encourages the establishment of preserves to protect species in danger of extinction.

Legal Protection

International efforts to preserve biodiversity have involved various activities of the United Nations. Most countries (168 of 191) of the world have signed the Convention on Biological Diversity, commonly known as the international biodiversity treaty. The United States signed the treaty, but it was not ratified by the

U.S. Senate. Some of the key components of the Convention on Biological Diversity are that the signatory countries will:

- Develop national strategies for the conservation and sustainable use of biological diversity.
- Identify components of biological diversity important for its conservation.
- Monitor biological diversity.
- Identify activities that have adverse impacts on the conservation and sustainable use of biological diversity.

- Establish a system of protected areas.
- Rehabilitate and restore degraded ecosystems and promote the recovery of threatened species.
- Develop or maintain necessary legislation for the protection of threatened species and populations.
- Integrate consideration of the conservation and sustainable use of biological resources into national decision making.

Awareness and concern about loss of biodiversity are high in many of the developed countries of the world. However, these parts of the world have few undisturbed areas and many vulnerable species have already been eliminated. The potential for the loss of biodiversity is greatest in tropical, developing countries. Many biologists estimate that there may be as many species in the tropical rainforests of the world as in the rest of the world combined. Unfortunately, loss of biodiversity is not a high priority for the general public in developing countries, even though their national governments have ratified the biodiversity treaty. This difference in level of concern is understandable since the developed world has surplus food, higher disposable income, and higher education levels, while people in many of the less-developed countries, where population growth is high, are most concerned with immediate needs for food and shelter, not with long-term issues such as species extinction. The developed world should not be held blameless, however, because our activities may indirectly be involved in extinction. When we purchase lumber, agricultural products, and fish from the less-developed world at low prices, we often are indirectly encouraging people in the developing world to overexploit their resources and are thus affecting the rates of extinction.

Many countries have established parks and other special designations of land use to protect areas of natural beauty or communities of organisms thought worthy of protection. Parks and protected lands allow a variety of uses, depending on the country and its laws. Some are used primarily as tourist attractions, which generate funds for the country. Others are set up primarily to protect a species or community of organisms, while others simply seek to restrict use so that certain environmental standards are met. For example, harvesting in a forest might be restricted to a certain number of trees each year, and the number of people allowed to visit a particular scenic site might be limited. Designating an area as protected does not solve the problem, however, since the areas must be protected from poachers and unauthorized agricultural activity of people who trespass on these protected areas.

These people are responding to basic biological and economic pressures to provide food for their families. In many countries, hunting for bush meat is an important source of income and much needed protein for the people of the area. Furthermore, protected lands generally do not generate any income for the government. Hiring security forces to patrol such areas is expensive for many of these countries, so protection is often inadequate to prevent trespass and poaching. Providing tourism opportunities that generate income for local people and the government is one way that some countries try to offset the costs of protection.

In the United States, the primary action related to the preservation of biodiversity involved the passage of the U.S. Endangered Species Act in 1973. This legislation designates species as endangered or threatened and gives the federal government jurisdiction over any species designated as endangered. **Endangered species** are those that have such small numbers that they are in immediate jeopardy of becoming extinct. (See figure 12.15.) **Threatened species** could become extinct if a critical factor in their environment were changed. About 300 U.S. species and subspecies have been so designated by the Office of Endangered Species of the Department of the Interior. The Endangered Species Act directs that no activity by a governmental agency should lead to the extinction of an endangered species and that all governmental agencies must use whatever measures are necessary to preserve these species.

Since one key to preventing extinctions is preservation of the habitat required by the endangered species, many U.S. governmental agencies and private organizations have purchased sensitive habitats or have managed areas to preserve suitable habitats for endangered species. Setting aside certain land areas or bodies of water forces government and private enterprise to confront the issue of endangered species. The question eventually becomes one of assigning a value to the endangered species. This is not an easy task; often politics become involved, and the endangered species does not always win. Since the protection of a

species requires the preservation of its habitat, inevitably landowners face losing the use of their land. Many lawsuits have been filed by landowners claiming that the federal government has taken their land from them by preventing them from using it for development purposes. As a result of these pressures, the rules were changed to allow some of the land to be used if a species conservation plan is in place. Another modification to the Endangered Species Act occurred in 1978, when Congress amended it so that exemptions to the act could be granted for federally declared major disaster areas or for national defense, or by a seven-member Endangered Species Review Committee. Because this group has the power to sanction the extinction of an organism, it has been nicknamed the "God Squad." If the committee found that the economic benefits of a project outweighed the harmful ecological effects, it would exempt a project from the Endangered Species Act.

The amendments to the Endangered Species Act also weakened the ability of the U.S. government to add new species to the endangered and threatened lists. Before a species can be listed, it is now necessary to determine the boundaries of its critical habitat, prepare an economic impact study, and hold public hearings—all within two years of the proposal of the listing. The political aspects of endangered species protection continue to be important in the U.S. Congress. Intense lobbying by environmental and business interests seeks to shape the reauthorization of the act to suit their interests.

A particularly sensitive issue is the designation of certain areas as wilderness. Although definitions of *wilderness* vary from country to country, the U.S. Congress, in the Wilderness Act of 1964, defined **wilderness** as "an area where the Earth and its community of life are untrammeled by man, where man himself is a visitor who does not remain." Recently, there has been intense pressure on members of Congress from oil-producing states such as Alaska to allow oil exploration in areas currently designated as wilderness, such as the Arctic National Wildlife Refuge. (See The Arctic National Wildlife

Blackfooted ferret

Whooping crane

Mission blue butterfly

Galápagos tortoise

Giant panda

Persistent trillium

Figure 12.15 **Endangered Species** Many species have been placed on the endangered species list. These are plants and animals that are present in such low numbers that they are in immediate danger of becoming extinct.

Refuge and Oil in chapter 10.) As population increases and resources become scarcer, pressures on wilderness areas will become greater throughout the world.

The designation of a species as endangered or threatened is biased toward large organisms that are relatively easy to identify. Many groups of organisms are very poorly known, so it is difficult to assess whether specific species are endangered or not. Table 12.5 compares the number of existing species in different taxonomic categories and the numbers and percentages considered endangered. Most endangered or threatened species are birds, mammals, certain categories of plants, some insects (particularly butterflies), a few mollusks, and fish. Bacteria, fungi, most insects, and many other inconspicuous organisms rarely show up on endangered species lists, even though they play vital roles in the nitrogen and carbon cycles, and as decomposers.

Table 12.5 Endangered and Threatened Species by Taxonomic Group

Taxonomic Group	Approximate Number of Described Species in the World	Number of U.S. Species Endangered or Threatened	World Conservation Union Threatened List	Percent of Total Species on World Conservation Union Threatened List
Insects	751,000	44	557	0.0076%
Plants	248,428	744	5,714	2.3%
Invertebrates other than insects and mollusks	180,000	33	436	0.24%
Mollusks	50,000	102	939	1.9%
Fishes	19,056	115	742	3.9%
Birds	9,040	92	1,192	13.2%
Reptiles	6,300	36	293	4.65%
Amphibians	4,184	21	157	3.75%
Mammals	4,000	74	1,137	28.4%
Fungi, bacteria, algae, protozoa	109,400	2	0	0%

Sources of data: The IUCN Red List of Threatened Species, and The U.S. Fish and Wildlife Service website.

Sustainable Management of Wildlife Populations

Many kinds of terrestrial wild animals are managed as game animals, to protect them from extinction, or for other purposes, and efforts are made to improve conditions for these species. Several techniques, some of which result in ecosystem modifications, are used to enhance certain wildlife populations. These techniques include habitat analysis and management, population assessment and management, and establishing refuges.

Habitat Analysis and Management

Managing a particular species requires an understanding of the habitat needs of that species. An animal's habitat must provide the following: food, water, and cover. **Cover** refers to any set of physical features that conceals or protects animals from the elements or enemies. Several kinds of cover are important and include places where the animal can rest and raise young, where it can escape from enemies, and where it is protected from the elements.

Animals are highly specific in their habitat requirements. For example, bob-

white quail must have a winter supply of food that protrudes above the snow. They require a supply of small rocks called grit, which the birds use in their gizzards to grind food. During most of the year, they need a field of tall grass and weeds as cover from natural predators. However, such protection is not available during the winter, so a thicket of brush is necessary. The thicket serves as cover from predators and shelter from the cold winter weather. Most areas provide suitable cover for resting, but for sleeping, quail prefer an open, elevated location, which allows the birds to quickly take flight if attacked at night. When raising young, they require grassy areas with some patches of bare ground where the young can sun themselves and dry out if they get wet. All of these requirements must be available within a radius of 400 meters (1300 feet), because this is the extent of a quail's normal daily travels. (See figure 12.16.)

Once the critical habitat requirements of a species are understood, steps can be taken to alter the habitat to improve the success of the species. The habitat modifications made to enhance the success of a species are known as **habitat**

management. The endangered Kirtland's warbler builds nests near the ground in dense stands of young jack pine. The density is important since their nests are vulnerable to predators and are better hidden in dense cover. Jack pine is a fire-adapted species that releases seeds following forest fires and naturally reestablishes dense stands following fire. Therefore, planned fires to regenerate new dense jack pine stands is a technique that has been used to provide appropriate habitat for this endangered bird.

Habitat management also may take the form of encouraging some species of plants that are the preferred food of the game species. For example, habitat management for deer may involve encouraging the growth of many young trees, saplings, and low-growing shrubs by cutting the timber in the area and allowing the natural regrowth to supply the food and cover the deer need. Both forest management and deer management may have to be integrated in this case because some other species of animals, such as squirrels, will be excluded if mature trees are cut, since they rely on the seeds of trees as a major food source.

Population Assessment and Management

Population management is another important activity that requires planning. Species of game animals are often managed so that they do not exceed the carrying capacity of their habitat. Wildlife managers use several techniques to establish and maintain populations at an appropriate level. Population censuses are used to check the population regularly to see if it is within acceptable limits. These census activities include keeping records of the number of animals killed by hunters, recording the number of singing birds during the breeding season, counting the number of fecal pellets, direct counting of large animals from aircraft, and a variety of other techniques.

Given suitable habitats and protection, most wild animals can maintain a sizable population. In general, organisms produce more offspring than can survive. High reproductive potential, protection from hunting, and the management and restoration of suitable habitats have resulted in large populations of once-rare animals. In Pennsylvania, where deer were once extinct, the number is now over 1 million. The wild turkey population in the United States has increased from about 20,000 birds in 1890 to over 5 million birds today. In Zimbabwe, unlike in most African countries, the number of elephants increased from about 200 in 1900 to 89,000 in 2003. This is well above the carrying capacity of about 45,000, and the government would like to reduce the number to about 30,000.

Since wildlife management often involves the harvesting of animals by hunting for sport and for meat, regulation of hunting activity is an important population management technique. Seasons are usually regulated to ensure adequate reproduction and provide the largest possible healthy population during the hunting season. Hunting seasons usually occur in the fall so that surplus animals are taken before the challenges of winter. Winter taxes the animal's ability to stay warm and is also a time of low food supplies in most temperate regions. A well-managed wildlife resource allows for a large number of animals to be harvested in

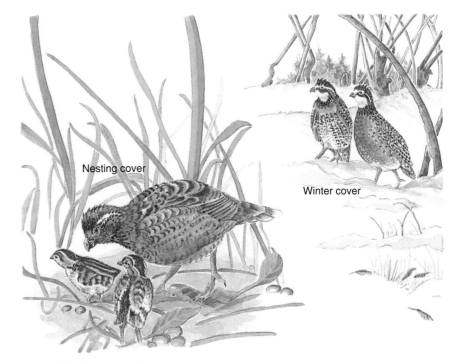

Figure 12.16 **Cover Requirements for Quail** Quail need several kinds of cover to be successful in an area. When raising young, they need areas of tall grass and weeds to provide protection from predators. In the winter, they need thickets to provide cover from weather and predators.

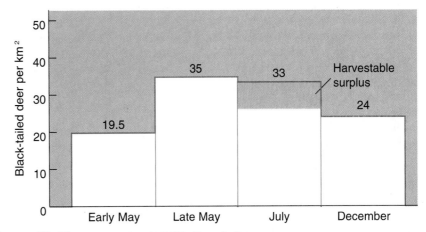

Figure 12.17 **Managing a Wildlife Population** The seasonal changes in this population of black-tailed deer are typical of many game species. The hunting season is usually timed to occur in the fall so that surplus animals will be harvested before winter, when the carrying capacity is lower.

Source: Data from R. D. Taber and R. F. Dasmann, "The Dynamics of Three Natural Populations of the Deer *Odocoileus hemionus columbianus*," *Ecology* 38, no. 2 (1957):233–46.

the fall and still leaves a healthy population to survive the winter and reproduce during the following spring. (See figure 12.17.)

In some cases, when a population of a species is below the desired number, organisms may be artificially introduced. Introductions are most often made when a species is being reintroduced to an area where it had become extinct. Wolves have been successfully reintroduced to Yellowstone National Park and large tracts of wooded land in the northern Midwest (Michigan, Wisconsin, Minnesota). Similarly an exchange between the state of Michigan and the province of Ontario reintroduced moose to Michigan and turkeys to parts of Ontario.

Cinnamon teal

Black duck

Mallard—male and female

Pintail (drake)

Atlantic flyway
Mississippi flyway
Central flyway
Pacific flyway

Figure 12.18 **Migration Routes for North American Waterfowl** Migratory waterfowl follow traditional routes when they migrate. These have become known as the Atlantic, Mississippi, Central, and Pacific flyways. Many of these waterfowl are hatched in Canada, migrate through the United States, and winter in the southern United States or Mexico.

Some game species are non-native species that have been introduced for sport hunting purposes. The ring-necked pheasant was originally from Asia but has been introduced into Europe, Great Britain, and North America. All of the large game animals of New Zealand are introduced species, since there was none in the original biota of these islands. Several species of deer have been introduced into Europe from Asia. Many of these are raised in deer parks, where the animals are similar to free-ranging domestic cattle. They may be hunted for sport or slaughtered to provide food. Introductions of exotic species were once common. However, today it is

recognized that these introductions often result in the decreased viability of native species or in the introduced species becoming pests.

Special Issues with Migratory Animals

Waterfowl (ducks, geese, swans, rails, etc.) present some special management problems because they are migratory. **Migratory birds** can fly thousands of kilometers and, therefore, can travel north in the spring to reproduce during the summer months and return to the south when cold weather freezes the ponds, lakes, and

streams that serve as their summer homes. (See figure 12.18.) Because many waterfowl nest in Canada and the northern United States and winter in the southern United States and Central America, an international agreement between Canada, the United States, and Mexico is necessary to manage and prevent the destruction of this wildlife resource. Habitat management has taken several forms. In Canada, where much of the breeding occurs, government and private organizations such as Ducks Unlimited have worked to prevent the draining of small ponds and lakes that provide nesting areas for the birds. In addition, new impound-

The California Condor

The California condor (*Gymnogyps californianus*) is thought to have been adapted to feed on the carcasses of large mammals found in North America during the Ice Age. With the extinction of the large, Ice Age mammals, the condors' major food source disappeared. By the 1940s, their range had shrunk to a small area near Los Angeles, California. Further fragmentation of their habitat caused by human activity, death by shooting, and death by eating animals containing lead shot reduced the wild population to about 17 animals by 1986.

A low reproductive potential makes it difficult for the species to increase in numbers. They do not become sexually mature until six years of age, and females typically lay one egg every two years. Because of concerns about the survival of the species, in 1987, all the remaining wild condors were captured to serve as breeding populations. The total population was 27 individuals. The plan was to raise young condors in captivity, leading to a large population that would ultimately allow for the release of animals back into the wild. Captive breeding is a very involved activity. Extra eggs can be obtained by removing the first egg laid and incubating it artificially. The female will lay a second egg if the first is removed. Raising the young requires careful planning. Although the young are fed by humans, they must not associate food with humans. This would result in inappropriate behaviors in animals eventually released to the wild. Therefore, puppets that resemble the parent birds are used to feed the young birds.

These efforts increased the number of offspring produced per female and resulted in a captive population of 54 individuals by 1991, when two condors were released into the wild north of Los Angeles. Six more were released in 1992. In 1996, a second population of condors was introduced in Arizona north of the Grand Canyon near the Utah border. The goal is to have two populations of up to 150 individuals each. In addition to natural mortality from predation, there has been unanticipated mortality from contact with power lines and lead poisoning from ingesting lead shot with food. To help reduce these mortality problems, the birds that are currently being released go through a period of training in which they are taught to avoid power lines. In addition, several condors have been recaptured to be treated for lead poisoning. These will eventually be released back to the wild.

In 2003, there were 220 California condors in existence. Thirty-six were in the wild in Arizona, 47 were in the wild in California and Baja California, 11 were awaiting release, and the rest were in captive breeding programs at three locations. The behavior of the animals seems natural, and there is hope that they may eventually reproduce in the wild.

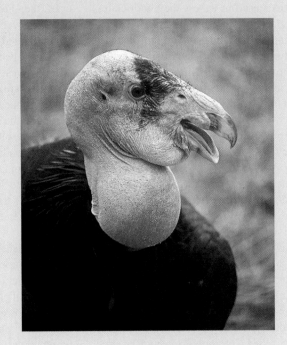

ments have been created where it is practical. Because birds migrate southward during the fall hunting season, a series of wildlife refuges provide resting places, food, and protection from hunting. In addition, these refuges may be used to raise local populations of birds. During the winter, many of these birds congregate in the southern United States. Refuges in these areas are important overwintering areas where the waterfowl can find food and shelter.

A similar management problem exists in Africa and Eurasia. Many birds that rely on wetlands migrate between Eurasia and Africa. In 1995, an international Agreement on the Conservation of African-Eurasian Waterbirds was developed. Its development was a complicated process, since it involves about 40 percent of the world's land surface and includes 117 countries. Several important countries have signed the agreement, but many have yet to do so.

In sub-Saharan Africa, many species of wildlife migrate across large expanses of land that may involve several countries. Often different countries have differing rules for harvest, and often there are barriers to the movement of migratory animals. For example, fences are commonly used to designate the boundaries between nations. A relatively new concept of transboundary parks has been developed to accommodate the movements of these animals. One typical action that occurs with the formation of transboundary parks across national boundaries is the removal of structures that interfere with the movement of the animals.

Sustainable Management of Fish Populations

For several reasons, the most productive areas of the ocean are those close to land. In shallow water, the entire depth (water column) is exposed to sunlight. Plants and algae can carry on photosynthesis, and biological productivity is high. The nutrients washed from the land also tend to make these waters more fertile. Furthermore, land masses modify currents that bring nutrients up from the ocean bottom. Many of the commercially important fish and other seafood species are bottom dwellers, but fishing for them at great depths is not practical. Therefore, fishing pressure is concentrated on areas where the water is shallow and relatively nutrient rich. Countries have taken control over the waters near them by establishing a 200-nautical-mile limit (approximately 300 kilometers) within which they control the fisheries. This has not solved conflicts, however, as neighboring countries dispute the fishing practices used in waters they both claim. Canada and the United States have had continuing arguments over the management of the lobster fishery and argued as well over the North Atlantic cod fishery until its collapse in the early 1990s.

One of the major problems associated with the management of marine fisheries resources is the difficulty in achieving agreement on limits to the harvest. The oceans do not belong to any country; therefore, each country feels it has the right to exploit the resource wherever it wants. Since each country seeks to exploit the resource for its advantage without regard for the sustainable use of the resource, international agreements are required to manage the resource. Such agreements are difficult to achieve because of political differences and disregard for the nature of the resource. (See Common Property Resource Problems—The Tragedy of the Commons in chapter 3.)

In recent years, several countries have designated areas within their territorial waters where fishing is not allowed. Several studies suggest that these reserves are effective ways to protect a portion of the fishery resource and serve to repopulate surrounding areas that are open to fishing. This approach is particularly effective for fish species that live on the bottom or are associated with specific structures on the bottom.

Although freshwater fisheries face some of the same problems as marine fisheries, they are typically easier to regulate because rivers, lakes, and streams are often contained within a particular country. Although people in many places in the world harvest freshwater fish for food, the quantities are relatively small compared to those produced by marine fisheries. Furthermore, the management of the freshwater fish resource is much more intense since the bodies of water are smaller and human populations have greater access to the resource. Freshwater fish management in much of the world includes managing for both recreational and commercial food purposes.

The fisheries resource manager must try to satisfy two interest groups. Both sports fishers and those who harvest for commercial purposes must adhere to regulations. However, the regulations usually allow the commercial fisher to use different harvesting methods, such as nets. In much of North America, freshwater fisheries are primarily managed for sport fishing. The management of freshwater fish populations is similar to that of other wild animals. Fish require cover, such as logs, stumps, rocks, and weed beds, so they can escape from predators. They also need special areas for spawning and raising young. These might be a gravel bed in a stream for salmon, a sandy area in a lake for bluegills or bass, or a marshy area for pike. A freshwater fisheries biologist tries to manipulate some of the features of the habitat to enhance them for the desired species of game fish. This might take the form of providing artificial spawning areas or cover. Regulation of the fishing season so that the fish have an opportunity to breed is also important.

In addition to these basic concerns, the fisheries biologist pays special attention to water quality. Whenever people use water or disturb land near the water, water quality is affected. For example, toxic substances kill fish directly, and organic matter in the water may reduce the oxygen. The use of water by industry or the removal of trees lining a stream warms the water and makes it unsuitable for certain species. Poor watershed management results in siltation, which covers spawning areas, clogs the gills of young fish, and changes the bottom so food organisms cannot live there. The fisheries biologist is probably as concerned about what happens outside the lake or stream as what happens within it.

The loss of biodiversity has become a major concern. Biodiversity can be measured in several ways. Genetic diversity is reduced when populations are reduced in size. Species diversity is reduced when a species becomes extinct from a local population or worldwide. Ecosystems involve the interactions of organisms and their physical environment. The loss of key species can result in the loss of a type of ecosystem either locally or worldwide.

Concern about the loss of biodiversity is based on several ways of looking at the value of a species. Functioning ecosystems and their component organisms provide many valuable services that are often overlooked because they are not easily measured in economic terms. For example, nitrogen-fixing bacteria provide nitrogen for crops, photosynthesis is responsible for food production, and decomposers recycle organic materials. Other biotic resources can be measured in economic terms. Meat and fish harvested from the wild have great economic value, many kinds of drugs are derived from plants and animals, and ivory or animal hides are frequently traded. Many people also consider the loss of biodiversity to be an ethical problem. They ask the question, Do humans have the right to drive other species to extinction?

The primary causes of biodiversity loss are habitat loss by humans converting ecosystems to agriculture and grazing, overexploitation by harvesting species at unsustainable levels, the introduction of exotic species that disrupt ecosystems and compete or prey on native organisms, and purposeful killing of pest organisms such as large predators.

Protection of biodiversity typically involves two kinds of activities: legal protections by national laws and international agreements, and management of the use of species and ecosystems at sustainable levels. These management activities typically include the establishment of regulations for hunting, fishing, and other uses.

Issues – Analysis

The Problem of Image

When we think of endangered species, we almost always visualize a mammal or a bird. In North America, we identify wolves, grizzly bears, various whales, bald eagles, whooping cranes, and similar species as endangered. We rarely think about clams, fish, plants, or insects. Because certain species are able to grab the attention of the public, they are called charismatic species. In addition, most of the charismatic species are carnivores at high trophic levels. Groups of people will organize to save the whales, whooping cranes, elephants, or osprey, but little interest is generated to save the Tar River spiny mussel, cave crayfish, or San Diego fairy shrimp. The graph shows the percentage of species in selected groups that are in various categories of concern in the United States. By far the most vulnerable category of organisms consists of freshwater species of mussels (clams), crayfish, fish, and stoneflies. The least vulnerable are birds and mammals, yet they capture most of the public's interest and are highlighted by the U.S. Fish and Wildlife Service on its endangered species website.

- What factors cause us to rank birds and mammals higher than clams and crayfish?
- Should we spend as much money to save the Tar River spiny mussel as we are spending on wolves or California condors?
- Since money is limited, how would you decide which species to spend limited resources on?

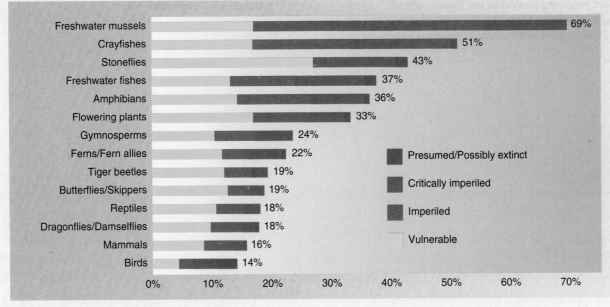

Freshwater mussels	69%
Crayfishes	51%
Stoneflies	43%
Freshwater fishes	37%
Amphibians	36%
Flowering plants	33%
Gymnosperms	24%
Ferns/Fern allies	22%
Tiger beetles	19%
Butterflies/Skippers	19%
Reptiles	18%
Dragonflies/Damselflies	18%
Mammals	16%
Birds	14%

Legend:
- Presumed/Possibly extinct
- Critically imperiled
- Imperiled
- Vulnerable

Source: Data from http://www.natureserve.org/images/precious/4-2.gif.

13 Land-Use Planning

Chapter Outline

Objectives

After reading this chapter, you should be
able to:

• Explain why most major cities are located
  on rivers, lakes, or the ocean.

• Describe the forces that result in farmland
  adjacent to cities being converted to urban
  uses.

• Explain why floodplains and wetlands are
  often mismanaged.

• Describe the economic and social values
  involved in planning for outdoor recreation
  opportunities.

• Explain why some land must be designated
  for particular recreational uses, such as
  wilderness areas, and why that decision
  sometimes invites disagreement from those
  who do not desire to use the land in the
  designated way.

• List the steps in the development and
  implementation of a land-use plan.

• Describe methods of enforcing compliance
  with land-use plans.

• Describe the advantages and disadvantages
  of both local and regional land-use
  planning.

• Describe the concept of smart growth.

PAST PRESENT FUTURE

1983 There were 350 million cars registered
globally.

1982 The average size of a new home in
North America was 159 square meters
(1710 square feet).

1980 The average number of persons per
household in North America was 2.75.

1982 The urban population in North America
was roughly 200 million.

2003 There were 600 million cars registered
globally.

2002 The average size of a new home in
North America was 216 square meters
(2320 square feet; a 36 percent
increase).

2002 The average number of persons per
household in North America was 2.58
(a 6 percent reduction).

2003 The urban population in North America
exceeded 260 million.

The Need for Planning

A large proportion of the land surface of the world (about one-third to one-half) has been changed by human activity. Most of this change occurred as people converted the land to agriculture and grazing, but, in our modern world, significant amounts have been covered with buildings, streets, highways, and other products of society. In many cases, cities grew without evaluating and determining the most logical use for the land. Los Angeles and Mexico City have severe air pollution problems because of their geographic location, Venice and New Orleans are threatened by high sea levels, and San Francisco and Tokyo are subject to earthquakes. Currently, most land-use decisions are still based primarily on economic considerations or the short-term needs of a growing population rather than on careful analysis of the capabilities and unique values of the land and landscape. Each piece of land has specific qualities based on its location and physical makeup. Some is valued for the unique species that inhabit it, some is valued for its scenic beauty, and some has outstanding potential for agriculture or urban uses. Since land and the resources it supports (soil, vegetation, elevation, nearness to water, watersheds) are not being created today (except volcanoes, river deltas, etc.), land should be considered a nonrenewable resource.

Once land has been converted from natural ecosystems or agriculture to intensive human use, it is generally unavailable for other purposes. As the population of the world grows, competition for the use of the land will increase, and systematic land-use planning will become more important. Furthermore, as the population of the world becomes more urbanized and cities grow, urban planning becomes critical.

Historical Forces That Shaped Land Use in North America

Today, most of the North American continent has been significantly modified by human activity. In the United States, about 47 percent of the land is used for crops and livestock, about 45 percent is forests and natural areas (primarily west of the Mississippi River), and nearly 5 percent is used intensively by people in urban centers and as transportation corridors. Canada is 54 percent forested and wooded and uses only 8 percent of its land for crops and livestock. Less than 1 percent of the land is in urban centers and transportation corridors. A large percentage of its remaining land is wilderness in the north.

This pattern of land use differs greatly from the original conditions experienced by the early European colonists who immigrated to the New World. The first colonists converted only small portions of the original landscape to farming, manufacturing, and housing, but as the population increased, more land was converted to agriculture, and settlements and villages developed into towns and cities. Although most of this early development was not consciously planned, it was not haphazard. Several factors influenced where development took place and the form it would take.

The Importance of Waterways

Waterways were the primary method of transportation, which allowed exploration and the development of commerce in early North America. Thus, early towns were usually built near rivers, lakes, and oceans. Typically, cities developed as far inland as rivers were navigable. Where abrupt changes in elevation caused waterfalls or rapids, goods being transported by boat or barge needed to be offloaded, transported around the obstruction, and loaded onto other boats. Cities often developed at these points. Buffalo, New York, and Sault Sainte Marie, Ontario, are examples of such cities. In addition to transportation, bodies of water provided drinking water, power, and waste disposal for growing villages and towns. Those towns and villages with access to waterways that provided easy transportation could readily receive raw materials and distribute manufactured goods. Some of these grew into major industrial or trade centers. Without access to water, St. Louis, Montreal, Chicago, Detroit, Vancouver, and other cities would not have developed. (See figure 13.1.) The availability of other natural resources, such as minerals, good farmland, or forests was also important in determining where villages and towns were established. Industrial development began on the waterfront since water supplied transportation, waste disposal, and power. As villages grew into towns and cities, large factories replaced small gristmills, sawmills, and blacksmith shops. The waterfront became a center of intense industrial activity. As industrial activity increased in the cities, people began to move from rural to urban centers for the job opportunities these centers presented.

The Rural-to-Urban Shift

North America remained essentially rural until industrial growth began in the last third of the 1800s and the population began a trend toward greater urbanization. (See figure 13.2.) Several forces led to this rural-to-urban transformation. First, the Industrial Revolution led to improvements in agriculture that required less farm labor at the same time industrial jobs became available in the city. Thus, people migrated from the farm to the city. The average person was no longer a farmer but, rather, a factory worker, shopkeeper, or clerk (with a regular paycheck) living in a tenement or tiny apartment near where they worked. This pattern of rural-to-urban migration occurred throughout North America, and it is still occurring in developing countries today. A second factor that affected the growth of cities was the influx of immigrants from Europe. Although some became farmers, many of these new citizens settled in towns and cities, where jobs were available. A third reason for the growth of cities was that they offered a greater variety of cultural, social, and artistic opportunities than did rural communities. Thus, cities were attractive for cultural as well as economic reasons. San Antonio, Texas, and Las Vegas, Nevada, are examples of cities developing around unique cultural attractions.

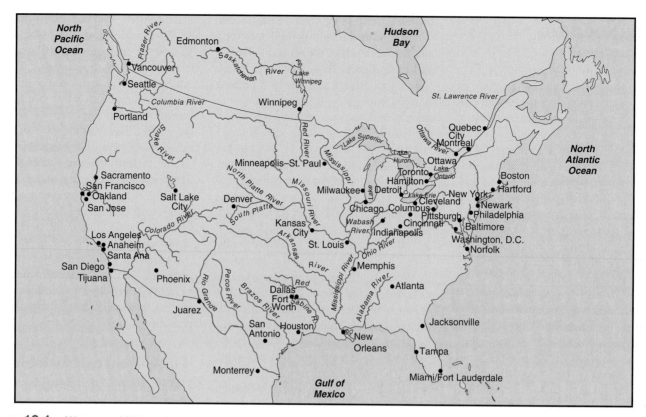

Figure 13.1   Water and Urban Centers Note that most of the large urban centers are located on water. Water is an important means of transportation and was a major determining factor in the growth of cities. The cities shown have populations of 1 million or more (except Edmonton, Winnipeg, and Quebec City).

Figure 13.2   Rural-to-Urban Population Shift In 1800, the United States was essentially a rural country. Industrialization in the late 1800s began the shift to an increasing urban population. In 2000, 75 percent of the U.S. population was urban.

Sources: Data from the *Statistical Abstract of the United States;* 2000 data from Population Reference Bureau, Inc., Washington, D.C.

GLOBAL PERSPECTIVE

Urbanization in the Developing World

Traditionally, most of the population of the developing world has been rural. However, in recent years, the number of people migrating to the cities has grown rapidly. By 2025, about 5 billion people are expected to be living in urban centers. Most of the increase in urban areas will be in the developing world. Many migrate to the cities because they feel they will have greater access to social services and other cultural benefits than are available in rural areas. Many also feel there are more employment opportunities. However, the increase in the urban population is occurring so rapidly that it is very difficult to provide the services needed by the population, and jobs are not being created as fast as the urban population is growing. Thus, many of the people live in poverty on the fringes of the city in shantytowns that lack water, sewer, and other services. Often these shantytowns are constructed without permission only a short distance from affluent urban dwellers. Because the poor lack safe drinking water and sewer services, they pollute the local water sources and disease is common. Because they burn wood and other poor-quality fuels in inefficient stoves, air pollution is common. The additional people also create traffic problems of staggering proportions. The chart shows the seven cities with the largest rate of population increase. All are in the developing world.

Cities Expected to Grow by More Than 50 Percent between 1995 and 2015

City	1995 Population (Millions)	Projected 2015 Population (Millions)
Bombay, India	15.14	26.22
Lagos, Nigeria	10.29	24.61
Delhi, India	9.95	16.86
Karachi, Pakistan	9.73	19.38
Metro Manila, Philippines	9.29	14.66
Jakarta, Indonesia	8.62	13.92
Dhaka, Bangladesh	8.55	19.49

Source: Data from *World Resources* 2002–2003.

Migration from the Central City to the Suburbs

During the early stages of industrial development, there was little control of industry activities, so the waterfront typically became a polluted, unhealthy, undesirable place to live. As roads and rail transport became available, anyone who could afford to do so moved away from the original, industrial city center. The more affluent moved to the outskirts of the city, and the development of suburban metropolitan regions began. Thus, the agricultural land surrounding the towns was converted to housing. Most cities originally had good farmland near them, since the floodplain near rivers typically has a deep, rich soil and agricultural land adjacent to the city was one of the factors that determined whether

the city grew or not. This was true because until land transportation systems became well developed, farms needed to be close to the city so that farmers could transport their produce to the markets in the city. This rich farmland adjacent to the city was ideal for the expansion of the city. As the population of the city grew, demand for land increased. As the price of land in the city rose, people and businesses began to look for cheaper land farther away from the city. Developers and real estate agents were quick to respond and to help people acquire and convert agricultural land to residential or commercial uses. Land was viewed as a commodity to be bought and sold for a profit, rather than as a nonrenewable resource to be managed. As long as money could be made by converting agricultural land to other purposes, it was impossible to prevent such conversion. There were no counteracting forces strong enough to prevent it.

The conversion of land around cities in North America to urban uses destroyed many natural areas that people had long enjoyed. The Sunday drive from the city to the countryside became more difficult as people had to drive farther to escape the ever-growing suburbs. The unique character of neighborhoods and communities was changed by the erection of shopping malls, apartment complexes, and expressways. Most of these alterations occurred without considering how they would affect the biological community or the lives of the people who lived in the area.

As cities continued to grow, certain sections within each city began to deteriorate. Industrial activity continued to be concentrated near water in the city's center. Industrial pollution and urban crowding turned the core of many cities into undesirable living areas. In the early 1900s, people who could afford to leave began to move to the outskirts. This trend continued after World War II, in the

1940s, 1950s, and 1960s, as a strong economy and government policies that favored new home purchases (tax deductions and low-interest loans) allowed more people to buy homes.

In 1947, William Levitt, a developer from the east coast of the United States, put forth a new plan for building affordable homes away from congested cities. Levitt bought thousands of hectares of potato fields on New York's Long Island. For no money down and $65 a month, a $6,900 Cape Cod home could be bought—and many were! Levitt's dream was so successful that he built and sold thousands of houses, completing as many as 36 in a day. Levittown, New York, and the suburban tract home and lifestyle were born—and a new form of "the American dream" took hold.

At first, this form of land planning and building worked. Large numbers of people found themselves homeowners, with a yard where they could plant a garden reminiscent of the farms and country villages from which many of them had come. Land was plentiful, and with cheap oil and labor, it appeared the answer had been found: A person could work in the city and "live in the country." Cars were also cheap, allowing workers to drive to town and be home by sundown.

In 1950, about 60 percent of the urban population lived in the central city; by 1990, this number was reduced to about 30 percent. Most of these new single-family homes were in attractive suburbs, away from the pollution and congestion of the central city. These collections of houses were built on large lots that provided outdoor space for family activities. The blocks of homes were added to the periphery of the city in a decentralized pattern in which single-family homes were separated from multifamily structures and both were some distance from places of work, shopping, and other service needs. This pattern also made it very difficult to establish efficient public transportation networks, which decreased energy efficiency and increased the cost of supplying utility services. However, public transport was not considered important because rising automobile ownership and improved highway systems would accommodate the transportation needs of the

suburban population. The convenience of a personal automobile escalated decentralized housing patterns, which, in turn, required better highways, which led to further decentralization.

By 1960, unplanned suburban growth had become known as urban sprawl. **Urban sprawl** is a pattern of unplanned, low-density housing and commercial development outside of cities that usually takes place on previously undeveloped land. In addition, blocks of housing are separated from commercial development, and the streets typically form branching patterns and often include cul-de-sacs. These large housing tracts surrounded cities, which made it difficult for people to find open space. A city dweller could no longer take a bus to the city limits and enjoy the open space of the countryside.

The way we plan the physical layout, or land use, of our communities is fundamental to sustainability. Two main features of our land-use practices over the past several decades have converged to generate haphazard, often inefficient, and unsustainable urban sprawl:

- Zoning ordinances that isolate employment locations, shopping, services, and housing locations from each other.
- Low-density growth planning aimed at creating automobile access to increasing expanses of land.

The complex problems shared by many cities are evidence of the impacts of urban sprawl, such as increasing traffic congestion and commute times, air pollution, inefficient energy consumption and greater reliance on oil, loss of open space and habitat, inequitable distribution of economic resources, and loss of sense of community.

Community sustainability requires a transition from poorly managed sprawl to land-use planning practices that create and maintain efficient infrastructure, ensure close-knit neighborhoods and a sense of community, and preserve natural systems.

Urban sprawl has been defined as auto-dependent development outside of the compact urban and village center, along highways, and in the rural countryside. Sprawl is typically characterized by the following:

- Excessive land consumption
- Low densities in comparison with older centers
- Lack of choice in ways to travel
- Fragmented open space, wide gaps between development, and a scattered appearance
- Lack of choice in housing types and prices
- Separation of uses into distinct areas
- Repetitive one-story development
- Commercial buildings surrounded by large areas for parking
- Lack of public spaces and community centers

Urban sprawl occurs in three ways. (See figure 13.3.) One type of growth involves the development of exclusive, wealthy suburbs adjacent to the city. These homes are usually on large individual lots in the more pleasing geographic areas surrounding the city. Often they are located along water, on elevated sites, or in wooded settings. A second development pattern is tract development. **Tract development** is the construction of similar residential units over large areas. Initially, these tracts are often separated from each other by farmland. New roads are constructed to link new housing to the central city and other suburbs, which stimulates the development of a third form of urban sprawl along transportation routes. This is referred to as **ribbon sprawl** and usually consists of commercial and industrial buildings that line each side of the highway that connects housing areas to the central city and shopping and service areas. Ribbon sprawl results in high costs for the extension of utilities and other public services. It also makes the extent of urbanization seem much larger than it actually is, since the undeveloped land is hidden by the storefronts that face the highway.

As suburbs continued to grow, cities (once separated by farmland) began to merge, and it became difficult to tell where one city ended and another began. This type of growth led to the development of regional cities. Although these cities maintain their individual names, they are really just part of one large urban

Figure 13.3 **Types of Urban Sprawl** Note the three different types of growth depicted in these photos. (*a*) The wealthy suburbs with large lots are adjacent to the city. (*b*) Ribbon sprawl develops as a commercial strip along highways. (*c*) Tract development results in neighborhoods consisting of large numbers of similar houses on small lots.

area called a **megalopolis.** (See figure 13.4.) The eastern seaboard of the United States, from Boston, Massachusetts, to Washington, D.C., is an example of a continuous city. In the midwestern United States, the area from Milwaukee, Wisconsin, to Chicago, Illinois, is a further example. Other examples are London to Dover in England, the Toronto-Mississauga region of Canada, and the southern Florida coast from Miami northward.

In some areas throughout North America, the growth of suburbs has been slowed due to the increased cost of housing and transportation. People have migrated back to some cities on a limited scale because of the lower cost of urban houses and the fact that public transportation is generally more efficient in the city than in the suburbs, thus freeing urban residents from the cost of daily commuting. This reverse migration, however, is still greatly offset by the continual growth of the suburban communities.

Factors That Contribute to Sprawl

As we enter the twenty-first century, many, including government policy makers, are beginning to question the prevailing pattern of sprawling urban development. However, powerful social, economic, and policy forces have been behind the development of the current decentralized urban centers typical of North

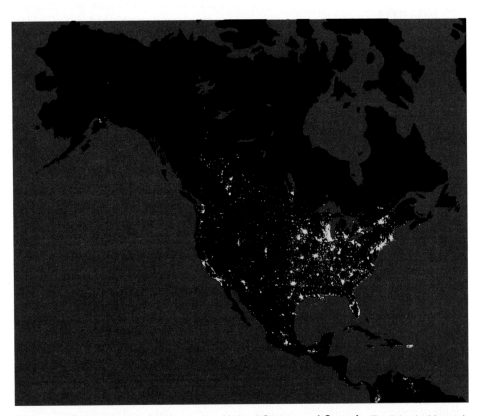

Figure 13.4 **Regional Cities in the United States and Canada** The lights in this satellite image show population concentrations. More than 30 major regional cities have developed, with each having more than 1 million people. Many of these cities merge with their neighbors to form huge regional cities. Major urban regions are the northeast coast of the United States (Boston to Washington, D.C.), the region south of the Great Lakes (Chicago to Pittsburgh), south Florida (Jacksonville to Miami), the Toronto and Montreal regions of Canada, and the west coast of California (San Francisco to San Diego).

America. If this pattern is to change, these underlying factors must be understood.

Among the causes of sprawl are:

- Public investments in roads, public buildings, water, sewer, and other infrastructure in peripheral areas; disinvestments in existing centers

- Land-use regulations that promote spread-out, land-consumptive development

- Consumer desire for a rural lifestyle with large homes and large yards, a safe environment, and less traffic congestion
- Preference of business and industry for easy highway access, plenty of free parking
- Demands of commercial tenants for particular locations and designs for buildings and sites
- Other public policies, including tax and utility rate policies
- Higher costs of development in older, traditional centers
- Lower land prices in peripheral areas
- Telecommunications advances
- Commercial lending practices that favor suburban development

Lifestyle Factors

One of the factors that has supported urban sprawl is the relative wealth of the population. This wealth is reflected in material possessions, two of which are automobiles and homes. Based on 2003 census data, in the United States, there are 78 motor vehicles for every 100 people. Since over 20 percent of the population is too young to drive, this means that there is essentially one motor vehicle for every licensed driver in the United States. In addition, about two-thirds of the population live in family-owned homes. With this level of wealth, people are free to choose where and how they want to live. Many are attracted to a lifestyle that includes low-density residential settings, with easy access to open space, isolated from the problems of the city. They are also willing to drive considerable distances to live in these settings. Thus, a decentralized housing pattern is possible because of the high rate of automobile ownership, which allows for ease of movement.

Sprawl impacts our quality of life in several ways, including:

- Increased dependency on the auto, fuel consumption, and air pollution
- Increased commuting times and costs
- Reduced opportunity for public transportation services
- Increased health problems in children and adults due to a sedentary lifestyle

- More time in cars and less time for family, friends, and recreation
- Loss of a sense of place and community decline, resulting in fragmented and dispersed communities, and a decline in social interaction and the isolation of some populations, such as the poor and the elderly, in urban areas.

Economic Factors

Several economic forces operate to encourage sprawl development. First of all, it is less expensive to build on agricultural and other nonurban land than it is to build within established cities. The land is less expensive, the regulations and permit requirements are generally less stringent, and there are fewer legal issues to deal with. An analysis of building costs in the San Francisco Bay area of California determined that it costs between 25 percent and 60 percent more to build in the city than it does to build in the suburbs.

Several tax laws also contributed to encouraging home ownership. The interest on home loans is deductible from income taxes, and, in the past, people could avoid paying capital gains taxes on homes they sold if they bought another home of equal or higher value. Since homes usually increased in value, this created a market for increasingly expensive homes.

Sprawl impacts the economy in several ways, including:

- Excessive public costs for roads, utility line extensions, and service delivery to dispersed development
- Decline in economic opportunity in traditional centers
- Premature disinvestments in existing buildings, facilities, and services in urban and village centers
- Relocation of jobs to peripheral areas at some distance from population centers
- Decline in the number of jobs in some sectors, such as retail
- Isolation of employees from activity centers, homes, daycare facilities, and schools

- Reduced ability to finance public services in urban centers

Planning and Policy Factors

Many planning and policy issues have contributed to sprawl development. First of all, until recently, little coordinated effort has been given to planning how development should occur in metropolitan areas. There are several reasons for this. First of all, most metropolitan areas include hundreds of different political jurisdictions (the New York City metropolitan area includes about 700 separate government units and involves the states of New York, New Jersey, and Connecticut). It is very difficult to integrate the activities of these separate jurisdictions in order to achieve coordinated city planning. In addition, it is very difficult for a small local unit of government to see the "big picture," and many are unwilling to give up their autonomy to a regional governmental body.

Local zoning ordinances have often fostered sprawl by prohibiting the mixing of different kinds of land use. Single-family housing, multiple-family housing, commercial, and light and heavy industry were restricted to specific parts of the community. In addition, many ordinances specify minimum lot sizes and house sizes. This tends to result in a decentralized pattern of development that is supported by the heavy use of automobiles and the roads and parking facilities necessary to support them.

In addition, many government policies actually subsidize the development of decentralized cities. For example, developers and the people who buy the homes and businesses they build are able to avoid paying for the full cost of extending services to new areas. The local unit of government picks up the cost of these improvements, and the cost is divided among all the taxpayers rather than just those who will benefit. The roads that are needed to support new developments are usually paid for with federal and state monies, so again the cost is not borne by those who benefit most but by the taxpayers of the state or nation. Furthermore, federal and state governments have not supported public transport in an equivalent way.

Sprawl impacts the environment in several ways, including:

- Fragmentation of open space and wildlife habitat
- Loss of productive farmland and forestland
- Decline in water quality from increased urban runoff, shoreline development, and loss of wetlands
- Inability to capitalize on unique cultural, historic, and public space resources (such as waterfronts) in urban and village centers

Problems Associated with Unplanned Urban Growth

As the population increased and metropolitan areas grew, several kinds of problems were recognized. They fall into several broad categories.

Transportation Problems

Most cities experience continual problems with transportation. This is primarily because, as cities grew, little thought was given to how people were going to move around and through the city. Furthermore, when housing patterns and commercial sectors changed, transportation mechanisms had to be changed to meet the shifting needs of the public. This often involved the abandonment of old transportation corridors and the establishment of new ones. Paradoxically, the establishment of new transportation corridors stimulates growth in the areas served, and the transportation corridors soon become inadequate. The reliance on the automobile as the primary method of transportation has required the constant building of new highways. A strong automobile-based bias exists in the culture in the United States. In Los Angeles, for example, 70 percent of surface area in the city is dedicated in some way to the automobile (roads, parking garages, etc.), compared with only 5 percent devoted to parks and open space.

According to the U.S. Department of Transportation, the average person in the

Figure 13.5 **Traffic and Suburbia** Traffic congestion is a common experience for people who work in cities but live in the suburbs. The popularity of automobiles and the desire to live in the suburbs are closely tied.

United States spends nine hours per week traveling in a car. In many metropolitan areas, the amount of time is much greater, and a significant amount of that time involves traffic jams. A study of traffic in the Washington, D.C., area concluded that the average commuter spent 80 hours per year stuck in traffic, in addition to the time normally needed to make the commute. (See figure 13.5.)

The local transportation commission in Ventura County, California (a semirural county with a population of approximately 650,000 located just north and west of Los Angeles County), found that the county will need to spend $1.35 billion over the next eight years to widen roads and build interchanges just to maintain and slightly improve the current level of service. That equals over $168 million per year. This example is not unique to southern California but rather the norm now for most urban and semiurban areas.

Air Pollution

Reliance on the automobile as the primary method of transportation has resulted in significant air pollution problems in many cities. Most of the large industrial sources of air pollution have been contained.

However, the individual car with its single occupant going to work, to shop, or to eat a meal is a constant source of air pollution. A simple solution to this problem is a centralized, efficient public transportation system. However, this is difficult to achieve with a highly dispersed population.

Low Energy Efficiency

Energy efficiency is low for several reasons. First of all, automobiles are the least energy-efficient means of transporting people from one place to another. Second, the separation of blocks of homes from places of business and shopping requires that additional distances be driven. Third, congested traffic routes result in hours being spent in stop-and-go traffic, which wastes much fuel. Finally, single-family homes require more energy for heating and cooling than multifamily dwellings.

Loss of Sense of Community

Although it is difficult to measure, general agreement exists that in dispersed suburban developments, there is a loss of a sense of community. In many places, people stay within their homes and yards and do not routinely walk through the productive and excellent places for aquatic species to grow rapidly. Wetlands are frequently critical to the reproduction of many kinds of animals. Many fish use estuaries and marshes for spawning. Wetlands also provide nesting sites for many kinds of birds and serve as critical habitats for many other species. Waterfowl hunters and commercial and sport fisheries depend on these habitats to produce and protect the young of the species they harvest. Human impact on wetlands has severely degraded or eliminated these spawning and nursery habitats.

Besides providing a necessary habitat for fish and other organisms, wetlands provide natural filters for sediments and runoff. This filtration process allows time for water to be biologically cleaned before it enters larger bodies of water, such as lakes and oceans, and reduces the sediment load carried by runoff. Wetlands also protect shorelines from erosion. When destroyed, the natural erosion protection provided by wetlands must be replaced by costly artificial measures, such as breakwalls.

Other Land-Use Considerations

The geologic status of an area must also be considered in land-use decisions. Building cities on the sides of volcanoes or on major earthquake-prone faults has led to much loss of life and property. Building homes and villages on unstable hillsides or in areas subject to periodic fires is also unwise. Yet, every year more houses slide down California hillsides and wildfires consume homes throughout the dry West.

Another problem in some locations is lack of water. Southern California and metropolitan areas in Arizona must import water to sustain their communities. Wise planning would limit growth to whatever could be sustained by available resources. The risk of serious water shortages in Beijing, China, has forced the government to consider a massive 1200-kilometer (745-mile) diversion of water from the Yangtze River to the city. As the population of cities in such locations continues to grow,

the strain on regional water resources will increase.

Water-starved cities often cause land-use dilemmas far from the city boundaries. Supplying water and power to cities often involves the construction of dams that flood valleys that may have significant agricultural, scenic, or cultural value. See chapter 16 on water use for a more extensive development of this topic.

Land-Use Planning Principles

Land-use planning is a process of evaluating the needs and wants of the population, the characteristics and values of the land, and various alternative solutions to the use of a particular land surface before changes are made. Planning land use brings with it the desires of many competing interests. The economic and personal needs of the population are a central driving force that requires land-use decisions to be made. However, the unique qualities of particular portions of the land surface prevent some uses, poorly accommodate others, but are highly suitable for others. For example, the floodplain beside a river is unsuitable for building permanent structures, can easily accommodate recreational uses such as parks, but may be most useful as a nature preserve. Agricultural land near cities can be easily converted to housing but may be more valuable for growing fruits and vegetables that are needed by the people of the city. This is particularly true when agricultural land is in short supply near urban centers. To make good land-use decisions, each piece of land must be evaluated and one of several competing uses assigned to it. When land-use decisions are made, the decision process usually involves the public, private landowners, developers, government, and special interest groups. Each interest has special wants and will argue that its desires are most important. Since this is the case, people must have principles and processes to guide land-use decisions, irrespective of their personal wants and interests.

A basic rule should be to make as few changes as possible, but when changes are

suggested or required, several things should be considered.

1. *Evaluate and record any unique geologic, geographic, or biologic features of the land.*

   Some land has unique features that should be preserved because of their special value to society. The Grand Canyon, Yellowstone National Park, and many wilderness areas have been set aside to preserve unique physical structures, scenic characteristics, special ecosystems, or unusual organisms. On a more local level, a stream may provide fishing opportunities near a city, or land may have excellent agricultural potential that should take precedence over other uses. New York City purchased and protects a watershed that provides water for the city at a much lower cost than treating water in the Hudson River.

2. *Preserve unique cultural or historical features.*

   Some portions of the landscape, areas within cities, and structures have important cultural, historic, or religious importance that should not be compromised by land-use decisions. In many cities, historic buildings have been preserved and particular sections set aside as historic districts. Sacred sites, many battlefields, and places of unique historic importance are usually protected from development.

3. *Conserve open space and environmental features.*

   It must be recognized that open space and natural areas are not unused, low-value areas. Many studies of human behavior have shown that when people are given choices, they will choose settings that provide a view of nature or that allow one to see into the distance. Some have argued that this is a deep-seated biological need, while others suggest that it is a culturally derived trait. Regardless of its origin, urban planners know that access to open space and natural areas is an important consideration when determining how to use land. Therefore, it makes sense to protect open space within and near centers of population.

Wetlands Loss in Louisiana

Louisiana has extensive coastal wetlands along the Gulf of Mexico that constitute about 40 percent of the wetlands in the United States, excluding Alaska and Hawaii. About 3500 square kilometers (1300 square miles) have been lost since 1956. This is an area approximately the size of the state of Rhode Island. Recent rates of loss have ranged from 65 to 90 square kilometers (25–35 square miles) per year. There are several reasons for the loss, but human activity is a primary cause.

Much of the Louisiana coastline consists of poorly compacted muds that are easily eroded if exposed to wind and wave action. Faults in the rock that lies below the mud are causing the land to tilt downward, which allows the ocean to invade. This is the basic cause for about 60 percent of the wetlands loss. In the past, the two actions—loss by tilting and addition of sediments from the Mississippi River—were in balance. The loss due to tilting was replaced by additional sediments from the river. However, two human actions have reduced the amount of sediments being added. Dams and reservoirs upstream on the Missouri and Mississippi Rivers have reduced the amount of sediment delivered to the region, making replacement of lost sediments impossible. In addition, levees on the Mississippi River have prevented the flooding that was a normal part of the behavior of the lower Mississippi River for centuries. The sediments, therefore, are carried out to the Gulf of Mexico and do not contribute to the building of wetlands near the shore. Without the constant addition of sediments, the muds settle and compact, which, along with the subsiding of the land, results in erosion and wetlands loss.

Shipping channels and canals associated with oil and gas production are responsible for about 30 percent of the current wetlands loss. There are nine major shipping channels and 13,000 kilometers (8000 miles) of canals. Typically, production companies cut canals through the marshes to give barges and other equipment easy access to the drilling platforms and production facilities. When oil or gas is found, it must be transported to land through pipelines. Laying pipelines often requires cutting additional canals through the wetlands. These canals break the wall of vegetation that protects the soft muds and allow salt water to penetrate into the marshes, killing the vegetation. Both of these activities open up the wetlands to wave action and accelerate the rate of erosion.

The nutria, a rodent introduced from South America, is also contributing to the problem. These animals reproduce rapidly and form such large populations that they eat all the vegetation in a local area. This exposes the muds to erosion and contributes to the loss of wetlands. Trapping these animals for their fur was at one time profitable for the local people and helped to control the animals. However, a shift in public opinion has reduced the acceptability of fur for clothing and depressed the fur market, making the trapping of nutria uneconomical.

A study led by the Louisiana Department of Natural Resources was released in 1998. It called for billions of dollars of federal investment to protect the wetlands. A logical source of these funds would be royalties paid to the federal government by offshore oil production companies. Major features of the plan include changes to current shipping practices, construction of new shipping facilities downstream from New Orleans, and controlled opening of levees along the lower Mississippi to restore some of the flooding that was at one time a normal part of the ecosystem. Powerful forces are in conflict over this plan. Shipping interests feel changes in shipping will cost them money. People who live in low-lying areas currently protected by levees may need to relocate even if controlled flooding is initiated. Changing the water quality by adding freshwater to a brackish water system will change the kinds of organisms present and hurt some traditional fishing activities. However, there is a plan and Congress will need to decide which elements of the plan to authorize and the amount of money to allocate for their implementation.

---

4. *Recognize and calculate the cost of additional changes that will be required to accommodate altered land use.*

Whenever land use is altered, additional modifications will be required to accommodate the change. For example, when a new housing development is constructed, schools and other municipal services will be required, roads will need to be improved, and the former use of the land is lost. Frequently, the cost of these changes is not borne by the developer or the homeowner but becomes the responsibility of the entire community; everyone pays for the additional cost through tax increases. In many areas, basic services, such as provision of water, are severe problems. It makes no sense to build new housing when there is not enough water to support the existing population.

5. *Plan for mixed housing and commercial uses of land in proximity to one another.*

One of the major problems associated with development in North America is segregation of different kinds of housing from one another and from shopping and other service necessities. Mixing various kinds of uses together (single-family housing, apartments, shopping and other service areas, and offices) allows easier connection between uses without reliance on the automobile. Walking and biking become possible when these different uses are within a short distance of one another.

6. *Plan for a variety of transportation options.*

Plan for transportation options other than the automobile. Currently, most urban and rural areas do not accommodate bicycles. Although bicycles can legally be ridden on streets, it is generally unsafe because of the high speeds of vehicular traffic and poor road surface conditions near the edge of the pavement. Special bike lanes are available in some areas, but this is rare in much of North America. Walking is a healthy, pleasant way to get

Farmland Preservation in Pennsylvania

The American Farmland Trust, a nongovernmental organization, recognizes Pennsylvania as the leading state in the number of farms and acres of farmland protected. The state bureau that oversees the program to purchase agricultural easements and administers the legislative programs designed to preserve farmland is the Agricultural Preservation Board (APB).

The Pennsylvania Agricultural Conservation Easement Purchase Program was developed in 1988 to help slow the loss of prime farmland to nonagricultural uses. The program enables state, county, and local governments to purchase conservation easements (sometimes called development rights) from owners of quality farmland. The first easements were purchased in 1989. Counties participating in the program have appointed agricultural land preservation boards. Overseeing the program, the APB is responsible for distributing state funds, approving and monitoring county programs, and making specific easement purchases.

To be chosen for easement purchase, a farm is rated against other eligible parcels according to the following criteria:

- Quality of the farmland. State regulations require that easements be purchased on farms of a minimum of 50 acres in size. Parcels as small as 10 acres may be preserved if they are adjacent to existing preserved farmland or used for the production of crops unique to the area. At least half the tract must either be harvested cropland, pasture, or grazing land.

- Stewardship. Farms are rated on the use of conservation and best-management practices in terms of nutrient management and control of soil erosion and sedimentation.

- Likelihood of conversion. Easements offered for sale to counties are scored and ranked for acquisition based on such factors as the proximity of the farm to sewer and water lines, the amount and type of agricultural use in the vicinity, and the amount of other preserved farmland in close proximity.

Farmers may choose to receive the proceeds from easement sales in a lump-sum payment or installments. Many farmers use the proceeds from easement sales to reduce debt loads and expand operations, and as a way to pass on their farms to the next generation.

---

from one task to another. However, crossing wide, busy streets is difficult, many areas lack sidewalks, and related service areas are often far apart. This tends to discourage this mode of transport. Clustering housing and service areas allows for easier planning of bus and rail routes to allow people to get from one place to another without relying on automobiles.

7. *Set limits and require managed growth with compact development patterns.*

Much of unplanned growth occurs because there is no plan or the land-use plan is not enforced. One very effective tool that promotes efficient use of the land is to establish an urban growth limit for a municipality. An **urban growth limit** establishes a boundary within which development can occur. Development outside the boundary is severely restricted. One of the most important outcomes of setting urban growth boundaries is that a great deal of planning must precede the establishment of the limit. This lets all in the community know

what is going on and can allow development to occur in logical stages that do not stress the community's ability to supply services. This mechanism also stimulates higher density uses of the urban land.

8. *Encourage development within areas that already have a supportive infrastructure so that duplication of resources is not needed.*

Because all development of land for human activity requires that services be provided, it makes sense that housing and commercial development occur where the infrastructure is already present. This includes electric, phone, sewer, water, and transportation systems. It includes service industries, such as shopping, banking, restaurants, hotels, and entertainment. It includes schools, hospitals, and police protection. If development occurs far from these services, it is very costly to extend or duplicate those services in the new location. Furthermore, all large cities and most smaller ones have vacant lots or abandoned buildings that have outlived their usefulness

or are vacant because of changing business or housing needs. This land is already urban and close to municipal services, and the buildings can be readily renovated or demolished and replaced. Many will argue that this is too expensive, but if there are no options, these spaces will be used, and their redevelopment can be important in revitalizing inner-city spaces.

Mechanisms for Implementing Land-Use Plans

Land-use planning is the construction of an orderly list of priorities for the use of available land. Developing a plan involves gathering data on current use and geological, biological, and sociological information. From these data, projections are made about what human needs will be. All of the data collected are integrated with the projections, and each parcel of land is evaluated and assigned a best use under the circumstances. There are basically three components that contribute to the successful implementation of a land-use development plan: land-use decisions can be assigned to a regional governmental body, the land or its development rights can be purchased, and laws or ordinances can be used to regulate land use. (See table 13.1.)

Establishing State or Regional Planning Agencies

National and regional planning is often more effective than local land-use planning because political boundaries seldom reflect the geological and biological database used in planning. Larger units contain more diverse collections of landscape resources and can afford to hire professional planners. A regional approach is also likely to prevent duplication of facilities and lead to greater efficiency. For example, airport locations should be based on a regional plan that incorporates all local jurisdictions. Three cities only 30 kilometers (20 miles) apart should not build three separate airports when one regional airport could serve their needs better and at a lower cost to the taxpayers.

In North America, the majority of regional governmental bodies are presently voluntary and lack any power to implement programs. Their only role is to advise the member governments. Members of local governments often are unwilling to give up power. They may view policy from a narrow perspective and put their own interests above the goals of the region.

One way to encourage regional planning is to develop policies at the state, provincial, or national level. The first state to develop a comprehensive statewide land-use program was Hawaii. During the early 1960s, much of Hawaii's natural beauty was being destroyed to build houses and apartments for the increasing population. The same land that attracted tourists was being destroyed to provide hotels and supermarkets for them. Local governments had failed to establish and enforce land-use controls. Consequently, in 1961, the Hawaii State Land-Use Commission was founded. This commission designated all land as urban, agricultural,

 Table 13.1 Land-Use Planning Strategies

Land-use planning for sustainability requires consideration of a wide spectrum of factors, including transportation, development density, energy efficiency, natural corridors and open space, and growth management. The following strategies are critical components of comprehensive planning to address the complex land-use issues facing our communities.

Transit-Oriented Design

Planning and design strategies for the development of mixed-use, walkable communities sited adjacent to transit access.

Mixed-Use Strategies

Development that promotes the coexistence of many community locales and services within close proximity, to reduce automobile dependency.

Urban Growth Boundaries

A regulatory strategy for limiting urban sprawl by creating a geographical boundary for new development over a period of time.

Infill Development

A strategy to promote greater development density and efficiency within existing urban boundaries.

Greenways

A strategy to preserve open spaces and natural systems, and provide recreation opportunities, by connecting cities, suburbs, and rural areas through linear corridors such as parks and trails.

Brownfield Redevelopment

A strategy for returning idle and often contaminated urban lands referred to as "brownfields" into productive use.

Transfer of Development Rights

A method of exchanging development rights among property developers to increase development density and protect open space and existing land uses.

Open Space Protection

Ways to protect a community's urban open space, farmland, wetlands, riparian lands, rangeland, forests and woodlands, and coastal lands.

Urban Forestry

Planting and maintenance of trees within a city or community as a strategy for reducing both carbon emissions and energy expenditures for heating and cooling.

Land Trusts

Local, regional, or statewide nonprofit organizations directly involved in protecting important land resources over the long term.

Agricultural Land Protection

Strategies for preserving the land that feeds and clothes us, provides open space, food, and habitat for diverse wildlife, and maintains a link to our nation's agricultural heritage.

Solar Access Protection

Regulatory measures to provide legal protection to property owners investing in solar energy systems through solar access ordinances.

Source: U.S. Department of Energy.

or conservational. Each parcel of land could be used only for its designated purpose. Other uses were allowed only by special permit. To date, the record for Hawaii's action shows that it has been successful in controlling urban growth and preserving the islands' natural beauty, even though the population continues to grow. (See figure 13.8.)

Several states and cities are attempting to follow Hawaii's lead in land-use regulation. (See tables 13.2 and 13.3.) Some have passed legislation dealing with special types of land use. Examples include wetland preservation, floodplain protection, and scenic and historic site preservation. Although direct state involvement in land-use regulation is relatively new, it is expected to grow. Only large, well-financed levels of government can afford to pay for the growing cost of adequate land-use planning. State, provincial, and regional governments are also more likely to have the power to counter the political and economic influences of land developers, lobbyists, and other special-interest groups when conflicts over specific land-use policies arise.

National governments also have a role to play. Since national governments own and administer the use of large amounts of land, national policy will determine how those lands are used. The designation of lands as wilderness, forests, rangelands, or parks at the federal level often involves a balancing of national priorities with local desires. As with local land-use issues, the conflicts over the use of federal lands are often economic and involve compromise.

Purchasing Land or Use Rights

Probably the simplest way to protect desirable lands is to purchase them. When privately owned land is desired for special purposes, it must be purchased from the owner, and the owner has a right to expect to get a fair price for the property. When it is determined that land has a high public value, then either the land or the rights to use it must be purchased. Many environmental organizations purchase lands of special historic, scenic, or environmental value.

In many cases, the owners may not be willing to sell the land but are willing to limit the uses to which the land can be put in the future. Therefore, landowners may sell the right to develop the land or may agree to place restrictions on the uses any future owners might consider.

Regulating Use

Many communities are not in a financial position to purchase lands; therefore, they attempt to regulate land use by zoning laws.

Zoning is a common type of land-use regulation that restricts the kinds of uses

Figure 13.8   Hawaii Land-Use Plan Hawaii was the first state to develop a comprehensive land-use plan. This development in an agricultural area was stopped when the plan was implemented in the 1960s.

Table 13.2   Examples of State Land-Use Planning Legislation

State	Legislation
California	Coastal Act (1976)
	Coastal Zone Conservation Act (1972)
	Tahoe Regional Planning Compact (1969)
Florida	Omnibus Growth Management Act (1985)
	State Comprehensive Plan (1985)
	State and Regional Planning Act (1984)
	Environmental Land and Water Management Act (1972)
Georgia	Coordinated Planning Legislation (1989)
Hawaii	Hawaii State Plan (1978)
	Hawaiian Land Use Law (1961)
Maine	Comprehensive Planning and Land-Use Regulation Act (1988)
Maryland	Economic Growth, Resource Protection and Planning Act (1992)
Massachusetts	Cape Cod Commission Act (1989)
	Martha's Vineyard Commission Act (1974)
New Jersey	State Planning Act (1985)
	State Pinelands Protection Act (1974)
New York	Adirondack Park Agency Act (1971)
Oregon	Land Conservation and Development Act (1973)
Rhode Island	Comprehensive Planning and Land Use Regulation Act (1988)
Vermont	Growth Management Act (1988)
Washington	Growth Management Act (1990)
	Amendments to the Growth Management Act (1991)

to which land in a specific region can be put. When land is zoned, it is designated for specific potential uses. Common designations are agricultural, commercial, residential, recreational, and industrial. (See figure 13.9.)

Most local zoning boards are elected or appointed and often lack specific training in land-use planning. As a result, zoning regulations are frequently made by people who see only the short-term gain and not the possible long-term loss. Often

PART FOUR   Resource Management

Table 13.3 Examples of Local Growth Management Actions

Locality	Action
Livermore, California	Established limits on the number of building permits issued
Petaluma, California	Limited the number of building permits to 500 per year and established criteria for awarding permits that emphasized planning
San Diego, California	Adopted a growth management plan
Boulder, Colorado	Established a growth management plan and limited the number of new buildings
Larimer County, Colorado	Established regional urban growth areas
Westminster, Colorado	Limited the number of building permits issued
Boca Raton, Florida	Established a cap on growth at 40,000 dwellings in 1972. The ordinance was eventually overturned by the U.S. Supreme Court.
Broward County, Florida	Public facilities must be developed along with new housing.
Palm Beach County, Florida	Growth control to provide affordable housing
Sanibel, Florida	Established growth limits based on environmental carrying capacity. The city is adjacent to an important wildlife refuge.
Hardin County, Kentucky	Eliminated rural zoning through growth management techniques
Montgomery County, Maryland	Established growth management controls
Minneapolis, Minnesota	Containment of development within city boundaries
Ramapo, New York	Does not permit development until municipal services are established in an area
Austin, Texas	Neighborhood-based conservation overlay districts

the land is simply zoned so that its current use is sanctioned and is rezoned when another use appears to have a higher short-term value to the community. Even when well-designed land-use plans exist, they are usually modified to encourage local short-term growth rather than to provide for the long-range needs of the community. The public needs to be alert to variances from established land-use plans, because once the plan is compromised, it becomes easier to accept future deviations that may not be in the best interests of the community. Many times, individuals who make zoning decisions are real-estate agents, developers, or local businesspeople. These individuals wield significant local political power and are not always unbiased in their decisions. Concerned citizens must try to combat special interests by attending zoning commission meetings and by participating in the planning process.

Special Urban Planning Issues

Urban areas present a large number of planning issues. Transportation, open space, and improving the quality of life in the inner city are significant problems.

Urban Transportation Planning

A growing concern of city governments is to develop comprehensive urban transportation plans. While the specifics of such plans might vary from region to region, urban transportation planning usually involves four major goals:

1. Conserve energy and land resources.
2. Provide efficient and inexpensive transportation within the city, with special attention to people who are unable to drive, such as many elderly, young, handicapped, and financially disadvantaged persons.
3. Provide suburban people opportunities to commute efficiently.
4. Reduce urban pollution.

Any successful urban transportation plan should integrate all of these goals, but funding and intergovernmental cooperation are needed to achieve this. The problems associated with current urban transportation will certainly not disappear overnight, but comprehensive

Figure 13.9   Zoning Most communities have a zoning authority that designates areas for particular use. This sign indicates that decisions have been made about the "best" use for the land.

planning is the first step to solving them.

Since automobiles are heavily used, transportation corridors and parking facilities must be included in any urban transportation plan.

However, many urban planners recognize that the automobile's disadvantages may outweigh its advantages, so some cities, such as Toronto, London, San Francisco, and New York, have attempted to dissuade automobile use by developing mass transit systems and by allowing automobile parking costs to increase substantially.

Environmental Close-Up

Land-Use Planning and Aesthetic Pollution

Unpleasant odors, disagreeable tastes, annoying sounds, and offensive sights can be aggravating. Yet it is difficult to get complete agreement on what is acceptable and when some aesthetic boundary has been crossed that is unacceptable. Furthermore, many useful activities generate stimuli that are offensive while the activity itself may be essential or at least very useful. Many of these do not harm us physically but may be harmful from an aesthetic point of view.

Odors are caused by various airborne chemicals. People who live near dairy farms, livestock-raising operations, paper mills, chemical plants, steel mills, and other industries may be offended by the odors originating from these sources, but the products of these activities are needed. Many of these industries discharge wastewater that contains materials that decompose or evaporate and cause odor pollution. People who are constantly exposed to an odor are usually not as offended by it as are people who are newly exposed to it. When an odor is constantly received, the brain ceases to respond to the stimulus. In other words, the person is not aware of the odor.

Chemicals can also affect the taste of things we eat or drink. Some naturally produced chemicals, as well as those discharged into waterways, can affect the taste of food or drinking water. Minute quantities of certain chemicals can affect the taste of our drinking water and our food. Algae in water produce flavors that are offensive to many. Some groundwater sources have sulfur or salts that cause unwanted flavors. In fish, a concentration of 100 ppm of phenol can be tasted. Although this small amount of chemical is not harmful biologically, it does make the fish very unappetizing.

Noise is unwanted sound. It is produced as an incidental by-product of industry, traffic, and other human activity. People vary considerably in their tolerance of unwanted sound. Many in cities adapt to the constant noise of the city, while those who live in more rural areas would find city noises annoying.

Visual pollution is a sight that offends us. This type of pollution is highly subjective and is, therefore, difficult to define or control. To most people, a dilapidated home or building is offensive, especially if located in an area of higher-priced homes. A heavily littered highway or street is aesthetically offensive to most people, and litter along a wilderness trail is even more unacceptable. Some sources of visual pollution are not so clear-cut, however. To many people, roadside billboards are offensive, but they can be helpful to advertisers and to travelers looking for information.

Since aesthetic pollutants—odors, tastes, sounds, and sights—are extremely difficult to define, it is difficult to establish aesthetic pollution standards. One of the simplest ways to eliminate many of these annoyances is to separate the generator of the offensive stimulus from the general public. Proper land-use planning can greatly reduce the amount of annoying aesthetic pollution. If uses that produce annoying stimuli can be clustered together rather than dispersed, the effect on the public is reduced. It does not make sense to allow new home construction in the vicinity of farming operations that will produce odors or industrial operations or airports that will generate noise. Similarly, allowing homes to be built in areas that have poor quality groundwater that will be used for drinking is a problem—unless an additional decision is made to provide a different source of drinking water. The roadside advertising that many find offensive can be regulated by allowing such signs only in specific areas rather than allowing them to be built anywhere along the roadway. None of these land-use remedies eliminates the aesthetic pollutant, but by segregating the public from the source of the annoyance, the impact is reduced.

Billboards line the landscape along Interstate Highway 40 in New Mexico.

The major urban mass transit systems are railroads, subways, trolleys, and buses. In many parts of the world, mass transportation is extremely efficient and effective. However, in the United States, where the automobile is the primary method of transportation, mass transportation systems are often underfunded and difficult to establish because mass transit is:

1. Economically feasible only along heavily populated routes
2. Less convenient than the automobile
3. Extremely expensive to build and operate
4. Often crowded and uncomfortable

Although mass transit meets a substantial portion of the urban transportation needs in some parts of the world, such as in Europe, its use in North America has grown slightly, while automobile and air travel has grown tremendously. One reason for this difference is the size of the region. Unlike North America, Japan and countries in Europe that have extensive public transportation are much smaller geographically. In addition, they have more uniformly dense populations. Thus, it is easier to make public transportation work in those settings. (See figure 13.10.) A variety of forces has contributed to this situation. As people became more affluent, they could afford to own automobiles, which are a convenient, individualized method of transportation. Governments in North

America encourage automobile use by financing highways and expressways, by maintaining a cheap energy policy, and by withdrawing support for most forms of mass transportation. Thus, they encourage automobile transportation with hidden subsidies (highway construction and cheap gasoline) but maintain that rail and bus transportation should not be subsidized. North Americans will seek alternatives to private automobile use only when the cost of fuel, the cost of parking, or the inconvenience of driving becomes too high.

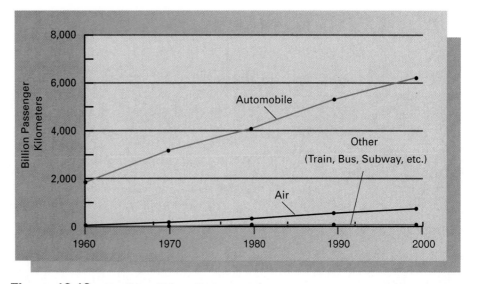

Figure 13.10 **Decline of Mass Transportation** Automobile use has increased consistently since 1960, while rail and bus transport use has remained low.

Source: Data from U.S. Department of Transportation, *National Transportation Statistics 2003*.

Urban Recreation Planning

About three-fourths of the population of North America live in urban areas. These urban dwellers value open space because it breaks up the sights and sounds of the city and provides a place for recreation. Inadequate land-use planning in the past has rapidly converted urban open spaces to other uses. Until recently, creating a new park within a city was considered an uneconomical use of the land, but people are now beginning to realize the need for parks and open spaces.

Some cities recognized the need for open space a long time ago and allocated land for parks. (See figure 13.11.) London, Toronto, and Perth, Australia, have centrally located and well-used parks. New York City set aside approximately 200 hectares (500 acres) for Central Park in the late 1800s. Boston has developed a park system that provides a variety of urban open spaces. Other cities have not dealt with this need for open space because they have lacked either the foresight or the funding. Recreation is a basic human need. The most primitive tribes and cultures all engaged in games or recreational activities. New forms of recreation are continually being developed. In the congested urban center, cities often must construct special areas where recreation can take place.

A major problem with urban recreation is locating recreational facilities near residential areas. Facilities that are not conveniently located may be infrequently used. For example, the hundreds of thousands of square kilometers of national parks in Alaska and the Yukon will be visited every year by a tiny proportion of the population of North America. Large urban centers are discovering that they must provide adequate, low-cost recreational opportunities within their jurisdiction. Some of these opportunities take the form of commercial establishments, such as bowling centers, amusement parks, and theaters. Others must be subsidized by the community. (See figure 13.12.) Playgrounds, organized recreational activities, and open space have usually been combined into an arm of the municipal government known as the parks and recreation department. Cities spend millions of dollars to develop and maintain recreation programs. Often, conflict arises over the allocation of financial and land resources. These are closely tied because open land is scarce in urban areas, and it is expensive. Riverfront property is ideal for park and recreational use, but it is also prime land for industry, commerce, or high-rise residential buildings. Although conflict is inevitable, many metropolitan areas are beginning to see that recreational resources may be as important as economic growth for maintaining a healthy community.

An outgrowth of the trend toward urbanization is the development of **nature centers.** In many urban areas, so little natural area is left that the people who live there need to be given opportunities to learn about nature. Nature centers are basically teaching institutions that provide a variety of methods for people to learn about and appreciate the natural world. Zoos, botanical gardens, and some urban parks, combined with interpretive centers, also provide recreational experiences. Nature centers are usually located near urban centers, in places where some appreciation of the natural processes and phenomena can be developed. They may be operated by municipal governments, school systems, or other nonprofit organizations.

Redevelopment of Inner-City Areas

As people moved to the suburbs during the past 50 years, the inner city was abandoned. Many old industrial sites sit vacant. Businesses have moved to the suburban malls. The quality of housing has declined. Services have been reduced. To improve the quality of life of the residents of the city, special efforts must be made to revitalize the city. Although activities that will improve the quality of life in the inner city vary from city to city, several land-use processes can help. One problem that has plagued industrial cities is vacant industrial and commercial sites.

a.   b.

Figure 13.11   **Urban Open Space** These photos show open space in two urban areas: (a) Stanley Park in Vancouver, British Columbia, and (b) Echo Park Lake in Los Angeles, California. If the land had not been set aside, it would have been developed.

Figure 13.12   **Urban Recreation** In urban areas, recreation often takes the form of sports programs, playgrounds, and walking. Most cities recognize the need for such activities and facilities, and develop extensive recreation programs for their citizens.

Many of these buildings have remained vacant because the cost of cleanup and renovation is expensive. Such sites have been called **brownfields.** Many of these sites involved environmental contamina-tion, and since the EPA required that they be cleaned up to a pristine condition, no one was willing to do so.

A new approach to utilizing these sites is called **brownfields development.**

This involves a more realistic approach to dealing with the contamination at these sites. Instead of requiring complete cleanup, the degree of cleanup required is matched to the intended use of the site. Although an old industrial site with specific contamination problems may not be suit-able for housing, it may be redeveloped as a new industrial site, since access to the contamination can be controlled. An old industrial site that has soil contamination may be paved to provide parking.

Another important focus in urban re-development is the remodeling of aban-doned commercial buildings for shopping centers, cultural facilities, or high-density housing. Chattanooga, Tennessee, has achieved a reputation for revitalizing its inner-city area. The process of revitaliza-tion involved extensive planning activities that included the public, public and private funding of redevelopment activities, es-tablishment of an electric bus system to al-leviate air pollution, renovation of existing housing, redevelopment of old ware-houses into a shopping center, and incor-poration of a condemned bridge over the Tennessee River into a portion of a park that is also an important pedestrian connection between a residential area and the downtown business district.

Smart Growth

The current pattern of growth throughout most of North America is commonly referred to as "sprawl." Concern about sprawl is growing because of the sheer pace of land development, which, according to the U.S. Department of Agriculture, is roughly double what it was only a decade ago. Such development has had a number of negative cultural, economic, environmental, and social consequences. In central cities and older suburbs, these include deteriorating infrastructure, poor schools, and a shortage of affordable, quality housing. In newer suburban areas, these problems may include increased traffic congestion, declining air quality, and the loss of open space. Smart growth is an approach that argues that these problems are two sides of the same coin and that the neglect of our central cities is fueling the growth and related problems of the suburbs.

To address these problems, smart growth advocates emphasize the concept of developing "livable" cities and towns. Livability suggests, among other things, that the quality of our built environment and how well we preserve the natural environment both directly affect our quality of life.

Though they support growth, communities are questioning the economic costs of abandoning infrastructure in the city and rebuilding it farther out. They are questioning the necessity of spending increasing time in cars, locked in traffic or traveling miles to the nearest store. They are questioning the practice of abandoning older communities while developing open space and prime agricultural lands at the suburban fringe.

Smart growth recognizes the benefits of growth. It invests time, attention, and resources in restoring a sense of community and vitality to center cities and older suburbs. In new developments, smart growth practices create communities that are more town-centered, transit- and pedestrian-oriented, and have a greater mix of housing, commercial, and retail uses. These practices also preserve open space and other environmental amenities. Smart growth recognizes connections between development and quality of life that can be presented as a list of principles.

Smart growth principles

1. Mix land uses.
2. Take advantage of compact building design.
3. Create a range of housing opportunities and choices.
4. Create walkable neighborhoods.
5. Foster distinctive, attractive communities with a strong sense of place.
6. Preserve open space, farmland, natural beauty, and critical environmental areas.
7. Strengthen and direct development toward existing communities.
8. Provide a variety of transportation choices.
9. Make development decisions predictable, fair, and cost-effective.
10. Encourage community and stakeholder collaboration in development decisions.

The features that distinguish smart growth vary by community. No two streets, neighborhoods, or cities are identical. There is no "one-size-fits-all" solution. Smart growth in Portland, Oregon, has different characteristics than smart growth in Austin, Texas. For that reason, smart growth does not prescribe solutions. Rather, it provides choices and seeks to build on proven successes. The smart growth principles listed above reflect the experience of localities that have successfully created smart growth communities. These communities had a vision of where they wanted to go and of what things they valued in their community.

An example of the principles of smart growth being applied is Suisun City, California. In 1989, the *San Francisco Chronicle* rated Suisun City, a town of 25,000 people between San Francisco and Sacramento, the worst place to live in the Bay Area. At that time, Suisun City's historic Main Street was a strip of boarded-up storefronts and vacant lots. Several blocks away, an oil refinery sat at the head of the polluted, silt-laden Suisun Channel. To-day, Suisun's harbor is filled with boats and lined with small businesses. A train and bus station that connects the city to the rest of northern California is located nearby. The town is diverse, walkable, and picturesque. Its crime rate is low and its housing affordable. This dramatic change occurred because Suisun City's residents, businesses, and elected officials agreed on a common vision for their town's future. Cleaning up the polluted Suisun Channel and making the waterfront a focal point of their town was a common goal. The citizens also wanted to reestablish the historic Main Street as a social and retail gathering place. It was recognized that by encouraging tax-generating commercial development such as retail shops and restaurants along Main Street and the waterfront, municipal finances would be strengthened.

In its turnaround, Suisun City avoided large-scale redevelopment projects such as shopping centers and industrial parks that would have altered its historic small-town character. Suisun City is still a work in progress, but it is now invigorated with new businesses and residents, a stronger community spirit, and optimism about its future.

Federal Government Land-Use Issues

Since the federal government manages large amounts of land, the laws and regulations that shape land-use policy are important. For example, the 1960 Multiple Use Sustained Yield Act divided use of national forests into four categories: wildlife habitat preservation, recreation, lumbering, and watershed protection. This act was designed to encourage both economic and recreational use of the forests. However, specific users of this public land are often in conflict, particularly recreational users with timber harvesters.

An 1872 mining law has also been important in federal land management. The law allows anyone to prospect for minerals on public lands and to establish a claim if such minerals are discovered. The miner is then allowed to purchase the rights to extract the mineral for $5 per acre. Many feel that the law is obsolete, but it is still in force and public land is still being sold to mining interests at ridiculously low prices. These laws are examples of past policy decisions that are still in effect. Today, one of the major uses of public lands is outdoor recreation.

Many people want to use the natural world for recreational purposes because nature can provide challenges that may be lacking in their day-to-day lives. Whether the challenge is hiking in the wilderness, exploring underwater, climbing mountains, or driving a vehicle through an area that has no roads, these activities offer a sense of adventure. Look at table 13.4. All of these activities use the out-of-doors but not in the same way. Conflicts develop because some of these activities cannot occur in the same place at the same time. For example, wilderness camping and backpacking often conflict with off-road vehicles.

A basic conflict exists between those who prefer to use motorized vehicles and those who prefer to use muscle power in their recreational pursuits. (See figure 13.13.) This conflict is particularly strong because both groups would like to use the same public land. Both have paid taxes, and both feel that it should be available for them to use as they wish.

Finally, as more rangelands and forests have had vehicular access controlled or eliminated, those who want to use public lands for motorized recreation have become upset.

Land-use conflicts also arise between business interests and recreational users of public lands. Federal and state governments give special use permits to certain users of public lands. Many ski resorts in the West make use of public lands. Grazing is also an important use of federal land. Based on "Animal Use Months" established by the Bureau of Land Management or the Forest Service, ranchers are allowed to graze cattle on certain public lands. Technically, failure to comply can mean a loss of grazing rights. However, since the establishment of regulations is highly political, many maintain that the political influence of ranchers allows them to use a public resource without adequately compensating the government. In addition, the regulatory agencies are un-

Table 13.4	Number of People Who Participated in Selected Outdoor Recreational Activities in 2003
Activity	**Percent of Population over Seven Years of Age Participating**
Exercise walking	33
Swimming	27
Bicycling	24
Fishing	15
Camping	18
Golf	14
Hiking	10
Running	8
Hunting	6
Backpacking	4
Skiing	4

Source: Data from *Statistical Abstract of the United States*, 2003.

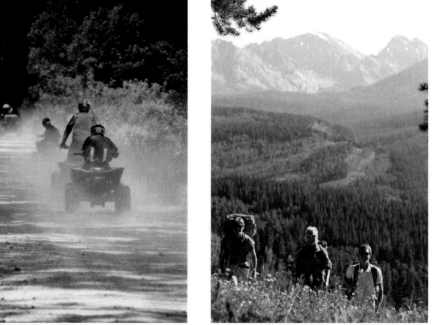

Figure 13.13 **Conflict over Recreational Use of Land** The people who use land for motorized and nonmotorized activities are often antagonists over the allocation of land for recreational use.

derstaffed and find it difficult to adequately regulate the actions of individual ranchers. As a result, some lands are overgrazed. Many people who want to use the publicly owned rangelands for outdoor recreation resent the control exercised by grazing interests. On the other hand, ranchers resent the intrusion of hikers and campers on land they have traditionally controlled.

An obvious solution to this problem is to allocate land to specific uses and to regulate the use once allocations have been made. Several U.S. governmental agencies, such as the National Park Service, the Bureau of Land Management, the Forest Service, and the Fish and Wildlife Service, allocate and regulate the lands they control. However, these agencies have conflicting roles. The Forest Service has a mandate to manage forested public lands for timber production. This mandate often comes in conflict with recreational uses. Similarly, the Bureau of Land Management has huge tracts of land that can be used for recreation, but it traditionally has been mandated to manage grazing rights.

A particularly sensitive issue is the designation of certain lands as wilderness areas. Obviously, if an area is to be wilderness, human activity must be severely restricted. This means that the vast majority of Americans will never see or make use of it. Many people argue that this is unfair because they are paying taxes to provide recreation for a select few. Others argue that if everyone were to use these areas, their charm and unique character would be destroyed and that, therefore, the cost of preserving wilderness is justifiable.

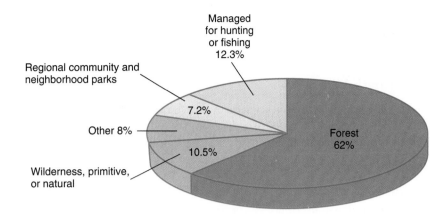

Figure 13.14 **U.S. Federal Recreational Lands** Of the approximately 108 million hectares (267 million acres) of federal recreational lands in the United States, approximately 10 percent is designated as wilderness, primitive, or natural.

Source: Data from the *Statistical Abstract of the United States*, 2003.

Areas designated as wilderness make up a very small proportion of the total public land available for recreation. (See figure 13.14.) Such areas, however, are becoming increasingly visible due to their unique character.

Summary

Historically, waterways served as transportation corridors that allowed for the exploration of new land and for the transport of goods. Therefore, most large urban centers began as small towns located near water. Water served the needs of the towns in many ways, especially as transportation. Several factors resulted in the shift of the population from rural to urban. These included the Industrial Revolution, which provided jobs in cities, and the addition of foreign immigrants to the cities. As towns became larger, the farmland surrounding them became suburbs surrounding industrial centers. Unregulated industrial development in cities resulted in the degradation of the waterfront and stimulated the development of suburbs around the city as people sought better places to live and had the money to purchase new homes. The rise in automobile ownership further stimulated the movement of people from the cities to the suburbs.

Many problems have resulted from unplanned growth. Current taxation policies encourage residential development of farmland, which results in a loss of valuable agricultural land. Floodplains and wetlands are often mismanaged. Loss of property and life results when people build on floodplains. Wetlands protect our shorelines and provide a natural habitat for fish and wildlife. Transportation problems and lack of open space are also typical in many large metropolitan areas.

Land-use planning involves gathering data, projecting needs, and developing mechanisms for implementing the plan. Good land-use planning should include assessment of the unique geologic, geographic, biological, and historic and cultural features of the land; the costs of providing additional infrastructure; preservation of open space; provision for a variety of transportation options; a mixture of housing and service establishments; redevelopment of disused urban land; and establishment of urban growth limits. Establishing regional planning agencies, purchasing land or its development rights, and zoning are ways to implement land-use planning. The scale of local planning is often not large enough to be effective because problems may not be confined to political boundaries. Regional planning units can afford professional planners and are better able to withstand political and economic pressures. A growing concern of urban governments is to develop comprehensive urban transportation plans that seek to conserve energy and land resources, provide efficient and inexpensive transportation and commuting, and help to reduce urban pollution. Urban areas must also provide recreational opportunities for their residents and seek ways to rebuild decaying inner cities.

Federal governments own and manage large amounts of land, therefore national policy must be developed. This usually involves designating land for particular purposes, such as timber production, grazing land, parks, or wilderness. The recreational use of public land often requires the establishment of rules that prevent conflict between potential users who have different ideas about what appropriate uses should be. Often federal policy is a compromise between competing uses and land is managed for multiple uses.

Issues – Analysis

Smart Growth Winners

The U.S. Environmental Protection Agency annually awards a series of smart growth awards to communities throughout the United States. The recipients of the awards must incorporate the principles of smart growth to create places that respect community culture and the environment, foster economic development, and enhance quality of life.

Minneapolis–St. Paul Programs

The 2003 award for overall excellence in smart growth went to the Minneapolis–St. Paul metropolitan area of Minnesota. Consistently ranked among the top locations in the United States to raise a family or establish a business, the Minneapolis–St. Paul region is experiencing rapid population growth. The metropolitan area is showing signs of growth-related stress: increasing traffic congestion, rising housing prices, and dwindling open space. Instead of trying to limit growth, the Minnesota Legislature provided the Metropolitan Council, a regional planning agency, with a voluntary, incentive-based approach to help communities grow in a way that addresses a variety of the region's issues. In 1995, the legislature passed the Livable Communities Act (LCA) to get innovative projects off the ground.

The LCA underwrites three distinct grant programs: Tax Base Revitalization (brownfield cleanup), Local Housing Incentives (affordable housing), and the Livable Communities Demonstration Account (mixed-use projects). Supported by a metropolitan-area property tax authorized by the legislature, the LCA uses a straightforward grant application with well-established selection criteria guiding the evaluation process. To be eligible for funding from any of the three grant programs, municipalities must first develop a housing action plan that addresses affordable housing needs and provides for a variety of housing types.

From 1996 through 2003, the Metropolitan Council awarded 292 LCA grants totaling nearly $100 million. The grants are expected to help in obtaining more than $3.4 billion in private and other public investments. LCA funding has helped create an anticipated 7267 new or rehabilitated housing units, 1911 new or rehabilitated rental units, 11,500 new or retained jobs, and 600 hectares (1483 acres) of reclaimed polluted land.

The voluntary program has engaged 106 communities in the seven-county metropolitan area. Annual requests for grants consistently exceed the funds available. The range of funded projects reflects the diverse urban, suburban, and rural communities in the region. In Hastings, a mid-sized town south of St. Paul, the Guardian Angels project used LCA funds to construct affordable apartments and to redevelop a church and school into a community center. An award to St. Louis Park, a suburb of Minneapolis, created a walkable town center with 10 new buildings and 660 housing units along an aging commercial corridor.

Georgia Quality Growth Program

More-informed communities make more-informed decisions about how and where they want to grow. Since 2000, the Georgia Office of Quality Growth (OQG) has helped communities implement smart growth principles. The OQG has focused its efforts on the communities that are ready for implementation of smart growth and on educating communities about smart growth.

OQG offers services to Georgia communities, including a Web-based toolkit of ways to put smart growth principles into practice at the local level (e.g., improving strip commercial corridors, creating infill development, preserving open space), model ordinances, a clearinghouse of George examples of quality growth projects, and direct technical assistance. OQG partners with over 40 organizations through the Georgia Quality Growth Partnership to provide these services.

Several new programs are being piloted by OQG: (1) Advisory clinics, which are expert panels made available at major in-state conferences for one-on-one consultation and advice about local smart growth projects; (2) Special Issue Retreats with key local officials for hands-on work with a recognized expert in a particular approach; (3) Quality Growth Assessments that provide a comprehensive review of a local government's regulations, plans, policies, and development patterns to identify barriers to implementation of smart growth; and (4) Consultants' Bureau, in which professional facilitators lead local planning and visioning meetings while bringing smart growth principles into the discussion.

Many communities follow up by applying for grants to implement the ideas put forward by OQG experts. Grants have included new projects such as detailed master plans for communities as well as infill design guidelines and development regulations for other communities. Other projects have included corridor studies, demonstration projects, ordinance development, and review of regulatory barriers to smart growth.

- Would your community be eligible for an EPA Smart Growth Award? Why or why not?
- Can you think of innovative ways to help your community grow smarter?
- Would you be willing to submit your ideas to your local planning agency?

Key Terms

brownfields *298*
brownfields development *298*
floodplains *288*
floodplain zoning ordinances *289*
infrastructure *288*

land-use planning *290*
megalopolis *285*
nature centers *297*
ribbon sprawl *284*
tract development *284*

urban growth limit *292*
urban sprawl *284*
wetlands *289*
zoning *294*

Review Questions

1. Why did urban centers develop near waterways? Are they still located near water?
2. Describe the typical changes that have occurred in cities from the time they were first founded until now.
3. Why do people move to the suburbs?
4. Why do some farmers near urban areas sell their land for residential or commercial development? If you were in this position, would you sell?
5. What is a megalopolis?
6. What land uses are suitable on floodplains?
7. What is multiple land use? Can land be used for multiple purposes?
8. Why is it important to provide recreational space in urban planning?
9. How can recreational activities damage the environment? Do you engage in any of those activities?
10. What is the monetary impact of recreational activities?
11. What are some strictly urban-related recreational activities?
12. List some conflicts that arise when an area is designated strictly as wilderness.
13. Describe the steps necessary to develop a land-use plan.
14. What are the advantages of regional or state planning?
15. List three benefits of land-use planning.
16. Describe the concept of smart growth.

Critical Thinking Questions

1. Choose the city where you live. Interview local residents and look at old city maps. What did the city look like 75 years ago? What were the city's boundaries? Where did people do their shopping? How did they get around? How does this compare with the current situation in the city?
2. What historical factors brought members of your family to the city? How does this compare to the factors that are currently contributing to the growth of cities in the developing world?
3. Consider the outer rim of the city closest to you. Which, if any, of the problems associated with unplanned growth are associated with your city? What factors make them a problem? What do you think can be done about them?
4. There has been tremendous development in the arid West of the United States over the past few decades, creating demands for water. How should these demands be met? Should there be limits to this type of development? What kinds of limits, if any?
5. Imagine you are a U.S. Forest Service supervisor who is creating a 10-year plan that is in the public comment stage. What interests would be contacting you? What power would each interest have? How would you manage the competing interests of timber, mining, grazing, and recreation or between motorized and nonmotorized recreation? What values, beliefs, and perspectives help you form your recommendations?
6. Imagine that you lived in an area of the country that has the potential to be named a wilderness area. What conflicts do you think would arise from such a declaration? Who might be some of the antagonists? Which perspective do you think is most persuasive? How would you answer the objections of the other perspective?
7. After reading the Environmental Close-Up in this chapter concerning wetland loss in Louisiana, what kinds of recommendations would you make to help preserve wetlands? What do you suppose might happen if nothing is done? What resistance might wetland preservation generate?

Interactive Exploration

Check out the website at www.mhhe.com/enger10e and click on the cover of this textbook for quizzing, career information, case studies, and hot links for the following topics:

Land-Use Planning

Urbanization and Sustainable Cities

Environmental Policy and Decision Making

Air Pollution

Refuges and Sanctuaries

Restoration Ecology

Conservation and Management of Habitats and Species

Landscape Ecology

Soil and Its Uses

14

Objectives

After reading this chapter, you should be able to:

- Describe the geologic processes that build and erode the Earth's surface.
- List the physical, chemical, and biological factors involved in soil formation.
- Explain the importance of humus to soil fertility.
- Differentiate between soil texture and soil structure.
- Explain how texture and structure influence soil atmosphere and soil water.
- Explain the role of living organisms in soil formation and fertility.
- Describe the various layers in a soil profile.
- Describe the processes of soil erosion by water and wind.
- Explain how contour farming, strip farming, terracing, waterways, windbreaks, and conservation tillage reduce soil erosion.
- Understand that the misuse of soil reduces soil fertility, pollutes streams, and requires expensive remedial measures.
- Explain how land not suited for cultivation may still be productively used for other purposes.

Chapter Outline

Geologic Processes
Soil and Land
Soil Formation
Soil Properties
Soil Profile
Soil Erosion
Soil Conservation Practices
   Contour Farming
   Strip Farming
   Terracing
   Waterways
   Windbreaks

Conventional Versus Conservation Tillage
Protecting Soil on Nonfarm Land
Issues—Analysis: Soil Fertility and Hunger in Africa

PAST PRESENT FUTURE

1983 The amount of arable land in the world was 0.29 hectares (0.72 acres) per capita.

1983 The amount of irrigated land in the world was 213 million hectares (526 million acres).

1982 The per capita food supplies in the developing world were 2,000 calories per day.

2001 The amount of arable land in the world was 0.22 hectares (0.54 acres) per capita.

2002 The amount of irrigated land in the world was 285 million hectares (704 million acres).

2003 The per capita food supplies in the developing world were 2,600 calories per day.

Figure 14.3 Physical Fragmentation by Freezing and Thawing The crack in the rock fills with water. As the water freezes and becomes ice, it expands. The pressure of the ice enlarges the crack. The ice melts, and water again fills the crack. The water freezes again and widens the crack. Alternate freezing and thawing splits the rock into smaller fragments.

Figure 14.4 An Eroded Landscape This landscape in Bryce Canyon National Park in Utah was created by the action of wind and moving water. The particles removed by these forces were deposited elsewhere and may have become part of the soil in that new location.

The physical breakdown of rock is also caused by forces that move and rub rock particles against each other (abrasion). For example, a glacier causes rock particles to grind against one another, resulting in smaller fragments and smoother surfaces. These particles are deposited by the glacier when the ice melts. In many parts of the world, the parent material from which soil is formed consists of glacial deposits. Wind and moving water also cause small particles to collide, resulting in further weathering. The smoothness of rocks and pebbles in a stream or on the shore is evidence that moving water has caused them to rub together, removing their sharp edges. Similarly, particles carried by wind collide with objects, fragmenting both the objects and the wind-driven particles.

Wind and moving water also remove small particles and deposit them at new locations, exposing new surfaces to the weathering process. For example, the landscape of the Bryce Canyon in the southwest United States was created by a combination of wind and moving water that removed easily transported particles, while rocks more resistant to weathering remained. (See figure 14.4.)

The activities of organisms can also assist mechanical weathering. The roots of plants can exert considerable force and move particles apart from one another. The burrows of animals expose new surfaces that can be altered by freezing and thawing.

Chemical weathering involves the chemical alteration of the rock in such a manner that it is more likely to fragment or to be dissolved. Some small rock fragments exposed to the atmosphere may be oxidized; that is, they combine with oxygen from the air and chemically change to different compounds. Other kinds of rock may combine with water molecules in a process known as hydrolysis. Often, the oxidized or hydrolyzed molecules are more readily soluble in water and, therefore, may be removed by rain or moving water. Rain is normally slightly acidic, and the acid content helps dissolve rocks.

Because of gravity, the prevailing movement of particles is from high elevations to lower ones. This process of loosening and redistributing particles is known as **erosion.** Wind can move sand and dust and can cause the wearing away of rocky surfaces by sandblasting their surfaces. Glaciers can move large rocks and cause

their surfaces to be rounded by being rubbed against each other and the surface of the Earth. Moving water transports much material in streams and rivers. In addition, wave action along the shores of lakes and the coasts of oceans constantly wears away and transports particles.

Soil and Land

The geologic processes just discussed are involved in the development of both soil and land; however, soil and land are not the same. **Land** is the part of the world not covered by the oceans. **Soil** is a thin covering over the land consisting of a mixture of minerals, organic material, living organisms, air, and water that together support the growth of plant life. The proportions of the soil components vary with different types of soils, but a typical, "good" agricultural soil is about 45 percent mineral, 25 percent air, 25 percent water, and 5 percent organic matter. (See figure 14.5.) This combination provides good drainage, aeration, and organic matter. Farmers are particularly concerned with soil because the nature of the soil

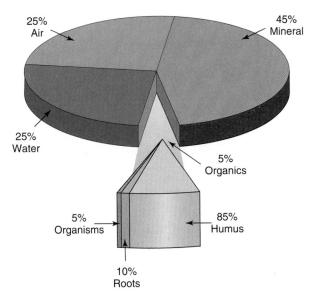

Figure 14.5 **The Components of Soil** Although soils vary considerably in composition, they all contain the same basic components: mineral material, air, water, and organic material. The organic material can be further subdivided into humus, roots, and other living organisms. The percentages shown are those that would be present in a good soil.

determines the kinds of crops that can be grown and which farming methods must be employed. Urban dwellers should also be concerned about soil because its health determines the quality and quantity of food they will eat. If the soil is so abused that it can no longer grow crops or if it is allowed to erode, degrading air and water quality, both urban and rural residents suffer. To understand how soil can be protected, we must first understand its properties and how it is formed.

Soil Formation

A combination of physical, chemical, and biological events acting over time is responsible for the formation of soil. Soil building begins with the fragmentation of the **parent material,** which consists of ancient layers of rock or more recent geologic deposits from lava flows or glacial activity. The kind and amount of soil developed depends on the kind of parent material present, the plants and animals present, the climate, the time involved, and the slope of the land. As was discussed earlier, the breakdown of parent material is known as weathering. The climate and chemical nature of the rock

material greatly influence the rate of weathering. Similarly, the size and chemical nature of the particles have a great impact on the nature of the soil that will develop in an area.

The role of organisms in the development of soil is also very important. The first organisms to gain a foothold in this modified parent material also contribute to soil formation. Lichens often form a pioneer community that grows on the surface of rocks and traps small particles. The decomposition of dead lichens and other organic matter releases acids that chemically alter the underlying rock, causing further fragmentation. The release of chemicals from the roots of plants causes further chemical breakdown of rock particles. As other kinds of organisms, such as plants and small animals, become established, they contribute, through their death and decay, increasing amounts of organic matter, which are incorporated with the small rock fragments.

The organic material resulting from the decay of plant and animal remains is known as **humus.** It is a very important soil component that accumulates on the surface and ultimately becomes mixed with the top layers of mineral particles. This material contains nutrients that are taken up by plants from the soil. Humus

also increases the water-holding capacity and the acidity of the soil so that inorganic nutrients, which are more soluble under acidic conditions, become available to plants. Humus also tends to stick other soil particles together and helps to create a loose, crumbly soil that allows water to soak in and permits air to be incorporated into the soil. Compact soils have few pore spaces, so they are poorly aerated, and water has difficulty penetrating, so it runs off.

Burrowing animals, soil bacteria, fungi, and the roots of plants are also part of the biological process of soil formation. One of the most important burrowing animals is the earthworm. One hectare (2.47 acres) of soil may support a population of 500,000 earthworms that can process as much as 9 metric tons of soil a year. These animals literally eat their way through the soil, resulting in further mixing of organic and inorganic material, which increases the amount of nutrients available for plant use. They often bring nutrients from the deeper layers of the soil up into the area where plant roots are more concentrated, thus improving the soil's fertility. Soil aeration and drainage are also improved by the burrowing of earthworms and other small soil animals, such as nematodes, mites, pill bugs, and tiny insects. Small soil animals also help to incorporate organic matter into the soil by collecting dead organic material from the surface and transporting it into burrows and tunnels. When the roots of plants die and decay, they release organic matter and nutrients into the soil and provide channels for water and air.

Fungi and bacteria are decomposers and serve as important links in many mineral cycles. (See chapter 5.) They, along with animals, improve the quality of the soil by breaking down organic material to smaller particles and releasing nutrients.

The position on the slope also influences soil development. Soil formation on steep slopes is very slow because materials tend to be moved downslope with wind and water. Conversely, river valleys often have deep soils because they receive materials from elsewhere by these same erosive forces.

Climate and time are also important in the development of soils. In general, extremely dry or cold climates develop soils very slowly, while humid and warm climates develop them more rapidly. Cold and dry climates have slow rates of accumulation of organic matter needed to form soil. Furthermore, chemical weathering proceeds more slowly at lower temperatures and in the absence of water. Under ideal climatic conditions, soft parent material may develop into a centimeter (less than ½ inch) of soil within 15 years. Under poor climatic conditions, a hard parent material may require hundreds of years to develop into that much soil. In any case, soil formation is a slow process.

The amount of rainfall and the amount of organic matter influence the pH of the soil. In regions of high rainfall, basic ions such as calcium, magnesium, and potassium are leached from the soils, and more acid materials are left behind. In addition, the decomposition of organic matter tends to increase the soil's acidity. Soil pH is important since it influences the availability of nutrients, which affects the kinds of plants that will grow, which affects the amount of organic matter added to the soil. Since calcium, magnesium, and potassium are important plant nutrients, their loss by leaching reduces the fertility of the soil. Excessively acidic soils also cause aluminum ions to become soluble, which in high amounts are toxic to many plants. (See the discussion of acid rain in chapter 17.) Most plants grow well in soils with a pH between 6 and 7, although some plants such as blueberries and potatoes grow well in acidic soils. In most agricultural situations, the pH of the soil is usually adjusted by adding chemicals to the soil. Lime can be added to make soils less acid, and acid-forming materials such as sulfates can be added to increase acidity.

Soil Properties

Soil properties include soil texture, structure, atmosphere, moisture, biotic content, and chemical composition. **Soil texture** is determined by the size of the mineral particles within the soil. The largest soil particles are gravel, which consists of fragments larger than 2.0 millimeters in diameter. Particles between 0.05 and

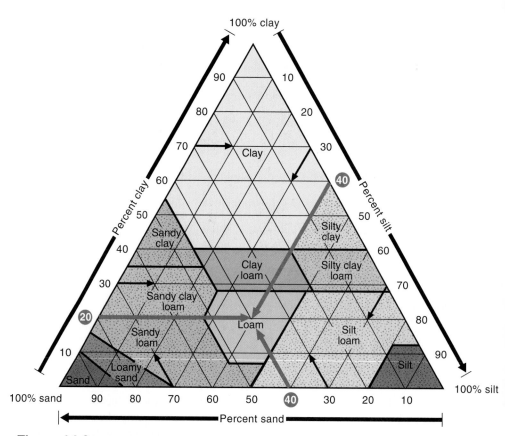

Figure 14.6 **Soil Texture** Texture depends on the percentage of clay, silt, and sand particles in the soil. A loam soil has the best texture for most crops. As shown in the illustration, if a soil were 40 percent sand, 40 percent silt, and 20 percent clay, it would be a loam.

Source: Data from U.S. Soil Conservation Service.

2.0 millimeters are classified as sand. Silt particles range from 0.002 to 0.05 millimeters in diameter, and the smallest particles are clay particles, which are less than 0.002 millimeters in diameter.

Large particles, such as sand and gravel, have many tiny spaces between them, which allow both air and water to flow through the soil. Water drains from this kind of soil very rapidly, often carrying valuable nutrients to lower soil layers, where they are beyond the reach of plant roots. Clay particles tend to be flat and are easily packed together to form layers that greatly reduce the movement of water through them. Soils with a lot of clay do not drain well and are poorly aerated. Because water does not flow through clay very well, clay soils tend to stay moist for longer periods of time and do not easily lose minerals to percolating water.

However, rarely does a soil consist of a single size of particle. Various particles are mixed in many different combinations,

resulting in many different soil classifications. (See figure 14.6.) An ideal soil for agricultural use is a **loam,** which combines the good aeration and drainage properties of large particles with the nutrient-retention and water-holding ability of clay particles.

Soil structure is different from its texture. **Soil structure** refers to the way various soil particles clump together. The particles in sandy soils do not attach to one another; therefore, sandy soils have a granular structure. The particles in clay soils tend to stick to one another to form large aggregates. Other soils that have a mixture of particle sizes tend to form smaller aggregates. A good soil is **friable,** which means that it crumbles easily. The soil structure and its moisture content determine how friable a soil is. Sandy soils are very friable, while clay soils are not. If clay soil is worked when it is too wet, it can stick together in massive blocks that will be difficult to break up.

A good soil for agricultural use will crumble and has spaces for air and water. In fact, the air and water content depends on the presence of these pore spaces. (See figure 14.7.) In good soil, about one-half to two-thirds of the spaces contain air after the excess water has drained. The air provides a source of oxygen for plant root cells and all the other soil organisms. The relationship between the amount of air and water is not fixed. After a heavy rain, most of the spaces may be filled with water and less oxygen is available to plant roots and other organisms. If some of the excess water does not drain from the soil, the plant roots may die from lack of oxygen. They are literally drowned. On the other hand, if there is not enough soil moisture, the plants wilt from lack of water. Soil moisture and air are also important in determining the numbers and kinds of soil organisms.

Protozoa, nematodes, earthworms, insects, algae, bacteria, and fungi are typical inhabitants of soil. (See figure 14.8.) The role of protozoa in the soil is not firmly established, but they seem to act as parasites and predators on other forms of soil organisms and, therefore, help to regulate the populations of those organisms.

Figure 14.7 **Pore Spaces and Particle Size** (a) Soil that is composed of particles of various sizes has spaces for both water and air. The particles have water bound to their surfaces (represented by the colored halo around each particle), but some of the spaces are so large that an air space is present. (b) Soil composed of uniformly small particles has less space for air. Since roots require both air and water, the soil in (a) would be better able to support crops than would the soil in (b).

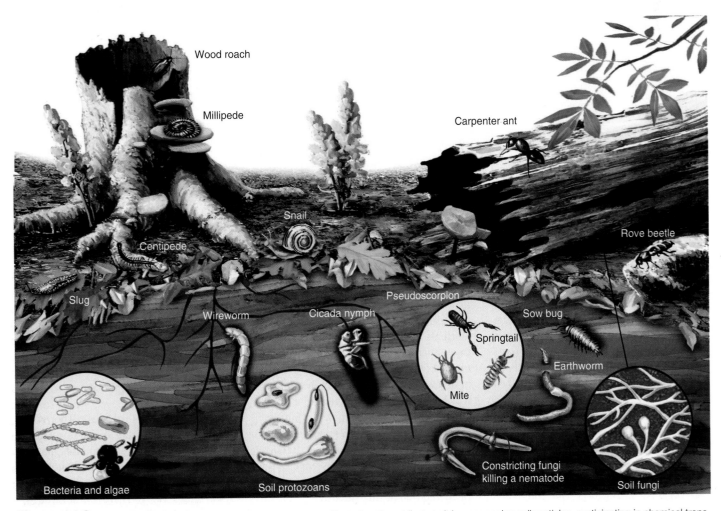

Figure 14.8 **Soil Organisms** All of these organisms occupy the soil and contribute to it by rearranging soil particles, participating in chemical transformation, and recycling dead organic matter.

CHAPTER 14 Soil and Its Uses

A horizon (topsoil) — Horizon of maximum biological activity

E horizon — Removal of materials dissolved or suspended in water occurs

B horizon (subsoil) — Horizon of accumulation of suspended materials from above

C horizon — Weathered parent material

R horizon — Bedrock

O_i — Loose leaves and organic debris, largely undecomposed

O_a — Organic debris partially decomposed

A — A dark-colored horizon with much organic matter mixed with mineral matter

E — Leaching removes some soluble minerals from this region

E/B — Transitional to horizon B

B/E — Transitional to horizon E

B — Maximum accumulation of silicate clay minerals or iron and organic matter

B/C — Transitional to horizon C

C

R

Figure 14.9 **Soil Profile** A soil has layers that differ physically, chemically, and biologically. The top layer is known as the *A* horizon and contains most of the organic matter. Organic matter that collects on the surface is known as the *O* horizon, which can be subdivided into an undecomposed top layer (*O*ᵢ) and a partially decomposed bottom layer (*O*ₐ). Many soils have a light-colored *E* horizon below the *A* horizon. It is light in color because dark-colored materials are leached from the layer. The *B* horizon accumulates minerals and particles as water carries dissolved minerals downward from the *A* and *E* to the *B* horizon. The *B* horizon is often called the subsoil. Below the *B* horizon is a *C* horizon of weathered parent material.

Nematodes, which are often called wire-worms or roundworms, may aid in the breakdown of dead organic matter. Some nematodes are parasitic on the roots of plants. Insects and other soil arthropods contribute to the soil by forming burrows and consuming and fragmenting organic materials, but they are also major crop pests that feed on plant roots. Several kinds of bacteria are able to fix nitrogen from the atmosphere. Algae carry on photosynthesis and are consumed by other soil organisms. Bacteria and fungi are particularly important in the decay and recycling of materials. Their chemical activ-

ities change complex organic materials into simpler forms that can be used as nutrients by plants. For example, some of these microorganisms can convert the nitrogen contained in the protein component of organic matter into ammonia or nitrate, which are nitrogen compounds that can be utilized by plants. The amount of nitrogen produced varies with the type of organic matter, type of microorganisms, drainage, and temperature. Finally, it is important to recognize that the soil contains a complicated food chain in which all organisms are subject to being consumed by others. All of these organisms are active within

distinct layers of the soil, known as the soil profile.

Soil Profile

The **soil profile** is a series of horizontal layers in the soil that differ in chemical composition, physical properties, particle size, and amount of organic matter. Each recognizable layer is known as a **horizon.** (See figure 14.9.) There are several systems for describing and classifying the horizons in soils. In general, the uppermost layer of the soil contains more nutrients and organic matter than do the deeper layers. The top layer is known as the *A* horizon, or topsoil. The *A* horizon consists of small mineral particles mixed with organic matter. Because of the relatively high organic content, it is dark in color. If there is a layer of **litter** (undecomposed or partially decomposed organic matter) on the surface, it is known as the *O* horizon. Forest soils typically have an *O* horizon. Many agricultural soils do not, since the soil is worked to incorporate surface crop residue. As the organic matter decomposes, it becomes incorporated into the *A* horizon. The thickness of the *A* horizon may vary from less than a centimeter (less than ½ inch) on steep mountain slopes to over a meter (over 40 inches) in the rich grasslands of central North America. Most of the living organisms and nutrients are found in the *A* horizon. As water moves down through the *A* horizon, it carries dissolved organic matter and minerals to lower layers. This process is known as **leaching.** Because of the leaching away of darker materials such as iron compounds, a lighter-colored layer develops below the *A* horizon that is known as the *E* horizon. Not all soils develop an *E* horizon. This layer usually contains few nutrients because water flowing down through the soil dissolves and transports nutrients to the underlying *B* horizon. The *B* horizon, often called the subsoil, contains less organic material and fewer organisms than the *A* horizon. However, it contains accumulations of nutrients that were leached from higher levels. Often, clay minerals that are leached from the topsoil are deposited in this layer. Because nutrients are deposited in this layer, the *B* horizon in many soils is a valuable source of nutrients for plants, and such subsoils support a well-developed root system. Because the amount of leaching depends on the available rainfall, grasslands soils, which develop under low rainfall, often have a poorly developed *B* horizon, while soils in woodlands that receive higher rainfall usually have a well-developed *B* horizon.

The area below the subsoil is known as the *C* horizon, and it consists of weathered parent material. This parent material contains no organic materials, but it does contribute to some of the soil's properties. The chemical composition of the minerals of the *C* horizon helps to determine the pH of the soil. If the parent material is limestone, the soil will tend to neutralize acids; whereas, if the parent material is granite rock, the soil will not be able to do so. The characteristics of the parent material in the *C* horizon may also influence the soil's rate of water absorption and retention. Ultimately, the *C* horizon rests on bedrock, which is known as the *R* horizon.

Soil profiles and the factors that contribute to soil development are extremely varied. Over 15,000 separate soil types have been classified in North America, but the three major classifications are grassland, forest, and desert. (See figure 14.10.) Most of the cultivated land in the world can be classified as either grassland soil or forest soil.

Because the amount of rainfall in grassland areas is relatively low, it does not penetrate into the soil layers very far. Most of the roots of the grasses and other plants remain near the surface, and little leaching of minerals from the topsoil to deeper layers occurs. Since the roots of the plants rot in place when the grasses die, a deep layer of topsoil develops. This lack of leaching also results in a thin layer of subsoil, which is low in mineral and organic content and supports little root growth.

Forest soils develop in areas of more abundant rainfall. Water moves down through the soil so that deeper layers of the soil have a great deal of moisture. The roots of the trees penetrate to this layer and extract the water they need. The leaves and other plant parts that fall to the soil surface form a thin layer of organic matter on the surface. This organic matter decomposes and mixes with the mineral material of the top layers of the soil. The water that moves through the soil tends to carry material from the topsoil to the subsoil, where many of the roots of the plants are located. One of the materials that accumulates in the *B* horizon is clay. In some soils, particularly forest soils, clay or other minerals may accumulate and form a relatively impermeable "hardpan" layer that limits the growth of roots and may prevent water from reaching the soil's deeper layers.

Desert soils have very poorly developed horizons. Since there is little rainfall, deserts do not support a large amount of plant growth, and much of the soil is exposed. Therefore, little organic matter is added to the soil, and little leaching of materials occurs from upper layers to lower layers. Since much of the soil is exposed to wind and water erosion, much of the organic material and smaller particles are carried away by wind or by flash floods when it rains.

In cold, wet climates, typical of the northern parts of Europe, Russia, and Canada, there may be considerable accumulations of organic matter since the rate of decomposition is reduced. The extreme acidity of these soils also reduces the rate of decomposition. Hot, humid climates also tend to have poorly developed soil horizons since the organic matter decays very rapidly, and soluble materials are carried away by the abundant rainfall.

Because tropical rainforests support such a vigorous growth of plants and an incredible variety of plant and animal species, it is often assumed that tropical soils must be very fertile. Consequently, many people have tried to raise crops on tropical soils. It is possible to grow certain kinds of crops that are specially adapted to tropical soils, but raising of most traditional crop species is not successful. To understand why this is so, it is important to understand the nature of tropical rainforest soils. Two features of the tropical rainforest climate have a great influence on the nature of the soil. High temperature results in rapid decomposition of organic matter, so that the soils have very little litter and humus. High rainfall tends to leach nutrients from the upper layers of the soil, leaving behind a soil that is rich in iron and aluminum. The high iron content results in a reddish color for most of these soils. Because the nutrients are quickly removed, these soils are very infertile. Furthermore, when the vegetation is removed, the soil is quickly eroded.

In addition to the differences caused by the kind of vegetation and rainfall, topography influences the soil profile. (See figure 14.11.) On a relatively flat area, the topsoil formed by soil-building processes will collect in place and gradually increase in depth. The topsoil formed

Figure 14.10 **Major Soil Types** There are thousands of different soil types, but many of them can be classified into three broad categories. Soils formed in the grasslands have a deep *A* horizon. The shallow *B* horizon does not have sufficient nutrients to support root growth. In forest soils, the *A* horizon is thinner, and leaching transfers many nutrients to the *B* horizon. Thus, roots are found in both the *A* and *B* horizons. Desert soils have very thin *A* horizons.

A horizon	} 15 cm (6 inches)
B horizon	} 25 cm (10 inches)
A horizon	} 150 cm (60 inches)
B horizon	} 15 cm (6 inches)

Figure 14.11 The Effect of Slope on a Soil Profile The topsoil formed on a large area of the hillside is continuously transported down the slope by the flow of water. It accumulates at the bottom of the slope and results in a thicker *A* horizon. The resulting "bottomland" is highly productive because it has a deep, fertile layer of topsoil, while the soil on the slope is less productive.

Figure 14.12 Water Erosion The force of moving water is able to pick up soil particles and remove them. In cases of prolonged erosion, gullies (such as the one shown here) are likely to form.

on rolling hills or steep slopes is often transported down the slope as fast as it is produced. On such slopes, the accumulation of topsoil may not be sufficient to support a cultivated crop. The topsoil removed from these slopes is eventually deposited in the flat floodplains. These regions serve as collection points for topsoil that was produced over extensive areas. As a result, these river-bottom and delta regions have a very deep topsoil layer and are highly productive agricultural land.

Soil Erosion

Erosion is the wearing away and transportation of soil by water, wind, or ice.

The Grand Canyon of the Colorado River, the floodplains of the Nile in Egypt, the little gullies on hillsides, and the deltas that develop at the mouths of rivers all attest to the ability of water to move soil. Anyone who has seen muddy water after a rainstorm has observed soil being moved by water. (See figure 14.12.) The force of moving water allows it to carry large amounts of soil. While erosion is a natural process, it is greatly accelerated by agricultural practices that leave the soil exposed. Each year, the Mississippi River transports over 325 million metric tons of soil from the central regions of North America to the Gulf of Mexico. This is equal to the removal of a layer of topsoil approximately 1 millimeter (0.04 inches) thick from the entire region. Although the rate of erosion varies from place to place, movement of soil by water occurs in every stream and river in the world. Dry Creek, a small stream in California, has only 500 kilometers (310 miles) of mainstream and tributaries; however, each year, it removes 180,000 metric tons of soil from a 340-square-kilometer (130-square-mile) area.

Soil erosion takes place everywhere in the world, but some areas are more exposed than others. Erosion occurs wherever grass, bushes, and trees are disappearing. Deforestation and desertification both leave land open to erosion. In deforested areas, water washes down steep, exposed slopes, taking the soil with it. In desertified regions, exposed soils, cleared for farming, building, or mining, or overgrazed by livestock, simply blow away. Wind erosion is most extensive in Africa and Asia. Blowing soil not only leaves a degraded area behind but can bury and kill vegetation where it settles. It will also fill drainage and irrigation ditches. When high-tech farm practices are applied to poor lands, soil is washed away and chemical pesticides and fertilizers pollute the runoff. Every year, erosion carries away far more topsoil than is created, primarily because of agricultural practices that leave the soil exposed. (See figure 14.13.)

Worldwide, erosion removes about 25.4 billion metric tons of soil each year. In Africa, soil erosion has reached critical levels, with farmers pushing farther onto deforested hillsides. In Ethiopia, for example, soil loss occurs at a rate of between 1.5 billion and 2 billion cubic meters (53–70 billion cubic feet) a year, with some 4 million hectares (about 10 million acres) of highlands considered irreversibly degraded. In Asia, in the eastern hills of Nepal, 38 percent of the land area is fields that have been abandoned because the topsoil has washed away. In the Western Hemisphere, Ecuador is losing soil at 20 times the acceptable rate.

According to the International Fund for Agricultural Development (IFAD), traditional labor-intensive, small-scale soil conservation efforts that combine maintenance of shrubs and trees with crop growing and cattle grazing work best at controlling erosion. In parts of Pakistan, a program begun by IFAD in 1980 to control rainfall runoff, erosion, and damage to rivers from siltation has increased crop yields and livestock productivity by 20 percent to 30 percent.

Badly eroded soil has lost all of the topsoil and some of the subsoil and is no longer productive farmland. Most current agricultural practices lose soil faster than it is replaced. Farming practices that reduce erosion, such as contour farming and terracing, are discussed in the Soil Conservation Practices section of this chapter.

Wind is also an important mover of soil. Under certain conditions, it can move large amounts. (See figure 14.14a.) Wind erosion may not be as evident as water erosion, since it does not leave gullies.

a.

Silt-laden river
b.

Figure 14.13 **Worldwide Soil Erosion** (a) Soil erosion is widespread throughout the world. (b) Silty rivers are evidence of poor soil conservation practices upstream.

Source: (a) U.S. Soil Conservation Service.

Nevertheless, it can be a serious problem. Wind erosion is most common in dry, treeless areas where the soil is exposed. In the Sahel region of Africa, much of the land has been denuded of vegetation because of drought, overgrazing, and improper farming practices. This has resulted in extensive wind erosion of the soil. (See figure 14.14b.) In the Great Plains region of North America, there have been four serious periods of wind erosion since European settlement in the 1800s. If this area receives less than 30 centimeters (12 inches) of rain per year, there is not enough moisture to support crops. When this occurs for several years in a row, it is called a drought. Farmers plant crops, hoping for rain. When the rain does not come, they plow their fields again to prepare them for another crop. Thus, the loose, dry soil is left exposed, and wind erosion results. Because of the large amounts of dust in the air during those times, the region is known as the Dust Bowl. During the 1930s, wind destroyed 3.5 million hectares (over 8.5 million acres) of farmland and seriously damaged an additional 30 million hectares (75 million acres) in the Dust Bowl.

Fortunately, many soil conservation practices have been instituted that protect soil. However, much soil is being lost to erosion, and more protective measures should be taken.

Soil Conservation Practices

The kinds of agricultural activities that land can be used for are determined by soil structure, texture, drainage, fertility, rockiness, slope of the land, amount and nature of rainfall, and other climatic conditions. A relatively large proportion— about 20 percent—of U.S. land is suitable for raising crops. However, only 2 percent of that land does not require some form of soil conservation practice. (See figure 14.15.) This means that nearly all of the soil in the United States must be managed in some way to reduce the effects of soil erosion by wind or water.

Not all parts of the world are as well supplied as the United States with land that has agricultural potential. (See table 14.1.) For example, worldwide, approximately 11 percent of the land surface is suitable for crops, and an additional 24 percent is in permanent pasture. In the United States, about 20 percent is cropland, and 25 percent is in permanent pasture. Contrast this with the continent of Africa, in which only 6 percent is suitable for crops and 29 percent can be used for pasture. Canada has only 5 percent

a.

b.

Figure 14.14 **Wind Erosion** (*a*) The dry, unprotected topsoil from this field is being blown away. The force of the wind is capable of removing all the topsoil and transporting it several thousand kilometers. (*b*) The semiarid region just south of the Sahara Desert is in an especially vulnerable position. The rainfall is unpredictable, which often leads to crop failure. In addition, population pressure forces people in this region to try to raise crops in marginal areas. This often results in increased wind erosion.

suitable for crops and 3 percent for pasture. Europe has the highest percentage of cropland with 30 percent, but it has only 17 percent in permanent pasture.

Since very little land is left that can be converted to agriculture, we must use what we have wisely. A study by the Food and Agricultural Organization states that up to 40 percent of the world's agricultural lands are seriously affected by soil degradation. The World Resources Institute in 2003 stated that land degradation affects around 70 percent of the world's rangelands, 40 percent of rain-fed agricultural lands, and 30 percent of irrigated lands. According to the United Nations Environment Programme, an estimated 500 million hectares (1.24 billion acres) of land in Africa have been affected by soil degradation since about 1950, including as much as 65 percent of agricultural land.

While the global expansion of agricultural area has been modest in recent decades, intensification has been rapid, as irrigated area increased and fallow time decreased to produce more output per hectare.

Many techniques can protect soil from erosion while allowing agriculture. Some of the more common methods are

Figure 14.15 **U.S. Land Used for Agricultural Purposes** Only 2 percent of the land in the United States can be cultivated without some soil conservation practices. This 2 percent is primarily flatland, which is not subject to wind erosion. On 42 percent of the remaining land, some special considerations for protecting the soil are required, or the kinds of crops are limited. Twelve percent of the land is marginal cropland that can provide crops only in years when rainfall and other conditions are ideal. Forty-four percent is not suitable for cultivating crops but may be used for other purposes, such as grazing cattle or as forests.

Environmental Close-Up

Desertification and Global Security

It is widely recognized that environmental degradation has a role to play in considerations of national security as well as international stability. Desertification is one of the major environmental issues that affects security and stability. The United Nations Convention to Combat Desertification (UNCCD) states that desertification threatens food security, health security, environmental security, and livelihood security. Environmental changes such as desertification put pressure on social, economic, political, and demographic dynamics, triggering insecurity.

The UNCCD Secretariat has long advocated the relevance of the convention's environmental concerns for a more comprehensive concept of international security. It organized an expert meeting in Spain in 1994 on the theme of desertification, migrations and conflicts, with the support of the Spanish and Dutch governments. According to research presented at that time, 50 armed conflicts in 1994 had environmental factors characteristic of drylands. The conflicts are triggered by the mass migration of people who have no choice but to leave their lands in search of livelihoods once their land becomes degraded and loses productivity, exposing them to the risk of conflicts with established populations in the new lands. Often, the environmental push element for migrations triggered by the environmental scarcity factors is hidden behind ethnic or socioeconomic considerations.

It is estimated that 135 million people—the equivalent of the population of Germany and France combined—are at risk of being displaced as a consequence of desertification. Some 60 million people are expected to eventually move from the desertified areas in sub-Saharan Africa toward northern Africa and Europe in the next 20 years. Long-term studies of West Africa project a constant migratory flow to coastal cities, whose population is expected to grow 3.5 times the numbers in 1997 to a total of 271 million in 2020.

Every year, between 700,000 and 900,000 Mexicans leave their rural dryland homes to find a living as migrant workers in the United States. In China, a considerable number of villages have been lost to expanding deserts, sand drifts, dune movement, and sandstorms in the last few decades. In Haiti, land degradation that reduced the per capita grain production to half of what it was 40 years ago, compounded with chronic political unrest, has resulted in 1.3 million Haitians fleeing their island over the past two decades.

Sixty percent of all land in Portugal and 75 percent of arable land in Turkey are at risk of desertification. Further, most of the territory in Northern Africa lies in arid and semiarid zones, which are prone to desertification and serve as transit zones toward Europe for migrants from sub-Saharan Africa. There is a growing awareness today that, in the context of global climate change, desertification is a survival issue that involves severe social and food security problems in these regions and must be addressed as one of the root causes of internal strife and public insecurity.

Table 14.1 Percentage of Land Suitable for Agriculture

Country	Percent Cropland	Percent Pasture
World	11.0	26.0
Africa	6.3	28.8
Egypt	2.8	5.0
Ethiopia	12.7	40.7
Kenya	7.9	37.4
South Africa	10.8	66.6
North America	13.0	16.8
Canada	4.9	3.0
United States	19.6	25.0
South America	6.0	28.3
Argentina	9.9	51.9
Venezuela	4.4	20.2
Asia	15.2	25.9
China	10.3	30.6
Japan	12.0	1.7
Europe	29.9	17.1

Source: Data from *World Resources*, 2003.

discussed here. Whenever soil is lost by water or wind erosion, the topsoil, the most productive layer, is the first to be removed. When the topsoil is lost, the soil's fertility decreases, and larger amounts of expensive fertilizers must be used to restore the fertility that was lost. This raises the cost of the food we buy. In addition, the movement of excessive amounts of soil from farmland into streams has several undesirable effects. First, a dirty stream is less aesthetically pleasing than a clear stream. Second, too much sediment in a stream affects the fish population by reducing visibility, covering spawning sites, and clogging the gills of the fish. Fishing may be poor because of unwise farming practices hundreds of kilometers upstream. Third, the soil carried by a river is eventually deposited somewhere. In many cases, this soil must be removed by dredging to clear shipping channels. We pay for dredging with our tax money, and it is a very expensive operation.

The Conservation Security Program

The 2002 U.S. farm legislation created an innovative environmental initiative called the Conservation Security Program (CSP). If implemented as designed, CSP will use three tiers to reward farmers who produce measurable environmental benefits on working lands. Farmers who qualify for the lowest tier (and the lowest payment) use basic conservation production systems that, for example, reduce soil erosion. Those in the highest tier (and the highest payment category) are taking a whole-farm approach to environmental problems—using a grass-based livestock pro-

duction system that integrates trees and wildlife habitat, for example. The CSP is a major departure from conventional farm policy, which pays farmers based on how many hectares of "program crops" such as corn and soybeans they raise, regardless of the environmental consequences.

CSP has yet to be fully used in the field. U.S. Department of Agriculture officials are still working out details of how it will be implemented. One area of concern is how "real environmental benefits" would be measured on-farm and over time. Monitoring systems developed by farmers and natural resource practitioners offer potential for providing the basis for such measurements.

For all of these reasons, proper soil conservation measures should be employed to minimize the loss of topsoil. Figure 14.16 contrasts poor soil conservation practices with proper soil protection. When soil is not protected from the effects of running water, the topsoil is removed and gullies result. This can be prevented by slowing the flow of water over sloping land.

Contour Farming

Contour farming, which is tilling at right angles to the slope of the land, is one of the simplest methods for preventing soil erosion. This practice is useful on gentle slopes and produces a series of small ridges at right angles to the slope. (See figure 14.17.) Each ridge acts as a dam to hold water from running down the incline. This allows more of the water to soak into the soil. Contour farming reduces soil erosion by as much as 50 percent and, in drier regions, increases crop yields by conserving water.

Strip Farming

When a slope is too steep or too long, contour farming alone may not prevent soil erosion. However, a combination of contour and strip farming may work. **Strip farming** is alternating strips of closely sown crops such as hay, wheat, or other small grains with strips of row crops such as corn, soybeans, cotton, or sugar beets. (See figure 14.18.) The closely sown crops retard the flow of water, which

a.　　　　　　　　　　　　b.

Figure 14.16　Poor and Proper Soil Conservation Practices (a) This land is no longer productive farmland since erosion has removed the topsoil. (b) This rolling farmland shows strip contour farming to minimize soil erosion by running water. It should continue indefinitely to be productive farmland.

Figure 14.17　Contour Farming Tilling at right angles to the slope creates a series of ridges that slows the flow of the water and prevents soil erosion. This soil conservation practice is useful on gentle slopes.

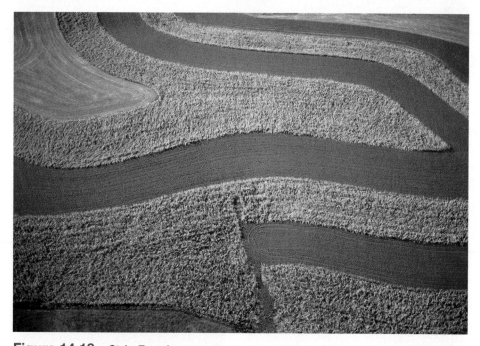

Figure 14.18 **Strip Farming** On rolling land, a combination of contour and strip farming prevents excessive soil erosion. The strips are planted at right angles to the slope, with bands of closely sown crops, such as wheat or hay, alternating with bands of row crops, such as corn or soybeans.

reduces soil erosion and allows more water to be absorbed into the ground. The type of soil, steepness, and length of slope dictate the width of the strips and determine whether strip or contour farming is practical.

Terracing

On very steep land, the only practical method of preventing soil erosion is to construct terraces. **Terraces** are level areas constructed at right angles to the slope to retain water and greatly reduce the amount of erosion. (See figure 14.19.) Terracing has been used for centuries in nations with a shortage of level farmland. The type of terracing seen in figure 14.19a requires the use of small machines and considerable hand labor and is not suitable for the mechanized farming typical in much of the world. The modification shown in figure 14.19b allows for the use of large farm machines. Terracing is an expensive method of controlling erosion since it requires moving soil to construct the level areas, protecting the steep areas between terraces, and constant repair and maintenance. Many factors, such as length and steepness of slope, type of soil,

and amount of precipitation, determine whether terracing is feasible.

Waterways

Even with such soil conservation practices as contour farming, strip farming, and terracing, farmers must often provide protected channels for the movement of water. **Waterways** are depressions on sloping land where water collects and flows off the land. When not properly maintained, these areas are highly susceptible to erosion. (See figure 14.20.) If a waterway is maintained with a permanent sod covering, the speed of the water is reduced, the roots tend to hold the soil particles in place, and soil erosion is decreased.

Windbreaks

Contour farming, strip farming, terracing, and maintaining waterways are all important ways of reducing water erosion, but wind is also a problem with certain soils, particularly in dry areas of the world. Wind erosion can be reduced if the soil is protected. The best protection is a layer of vegetation over the surface. However, the process of preparing soil for planting and

the method of planting often leave the soil exposed to wind. **Windbreaks** are plantings of trees or other plants that protect bare soil from the full force of the wind. Windbreaks reduce the velocity of the wind, thereby decreasing the amount of soil that it can carry away. (See figure 14.21.) In some cases, rows of trees are planted at right angles to the prevailing winds to reduce their force, while in other cases, a kind of strip farming is practiced in which strips of hay or grains are alternated with row crops that leave large amounts of the soil exposed. In some areas of the world, the only way to protect the soil is to not cultivate it at all but to leave it in a permanent cover of grasses.

Conventional Versus Conservation Tillage

Conventional tillage methods in much of the world require extensive use of farm machinery to prepare the soil for planting and to control weeds. Typically, a field is plowed and then disked or harrowed one to three times before the crop is planted. The plowing, which turns the soil over, has several desirable effects: any weeds or weed seeds are buried, thus reducing the weed problem in the field. Crop residue from previous crops is incorporated into the soil, where it will decay faster and contribute to soil structure. Nutrients that had been leached to deeper layers of the soil are brought near the surface. And the dark soil is exposed to the sun so that it warms up faster. This last effect is most critical in areas with short growing seasons. In many areas, fields are plowed in the fall, after the crop has been harvested, and the soil is left exposed all winter.

After plowing, the soil is worked by disks or harrows to break up any clods of earth, kill remaining weeds, and prepare the soil to receive the seeds. After the seeds are planted, there may still be weed problems. Farmers often must cultivate row crops to kill the weeds that begin to grow between the rows. Each trip over the field costs the farmer money, while at the same time increasing the amount of

a.

b.

Figure 14.19 **Terraces** Since the construction of terraces requires the movement of soil and the protection of the steep slope between levels, terraces are expensive to build. (a) The terraces seen here are extremely important for people who live in countries that have little flatland available. They require much energy and hand labor to maintain but make effective agricultural use of the land without serious erosion. (b) This modification of the terracing concept allows the use of the large farm machines typical of farming practices in Canada, Europe, and the United States.

a.

b.

Figure 14.20 **Protection of Waterways Prevents Erosion** (a) An unprotected waterway has been converted into a gully. (b) A well-maintained waterway is not cultivated; a strip of grass retards the flow of water and protects the underlying soil from erosion.

time the soil is exposed to wind or water erosion.

In recent years, several new systems of tillage have developed, as innovations in chemical herbicides and farm equipment have taken place. These tilling practices protect the soil by leaving the crop residue on the soil surface, thus reducing the amount of time it is exposed to erosion forces. **Reduced tillage** is a method that uses less cultivation to control weeds and to prepare the soil to receive seeds but generally leaves 15 percent to 30 percent of the soil surface covered with crop residue after planting. **Conservation tillage** methods further reduce the amount of disturbance to the soil and leave 30 percent or more of the soil surface covered with crop residue following planting. Selective herbicides are used to kill unwanted vegetation before planting the new crop and to control weeds afterward. Several variations of conservation tillage are used:

1. Mulch tillage involves tilling the entire surface just before planting or as planting is occurring.

2. Strip tillage is a method that involves tilling only in the narrow strip that is to receive the seeds. The rest of the soil and the crop residue from the previous crop are left undisturbed.

3. Ridge tillage involves leaving a ridge with the last cultivation of the previous year and planting the crop on the

a.   b.

Figure 14.21 **Windbreaks** (a) In sections of the Great Plains, trees provide protection from wind erosion. The trees along the road protect the land from the prevailing winds. (b) In this field, temporary strips of vegetation serve as windbreaks.

Table 14.2 Comparison of Various Tillage Methods

Tillage Method	Fuel Use (Liters per Hectare)	Time Involved (Hours per Hectare)
Conventional plowing	8.08	3.00
Reduced tillage	5.11	2.20
Mulch tillage	4.64	2.07
Ridge tillage	4.12	2.25
No tillage	2.19	1.21

Source: Data from University of Nebraska, Institute of Agriculture and Natural Resources, 2001.

ridge with residue left between the ridges. The crop may be cultivated during the year to reduce weeds.

4. No-till farming involves special planters that place the seeds in slits cut in the soil that still has on its surface the residue from the previous crop.

Both reduced tillage and conservation tillage methods reduce the amount of time and fuel needed by the farmer to produce the crop and, therefore, represent an economic savings. By 2000, over half of the cropland in the United States was being farmed using reduced or conservation tillage methods. (See table 14.2.) For many kinds of crops, yields are comparable to those produced by conventional tillage methods.

Other positive effects of reduced tillage, in addition to reducing erosion, are:

1. The amount of winter food and cover available for wildlife increases, which can lead to increased wildlife populations.

2. Since there is less runoff, siltation in streams and rivers is reduced. This results in clearer water for recreation and less dredging to keep waterways open for shipping.

3. Row crops can be planted on hilly land that cannot be converted to such crops under conventional tilling methods. This allows a farmer to convert low-value pastureland into cropland that gives a greater economic yield.

4. Since fewer trips are made over the field, petroleum is saved, even when the petrochemical feedstocks necessary to produce the herbicides are taken into account.

5. Two crops may be grown on a field in areas that had been restricted to growing one crop per field per year. In some areas, immediately after harvesting wheat, farmers have planted soybeans directly in the wheat stubble.

6. Because conservation tillage reduces the number of trips made over the field by farm machinery, the soil does not become compacted as quickly.

However, there are also some drawbacks to conservation tillage methods:

1. The residue from previous vegetation may delay the warming of the soil, which may, in turn, delay planting some crops for several days.

2. The crop residue reduces evaporation from the soil and the upward movement of water and soil nutrients from deeper layers of the soil, which may retard the growth of plants.

3. The accumulation of plant residue can harbor plant pests and diseases that will require more insecticides and fungicides. This is particularly true if the same crop is planted repeatedly in the same location.

Conservation tillage is not the complete answer to soil erosion problems but may be useful in reducing soil erosion on well-drained soils. It also requires that farmers pay close attention to the condition of the soil and the pests to be dealt with.

In 1972, 12 million hectares (30 million acres) in the United States were under some form of conservation tillage. About 1.3 million hectares (3.2 million acres) were being farmed using no-till methods. By 1992, this had risen to 60 million hectares (150 million acres) of conservation-tilled land, of which 6.2 million hectares (15 million acres) were being farmed using no-till methods. It is estimated that by the year 2010, 95 percent of U.S. cropland will be under some form of reduced-tillage practice.

Figure 14.22 **Noncrop Use of Land to Raise Food** As long as it is properly protected and managed, this land can produce food through grazing, but it should never be plowed to plant crops because the topsoil is too shallow and the rainfall is too low.

Protecting Soil on Nonfarm Land

Each piece of land has characteristics, such as soil characteristics, climate, and degree of slope, that influence the way it can be used. When all these factors are taken into consideration, a proper use can be determined for each portion of the planet. Wise planning and careful husbandry of the soil is necessary if the land and its soil are to provide food and other necessities of life. Not all land is suitable for crops or continuous cultivation. Some has highly erodible soil and must never be plowed for use as cropland, but it can still serve other useful functions. By using appropriate soil conservation practices, much of the land not usable for crops can be used for grazing, wood production, wildlife production, or scenic and recreational purposes. Figure 14.22 shows land that is not suitable for cultivation because it is too arid but that, if it is used properly, can provide grazing for cattle or sheep.

The land shown in figure 14.23 is not suitable for either crops or grazing because of the steep slope and thin soil. However, it is still a valuable and productive piece of land, since it can be used to furnish lumber, wildlife habitats, and recreational opportunities.

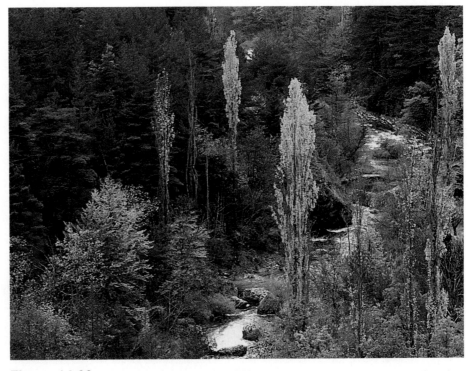

Figure 14.23 **Forest and Recreational Use** Although this land is not capable of producing crops or supporting cattle, it furnishes lumber, a habitat for wildlife, and recreational opportunities.

CHAPTER 14   Soil and Its Uses

Environmental Close-Up

Land Capability Classes

Not all land is suitable for raising crops or urban building. Such factors as the degree of slope, soil characteristics, rockiness, erodibility, and other characteristics determine the best use for a parcel of land. In an attempt to encourage people to use land wisely, the U.S. Soil Conservation Service has established a system to classify land-use possibilities. The table shows the eight classes of land and lists the characteristics and capabilities of each. The photo shows how these classes can coexist.

Unfortunately, many of our homes and industries are located on type I and II land, which has the least restrictions on agricultural use. This does not make the best use of the land. Zoning laws and land-use management plans should consider the land-use capabilities and institute measures to ensure that land will be used to its best potential.

- Can you provide examples from your community where you feel land has not been used properly?

- In your opinion, should there be more or less land-use planning?

- What are some of the consequences of building or farming on land that is not suitable?

	Land Class	Characteristics	Capability	Special Conservation Measures
Land suitable for cultivation	I	Excellent, flat, well-drained land	Cropland	Normal good practices adequate
	II	Good land; has minor limitations, such as slope, sandy soil, or poor drainage	Cropland Pasture	Strip cropping Contour farming
	III	Moderately good land with important limitations of soil, slope, or drainage	Cropland Pasture Watershed	Contour farming Strip cropping Terraces Waterways
	IV	Fair land with severe limitations of soil, slope, or drainage	Pasture Orchards Urban Industry Limited cropland	Crops on a limited basis Contour farming Strip cropping Terraces Waterways
Land not suitable for cultivation	V	Use for grazing and forestry; slightly limited by rockiness, shallow soil, or wetness	Grazing Forestry Watershed Urban Industry	No special precautions if properly grazed or logged; must not be plowed
	VI	Moderate limitations for grazing and forestry because of moderately steep slopes	Grazing Forestry Watershed Urban Industry	Grazing or logging may be limited at times.
	VII	Severe limitations for grazing and forestry because of very steep slopes vulnerable to erosion	Grazing Forestry Watershed Recreation Wildlife Urban Industry	Careful management is required when used for grazing or logging.
	VIII	Unsuitable for grazing and forestry because of steep slope, shallow soil, lack of water, or too much water	Watershed Recreation Wildlife Urban Industry	Not to be used for grazing or logging; steep slope and lack of soil present problems

Worldwide Soil Degradation

As humans have used the land and its soil for agriculture, grazing, fuelwood production, and forestry, we have had profound effects on the rates at which soils are formed and lost. Removal of the vegetation has accelerated soil erosion so that most of the world is losing soil faster than it is being formed. The degradation is most severe where the need for food is greatest, since starving people are forced to overexploit their already overused soils. The map and graphs illustrate the magnitude of the soil degradation problem and the relative importance of the causes.

Percent of Soil Degraded by Human Activity

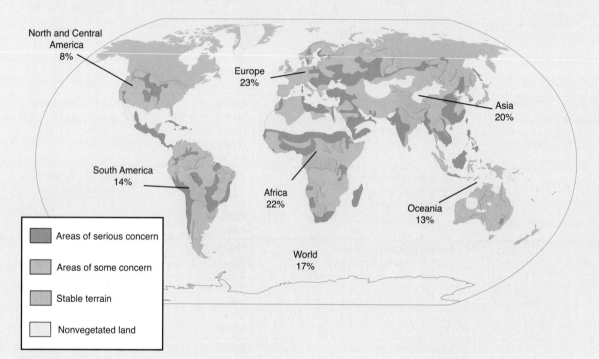

Causes of Soil Degradation

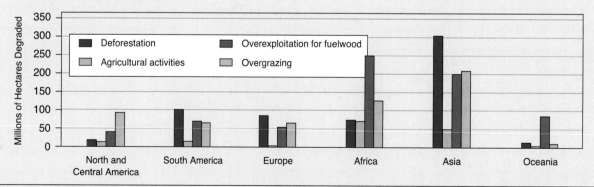

Source: World Resources Institute in collaboration with the United Nations Environment Programme and the United Nations Development Programme, *World Resources 1992-93* (Oxford University Press, New York, 1992), table 19.4, p. 290.

Summary

The surface of the Earth is in constant flux. The movement of tectonic plates results in the formation of new land as old land is worn down by erosive activity. Soil is an organized mixture of minerals, organic material, living organisms, air, and water. Soil formation begins with the breakdown of the parent material by such physical processes as changes in temperature, freezing and thawing, and movement of particles by glaciers, flowing water, or wind. Oxidation and hydrolysis can chemically alter the parent material. Organisms also affect soil building by burrowing into and mixing the soil, releasing nutrients, and decomposing.

Topsoil contains a mixture of humus and inorganic material, both of which supply soil nutrients. The ability of soil to grow crops is determined by the inorganic matter, organic matter, water, and air spaces in the soil. The mineral portion of the soil consists of various mixtures of sand, silt, and clay particles.

A soil profile typically consists of the *O* horizon of litter; the *A* horizon, which is rich in organic matter; an *E* horizon from which materials have been leached; the *B* horizon, which accumulates materials leached from above; and the *C* horizon, which consists of slightly altered parent material. Forest soils typically have a shallow *A* horizon and an *E* horizon and a deep, nutrient-rich *B* horizon with much root development. Grassland soils usually have a thick *A* horizon containing most of the roots of the grasses. They lack an *E* horizon and have few nutrients in the thin *B* horizon.

Soil erosion is the removal and transportation of soil by water or wind. Proper use of such conservation practices as contour farming, strip farming, terracing, waterways, windbreaks, and conservation tillage can reduce soil erosion. Misuse reduces the soil's fertility and causes air- and water-quality problems. Land unsuitable for crops may be used for grazing, lumber, wildlife habitats, or recreation.

Issues – Analysis

Soil Fertility and Hunger in Africa

Africa south of the Sahara is the only remaining region of the world where per capita food production has remained stagnant over the past 40 years. About 180 million Africans—up 100 percent since 1970—do not have access to sufficient food to lead healthy and productive lives.

Africa's food insecurity is directly related to insufficient total food production, in contrast to South Asia and other regions, where food insecurity is primarily due to poor distribution and lack of purchasing power. Depletion of soil fertility is a major cause of low per capita food production in Africa. Over decades, small-scale farmers have removed large quantities of nutrients from their soils without using sufficient quantities of manure or fertilizer to replenish the soil.

The traditional way to overcome nutrient depletion is the use of mineral fertilizers. But fertilizers in Africa cost two to six times as much as those in Europe, North America, or Asia.

A soil fertility replenishment approach has been developed during the past decade using resources naturally available in Africa. Participatory research trials were conducted in which farmers tested and adapted several new practices. These included the use of nitrogen-fixing leguminous trees and biomass transfer of the leaves of nutrient-accumulating shrubs.

Leguminous trees are interplanted into a young maize (corn) crop and allowed to grow as fallows during dry seasons. The quantities of nitrogen captured are similar to those applied as fertilizers by commercial farmers to grow maize in developed countries. After the wood is harvested from the trees, nitrogen-rich leaves, pods, and green branches are hoed into the soil before maize is planted at the start of a subsequent season. This litter decomposes with the tree roots, releasing nitrogen and other nutrients to the soil.

The transfer of the leaf biomass of nutrient-accumulating shrubs from roadsides and hedges into cropped fields also adds nutrients and can double maize yields without fertilizer additions. This organic source of nutrients is effective because it also adds other plant nutrients, particularly potassium and micronutrients. Because of the high labor requirements for cutting and carrying the biomass to fields, this is profitable with high-value crops such as vegetables but not with relatively low-valued maize.

Tens of thousands of farm families in Kenya, Tanzania, Malawi, Zambia, Zimbabwe, and Mozambique use various combinations of tree fallows and biomass transfers with good and consistent results. Adoption of this approach is taking place by the transfer of knowledge from farmer to farmer and village to village by community-based organizations and by a multitude of national research and extension institutes, universities, nongovernmental organizations, and development projects. The challenge, now, is to accelerate the adoption rate to reach tens of millions of farm families.

- Do you feel that such low-tech solutions as those discussed in this essay have a role in solving the challenge of feeding the world's growing population?
- What are the barriers to expanding the programs discussed?
- Are low-tech solutions equally applicable to the developed nations in the world? Why or why not?

Key Terms

asthenosphere *306*
chemical weathering *308*
conservation tillage *321*
contour farming *319*
crust *306*
erosion *308/315*
friable *310*
horizon *312*
humus *309*
land *308*

leaching *313*
lithosphere *306*
litter *313*
loam *310*
mantle *306*
mechanical weathering *307*
parent material *309*
plate tectonics *306*
reduced tillage *321*
soil *308*

soil profile *312*
soil structure *310*
soil texture *310*
strip farming *319*
terraces *320*
waterways *320*
weathering *307*
windbreaks *320*

Review Questions

1. How are soil and land different?
2. Name the five major components of soil.
3. Describe the process of soil formation.
4. Name five physical and chemical processes that break parent material into smaller pieces.
5. In addition to fertility, what other characteristics determine the usefulness of soil?
6. How does soil particle size affect texture and drainage?
7. Describe a soil profile.
8. Define erosion.
9. Describe three soil conservation practices that help to reduce soil erosion.
10. Besides cropland, what are other possible uses of soil?

Critical Thinking Questions

1. Minimum tillage soil conservation often uses greater amounts of herbicides to control weeds. What do you think about this practice? Why?
2. As populations grow, should we try to bring more land into food production, or should we use technology to aid in producing more food on the land we already have in production of food? What are the trade-offs?
3. Given what you know about soil formation, how might you explain the presence of a thick *A* horizon in soils in the North American Midwest?
4. Why should nonfarmers be interested in soil conservation?
5. Imagine that you are a scientist hired to consult on a project to evaluate land-use practices at the edge of a small city. The area in question has deep ravines and hills. What kinds of agricultural, commercial, and logging practices would you recommend in this area to help preserve the environment?
6. Look at your own community. Can you see examples of improper land use (urban or rural)? What are the consequences of these land-use practices? What recommendations would you make to improve land use?

Welcome to McGraw-Hill's Online Learning Center

Location: www.mhhe.com/enger10e

Interactive Exploration

Check out the website at www.mhhe.com/enger10e and click on the cover of this textbook for quizzing, career information, case studies, and hot links for the following topics:

Soil and Its Uses

Agricultural Methods

Nutrient Cycling

Sustainable Agriculture

Deserts

PART FOUR   Resource Management

Agricultural Methods and Pest Management

15

Objectives

After reading this chapter, you should be able to:

- Explain how mechanization encouraged monoculture farming.
- List the advantages and disadvantages of monoculture farming.
- Explain why chemical fertilizers are used.
- Understand how fertilizers alter soil characteristics.
- Explain why modern agriculture makes extensive use of pesticides.
- Differentiate between persistent pesticides and nonpersistent pesticides.
- List four problems associated with pesticide use.
- Define biomagnification.
- Define organic farming.
- Explain why integrated pest management depends on a complete knowledge of the pest's life history.
- Recognize that genetically modified crops are created by using biotechnological techniques to insert genes from one species into another.

Chapter Outline

The Development of Agriculture
  Shifting Agriculture
  Labor-Intensive Agriculture
  Mechanized Agriculture
Fossil Fuel Versus Muscle Power
The Impact of Fertilizer
Agricultural Chemical Use
  Insecticides
  Herbicides
  Fungicides and Rodenticides
  Other Agricultural Chemicals
Problems with Pesticide Use
  Persistence
  Bioaccumulation and Biomagnification
  Pesticide Resistance
  Effects on Nontarget Organisms
  Human Health Concerns

Why Are Pesticides So Widely Used?
Alternatives to Conventional Agriculture
  Techniques for Protecting Soil and Water
    Resources
  Integrated Pest Management
Issues—Analysis: What Are Organic Foods?

Environmental Close-Ups
DDT—A Historical Perspective, p. 334
A New Generation of Insecticides, p. 337
Industrial Production of Livestock, p. 343
Food Additives, p. 344

Global Perspectives
Economic Development and Food Production in China, p. 340
Contaminated Soils in the Former Soviet Union, p. 341

PAST PRESENT FUTURE

1983 No genetically modified crops were grown.	**1996** Genetically modified crops were introduced. By 2002, 22 percent of crops worldwide were genetically modified.
1983 U.S. farmland under conservation tillage amounted to 16 percent.	**2002** U.S. farmland under conservation tillage amounted to 36.6 percent.
1983 The average U.S. corn yield was 5100 kilograms per hectare (4550 pounds per acre).	**2002** The average U.S. corn yield was 8175 kilograms per hectare (7300 pounds per acre; a 60 percent increase).
1983 World fertilizer use was 126,291,700 metric tons.	**2000** World fertilizer use was 136,435,100 metric tons (an 8 percent increase).
1983 U.S. farmland equaled 414,340 hectares (1,023,800 acres).	**2002** U.S. farmland equaled 381,170 hectares (941,900 acres; a loss of 8 percent).

The Development of Agriculture

Our early ancestors obtained food from nature by hunting and gathering. The development of agriculture involved manipulating the natural environment to produce the kinds of foods humans want and allowed for an increase in the size of the human population. The history of the development of agriculture has involved various kinds of innovations. One of the simplest is shifting agriculture, also known as "slash and burn" agriculture.

Shifting Agriculture

Shifting agriculture involves cutting down the trees and burning the trees and other vegetation in a small area of the forest. (See figure 15.1.) Burning releases nutrients that were tied up in the biomass and allows a few crops to be raised before the soil is exhausted. Once the soil is no longer suitable for raising crops (within two or three years), the site is abandoned. The surrounding forest recolonizes the area, which will return over time to the original forest through the process of succession.

In some parts of the world with poor soil and low populations, shifting agriculture is still used successfully. This method is particularly useful on thin, nutrient-poor tropical soils and on steep slopes. The small size of the openings in the forest and their temporary existence prevent widespread damage to the soil, and erosion is minimized. While this system of agriculture is successful when human population densities are low, it is not suitable for large, densely populated areas. When populations become too large, the size and number of the garden plots increase and the time between successive uses of the same plot of land decreases. When a large amount of the forest is disturbed and the time between successive uses is decreased, the forest cannot return and repair the damage done by the previous use of the land, and the nature of the forest is changed.

The traditional practices of the people who engage in small-scale, shifting agriculture have been developed over hun-

Figure 15.1 **Shifting Agriculture** In many areas of the world where the soils are poor and human populations are low, crops can be raised by disturbing small parts of the ecosystem followed by several years of recovery. The burning of vegetation releases nutrients that can be used by crops for one or two years before the soil is exhausted. The return of the natural vegetation prevents erosion and repairs the damage done by temporary agricultural use.

dreds of years and often are more effective for their local conditions than other methods of gardening. Typically, these gardens are planted with a mixture of plants; a system known as **polyculture.** Mixing plants together in a garden often is beneficial, since shade-requiring species may be helped by taller plants, or nitrogen-fixing legumes may provide nitrogen for species that require it. In addition, mixing species may reduce insect pest problems because some plants produce molecules that are natural insect repellents. The small, isolated, temporary nature of the gardens also reduces the likelihood of insect infestations. While today we see this form of agriculture practiced most commonly in tropical areas, it is important to note that many Native American cultures used shifting agriculture and polyculture in temperate areas.

Labor-Intensive Agriculture

In many areas of the world with better soils, more intense forms of agriculture developed that involved a great deal of manual labor to till, plant, and harvest the crop. This style of agriculture is still practiced in much of the world today. Three situations favor this kind of farming: (1) when the growing site does not allow for mechanization, (2) when the kind of crop does not allow it, and (3) when the

economic condition of the people does not allow them to purchase the tools and machines used for mechanized agriculture. Crops or terrain that requires that fields be small discourages mechanization, since large tractors and other machines cannot be used efficiently on small, oddly shaped fields. Many mountainous areas of the world fit into this category. In addition, some crops require such careful handling in planting, weeding, or harvesting that large amounts of hand labor are required. The planting of paddy rice and the harvesting of many fruits and vegetables are examples. However, the primary reason for labor-intensive farming is economic. Many densely populated countries have numerous small farms of 1 to 2 hectares (a hectare is about the size of a soccer field) that can be effectively managed with human labor, supplemented by that of draft animals and a few small gasoline-powered engines. (See figure 15.2.) In addition, in the less-developed regions of the world, the cost of labor is low, which encourages the use of hand labor rather than relatively expensive machines to do planting, weeding, and other activities.

Mechanization requires large tracts of land that could be accumulated only by the expenditure of large amounts of money or the development of larger cooperative farms from many small units. Even if social and political obstacles to such large landholdings could be overcome, there is still the problem of obtaining the necessary capital to purchase the machines. Large parts of the developing world fit into this category, including much of Africa, many areas in Central and South America, and many areas in Asia. In countries such as China and India, which have a combined population of over 2 billion people, about 70 percent of the population is rural. Many of these people are engaged in nonmechanized agriculture.

Mechanized Agriculture

The development of various kinds of machines following the Industrial Revolution resulted in changes in agriculture. Horse-drawn farm implements and the subsequent development of tractors led to the modern mechanized agriculture typical of North America, much of Europe, and other parts of the world where money and land are available to support this form of agriculture. In large measure, machines and fossil-fuel energy replace the energy formerly supplied by human and animal muscles. Mechanization requires large expanses of fairly level land for the machines to operate effectively. In addition, large tracts of land must be planted in the same crop for efficient planting, cultivating, and harvesting, a practice known as **monoculture.** (See figure 15.3.) Small sections of land with many kinds of crops require many changes of farm machinery, which takes time. Also, many crops interspersed with one another reduces the efficiency of farming operations because farmers must skip parts of the field, which increases travel time and uses expensive fuel.

Even though mechanized monoculture is an efficient method of producing food, it is not without serious drawbacks. When large tracts of land are prepared for planting, they are often left uncovered by vegetation, and soil erosion increases. Because of problems with erosion, many farmers are now using methods that reduce the time the fields are left bare. (See chapter 14.)

Traditionally, mechanized farming has removed much of the organic matter each year when the crop was harvested. This tended to reduce the soil organic mat-

Figure 15.2 Labor-Intensive Agriculture In many of the less-developed countries of the world, the extensive use of hand labor allows for impressive rates of production with a minimal input of fossil fuels and fertilizers. This kind of agriculture is also necessary in areas that have only small patches of land suitable for farming.

Figure 15.3 Mechanized Monoculture This wheat field is an example of monoculture, a kind of agriculture that is highly mechanized and requires large fields for the efficient use of machinery. In this kind of agriculture, machines and fossil-fuel energy have replaced the energy of humans and draft animals.

ter. As agricultural scientists and farmers have recognized the need to improve the organic-matter content of soils, many farmers have been leaving increased amounts of organic matter after harvest, or they specifically plant a crop that is later plowed under to increase the soil's organic content.

To ensure that a crop can be planted, tended, and harvested efficiently by machines, farmers rely on seeds that are identical genetically and thus provide uniform plants with characteristics suitable for mechanized farming. These special seeds can ensure that all the plants germinate at the same time, resist the same pests, ripen at the same time, and grow to the same height. These are valuable characteristics to the farmer, but these plants have little genetic variety. When all the farmers in an area plant the same genetic varieties, pest control becomes a serious problem. If diseases or pests begin to spread, the magnitude of the problem becomes devastating because all the plants have the same characteristics and, thus, are susceptible to the same diseases. If genetically diverse crops are planted or crops are rotated from year to year, this problem is not as great.

Because farm equipment is expensive, farmers tend to specialize in a few crops. This means that the same crop may be planted in the same field several years in a row. This lack of crop rotation may deplete certain essential soil nutrients, thereby requiring special attention be paid to soil chemistry. In addition, planting the same crop repeatedly encourages the growth of insect and fungus pest populations because they have a huge food supply at their disposal. This requires the frequent use of insecticides and fungicides or other methods of pest control.

Even though there are problems associated with mechanized, monoculture agriculture, it has greatly increased the amount of food available to the world over the past 100 years. Yields per hectare of land being farmed have increased over much of the world, particularly in the developed world, which includes the United States. (See figure 15.4.) This increase has come about because of improved varieties of crops, irrigation, better farming methods, the use of agricultural chemicals, more efficient machines, and the use of energy-intensive as opposed to labor-intensive technology.

Throughout the 1950s, 1960s, and 1970s, the introduction of new plant varieties and farming methods resulted in increased agricultural production worldwide. This has been called the **Green Revolution.** Both the developed world, which uses highly mechanized farming methods, and the developing world, where labor-intensive farming is typical, have benefited from these advances, and food

Figure 15.4 **Increased Yields Resulting from Modern Technology** The increased crop yields in the United States and many other parts of the world are the result of a combination of factors, including the development of high-yielding varieties, changed agricultural methods, the application of fertilizers and pesticides, and more efficient machinery.

Source: Data from U.S. Department of Agriculture, *Agriculture Statistics*.

production has increased significantly. The Green Revolution has also had some negative effects. For example, many modern varieties of plants require fertilizer and pesticides that the traditional varieties they replaced did not need. In addition, many of the crops require higher amounts of water and increase the demand for irrigation. Farmers also have become more dependent on the industries that provide the specialized seeds. When the positives and negatives are balanced, however, the end result is that food production per hectare has increased. This has not solved the world's food problem, however, because the population of the world continues to increase and more food is needed.

Fossil Fuel Versus Muscle Power

Mechanized agriculture has substituted the energy stored in petroleum products for the labor of humans. For example, in the United States in 1913, it required 135 hours of labor to produce 2500 kilograms (5500 pounds) of corn. In 1980, it required only about 15 hours of labor to produce

the same amount of corn. The energy supplied by petroleum products replaced the equivalent of 120 hours of labor. Energy is needed for tilling, planting, harvesting, and pumping irrigation water. The manufacture of fertilizer and pesticides also requires the input of large amounts of fossil fuels, both as a source of energy for the industrial process and as raw material from which these materials are made. For example, about 5 metric tons of fossil fuel are required to produce about 1 metric ton of fertilizer. In addition, the pesticides used in mechanized agriculture require the use of oil as a raw material. Since the developed world depends on oil to run machines and manufacture fertilizer and pesticides to support its agriculture, any change in the availability or cost of oil will have a major impact on the world's ability to feed itself.

The Impact of Fertilizer

Various experts estimate that approximately 25 percent of the world's crop yield can be directly attributed to the use

Estimated Growth in Worldwide Fertilizer Use, 1960–2020

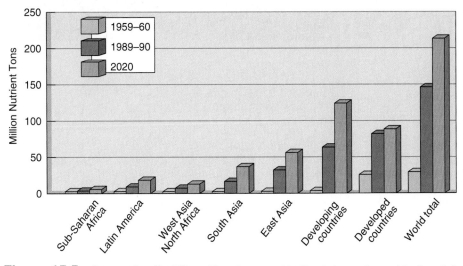

Figure 15.5 **Increasing Fertilizer Use** The use of fertilizer is increasing rapidly. Growth in use will be most rapid in the developing countries over the next 20 years.

Source: Balu Bumb and Carlos Baanante, *World Trends in Fertilizer Use and Projections to 2020,* 2020 Brief No. 38 (International Food Policy Research Institute, Washington, D.C., 1996), table 1.

a source of energy. Since these organisms serve as important links in the carbon and nitrogen cycles, the presence of organic matter is important to their function. Thus, total dependency on chemical fertilizers usually reduces the amount of organic matter and can change the physical, chemical, and biological properties of the soil.

As water moves through the soil, it dissolves soil nutrients (particularly nitrogen compounds) and carries them into streams and lakes, where they may encourage the growth of unwanted plants and algae. This is particularly true when fertilizers are applied at the wrong time of the year, just before a heavy rain, or in such large amounts that the plants cannot efficiently remove them from the soil before they are lost. These ideas are covered in greater detail in chapter 16 on water management.

of chemical fertilizers. The use of fertilizer has increased significantly over the last few decades and is projected to increase even more. (See figure 15.5.) However, since fertilizer production relies on energy from fossil fuels, the price and availability of chemical fertilizers are strongly influenced by world energy prices. If the price of oil increases, the price of fertilizer goes up, as does the cost of food. This is felt most acutely in parts of the world where money is in short supply, since the farmers are unable to buy fertilizer and crop yields fall accordingly.

Fertilizers are valuable because they replace the soil nutrients removed by plants. Some of the chemical building blocks of plants, such as carbon, hydrogen, and oxygen, are easily replaced by carbon dioxide from the air and water from the soil, but others are less easily replaced. The three primary soil nutrients often in short supply are nitrogen, phosphorus, and potassium compounds. They are often referred to as **macronutrients** and are the common ingredient of chemical fertilizers. Their replacement is important because when the crop is harvested, the chemical elements that are a part of the crop are removed from the field. Since many of those elements originated from the soil, they need to be replaced if

another crop is to be grown. Certain other elements are necessary in extremely small amounts and are known as **micronutrients.** Examples are boron, zinc, and manganese. As an example of the difference between macronutrients and micronutrients, harvesting 1 metric ton of potatoes removes 10 kilograms (22 pounds) of nitrogen (a macronutrient) but only 13 grams (0.03 pounds) of boron (a micronutrient). When the same crop is grown repeatedly in the same field, certain micronutrients may be depleted, resulting in reduced yields. These necessary elements can be returned to the soil in sufficient amounts by incorporating them into the fertilizer the farmer applies to the field.

Although chemical fertilizers replace inorganic nutrients, they do not replace soil organic matter. Organic material is important because it modifies the structure of the soil, preventing compaction and maintaining pore space, which allows water and air to move to the roots. The decomposition of organic matter produces humus, which helps to maintain proper soil chemistry because it tends to loosely bind many soil nutrients and other molecules and modifies the pH so that nutrients are not released too rapidly. Soil bacteria and other organisms use organic matter as

Agricultural Chemical Use

In addition to chemical fertilizers, mechanized monoculture requires large amounts of other agricultural chemicals, such as pesticides, growth regulators, and preservatives. These chemicals have specific scientific names but are usually categorized into broad groups based on their effects. A **pesticide** is any chemical used to kill or control populations of unwanted fungi, animals, or plants, often called **pests.** The term *pest* is not scientific but refers to any organism that is unwanted. Insects that feed on crops are pests, while others, such as bees, are beneficial for pollinating plants. Unwanted plants are generally referred to as **weeds.**

Pesticides can be subdivided into several categories based on the kinds of organisms they are used to control. **Insecticides** are used to control insect populations by killing them. Unwanted fungal pests that can weaken plants or destroy fruits are controlled by **fungicides.** Mice and rats are killed by **rodenticides,** and plant pests (weeds) are controlled by **herbicides.** Since pesticides do not kill just pests but can kill a large variety of living things, including humans, these

Environmental Close-Up

DDT—A Historical Perspective

The first synthetic organic insecticide to be used was DDT [1,1,1-trichloro-2,2-bis-(p-chlorophenyl)ethane] (see the figure). DDT was originally thought to be the perfect insecticide. It was inexpensive, long-lasting, relatively harmless to humans, and very deadly to insects. After its discovery, it was widely used in agriculture and to control disease-carrying insects. During the first 10 years of its use (1942–52), DDT is estimated to have saved 5 million lives, primarily because of its use in controlling disease-carrying mosquitoes.

However, DDT had several drawbacks. Although most of these drawbacks are not unique to DDT, they were first recognized with DDT because it was the first widely used insecticide. One problem involved the development of resistance in insect populations that were repeatedly subjected to spraying by DDT. Another was the fact that it affected many nontarget organisms, not just the disease-carrying insects that were the original targets. This was a particular problem because DDT is a persistent chemical, which is another major drawback. In temperate regions of the world, DDT has a half-life (the amount of time required for half of the chemical to decompose) of 10 to 15 years. This means that if 1000 kilograms (2200 pounds) of DDT were sprayed over an area, 500 kilograms (1100 pounds) would still be present in the area 10 to 15 years later; 30 years from the date of application, 250 kilograms (550 pounds) would still be present. The half-life of DDT varies depending on soil type, temperature, the kinds of soil organisms present, and other factors. In tropical parts of the world, the half-life may be as short as six months. An additional complication is that persistent pesticides may break down into products that are still harmful. Furthermore, because it is persistent, DDT tends to accumulate and reach higher concentrations in older animals and in animals at higher trophic levels.

When these effects were recognized, the United States and many other developed countries banned the use of DDT. A key force in bringing about this change in thinking was the book *Silent Spring*, by Rachel Carson. (See Environmental Close-Up: Naturalist Philosophers in chapter 2.) The results of this ban are clearly seen in the reduced levels of DDT in the environment and in the recovery of populations of such organisms as eagles, cormorants, and pelicans.

Because of these problems, the World Wildlife Fund asked delegates at a United Nations Environment Programme conference on persistent organic pollutants in 1999 to ban DDT from use throughout the world because it was showing up worldwide in the bodies of many kinds of animals that were great distances from sources of DDT and was present in the breast milk of women where DDT is used. This resulted in a great deal of protest from public health people concerned about controlling malaria, since DDT is still widely used in many parts of the world, where it is sprayed on the walls of houses to kill mosquitoes that spread malaria. The use of DDT is still effective and is much less expensive than materials that would be substituted for it. They feared that the total ban of DDT would result in less effective control of mosquitoes and increased deaths from malaria. Where the risk of dying from malaria is high and the risk of possible health effects from exposure to DDT is low, it makes sense to use an inexpensive, effective material such as DDT. In countries where malaria and other serious insect-transmitted diseases are rare (most of the developed world), it is possible to eliminate the use of DDT. Ultimately, continued use of DDT was sanctioned for the control of malaria and other diseases transmitted by insects, and it continues to be used for these purposes today.

The chemical structure of DDT.

---

chemicals might be more appropriately called **biocides.** They kill many kinds of living things. A perfect pesticide is one that kills or inhibits the growth of only the specific pest organism causing a problem. The pest is often referred to as the **target organism.** However, most pesticides are not very specific and kill many **nontarget organisms** as well. For example, most insecticides kill both beneficial and pest species, rodenticides kill other animals as well as rodents, and most herbicides kill a variety of plants, both pests and nonpests.

Many of the older pesticides were very stable and remained active for long periods of time. These are called **persistent pesticides.** Pesticides that break down quickly are called **nonpersistent pesticides.**

Insecticides

If insects are not controlled, they consume a large proportion of the crops produced by farmers. In small garden plots, insects can be controlled by manually removing them and killing them. However, in large fields, this is not practical, so people have sought other ways to control pest insects.

Nearly 3000 years ago, the Greek poet Homer mentioned the use of sulfur to control insects. For centuries, it was known that natural plant products could repel or kill insect pests. Plants with insect-repelling abilities were interplanted with crops to help control the pests. Nicotine from tobacco, rotenone from tropical legumes, and pyrethrum from chrysanthemums were extracted and used to control insects. In fact, these compounds are still used today. However, because plant products are difficult to extract and apply and have short-lived effects, other compounds were sought. In 1867, the first synthetic inorganic insecticide, Paris green, was formulated. It was a mixture of acetate and arsenide of copper and was used to control Colorado potato beetles.

In addition, many insects harm humans because they spread diseases, such as sleeping sickness, bubonic plague, and malaria. Mosquitoes are known to carry over 30 diseases harmful to humans.

Currently, the World Health Organization estimates that between 300 million and 500 million people have malaria and 1.1 million people die of the disease each year. Most of these deaths are children. Malaria is one of the top five causes of death in children. The discovery of chemicals that could kill insects was celebrated as a major advance in the control of disease and the protection of crops.

DDT was the first synthetic organic insecticide produced. (See Environmental Close-Up: DDT—A Historical Perspective on page 334.) Since then, over 60,000 different compounds that have potential as insecticides have been synthesized. However, most of these have never been put into production because cost, human health effects, or other drawbacks make them unusable. Several categories of these compounds have been developed. Three that are currently used are chlorinated hydrocarbons, organophosphates, and carbamates.

Chlorinated Hydrocarbons

Chlorinated hydrocarbons are a group of pesticides of complex, stable structure that contain carbon, hydrogen, and chlorine. DDT was the first such pesticide manufactured, but several others have been developed. Other chlorinated hydrocarbons are chlordane, aldrin, heptachlor, dieldrin, and endrin. It is not fully understood how these compounds work, but they are believed to affect the nervous systems of insects, resulting in their death.

One of the major characteristics of these pesticides is that they are very stable chemical compounds. This is both an advantage and a disadvantage. They can be applied once and be effective for a long time. However, since they do not break down easily, they tend to accumulate in the soil and in the bodies of animals in the food chain. Thus, they affect many nontarget organisms, not just the original target insects.

Because of their negative effects, most of the chlorinated hydrocarbons are no longer used in many parts of the world. DDT, aldrin, dieldrin, toxaphene, chlordane, and heptachlor have been banned in the United States and many other developed countries. However, many developing countries still use chlorinated hydrocarbons for insect control to protect crops and public health. Because of their persistence and continued use in many parts of the world, chlorinated hydrocarbons are still present in the food chain, although the level of contamination has dropped. These molecules continue to enter parts of the world where their use has been banned through the atmosphere and as trace contaminants of imported products.

Organophosphates and Carbamates

Because of the problems associated with persistent insecticides, nonpersistent insecticides that decompose to harmless products in a few hours or days were developed. However, like other insecticides, these are not species-specific; they kill beneficial insects as well as harmful ones. Although the short half-life prevents the accumulation of toxic material in the environment, it is a disadvantage for farmers, since more frequent applications are required to control pests. This requires more labor and fuel and, therefore, is more expensive.

Both **organophosphates** and **carbamates** work by interfering with the ability of the nervous system to conduct impulses normally. Under normal conditions, a nerve impulse is conducted from one nerve cell to another by means of a chemical known as a neurotransmitter. One of the most common neurotransmitters is acetylcholine. When this chemical is produced at the end of one nerve cell, it causes an impulse to be passed to the next cell, thereby transferring the nerve message. As soon as this transfer is completed, an enzyme known as cholinesterase destroys acetylcholine, so the second nerve cell in the chain is stimulated for only a short time. Organophosphates and carbamates interfere with cholinesterase, preventing it from destroying acetylcholine. This results in nerve cells being continuously stimulated, causing uncontrolled spasms of nervous activity and uncoordination that result in death.

Although these pesticides are less persistent in the environment than are chlorinated hydrocarbons, they are generally much more toxic to humans and other vertebrates because these insecticides affect their nerve cells as well. Persons who apply such pesticides must use special equipment and should receive special training because improper use can result in death. Since organophosphates interfere with cholinesterase more strongly than do carbamates, they are considered more dangerous and, for many applications, have been replaced by carbamates.

Common organophosphates are malathion, parathion, and diazinon. Malathion is widely used for such projects as mosquito control, but parathion is a restricted organophosphate because of its high toxicity to humans. Diazinon is widely used in gardens. Carbaryl, propoxur, and aldicarb are examples of carbamates. Carbaryl (Sevin) is widely used in home gardens and in agriculture to control many kinds of insects. Propoxur is not used on crops but is used to control insects around homes and farms. Aldicarb is a restricted-use insecticide that is used primarily on cotton, soybeans, and peanuts. It has been associated with groundwater contamination and has been discontinued for some uses, such as control of insects in potatoes.

Herbicides

Herbicides are another major class of chemical control agents. In fact, about 60 percent of the approximately 440 million kilograms (about 1 billion pounds) of pesticides used in U.S. agriculture are herbicides. They are widely used to control unwanted vegetation along power-line rights-of-way, railroad rights-of-way, and highways, as well as on lawns and cropland, where they are commonly referred to as weed killers.

Weeds are plants we do not want to have growing in a particular place. Weed control is extremely important for agriculture since weeds take nutrients and water from the soil, making them unavailable to the crop species. In addition, weeds may shade the crop species and prevent it from getting the sunlight it needs for rapid growth. At harvest time, weeds reduce the efficiency of harvesting machines. Also, weeds generally must be sorted from the crop before it can be sold, which adds to the time and expense of harvesting.

Traditionally, farmers have expended much energy trying to control weeds.

GLOBAL PERSPECTIVE

Economic Development and Food Production in China

When the economic conditions of a people improve, the change is reflected in the quality of their diet. In particular, meat becomes a bigger part of the diet. Therefore, people feed some of their grains to animals and consume animal products such as meat, fish, eggs, and butter. They eat more vegetables and use more edible oils. In short, they move a bit higher up the food chain.

A sizable percentage of China's nearly 1.3 billion citizens are moving up the food chain. In the early 1980s, the typical urban Chinese diet consisted of rice, porridge, and cabbage. By the middle 1990s, the diet had dramatically changed to include meat, eggs, or fish at least once a day.

China has about 21 percent of the world's population but only 7 percent of the arable land; in other words, it has less arable land than the United States but nearly five times the population. China's farm sector is simply unable to keep up with the surging demand for what people regard as better food.

Total meat consumption in China is growing by 10 percent a year; feed for animal consumption is growing by 15 percent. Demand for poultry, which requires 2 to 3 kilograms of feed per kilogram of bird, has doubled in five years. China's new diet will make it more dependent on the United States, Canada, and Australia for feed grains. By 1997, China had gone from being a net exporter of grain to importing 16 million metric tons. The switch in corn is even more dramatic. As recently as 1995, China was the second-largest corn exporter in the world. But with the chickens and pigs eating so much corn, in one year, China moved from exporting 12 million metric tons of corn to importing 4 million metric tons.

To support its increased agricultural needs, China has become the world's largest importer of fertilizer. In addition, the government is converting some marginal lands in the north for agriculture. Even so, its farmland is declining by at least 0.5 percent per year because of the loss of fertile, multiple-cropped farmland in the southern coastal provinces.

As in much of the world, irrigation is responsible for a major portion of agricultural productivity in China. In the Yellow River Valley's large grain belt, irrigation projects tripled crop yields during the 1950s and 1960s, resulting in severe overpumping of groundwater. Today, the water table is falling, aquifers are vanishing, and farmers now have to compete with industry and households for water.

China is only a part—but certainly the biggest part—of a larger story. What's true for China is also happening throughout much of Asia. Large, populous countries such as Indonesia, Thailand, and the Philippines are rapidly urbanizing. They are gaining purchasing power and losing farmland, and the people are adding more animal proteins and processed foods to their diets. Given the magnitude of the changes, the world's ability to feed itself in the future is difficult to predict.

KFC and McDonald's in Beijing. Demand for poultry and beef is soaring.

benefit in controlling the size of the insect population. However, the surviving individuals that are tolerant of the insecticide will constitute the majority of the breeding populations. Since these individuals possess genetic characteristics for tolerating the insecticide, so will many of their off-spring. Therefore, in the next generation, the number of individuals able to tolerate the insecticide will increase, and the second use of the insecticide will not be as effective as the first. Since some species of insect pests can produce a new generation each month, this process of selecting individuals capable of tolerating the insecticide can result in resistant populations in which 99 percent of the individuals are able to tolerate the insecticide within five years. As a result, that particular insecticide is no longer as effective in controlling insect pests, and increased dosages or

GLOBAL PERSPECTIVE

Contaminated Soils in the Former Soviet Union

The independent republics of the former Soviet Union have inherited many kinds of environmental problems generated by the failed policies of the Soviet government. One of these policies was a push for increased agricultural production. New land was brought under cultivation, irrigation projects were developed, and extensive use of fertilizer and pesticides stimulated production.

This single-minded approach has resulted in serious environmental damage. In many areas, the soil has been contaminated by excessive fertilizer and pesticide use, which has contaminated groundwater and food products. It is estimated that 30 percent of the food produced throughout the former Soviet Union is contaminated and should not be eaten. Forty-two percent of all baby food is contaminated by nitrates and pesticides. Latvian ecologists estimate that excessive pesticide use currently results in 14,000 deaths per year and that 700,000 people become ill from pesticides throughout the former Soviet Union annually.

One problem pesticide is DDT. It was officially banned in the Soviet Union in the 1970s, but the drive to improve agricultural production allowed its continued use. Waivers to the ban were allowed, or DDT was simply renamed by government bureaucrats and used under its new label. It is estimated that 10 million hectares (25 million acres) of agricultural land in the former Soviet Union are contaminated with DDT.

The widespread use of agricultural chemicals has also affected the groundwater. Persistent pesticides, nitrates, and heavy metals have entered the groundwater, which many rely on for drinking water. Some areas have cadmium levels three to 10 times higher than the amount permitted.

It appears that the fall of communism in the former Soviet Union did not solve all the problems of the people. These environmental problems are surfacing after years of secrecy and denial under the shortsighted, production-oriented agricultural policy of the failed political system.

Table 15.1 Insecticide Resistance of Houseflies to the Insecticide Cyfluthrin

	Houseflies from a Susceptible Population	Houseflies from a Poultry Facility Population		
Concentration (ng/cm²)	8.3	8.3 (1×)	83 (10×)	830 (100×)
Percent survival	0	100	90	38

Source: Data from Scott, Jeffrey G., et. al., *"Insecticide Resistance in Houseflies from Caged-Layer Poultry Facilities,"* *Pest Management Science* 56 (2000): 147–53 (2000).

more frequent spraying may be necessary. Over 500 species of insects have populations resistant to insecticides.

For example, a recent study of insecticide resistance of houseflies in poultry facilities used for egg production shows that these populations of houseflies are resistant to a wide variety of insecticides compared to a nonresistant test population. Poultry facilities are an excellent situation for developing resistance since the flies spend their entire lives within the facility and are subjected to repeated applications of insecticide. Table 15.1 shows the effect of the insecticide, Cyfluthrin, on resistance in a particular housefly population. The standard dose that killed susceptible houseflies did not kill the resistant population. Even at a dosage of 100 times the standard dose, nearly 40 percent of the houseflies in the resistant population survived.

Effects on Nontarget Organisms

Most pesticides are not specific and kill beneficial species as well as pest species. With herbicides, this is usually not a problem because an herbicide is chosen that does not harm the desired crop plant, and generally all other plants are competing pests. However, with insecticides, several problems are associated with the effects on nontarget organisms. The use of insecticides can harm populations of birds, mammals, and harmless or beneficial insects. In general, insecticides that harm vertebrates are restricted in their use. However, it is difficult to apply insecticides in such a way that only the harmful species are affected. When beneficial insects are killed by insecticides, pesticide use can be counterproductive. If an insecticide kills predator and parasitic insects that normally control the pest insects, there are no natural checks to control the population growth of the pest species. Additional applications of insecticides are necessary to prevent the pest population from rebounding to levels even higher than the initial one. Once the decision is made to use pesticides, it often becomes an irreversible tactic, because stopping their use would result in rapid increases in pest populations and extensive crop damage.

An associated problem is that the use of insecticides may change the population structures of the species present so that a species that was not previously a problem becomes a serious pest. For example, when synthetic organic insecticides came into common use with cotton in the 1940s, the insect parasites and predators were eliminated, and the bollworm and tobacco budworm became major pests. In the mid-1990s, a similar situation developed when repeated use of malathion removed predator insects and allowed beet army worms to become a major pest. The repeated use of insecticides caused a different pest problem to develop.

Human Health Concerns

Short-term and long-term health effects to persons applying pesticides and the public that consumes pesticide residues in food are also concerns. If properly applied, most pesticides can be used with little danger to the applicator. However, in many cases, people applying pesticides are unaware of how they work and what precautions should be used in their application. In many parts of the world, farmers may not be able to read the caution labels on the packages or do not have access to the protective gear specified for use with the pesticide. Therefore, many incidences of acute poisoning occur each year. In most cases, the symptoms disappear after a period free from exposure. Estimates of the number of poisonings are very difficult to obtain since many go unreported, but in the United States, pesticide poisonings requiring medical treatment are in the thousands per year. The World Health Organization estimates that each year, there are between 1 million and 5 million acute pesticide poisonings and that about 20,000 deaths occur from pesticide poisonings.

For most people, however, the most critical health problem related to pesticide use is inadvertent exposure to very small quantities of pesticides. Many pesticides have been proven to cause mutations, produce cancers, or cause abnormal births in experimental animals. Studies of farmers who were occupationally exposed to pesticides over many years show that they have higher levels of certain kinds of cancers than the general public. There also are questions about the effects of chronic, minute exposures to pesticide residues in food or through contamination of the environment.

Figure 15.10 **Changes in Pesticide Use** The use of pesticides is greatest in the developed world, comprising about 67 percent of the total. However, all parts of the world except Europe are experiencing increased use of pesticides. The use in Africa is low and relatively stable.

Source: Data from Food and Agriculture Organization of the United Nations.

Although the risks of endangering health by consuming tiny amounts of pesticides are very small compared to other risk factors, such as automobile accidents, smoking, or poor eating habits, many people find pesticides unacceptable and seek to prohibit their sale and use. The U.S. Environmental Protection Agency requires careful studies of the effectiveness and possible side effects of each pesticide licensed for use. It has banned several pesticides from further use because new information suggests that they are not as safe as originally thought. For example, it banned the use of dinoseb, an herbicide, because tests by the German chemical company Hoechst AG indicated that dinoseb causes birth defects in rabbits. Other studies indicate that dinoseb causes sterility in rats. Similarly, in 1987, Velsicol Chemical Corporation signed an agreement with the EPA to stop producing and distributing chlordane in the United States. Chlordane had been banned previously for all applications except for termite control. After it was shown that harmful levels of chlordane could exist in treated homes, its use was finally curtailed. Also in 1987, the Dow Chemical Company announced that it would stop producing the controversial

herbicide 2,4,5-T. In both of these cases, new pesticides were developed to replace the older types, so that discontinuing them did not result in an economic hardship for farmers or other consumers.

Finally, during the 1990s, the EPA required that many pesticides be reregistered. During this process, manufacturers needed to justify the continued production of the product. In many cases, companies simply did not want to spend the money to go through the registration process; therefore, the product was withdrawn from the market. In other cases, the variety of uses was modified. For example, chlorpyrifos, one of the most widely used insecticides, had its use altered. Through negotiations with the producers, its use in homes and schools will be eliminated, while it will still be used in most agricultural settings.

Why Are Pesticides So Widely Used?

Figure 15.10 shows changes in the use of pesticides throughout the world. Use is projected to continue to increase significantly. If pesticides have so many draw-

Industrial Production of Livestock

Over the last couple of decades, the trend has been toward the concentration of livestock production into large, factorylike operations. Dairy cattle, beef cattle, pigs, and chickens for both eggs and meat are raised in concentrations that are causing environmental problems. Many of these commercial operations feed thousands of animals, which produce equally large amounts of manure. (See photo.) Although such livestock-raising operations are supposed to contain and dispose of their animal waste properly, problems often result. Holding tanks or lagoons may fail and release animal waste into local streams.

One of the standard ways of dealing with the waste is to use it as fertilizer. This is a relatively simple and effective technology. However, the amount of waste produced is often greater than the land suitable and available for application of manure as fertilizer. The manure may not be applied on all crops, cannot be applied when the ground is frozen, and may not be applied at other times of the year for a variety of reasons. However, the animals produce manure every day.

Groundwater contamination is an additional problem. When concentrated animal wastes are in direct contact with the soil in feedlots or unlined sewage lagoons, or when they are applied to fields at too great a concentration, they can penetrate the soil and enter the groundwater. Other issues involve odors and dusts associated with large industrial livestock operations.

An additional concern related to this form of agriculture is the frequent use of antibiotics. Raising large numbers of animals together creates a ready environment for the spread of disease. Consequently, antibiotics are used regularly to prevent the spread of disease and foster faster growth of the animals. When antibiotics are used so freely, it is highly likely that the disease organisms being targeted will become resistant to the antibiotics. Since some of these disease organisms (particu-

A large-scale chicken operation outside of Raleigh, North Carolina, an area that has a lot of chicken and turkey farms.

larly those of pigs and chickens) can spread to humans, resistant disease organisms in livestock can become a human health problem as well.

There are also economic considerations. Since small farmers cannot compete with the large commercial operations, many have gone out of business or had to switch to other agricultural activities to replace the loss of income from raising livestock. While several organizations are concerned about the ethics, health implications, and social implications of such large livestock operations, the average consumer in the supermarket knows little about where the meat products originated or under what conditions the animals were raised.

---

backs, why are they used so extensively? There are three primary reasons. First, the use of pesticides has increased, at least in the short term, the amount of food that can be grown in many parts of the world. In the United States, pests are estimated to consume 33 percent of the crops grown. On a worldwide basis, pests consume approximately 35 percent of crops. This represents an annual loss of $18.2 billion in the United States alone. Farmers, grain-storage operators, and the food industry continually seek to reduce this loss. A retreat from dependence on pesticides would certainly reduce the amount of food produced. Agricultural planners in most countries are not likely to suggest changes in pesticide use that would result in malnutrition and starvation for many of their inhabitants.

The economic value of pesticides is the second reason they are used so extensively. The cost of pesticides is more than offset by increased yields and profits for the farmer. In addition, the production and distribution of pesticides is big business. Companies that have spent millions of dollars developing a pesticide are going to argue very strongly for its continued use. Since farmers and agrochemical interests have a powerful voice in government, they have successfully lobbied for continued use of pesticides.

A third reason for extensive pesticide use is that many health problems are currently impossible to control without insecticides. This is particularly true in areas of the world where insect-borne diseases would cause widespread public health consequences if insecticides were not used.

Alternatives to Conventional Agriculture

Before the invention of synthetic fertilizers, herbicides, fungicides, and other agrochemicals, animal manure and crop rotation provided soil nutrients; a mixture of crops prevented regular pest problems; and manual labor killed insects and weeds. With the development of mechanization, larger areas could be farmed, draft animals were no longer needed, and many farmers changed from mixed agriculture, in which animals were an important ingredient, to monoculture. Chemical fertilizers replaced manure as a source of soil nutrients, and crop rotation was no

---

CHAPTER 15   Agricultural Methods and Pest Management

Environmental Close-Up

Food Additives

Food additives are chemicals added to food before its sale. They have several purposes:

1. To prolong the storage life of the food.
2. To make the food more attractive by adding color or flavor.
3. To modify nutritive value.

Many kinds of molecules are added to foods to prolong shelf life. Calcium propionate is added to baked goods to retard the growth of air-borne spores of molds and bacteria on the food, which can spoil it. BHT, TBHQ, and other commonly seen alphabetic mixtures have a similar function.

Sometimes additives are used just to make the food appear more attractive. For years, red dye II was used to color a variety of foods. Its use was discontinued when it was found to be carcinogenic. Other food colorings are still widely used to increase appeal to the consumer. Many kinds of artificial flavors are added to products as well. Commonly used flavor enhancers are monosodium glutamate, table salt, and citric acid.

Other food additives are used to modify the nutritive value of the food product. Iodine in table salt is a good example. Iodine is a trace element required for proper thyroid functioning. Individuals suffering from a lack of iodine often develop an enlargement of the thyroid gland known as goiter. The addition of iodine to table salt has eliminated goiter in the United States. Most cereals and baked goods and many other products have various vitamins and minerals added to improve their nutritive value. Some additives, such as Nutrasweet, reduce the calories while giving the consumer the sensation of tasting something sweet.

Some additives are an unavoidable residue of some step of the food production process and could more properly be called contaminants. Pesticide residues are an example. Pesticides are used to grow foods, but they are also used to eliminate pests in the storing, processing, and transportation steps of food production.

Diethylstilbestrol (DES) was at one time used in the poultry industry to produce fatter birds. Because there were indications that DES is carcinogenic, the U.S. Food and Drug Administration banned the use of DES in chickens and declared that it was potentially hazardous to humans. Further studies were conducted to determine if DES was safe to use in raising beef. Cattle gain weight more rapidly with DES in their feed. Because DES was eventually linked to breast cancer in women, it has been banned from all animal feed use in the United States and Europe.

---

longer as important since hay and grain were no longer grown for draft animals and cattle. The larger fields of crops such as corn, wheat, and cotton presented opportunities for pest problems to develop, and chemical pesticides were used to "solve" this problem.

Today, many people feel current agricultural practices are not sustainable and that we should look for ways to reduce reliance on fertilizer and pesticides, while ensuring good yields and controlling the pests that compete with us for the food that we raise. Several alternative approaches have somewhat different but overlapping goals. Some of the terms used to describe these approaches are: *alternative agriculture, sustainable agriculture,* and *organic agriculture*. **Alternative agriculture** is the broadest term. It includes all nontraditional agricultural methods and encompasses sustainable agriculture, organic agriculture, alternative uses of traditional crops, alternative methods for raising crops, and producing crops for industrial use. The **sustainable agriculture** movement maintains that current practices are degrading natural resources and seeks methods to produce adequate, safe food in an economically viable manner while enhancing the health of agricultural land and related ecosystems. **Organic agriculture** advocates avoiding the use of chemical fertilizers and pesticides in the production of food, thus preventing damage to related ecosystems and the food-consuming public. The definition of organic agriculture by the National Organic Standards Board established by the 1990 Organic Foods Production Act points out a central theme of these three movements.

Organic agriculture is an ecological production management system that promotes and enhances biodiversity, biological cycles, and soil biological activity. It is based on minimal use of off-farm inputs and management practices that restore, maintain, and enhance ecological harmony.

See Issues—Analysis: What Are Organic Foods?, for more detailed listing of the requirements for foods that are labeled organic.

Techniques for Protecting Soil and Water Resources

In order to reduce negative impacts of conventional agriculture while making a profit, farmers must deal with the twin problems of protecting the quality and fertility of the soil while controlling pests of their crops. A wide variety of techniques can be used to reduce the negative impact of conventional agricultural activity and maintain economic viability for the farmer.

Conventional farming practices have several negative effects on soil and water. Soil erosion is a problem throughout the world. (See chapter 14.) However, two other problems are also important: compaction of the soil and reduction in soil organic matter. Several changes in agricultural production methods can help to reduce these problems. Reducing the number of times farm equipment travels over the soil will reduce the degree of compaction. Leaving crop residue on the soil and incorporating it into the soil reduces erosion and increases soil organic matter. In addition, introducing organic matter into the soil makes compaction less likely.

Conventional agriculture has several negative impacts on watersheds: fertilizer runoff stimulates aquatic growth and degrades water resources; pesticides can accumulate in food chains; and groundwater resources can be contaminated by fertilizer, pesticides, or animal waste. Reducing or eliminating these sources of contamination would enhance ecological harmony and reduce a threat to human health. Fertilizer runoff can be lessened by reducing the amount of fertilizer applied and the conditions under which it is applied. Applying fertilizer as plants need it will ensure that more of it is taken up by plants and less runs off. Increased organic matter in the soil also tends to reduce runoff. More careful selection, timing, and use of pesticides would decrease the extent to which these materials become environmental contaminants.

Precision agriculture is a new technique that addresses many of these concerns. With modern computer technology and geographic information systems, it is now possible, based on the soil and topography, to automatically vary the chemicals applied to the crop at different places within a field. Thus, less fertilizer is used, and it is used more effectively.

True organic agriculture that uses neither chemical fertilizers nor pesticides is the most effective in protecting soil and water resources from these forms of pollution, but it requires several adjustments in the way in which farming is done. Crop rotation is an effective way to enhance soil fertility, reduce erosion, and control pests. The use of nitrogen-fixing legumes, such as clover, alfalfa, beans, or soybeans, in crop rotation increases soil nitrogen but places other demands on the farmer. For example, it typically requires that cattle be a part of the farmer's operation in order to make use of forage crops and provide organic fertilizer for subsequent crops. Crop rotation also requires a greater investment in farm machinery, since certain crops require specialized equipment. Also, the raising of cattle requires additional expenditures for feed supplements and veterinary care. Critics say that organic farming cannot produce the amount of food required for today's population and that it

can be economically successful only in specific cases. Proponents disagree and stress that when the hidden costs of soil erosion and pollution are included, organic agriculture or some modification of it is a viable alternative approach to conventional means of food production. Furthermore, organic farmers are willing to accept lower yields because they do not have to pay for expensive chemical fertilizers and pesticides. In addition, organic farmers often receive premium prices for products that are organically grown. Thus, even with lower yields, they can still make a profit.

Integrated Pest Management

Integrated pest management uses a variety of methods to control pests rather than relying on pesticides alone. Integrated pest management is a technique that depends on a complete understanding of all ecological aspects of the crop and the particular pests to which it is susceptible to establish pest control strategies. It requires information about the metabolism of the crop plant, the biological interactions between pests and their predators or parasites, the climatic conditions that favor certain pests, and techniques for encouraging beneficial insects. It may involve the selective use of pesticides. Much of the information necessary to make integrated pest management work goes beyond the knowledge of the typical farmer. The metabolic and ecological studies necessary to pinpoint weak points in the life cycles of pests can usually be carried out only at universities or government research institutions. These studies are expensive and must be completed for each kind of pest, since each pest has a unique biology. Once a viable technique has been developed, an educational program is necessary to provide farmers with the information they need in order to use integrated pest management rather than the "spray and save" techniques that they used previously and that pesticide salespeople continually encourage.

Several methods are employed in integrated pest management. These include disrupting reproduction, using beneficial organisms to control pests, developing

resistant crops, modifying farming practices, and selectively using pesticides.

Disrupting Reproduction

In some species of insects, a chemical called a **pheromone** is released by females to attract males. Males of some species of moths can detect the presence of a female from a distance of up to 3 kilometers (nearly 2 miles). Since many moths are pests, synthetic odors can be used to control them. Spraying an area with the pheromone confuses the males and prevents them from finding females, which results in a reduced moth population the following year. In a similar way, a synthetic sex attractant molecule known as Gyplure is used to lure gypsy moths into traps, where they become stuck. Since the females cannot fly and the males are trapped, the reproductive rate drops, and the insect population may be controlled.

Another technique that reduces reproduction is male sterilization. In the southern United States and Central America, the screwworm fly weakens or kills large grazing animals, such as cattle, goats, and deer. The female screwworm fly lays eggs in open wounds on these animals, where the larvae feed. However, it was discovered that the female mates only once in her lifetime. Therefore, the fly population can be controlled by raising and releasing large numbers of sterilized male screwworm flies. Any female that mates with a sterile male fails to produce fertilized eggs and cannot reproduce. In Curaçao, an island 65 kilometers (40 miles) north of Venezuela, a program of introducing sterile male screwworm flies eliminated this disease from the 25,000 goats on the island. In parts of the southwestern United States, the sterile male technique has also been very effective. The screwworm fly has been eliminated from the United States and northern Mexico, and much of Central America may become free of them as well. In 1990, sterile males were released in Libya to begin eliminating screwworm flies that had been introduced with a South American cattle shipment.

During an epidemic of Mediterranean fruitflies in southern California and northern Mexico in the early 1980s, a similar technique was employed. Unfortunately, the X-ray technique used to sterilize the males was ineffective, and most of the flies released were not sterile, which made the problem worse rather than better. Pesticides were eventually used to control the fruitflies. Recent concern about the Mediterranean fruitfly in California has resulted in the controversial aerial spraying of malathion.

Using Beneficial Organisms to Control Pests

The manipulation of predator-prey relationships can also be used to control pest populations. For instance, the ladybird beetle, commonly called a ladybug, is a natural predator of aphids and scale insects. (See figure 15.11a.) Artificially increasing the population of ladybird beetles reduces aphid and scale populations. In California during the late 1800s, scale insects on orange trees damaged the trees and reduced crop yields. The introduction of a species of ladybird beetle from Australia quickly brought the pests under control. Years later, when chemical pesticides were first used in the area, so many ladybird beetles were accidentally killed that scale insects once again became a serious problem. When pesticide use was discontinued, ladybird beetle populations rebounded, and the scale insects were once again brought under control. (See figure 15.11b.)

Herbivorous insects can also be used to control weeds. Purple loosestrife (*Lythrum salicaria*) is a wetlands plant accidentally introduced from Europe in the mid-1800s. It takes over sunny wetlands and eliminates native vegetation such as cattails, thereby eliminating many species that rely on cattails for food, nesting places, or hiding places. (See figure 15.12.) The plant exists in most states and provinces in the United States and Canada. In Europe, the plant is not a pest because there are several insect species that attack it in various stages of the plant's life cycle. Since purple loosestrife is a European plant, a search was made to identify candidate species of insects that would control it. The criteria were that the insects must live only on purple loosestrife and not infest other plants and must have the capacity to do major damage to

a.

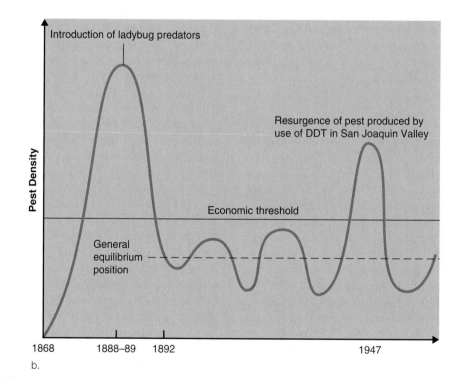

b.

Figure 15.11 **Insect Control with Natural Predators** (a) The ladybird beetle is a predator of many kinds of pest insects, including aphids. (b) In 1889, the introduction of ladybird beetles (ladybugs) brought the cottony cushion scale under control in the orange groves of the San Joaquin Valley. In the 1940s, DDT reduced the ladybird beetle population, and the cottony cushion scale population increased. Stopping the use of DDT allowed the ladybird population to increase, reducing the pest population and allowing the orange growers to make a profit.

Source: (b) *Man and the Environment*, 2d ed. by Arthur S. Boughey, 1975. Reprinted by permission of Prentice-Hall, Inc., Upper Saddle River, N.J.

a. Purple loosestrife

b. *Galerucella calmariensis*

c. *Galerucella pusilla*

d. *Hylobius transversovittatus*

Figure 15.12 **Biological Control of Purple Loosestrife** (*a*) Purple loosestrife is a European plant that invades wetlands and prevents the growth and reproduction of native species of plants. Several kinds of European beetles that feed on purple loosestrife have been released as biological control agents. (*b, c*) Two species (*Galerucella calmariensis* and *Galerucella pusilla*) feed on the leaves, shoots, and flowers of the newly growing purple loosestrife. (*d*) Another species (*Hylobius transversovittatus*) has larvae that feed on the roots. It appears that they are actually slowing the spread of purple loosestrife.

purple loosestrife. Five species of beetles have been identified that will attack the plant in various ways. After extensive studies to determine if introductions of these insects from Europe would be likely to cause other problems, several were selected as candidates to help control purple loosestrife in the United States and Canada. Some combination of these beetles has been released in a large proportion of the states and provinces infested with purple loosestrife. Two species (*Galerucella calmariensis* and *Galerucella pusilla*) feed on the leaves, shoots, and flowers of the newly growing purple loosestrife; one (*Hylobius transversovittatus*) has larvae that feed on the roots. Two other species (*Nanophyes brevis* and *Nanophyes marmoratus*) feed on the flowers. This multipronged attack by several species of beetles has been effective in reducing populations of purple loosestrife.

The use of specific strains of the bacterium *Bacillus thuringiensis* to control mosquitoes and moths is another example of the use of one organism to control another. A crystalline toxin produced by the bacterium destroys the lining of the gut of the feeding insect, resulting in its death. One strain of *B. thuringiensis* is used to control mosquitoes, while another is primarily effective against the caterpillars of leaf-eating moths, including the gypsy moth.

Naturally occurring pesticides found in plants also can be used to control pests. For example, marigolds are planted to reduce the number of soil nematodes, and garlic plants are used to check the spread of Japanese beetles.

Developing Resistant Crops

One outcome of the major advances in molecular genetics is the development of specialized organisms that have had specific, valuable genes inserted into their genetic makeup. Although this is a relatively new technology, it is an extension of an ancient art. Farmers have been involved in manipulating the genetic makeup of their plants and animals since these organisms were first domesticated. Initially, farmers either consciously or accidentally chose to plant specific seeds or breed certain animals that had specific desirable characteristics. This resulted in local varieties with particular characteristics. When the laws of genetics began to be understood in the early 1900s, scientists began to make precise crosses between carefully selected individuals to enhance the likelihood that their offspring would have certain highly desirable characteristics. This led to the development of hybrid seeds and specific breeds of domesticated animals. Controlled plant and animal breeding resulted in increased yields and better disease resistance in domesticated plants and animals. These activities are still the major driving force for developing improved varieties of domesticated organisms.

When the structure of DNA was discovered and it was determined that the DNA of organisms could be manipulated, an entirely new field of plant and animal breeding arose. **Genetic engineering** or **biotechnology** involves inserting specific pieces of DNA into the genetic makeup of organisms. The organism with the altered genetic makeup is known as a **genetically modified organism.** The DNA inserted could be from any source, even an entirely different organism. In agricultural practice, two kinds of genetically modified organisms have received particular attention. One involves the insertion of genes from a specific kind of bacterium called *Bacillus thuringiensis israeliensis* (Bti). Bti produces a material that causes the destruction of the lining of the gut of insects that eat it. It is a natural insecticide. To date, the gene has been inserted into the genetic makeup of several crop plants, including corn. In field tests, the genetically engineered corn was protected against some of its insect pests, but there was some concern that pollen grains from the corn might be blown to neighboring areas and affect nontarget insect populations. In particular, a study of monarch butterflies indicated that populations of butterflies adjacent to fields of this genetically engineered corn were negatively affected. One could argue that since the use of Bti corn results in less spraying of insecticides in cornfields, this is just a trade-off.

A second kind of genetically engineered plant involves inserting a gene for herbicide resistance into the genome of certain crop plants. The value of this to farmers is significant. For example, a farmer could plant cotton with very little preparation of the field to rid it of weeds. When both the cotton and the weeds began to grow, the field could be sprayed with a specific herbicide that would kill the weeds but not harm the herbicide-resistant cotton. This has been field-tested and it works. Critics have warned that the genes possibly could escape from the crop plants and become part of the genome of the weeds that we are trying to control, thus creating superweeds.

Many groups oppose the use of genetically modified organisms. They argue that this technology is going a step too far, that no long-term studies have been done to ensure their safety, that there are dangers we cannot anticipate, and that if such crops are grown, they should be labeled so that the public knows when they are consuming products from genetically modified organisms. The European Union (EU) will not buy genetically modified grains from the United States. This has resulted in farmers having to segregate their stores of grains so that they can guarantee to an EU buyer that a crop is not genetically modified and at the same time sell a genetically modified crop to other buyers.

Supporters argue that all plant and animal breeding involves genetic manipulation and that this is just a new kind of genetic manipulation. A great deal of evidence exists that genes travel between species in nature and that genetic engineering simply makes a common, natural process more frequent. Over the next 20 to 30 years, scientists hope to use biotechnology to produce high-yield plant strains that are more resistant to insects and disease, thrive on less fertilizer, make their own nitrogen fertilizer, do well in slightly salty soils, withstand drought, and use solar energy more efficiently during photosynthesis. These new kinds of genetically modified organisms will continue to be developed and tested, and the political arguments about their appropriateness will continue as well.

Modifying Farming Practices

Modification of farming practices can often reduce the impact of pests. In some cases, all crop residues are destroyed to prevent insect pests from finding overwintering sites. For example, shredding and plowing under the stalks of cotton in the fall reduces overwintering sites for boll weevils and reduces their numbers significantly, thereby reducing the need for expensive insecticide applications. Many farmers are also returning to crop rotation, which tends to prevent the buildup of specific pests that typically occurs when the same crop is raised in a field year after year.

Selective Use of Pesticides

Pesticides can also play a part in integrated pest management. Identifying the precise time when the pesticide will have the greatest effect at the lowest possible dose has these advantages: it reduces the amount of pesticide used and may still allow the parasites and predators of pests to survive. Such precise applications often require the assistance of a trained professional who can correctly identify the pests, measure the size of the population, and time pesticide applications for maximum effect. In several instances, pheromone-baited traps capture insect pests from fields, and an assessment of the number of insects caught can be a guide to when insecticides should be applied.

Integrated pest management will become increasingly popular as the cost of pesticides rises and knowledge about the biology of specific pests becomes available. However, as long as humans raise crops, there will be pests that will outwit the defenses we develop. Integrated pest management is just another approach to a problem that began with the dawn of agriculture.

Summary

Although small slash-and-burn garden plots are common in some parts of the world, most of the food in the world is raised on more permanent farms. In countries where population size is high and money is in short supply, much of the farming is labor-intensive, using human labor for many of the operations necessary to raise crops. However, much of the world's food is grown on large, mechanized farms that use energy rather than human muscle for tilling, planting, and harvesting crops, and for producing and applying fertilizers and pesticides.

Monoculture involves planting large areas of the same crop year after year. This causes problems with plant diseases, pests, and soil depletion. Although chemical fertilizers can replace soil nutrients that are removed when the crop is harvested, they do not replace the organic matter necessary to maintain soil texture, pH, and biotic richness.

Mechanized monoculture depends heavily on the control of pests by chemical means. Persistent pesticides are stable and persist in the environment, where they may biomagnify in ecosystems. Consequently, many of the older persistent pesticides have been quickly replaced by nonpersistent pesticides that decompose much more quickly and present less of an environmental hazard. However, most nonpersistent pesticides are more toxic to humans and must be handled with greater care than the older persistent pesticides.

Pesticides can be divided into several categories based on the organism they are used to control. Insecticides are used to control insects, herbicides are used for plants, fungicides for fungi, and rodenticides for rodents. Because of the problems of persistence, biomagnification, resistance of pests to pesticides, and effects on human health, many people are seeking pesticide-free alternatives to raising food. Several different philosophies that seek the same ends—less use of chemicals and better stewardship of soil—are alternative agriculture, sustainable agriculture, and organic agriculture. One ingredient in all of these approaches is the use of integrated pest management, which employs a complete understanding of an organism's ecology to develop pest-control strategies that use no or few pesticides.

Issues – Analysis

What Are Organic Foods?

The Organic Food Production Act of 1990 required that the U.S. Department of Agriculture (USDA) establish regulations on the production of organic foods. In 2002, the USDA published the final standards for organic foods. These standards set requirements for farmers who desire to have their crops and livestock marketed as organic. Farmers must submit an organic system plan to their certifying agent. Crops or livestock to be labeled organic must meet the following conditions:

Organic Crops	Organic Livestock	Organic Labels
Land must not have prohibited substances applied to it for at least three years.	Animals for slaughter must be raised under organic management from the last third of gestation, or no later than the second day of life for poultry.	Foods labeled "100 percent organic" and "organic" must contain (excluding water and salt) only organically produced ingredients.
Use of genetic engineering, ionizing radiation, and sewage sludge is prohibited.	Livestock must be fed products that are 100 percent organic. Vitamin and mineral supplements are permitted.	Products labeled "organic" must consist of at least 95 percent organically produced ingredients (excluding water and salt).
No chemical fertilizer can be used. Only animal and crop wastes and tillage methods may be used to provide fertility.	No antibiotics or hormones may be used.	Processed products that contain at least 70 percent organic ingredients can use the phrase "made with organic ingredients."
Synthetic pesticides must not be used, but pesticides made from plants and other natural sources may be used.	Animal welfare conditions require that: 1. Animals have access to the outdoors. (They may be temporarily confined only for reasons of health, safety, the animal's stage of production, or protection of soil or water quality.) 2. Preventive health management practices must be employed, including the use of vaccines to keep animals healthy. 3. Sick or injured animals must be treated; however, animals treated with a prohibited medication may not be sold as organic.	Processed products that contain less than 70 percent organic ingredients cannot use the term *organic* anywhere on the label but may list the ingredients that are organic.
Farms and handling operations that sell less than $5,000 annually of organic agricultural products are exempt from certification. But if they comply with all other national standards for organic products, they may label their products as organic.	A dairy herd can be converted to organic production if 80 percent organically produced feed is used for nine months, followed by three months of 100 percent organically produced feed.	Products with pesticide residues up to 5 percent of the EPA pesticide tolerance may be sold as organic.

- Many people feel there is no difference in food quality between foods that are labeled organic and those that are not. Do you think the nutritional quality of organic foods differs from that of conventionally produced foods?
- The rules permit small amounts of pesticide residues. Why do you think these are allowed?
- The rule that animals must have access to the outdoors has been challenged. Do you think being outdoors changes the quality of the meat, milk, or eggs produced?

Key Terms

alternative agriculture *344*
auxins *336*
bioaccumulation *338*
biocides *334*
biomagnification *338*
biotechnology *347*
carbamates *335*
chlorinated hydrocarbons *335*
fungicides *333*
genetically modified organisms *347*
genetic engineering *347*

Green Revolution *332*
herbicides *333*
insecticides *333*
integrated pest management *345*
macronutrient *333*
micronutrient *333*
monoculture *331*
nonpersistent pesticides *334*
nontarget organisms *334*
organic agriculture *344*
organophosphates *335*

persistent pesticides *334*
pests *333*
pesticides *333*
pheromone *345*
polyculture *330*
precision agriculture *345*
rodenticides *333*
sustainable agriculture *344*
target organism *334*
weeds *333*

Review Questions

1. What is monoculture?
2. List three reasons why fossil fuels are essential for mechanized agriculture.
3. Describe why pesticides are commonly used in mechanized agriculture.
4. Why are fertilizers used? What problems are caused by fertilizer use?
5. How do persistent and nonpersistent pesticides differ?
6. What is biomagnification? What problems does it cause?
7. How do organic farms differ from conventional farms?
8. Name three nonchemical methods of controlling pest populations.
9. What are the advantages and disadvantages of integrated pest management?
10. List three uses of food additives.
11. List three actions farmers could take to reduce the effect of pesticides on the environment.

Critical Thinking Questions

1. If you were a public health official in a developing country, would you authorize the spraying of DDT to control mosquitoes that spread malaria? What would be your reasons?
2. Look at table 15.1. What caused the changes in the effectiveness of the insecticide? If you were an agricultural extension agent, what alternatives to pesticides might you recommend?
3. Imagine that you are a scientist examining fish in Lake Superior and you find toxaphene in the fish you are studying. Toxaphene was used primarily in cotton farming and has been banned since 1982. How can you explain its presence in these fish?
4. Are the risks of pesticide use worth the benefits? What values, beliefs, and perspectives lead you to this conclusion?
5. Do you think that current agricultural practices are sustainable? Why or why not? What changes in agriculture do you think will need to happen in the next 50 years?
6. Imagine you are an EPA official who is going to make a recommendation about whether an agricultural pesticide can remain on the market or should be banned. What are some of the facts you would need to make your recommendation? Who are some of the interest groups interested in the outcome of your decision? What arguments might they present regarding their positions? What political pressures might they be able to bring to bear on you?
7. Why are few consumers demanding alternative methods of crop production, and why are farmers not using those methods?

Location: www.mhhe.com/enger10e

Mc
Graw
Hill

Interactive Exploration

Check out the website at www.mhhe.com/enger10e and click on the cover of this textbook for quizzing, career information, case studies, and hot links for the following topics:

Agricultural Methods

Food and Agriculture

Sustainable Agriculture

The Informed Consumer

Genetically Modified Foods

Economic and Ecological Importance of Angiosperms

Hazardous and Toxic Wastes

Food Webs

Pest Control

Integrated Pest Management

16

Water Management

Chapter Outline

Objectives

After reading this chapter, you should be able to:

- Explain how water is cycled through the hydrologic cycle.
- Explain the significance of groundwater, aquifers, and runoff.
- Explain how land use affects infiltration and surface runoff.
- List the various kinds of water use and the problems associated with each.
- List the problems associated with water impoundment.
- List the major sources of water pollution.
- Define biochemical oxygen demand (BOD).
- Differentiate between point and nonpoint sources of pollution.
- Explain how heat can be a form of pollution.

- Differentiate between primary, secondary, and tertiary sewage treatments.
- Describe some of the problems associated with storm-water runoff.
- List sources of groundwater pollution.
- Explain how various federal laws control water use and prevent misuse.
- List the problems associated with water-use planning.
- Explain the rationale behind the federal laws that attempt to preserve certain water areas and habitats.
- List the problems associated with groundwater mining.
- Explain the problem of salinization associated with large-scale irrigation in arid areas.
- List the water-related services provided by local governments.

PAST PRESENT FUTURE

1972 Sixty-four percent of streams and rivers in the United States were unsafe for swimming and fishing.

1991 Over 2000 of the nation's beaches were closed due to water pollution.

1990 Fifty-one percent of the world's people had access to sewage treatment, which is critical to controlling the spread of disease.

1995 Forty-one percent of the world's population experienced regular water shortages.

2003 Thirty-nine percent of streams and rivers were unsafe for swimming and fishing. This substantial improvement is due to the Clean Water Act.

2001 Beach closings had increased by sixfold, to 13,000.

2000 Sixty-one percent of the world's people had access to sewage treatment.

2025 By this time, it is predicted that 48 percent of the world's population will live in areas where regular water shortages occur.

The Water Issue

Water in its liquid form is the material that makes life possible on Earth. All living organisms are composed of cells that contain at least 60 percent water. Furthermore, their metabolic activities take place in a water solution. Organisms can exist only where they have access to adequate supplies of water. Water is also unique because it has remarkable physical properties. Water molecules are polar; that is, one part of the molecule is slightly positive and the other is slightly negative. Because of this, the water molecules tend to stick together, and they also have a great ability to separate other molecules from each other. Water's ability to act as a solvent and its capacity to store heat are a direct consequence of its polar nature. These abilities make water extremely valuable for societal and industrial activities. Water dissolves and carries substances ranging from nutrients to industrial and domestic wastes. A glance at any urban sewer will quickly point out the importance of water in dissolving and transporting wastes. Because water heats and cools more slowly than most substances, it is used in large quantities for cooling in electric power generation plants and in other industrial processes. Water's ability to retain heat also modifies local climatic conditions in areas near large bodies of water. These areas do not have the wide temperature-change characteristic of other areas.

For most human as well as some commercial and industrial uses, the quality of the water is as important as its quantity. Water must be substantially free of dissolved salts, plant and animal waste, and bacterial contamination to be suitable for human consumption. The oceans, which cover approximately 70 percent of the Earth's surface, contain over 97 percent of its water. However, saltwater cannot be consumed by humans or used for many industrial processes. Freshwater is free of the salt found in ocean waters. Of the freshwater found on Earth, only a tiny fraction is available for use. (See figure 16.1.) Unpolluted freshwater that is suitable for drinking is known as **potable water.** Early human migration routes and settlement sites were influenced by the availability of drinking water. At one time,

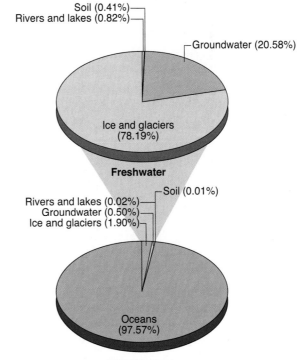

Figure 16.1 **Freshwater Resources** Although water covers about 70 percent of the Earth's surface, over 97 percent is saltwater. Of the less than 3 percent that is freshwater, only a tiny fraction is available for human use.

clean freshwater supplies were considered inexhaustible. Today, despite advances in drilling, irrigation, and purification, the location, quality, quantity, ownership, and control of potable waters remain important concerns.

Although the world's population has tripled in the past century and water use has risen sixfold, only recently have we begun to understand that we will probably exhaust our usable water supplies in some areas of the world. Some areas of the world have abundant freshwater resources, while others have few. In addition, demand is increasing for freshwater for industrial, agricultural, and personal needs.

Shortages of potable freshwater throughout the world can also be directly attributed to human abuse in the form of pollution. Water pollution has negatively affected water supplies throughout the world. In many parts of the developing world, people do not have access to safe drinking water. The World Health Organization estimates that about 25 percent of the world's people do not have access to

safe drinking water. Even in the economically advanced regions of the world, water quality is a major issue. According to the United Nations Environment Programme, 5 million to 10 million deaths occur each year from water-related diseases. The illnesses include cholera, malaria, dengue fever, and dysentery. The United Nations also reports that these illnesses have been increasing over the past decade and that without large economic investments in safe drinking-water supplies, the rate of increase will continue.

Unfortunately, the outlook for the world's freshwater supply is not very promising. According to studies by the United Nations and the International Joint Commission (a joint U.S.-Canadian organization that studies common water bodies between the two nations), many sections of the world are currently experiencing a shortage of freshwater and the problem will only intensify. A study by the United Nations in 2002 said that by 2025, one in three people globally could be threatened with a shortage of freshwater. The United Nations Third World Water Forum stated

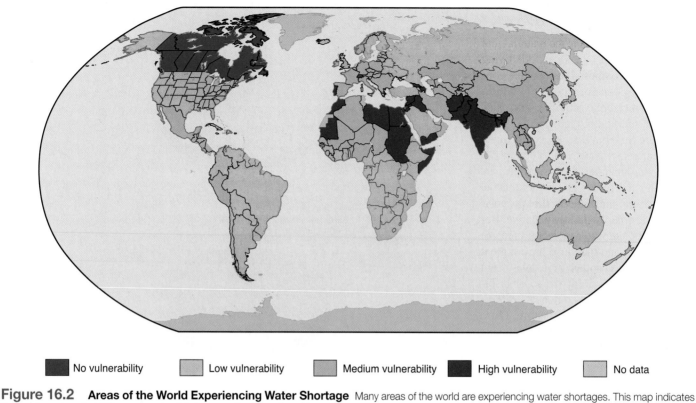

Figure 16.2 **Areas of the World Experiencing Water Shortage** Many areas of the world are experiencing water shortages. This map indicates that north and central Africa and parts of Asia are particularly vulnerable.

Source: Data from World Resources Institute.

that currently 450 million people in 29 countries lack water. (See figure 16.2.) The report went on to say that the Middle East, India, Pakistan, and China would struggle with serious water shortages. It was stated that water could become as important as oil as a major source of world conflict. Increasing scarcity, competition, and arguments over water in the first quarter of the twenty-first century could dramatically change the way we value and use water and the way we mobilize and manage water resources. Furthermore, changes in the amount of rain from year to year result in periodic droughts for some areas and devastating floods for others. However, rainfall is needed to regenerate freshwater and, therefore, is an important link in the cycling of water.

The Hydrologic Cycle

All water is locked into a constant recycling process called the **hydrologic cycle.** (See figure 16.3.) Two important processes involved in the cycle are the evaporation and condensation of water. Evaporation involves adding energy to molecules of a liquid so that it becomes a gas in which the molecules are farther apart. Condensation is the reverse process in which molecules of a gas give up energy, get closer together, and become a liquid. Solar energy provides the energy that causes water to evaporate from the ocean surface, the soil, bodies of freshwater, and from the surfaces of plants. The water evaporated from plants comes from two different sources. Some is water that has fallen on plants as rain, dew, or snow. In addition, plants take up water from the soil and transport it to the leaves, where it evaporates. This process is known as **evapotranspiration.**

The water vapor in the air moves across the surface of the Earth as the atmosphere circulates. As warm, moist air cools, water droplets form and fall to the land as precipitation. Although some precipitation may simply stay on the surface until it evaporates, most will either sink

into the soil or flow downhill and enter streams and rivers, which eventually return the water to the ocean. Surface water that moves across the surface of the land and enters streams and rivers is known as **runoff.** Water that enters the soil and is not picked up by plant roots moves slowly downward through the spaces in the soil and subsurface material until it reaches an impervious layer of rock. The water that fills the spaces in the substrate is called **groundwater.** It may be stored for long periods in underground reservoirs.

The porous layer that becomes saturated with water is called an **aquifer.** There are three basic kinds of aquifers: unconfined, semiconfined, and confined (see figure 16.4). An **unconfined aquifer** usually occurs near the land's surface where water enters the aquifer from the land above it. The top of the layer saturated with water is called the **water table.** The lower boundary of the aquifer is an impervious layer of clay or rock that does not allow water to pass through it. Unconfined aquifers are replenished (recharged) primarily by rain that falls on the ground

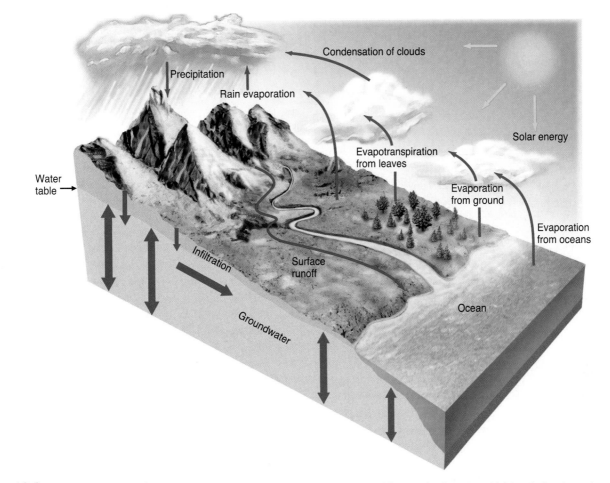

Figure 16.3   The Hydrologic Cycle The cycling of water through the environment follows a simple pattern. Moisture in the atmosphere condenses into droplets that fall to the Earth as rain or snow, supplying all living things with its life-sustaining properties. Water, flowing over the Earth as surface water or through the soil as groundwater, returns to the oceans, where it evaporates back into the atmosphere to begin the cycle again.

directly above the aquifer and infiltrates the layers below. The water in such aquifers is at atmospheric pressure and flows in the direction of the water table's slope, which may or may not be similar to the surface of the land above it. Above the water table and below the land surface is a layer known as the **vadose zone** (also known as the unsaturated zone or zone of aeration) that is not saturated with water.

A **confined aquifer** is bounded on both the top and bottom by layers that are impervious to water and is saturated with water under greater-than-atmospheric pressure. An impervious confining layer is called an **aquiclude.** If water can pass in and out of the confining layer, the layer is called an **aquitard.** A confined aquifer is primarily replenished by rain and surface water from a recharge zone (the area

where water is added to the aquifer) that may be many kilometers from where the aquifer is tapped for use. If the recharge area is at a higher elevation than the place where an aquifer is tapped, water will flow up the pipe until it reaches the same elevation as the recharge area. Such wells are called **artesian wells.** If the recharge zone is above the elevation of the top of the well pipe, it is called a flowing artesian well because water will flow from the pipe.

The nature of the substrate in the aquifer influences the amount of water the aquifer can hold and the rate at which water moves through it. **Porosity** is a measure of the size and number of the spaces in the substrate. The greater the porosity, the more water it can contain. The rate at which water moves through an aquifer is determined by the size of the pores, the

degree to which they are connected, and any cracks or channels present in the substrate. The rate at which the water moves through the aquifer determines how rapidly water can be pumped from a well per minute.

Human Influences on the Hydrologic Cycle

Human activities can significantly impact evaporation, runoff, and infiltration. When water is used for cooling in power plants or to irrigate crops, the rate of evaporation is increased. Water impounded in reservoirs also evaporates rapidly. This rapid evaporation can affect local atmospheric

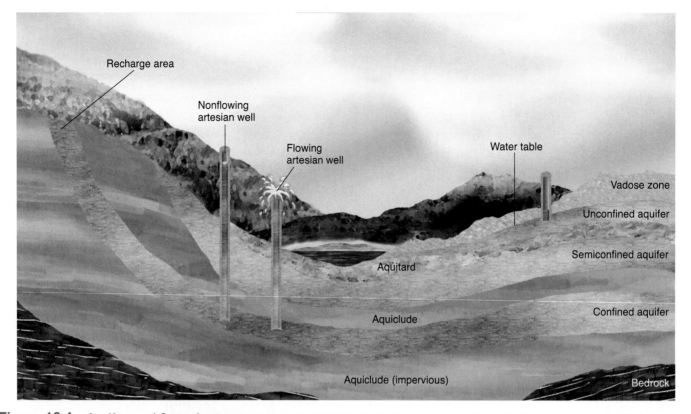

Figure 16.4 **Aquifers and Groundwater** Groundwater is found in the pores in layers of sediment or rock. The various layers of sediment and rock determine the nature of the aquifer and how it can be used.

conditions. Runoff and the rate of infiltration also are greatly influenced by human activity. Removing the vegetation by logging or agriculture increases runoff and decreases infiltration. Because there is more runoff, there is more erosion of soil. Urban complexes with a high percentage of impervious, paved surfaces have increased runoff and reduced infiltration. A major concern in urban areas is providing ways to carry storm water away rapidly. This involves designing and constructing surface waterways and storm sewers. Often cities have significant flooding problems when heavy rains overtax the ability of their storm-water management systems to remove excess water. Many cities combine their storm-sewer water with wastewater at their treatment plants, which can cause serious pollution problems after heavy rains. The treatment plants cannot process the increased volume of water and must discharge the combined untreated wastewater and sewer water into the receiving body of water.

Cities also have problems supplying water for industrial and domestic use. In many cases, cities rely on surface water sources for their drinking water. The source may be a lake or river, or an impoundment that stores water. Over one-third of the world's largest cities rely on large areas of protected forests to maintain a steady flow of clean drinking water. For example, New York City, with a population of over 8 million people, relies on a 998 square kilometer (385 square mile) watershed protected by the Catskill State Park for 90 percent of its drinking water. In addition, aquifers are extremely important in supplying water. Many large urban areas in the United States depend on underground water for their water supply. This groundwater supply can be tapped as long as it is not used faster than it can be replaced. Determining how much groundwater or surface water can be used and what the uses should be is a major concern, especially in water-poor areas of the world.

There are several ways to monitor water use from surface and groundwater sources. Water withdrawals are measurements of the amount of water taken from a source. This water may be used temporarily and then returned to its source and used again. For example, when a factory withdraws water from a river for cooling purposes, it returns most of the water to the river; thus, the water can be used later. Water that is incorporated into a product or lost to the atmosphere through evaporation or evapotranspiration cannot be reused in the same geographic area and is said to be consumed. Much of the water used for irrigation is lost to evaporation and evapotranspiration or is removed with the crop when it is harvested. Therefore, much of the water withdrawn for irrigation is consumed.

Kinds of Water Use

Water use varies considerably around the world, depending on availability of water and degree of industrialization. However, use can be classified into four broad categories: (1) domestic use, (2) agricultural use, (3) industrial use, and (4) in-stream use. It is important to remember that some uses of water are consumptive, while others are nonconsumptive.

Domestic Use of Water

Over 90 percent of the water used for domestic purposes in North America is supplied by municipal water systems, which typically include complex, costly storage, purification, and distribution facilities. Many rural residents, however, can obtain safe water from untreated private wells. Nearly 37 percent of municipal water supplies come from wells.

Regardless of the water source (surface or groundwater), water supplied to cities in the developed world is treated to ensure its safety. Treatment of raw water before distribution usually involves some combination of the following processes. The raw water is filtered through sand or other substrates to remove particles. Chemicals may be added to the water that will cause some dissolved materials to be removed. Then, before the water is released for public use, it is disinfected with chlorine, ozone, or ultraviolet light to remove any organisms that might still be present. Where no freshwater is available, expensive desalinization of saltwater may be the only option available.

Domestic activities in highly developed nations require a great deal of water. This domestic use includes drinking, air conditioning, bathing, washing clothes, washing dishes, flushing toilets, and watering lawns and gardens. On average, each person in a North American home uses about 400 liters (about 100 gallons) of **domestic water** each day. About 69 percent is used as a solvent to carry away wastes (nonconsumptive use; laundry, bathing, toilets, and dishes), about 29 percent is used for lawns and gardens, (consumptive use), and only a tiny amount (about 2 percent) is actually consumed by drinking or in cooking. (See figure 16.5.) Yet all water that enters the house has been purified and treated to make it safe for drinking. Natural processes cannot cope with the highly concentrated wastes typical of a large urban area. The unsightly and smelly wastewater also presents a potential health problem, so cities and towns must treat it before returning it

to a local water source. Until recently, the cost of water in almost every community has been so low that there was very little incentive to conserve, but shortages of water and increasing purification costs have raised the price of domestic water in many parts of the world, and it is becoming evident that increased costs do tend to reduce use. (See figure 16.6.)

Although domestic use of water is a relatively small component of the total water-use picture (see figure 16.7), urban

growth has created problems in the development, transportation, and maintenance of quality water supplies. In regions experiencing rapid population growth, such as Asia, domestic use is expected to increase sharply. Many cities in China are setting quotas on water use that are enforced by higher prices for larger users. In the coastal city of Dalian, a family that uses more than 8,000 liters (2,113 gallons) of water per month will pay four times more than a family that conserves water. In

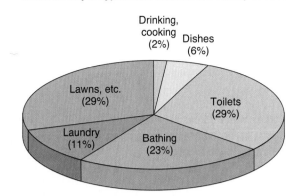

Water Use by a Typical North American Family of Four

Figure 16.5 **Urban Domestic Water Uses** Over 150 billion liters (40 billion gallons) of water are used each day for urban domestic purposes in North America. (Nonconsumptive uses are shown in blue, and consumptive uses are shown in green.)

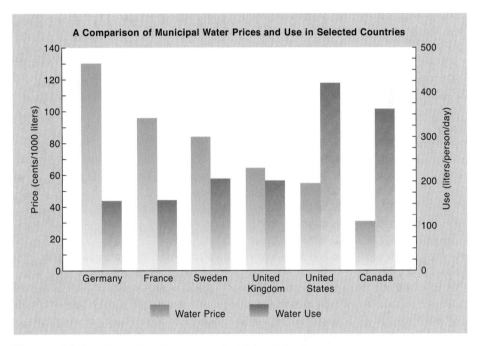

Figure 16.6 **Water Use Decreases As Water Price Increases** A general correlation exists between the amount of water that is used and its price.

North America, more than 36 states expect a water crisis in the next 10 years. Fast-growing cities in the West, especially those in arid areas, are experiencing water shortages. Demand for water in urban areas sometimes exceeds the immediate supply, particularly when the supply is local surface water. This is especially true during the summer, when water demand is high and precipitation is often low. Many communities have begun public education campaigns designed to help reduce water usage. Figure 16.8 shows some ways that people could reduce their water use. In addition, at the federal level, the U.S. Environmental Protection Agency (EPA) is in the process of developing a national consumer labeling program of water-efficient products modeled after its highly successful Energy STAR program. Mexico City, with a population of nearly 20 million and minimal access to surface water, has one of the most serious water management problems in the world. The water supply for the Mexico Basin is severely stressed by water management practices that do not ensure a sustainable supply of clean water. Since extractions from the Valley of Mexico aquifer began nearly 100 years ago, groundwater levels have dropped significantly. As the city's population continues to grow, managers have been forced to look at alternative sources of water from more than 160 kilometers (100 miles) away.

In addition to encouraging the public to conserve water, municipalities need to pay attention to losses that occur within the distribution system. Leaking water pipes and mains account for significant losses of water. Even in the developed world, such losses may be as high as 20 percent. Poorer countries typically exceed this, and some may lose over 50 percent of the water to leaks. Another major cause of water loss has been public attitudes. As long as water is considered a limitless, inexpensive resource, people will make little effort to conserve it. As the cost of water rises and attitudes toward water change, so will usage and efforts to conserve. For example, the California Urban Conservation Council, together with the EPA, has launched a highly innovative virtual home tour that allows Internet users to click on different areas of the

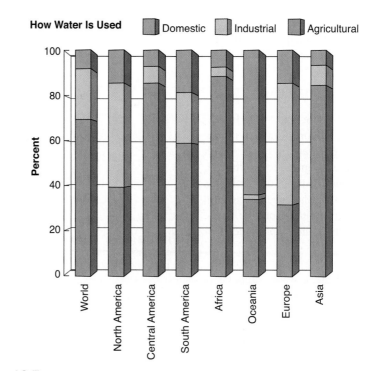

Figure 16.7 **World Uses of Water** Domestic, industrial, and agricultural uses dominate the allocation of water resources. However, there is considerable variety in different parts of the world in how these resources are used.

Source: World Resources Institute 2000.

H₂ouse floor-plan for facts and advice on saving water.

Agricultural Use of Water

In North America, groundwater accounts for about 37 percent of the water used in agriculture and surface water about 63 percent. **Irrigation** is the major consumptive use of water in most parts of the world and accounts for about 80 percent of all the water consumed in North America. About 500,000 million liters per day (134,000 million gallons per day or 150 million acre-feet per year) are used in irrigation in the United States. The amount of water used for irrigation and livestock continues to increase throughout the world. Future agricultural demand for water will depend on the cost of water for irrigation; the demand for agricultural products, food, and fiber; governmental policies; the development of new technology; and competition for water from a growing human population.

Since irrigation is common in arid and semiarid areas, local water supplies are often lacking, and it is often necessary to transport water great distances to water crops. This is particularly true in the western United States, where about 14 million hectares (35 million acres) of land are irrigated.

Four methods of irrigation are commonly used. Surface or flood irrigation involves supplying the water to crops by having the water flow over the field or in furrows. This requires extensive canals and is not suitable for all kinds of crops. Spray irrigation involves the use of pumps to spray water on the crop. Trickle irrigation uses a series of pipes with strategically placed openings so that water is delivered directly to the roots of the plants. Subirrigation involves supplying water to plants through underground pipes. Often this method is used where soils require draining at certain times of the year. The underground pipes can be used to drain excess water at one time of the year and supply water at others. Each of these methods has its drawbacks and advantages as well as conditions under which it works well.

Construction and maintenance of irrigation structures, such as dams, canals,

Water Savings Guide

Conservative use will save water		Normal use will waste water
Wet down, soap up, rinse off 15 liters (4 gal)	**Shower**	Regular shower 95 liters (25 gal)
May we suggest a shower?	**Tub bath**	Full tub 135 liters (36 gal)
Minimize flushing Each use consumes 20-25 liters (5-7 gal) New toilets use 6 liters (1.6 gal)	**Toilet**	Frequent flushing is very wasteful
Fill basin 4 liters (1 gal)	**Washing hands**	Tap running 8 liters (2 gal)
Fill basin 4 liters (1 gal)	**Shaving**	Tap running 75 liters (20 gal)
Wet brush, rinse briefly 2 liters (1/2 gal)	**Brushing teeth**	Tap running 38 liters (10 gal)
Take only as much as you require	**Ice**	Unused ice goes down drain
Please report immediately	**Leaks**	A small drip wastes 95 liters (25 gal) per week
Turn off light, TV, heaters, and air conditioning when not in room	**Energy**	Wasting energy also wastes water

Thank you for using this column...and not this one

Figure 16.8 **Ways to Conserve Water** Minor changes in the way people use water could significantly reduce domestic water use. Note: 1 gallon equals approximately 3.785 liters.

Reprinted by permission of the San Francisco Convention & Visitors Bureau.

pipes, and pumps, are expensive. Costs for irrigation water have traditionally been low since many of the dams and canals were constructed with federal assistance, and farmers have often used water wastefully. As competition has grown between urban areas and agriculture for scarce water resources, there has been pressure to raise the cost of water used for irrigation. Increasing the cost of water will stimulate farmers to conserve, just as it does homeowners. Another way to reduce the demand for irrigation water is to reduce the quantity of water-demanding crops grown in dry areas or change from high water-demanding to lower water-demanding crops. For example, wheat or soybeans require less water than do potatoes or sugar beets. It is also becoming increasingly important to modify irrigation practices to use less water. (See figure 16.9.) For example, the use of trickle irrigation and some variations of spray irrigation use water more efficiently than the more traditional flood irrigation methods.

Many forms of irrigation require a great deal of energy. This is particularly true when pumps are used to deliver the water to the crop. It is estimated that 40 percent of the energy devoted to agriculture in Nebraska is used for irrigation. Increasing energy costs may force some farmers to reduce or discontinue irrigation. In addition, much of western Nebraska relies on groundwater for irrigation, and the water table is dropping rapidly. If a water shortage develops, land values will decline. Land use and water use are interrelated and cannot be viewed independently. In the Salinas Valley, along the central coast of California, farmers are irrigating their fields with recycled water from the Monterey Regional Water Pollution Control Agency's (MRWPCA) local wastewater treatment plant. Because seawater had entered the groundwater as far as 10 kilometers (6 miles) inland, well water was becoming too salty for agricultural use. MRWPCA, aware that this problem may be caused by overpumping of groundwater, in the 1980s began studying the possibility of irrigating with treated municipal wastewater. Although California regulations allow the use of treated water for irrigation, health officials expressed concerns that its use on cold-season crops such as broccoli—many of which are consumed raw—may contaminate the crops with pathogens. The MRWPCA conducted a study that showed that crops irrigated with recycled water were not contaminated with pathogens. California then approved the plan, which required an upgrade of the local wastewater treatment plant. Sixty kilometers (40 miles) of pipeline were constructed to deliver the water to some 75 users throughout the system, which has the capacity to provide 24 billion liters (19,500 acre-feet) of water per year and irrigate 4900 hectares (12,000 acres) of coastal farmland. Such systems may become a common solution to the problem of deteriorating and diminishing groundwater supplies in California and perhaps throughout the world.

Industrial Use of Water

Industrial water use accounts for nearly half of total water withdrawals in the United States, about 70 percent in Canada, and about 23 percent worldwide. Since most industrial processes involve heat exchanges, 90 percent of the water used by industry is for cooling and is returned to the source, so only a small amount is actually consumed. Industrial use accounts for less than 20 percent of the water consumed in the United States. For example, electric-power generating plants use water to cool steam so that it changes back into water. Many industries, especially power plants, actually can use saltwater for cooling purposes. About 30 percent of the cooling water used by power plants is

a.

b.

c.

Figure 16.9 **Types of Irrigation** Many arid areas require irrigation to be farmed economically. (a) Surface or flood irrigation uses irrigation canals and ditches to deliver water to the crops. The land is graded so that water flows from the source into the fields. Water is siphoned from a canal into ditches between rows of crops. (b) Spray irrigation uses a pump to spray water into the air above the plants. This is an example of central-pivot spray irrigation, in which a long pipe on wheels slowly rotates about a central point. (c) Trickle irrigation conserves water by delivering water directly to the roots of the plants but requires an extensive network of pipes.

saltwater. If the water heated in an industrial process is dumped directly into a watercourse, it significantly changes the water temperature. This affects the aquatic ecosystem by increasing the metabolism of the organisms and reducing the water's ability to hold dissolved oxygen.

Industry also uses water to dissipate and transport waste materials. In fact, many streams are now overused for this purpose, especially in urban centers. The use of watercourses for waste dispersal degrades the quality of the water and may reduce its usefulness for other purposes, as well as directly harming fish and wildlife. This is especially true if the industrial wastes are toxic.

Historically, industrial waste and heat were major causes of pollution. However, most industrialized nations have passed laws that severely restrict industrial discharges of wastes or heated water into watercourses. In the United States, the federal government's role in maintaining water quality began in 1972, with the passage of the Federal Water Pollution Control Act (PL 92-500). This act provided federal funds and technical assistance to strengthen local, state, and interstate water-quality programs. The act (and subsequent amendments in 1977, 1981, 1987, and 1993) is commonly referred to as the "Clean Water Act." The Clean Water Act is a comprehensive and technically rigorous piece of legislation. It seeks to protect the waters of the United States from pollution. To do this, the act specifically regulates pollutant discharges into "navigable waters" by implementing two concepts: setting water-quality standards for surface water and limiting effluent discharges into the water. The policy objectives of the Clean Water Act are to restore and maintain the "chemical, physical and biological integrity of the nation's waters." Enforcement of this act has been extremely effective in improving surface-water quality. However, many countries in the developing world have done little to control industrial pollution, and water quality is significantly reduced by careless use. (See Global Perspective: Comparing Water Use and Pollution in Industrialized and Developing Countries on page 363.)

In-Stream Use of Water

In-stream water use does not remove water but makes use of it in its channels and basins. Therefore, all in-stream uses are nonconsumptive. Major in-stream uses of water are for hydroelectric power, recreation, and navigation. Although in-stream uses do not remove water, they may require modification of the direction, time, or volume of flow and can negatively affect the watercourse.

Electricity from hydroelectric power plants is an important energy resource.

The California Water Plan

The management of freshwater is often a controversial subject, involving social, ecological, and economic aspects. A good example is the California Water Plan.

In the early 1900s, it became clear that the growth of Los Angeles, which was then a small coastal town, would be encouraged by irrigating the surrounding land. Los Angeles looked to the Owens Valley, near Bishop, 400 kilometers (250 miles) north, for a source of water. The Los Angeles Aqueduct connecting these two areas was completed in 1913.

After the Owens Valley project, California developed a statewide water program known as the California Water Plan in 1957, and the plan has been further refined and modified in recent years. This plan is necessary because most of the state's population and irrigated land are found in the central and southern regions, but most of the water is in the north. The plan details the construction of aqueducts, canals, dams, reservoirs, and power stations to transport water from the north to the south. In addition to supplying water for southern regions, the aqueducts provide irrigation for the San Joaquin Valley. Eventually, the new land made available for agriculture amounted to about 400,000 hectares (1 million acres).

The California Water Plan has been one of the most controversial programs ever undertaken in California. Its adoption instigated a sectional feud between the moist "north" and the dry "south." Southern California was accused of trying to steal northern water, but because the large population in the southern part of the state carried the vote, the plan was adopted. Environmentalists still claim that the project has irreparably scarred the countryside and upset natural balances of streams, estuaries, vegetation, and wildlife. They argue that providing water to southern California promotes population growth, which leads to further urbanization and land development.

Many questions and controversies center on whether the water is really needed. Ninety percent of the water used in southern California is for irrigation; there is evidently abundant water for domestic and industrial use. A few of the crops raised in the San Joaquin Valley account for a large percentage of the irrigation water. Most notable is rice. Rice is not a native crop and demands intensive irrigation. The cost is borne by all the rate payers, most of whom are urban. California is now one of the nation's most productive agricultural areas.

Allocation of water resources is a matter of economics as well as technology. The California water project has been criticized for using public funds to increase the value of privately held farmland. Furthermore, technological advances in desalination plants may give a new dimension (unforeseen when the water plan was devised) to the problem of water resources. Although more aqueducts, canals, and pumping plants are planned, whether they will be completed is uncertain.

Southern California has been in a drought emergency for several years. Water rates have increased, certain municipal uses have been curtailed, and irrigation has been reduced. Increasing population creates a demand for water that is now causing people to recognize that they will need to choose between using water for municipal purposes or irrigation. There is not enough water to increase both.

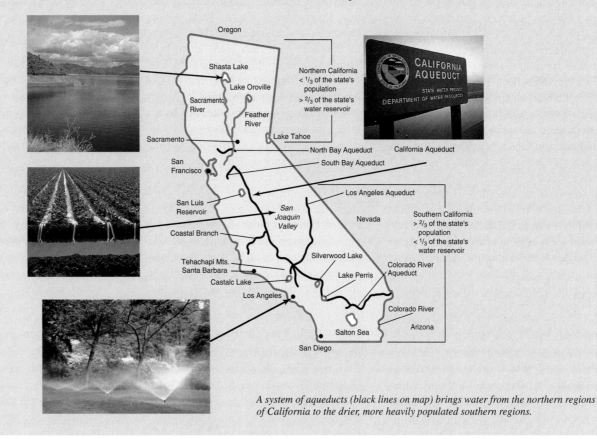

A system of aqueducts (black lines on map) brings water from the northern regions of California to the drier, more heavily populated southern regions.

a.   b.

Figure 16.10   Dams Interrupt the Flow of Water The flow of water in most large rivers is controlled by dams. Most of these dams provide electricity. In addition, they prevent flooding and provide recreational areas. (*a*) Large dams, however, destroy the natural river system, while smaller dams (*b*), do not always have a significant impact.

Presently, hydroelectric power plants produce about 13 percent of the total electricity generated in the United States. (See figure 16.10.) They do not consume water and do not add waste products to it. However, the dams needed for the plants have definite disadvantages, including the high cost of construction and the resulting destruction of the natural habitat in streams and surrounding lands. The sudden discharge of impounded water from a dam can seriously alter the downstream environment. If the discharge is from the top of the reservoir, the stream temperature rapidly increases. Discharging the colder water at the bottom of the reservoir causes a sudden decrease in the stream's water temperature. Either of these changes is harmful to aquatic life. The impoundment of water also reduces the natural scouring action of a flowing stream. If water is allowed to flow freely, the silt accumulated in the river is carried downstream during times of high water. This maintains the river channel and carries nutrient materials to the river's mouth. But if a dam is constructed, the silt is deposited behind the dam, eventually filling the reservoir.

Many other dams were constructed to control floodwaters. While dams reduce flooding, they do not eliminate it. In fact, the building of a dam often encourages people to develop the floodplain. As a result, when flooding occurs, the loss of property and lives may be greater.

Because dams create lakes that have a large surface area, evaporation is increased. In arid regions, the amount of water lost can be serious. This is particularly evident in hot climates. Furthermore, flow is often intermittent below the dam, which alters the water's oxygen content and interrupts fish migration. The populations of algae and other small organisms are also altered. Because of all these impacts, dam construction requires careful planning. It should also be noted that the ecological impact of dams varies considerably, depending on the size of the dam. Large-scale dams have significantly greater impact than small-scale dams.

Dam construction often creates new recreational opportunities because reservoirs provide sites for boating, camping, and related recreation. (See figure 16.11.) However, these opportunities come at the expense of a previously free-flowing river. Some recreational pursuits, such as river fishing, are lost. Sailing, waterskiing, swimming, fishing, and camping all require water of reasonably good quality. Water is used for recreation in its natural setting and often is not physically affected. Even so, it is necessary to plan for recreational use, because overuse or inconsiderate use can degrade water quality. For example, waves generated by powerboats can accelerate shoreline erosion and cause siltation. When large numbers of powerboats are used on water, they contribute significantly to water pollution since their exhaust contains unburned hydrocarbons. Newer boat motors are designed to produce less pollution.

Most major rivers and large lakes are used for navigation. North America currently has more than 40,000 kilometers (25,000 miles) of commercially navigable waterways. These waterways must have sufficient water depth to ensure passage of ships and barges. Canals, locks, and dams are used to ensure that adequate depths are provided. Often, dredging is necessary to maintain the proper channel depth.

GLOBAL PERSPECTIVE

Comparing Water Use and Pollution in Industrialized and Developing Countries

There are great differences in the way water is used in the industrialized areas of the world compared to the developing regions. In the industrialized world, money is available to support the technology to provide inexpensive, safe domestic water supplies and to treat wastewater. However, even industrialized countries have difficulty with nonpoint sources of pollution, such as fertilizer runoff, and must deal with damage caused by past practices. The tables summarize general differences in water use and pollution control in industrialized and developing regions of the world.

Children collect garbage for recycling from Manila Bay, the Philippines, in January 2000.

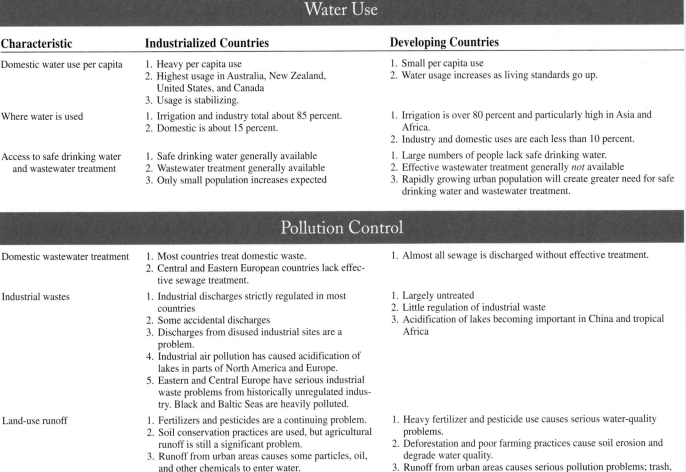

	Water Use	
Characteristic	**Industrialized Countries**	**Developing Countries**
Domestic water use per capita	1. Heavy per capita use 2. Highest usage in Australia, New Zealand, United States, and Canada 3. Usage is stabilizing.	1. Small per capita use 2. Water usage increases as living standards go up.
Where water is used	1. Irrigation and industry total about 85 percent. 2. Domestic is about 15 percent.	1. Irrigation is over 80 percent and particularly high in Asia and Africa. 2. Industry and domestic uses are each less than 10 percent.
Access to safe drinking water and wastewater treatment	1. Safe drinking water generally available 2. Wastewater treatment generally available 3. Only small population increases expected	1. Large numbers of people lack safe drinking water. 2. Effective wastewater treatment generally *not* available 3. Rapidly growing urban population will create greater need for safe drinking water and wastewater treatment.

	Pollution Control	
Domestic wastewater treatment	1. Most countries treat domestic waste. 2. Central and Eastern European countries lack effective sewage treatment.	1. Almost all sewage is discharged without effective treatment.
Industrial wastes	1. Industrial discharges strictly regulated in most countries 2. Some accidental discharges 3. Discharges from disused industrial sites are a problem. 4. Industrial air pollution has caused acidification of lakes in parts of North America and Europe. 5. Eastern and Central Europe have serious industrial waste problems from historically unregulated industry. Black and Baltic Seas are heavily polluted.	1. Largely untreated 2. Little regulation of industrial waste 3. Acidification of lakes becoming important in China and tropical Africa
Land-use runoff	1. Fertilizers and pesticides are a continuing problem. 2. Soil conservation practices are used, but agricultural runoff is still a significant problem. 3. Runoff from urban areas causes some particles, oil, and other chemicals to enter water.	1. Heavy fertilizer and pesticide use causes serious water-quality problems. 2. Deforestation and poor farming practices cause soil erosion and degrade water quality. 3. Runoff from urban areas causes serious pollution problems; trash, human waste, animal waste, chemical waste are all present.

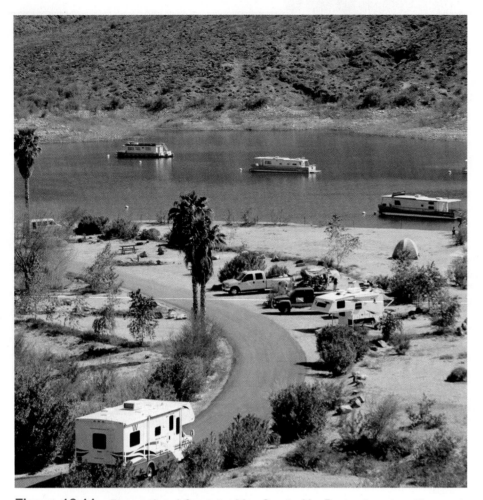

Figure 16.11 **Recreational Opportunities Created by Dams** Over nine million people annually enjoy camping and boating on Lake Mead, the largest man-made lake in the United States. This 550 mile shoreline was created when the Hoover Dam was completed on the Colorado River in 1935 to supply power and water to several western states.

Dredging can resuspend contaminated sediments. Another problem is determining where to deposit the contaminated sediments when they are removed from the bottom. In addition, the flow within the hydrologic system is changed, which, in turn, affects the water's value for other uses.

Most large urban areas rely on water to transport resources. During recent years, the inland waterway system in the United States has carried about 10 percent of goods such as grain, coal, ore, and oil. In North America, expenditures for the improvement of the inland waterway system have totaled billions of dollars.

In the past, almost any navigation project was quickly approved and funded, regardless of the impact on other uses. To-day, however, such decisions are not made until ecological impacts are analyzed.

Kinds and Sources of Water Pollution

Water pollution occurs when something enters water that changes the natural ecosystem or interferes with water use by segments of society. In an industrialized society, maintaining completely unpolluted water in all drains, streams, rivers, and lakes is probably impossible. (See table 16.1.) But we can evaluate the water quality of a body of water and take steps to preserve or improve its quality by eliminating sources of pollution. Some pollu-tants seriously affect the quality and possible uses of water. In general, water pollutants can be divided into several broad categories.

Toxic chemicals or acids may kill organisms and make the water unfit for human use. If these chemicals are persistent, they may bioaccumulate in individual organisms and biomagnify in food chains.

Dissolved organic matter is a significant water pollution problem because it decays in the water. As the microorganisms naturally present in water break down the organic matter, they use up available dissolved oxygen from the water. If too much dissolved oxygen is removed, aquatic organisms die. The amount of oxygen required to decay a certain amount of organic matter is called the **biochemical oxygen demand (BOD).** (See figure 16.12.) Measuring the BOD of a body of water is one way to determine how polluted it is. If too much organic matter is added to the water, all of the available oxygen will be used up. Then, anaerobic (not requiring oxygen) bacteria begin to break down wastes. Anaerobic respiration produces chemicals that have a foul odor and an unpleasant taste and generally interfere with the well-being of humans.

Disease-causing organisms are a very important pollution problem in most of the world. Untreated or inadequately treated human or domesticated animal waste is most often the source of these organisms. In the developed world, sewage treatment and drinking-water treatment plants greatly reduce this public health problem.

Nutrients are also a pollution problem. Additional nutrients in the form of nitrogen and phosphorus compounds from fertilizer, sewage, detergents, and animal waste increase the rate of growth of aquatic plants and algae. However, phosphates and nitrates are generally present in very limited amounts in unpolluted freshwater and, therefore, are a limiting factor on the growth of aquatic plants and algae. (A **limiting factor** is a necessary material that is in short supply, and because of the lack of it, an organism cannot reach its full potential growth. See chapter 5.) Thus, when phosphates or nitrates are added to the surface water, they act as a fertilizer and promote the growth of undesirable

Table 16.1 Sources and Impacts of Selected Pollutants

Pollutant	Source	Effects on Humans	Effects on Aquatic Ecosystem
Acids	Atmospheric deposition; mine drainage; decomposing organic matter	Reduced availability of fish and shellfish Increased heavy metals in fish	Death of sensitive aquatic organisms; increased release of trace metals from soils, rock, and metal surfaces, such as water pipes
Chlorides	Runoff from roads treated for removal of ice or snow; irrigation runoff; brine produced in oil extraction; mining	Reduced availability of drinking water supplies; reduced availability of shellfish	At high levels, toxic to freshwater organisms
Disease-causing organisms	Dumping of raw and partially treated sewage; runoff of animal wastes from feedlots	Increased costs of water treatment; death and disease; reduced availability and contamination of fish, shellfish, and associated species	Reduced survival and reproduction of aquatic organisms due to disease
Elevated temperatures	Heat trapped by cities that is transferred to water; unshaded streams; solar heating of reservoirs; warm-water discharges from power plants and industrial facilities	Reduced availability of fish	Elimination of cold-water species of fish and shellfish; less oxygen; heat-stressed animals susceptible to disease; inappropriate spawning behavior
Heavy metals	Atmospheric deposition; road runoff; discharges from sewage treatment plants and industrial sources; creation of reservoirs; acidic mine effluents	Increased costs of water treatment; disease and death; reduced availability and healthfulness of fish and shellfish; biomagnification	Lower fish population due to failed reproduction; death of invertebrates leading to reduced prey for fish; biomagnification
Nutrient enrichment	Runoff from agricultural fields, pastures, and livestock feedlots; landscaped urban areas; dumping of raw and treated sewage and industrial discharges; phosphate detergents	Increased water treatment costs; reduced availability of fish, shellfish, and associated species; color and odor associated with algal growth; impairment of recreational uses	Algal blooms occur. Death of algae results in low oxygen levels and reduced diversity and growth of large plants. Reduced diversity of animals; fish kills
Organic molecules	Runoff from agricultural fields and pastures; landscaped urban areas; logged areas; discharges from chemical manufacturing and other industrial processes; combined sewers	Increased costs of water treatment; reduced availability of fish, shellfish, and associated species; odors	Reduced oxygen; fish kills; reduced numbers and diversity of aquatic life
Sediment	Runoff from agricultural land and livestock feedlots; logged hillsides; degraded stream banks; road construction; and other improper land use	Increased water treatment costs; reduced availability of fish, shellfish, and associated species; filling in of lakes, streams, and artificial reservoirs and harbors, requiring dredging	Covering of spawning sites for fish; reduced numbers of insect species; reduced plant growth and diversity; reduced prey for predators; clogging of gills and filters
Toxic chemicals	Urban and agricultural runoff; municipal and industrial discharges; leachate from landfills and mines; atmospheric deposits	Increased costs of water treatment; increased risk of certain cancers; reduced availability and healthfulness of fish and shellfish	Reduced growth and survivability of fish eggs and young; fish diseases; death of carnivores due to biomagnification in the food chain

Source: Data, in part, from World Resources Institute 1994–95.

algae populations. The excessive growth of algae and aquatic plants due to added nutrients is called **eutrophication.** Algae and larger aquatic plants may interfere with the use of the water by fouling boat propellers, clogging water-intake pipes, changing the taste and odor of water, and causing the buildup of organic matter on the bottom. As this organic matter decays, oxygen levels decrease, and fish and other aquatic species die.

Physical particles also can negatively affect water quality. Particles alter the clarity of the water, can cover spawning sites, act as abrasives that injure organisms, and carry toxic materials.

Ideally, we should think in terms of eliminating all pollution, but any human use is going to have at least some minor negative impact. Even activities such as swimming and boating add particles and chemicals to the water. So, determining acceptable water quality involves economic considerations. Removing the last few parts per million of some materials from the water may not significantly improve water quality and may not be economically justifiable. This is certainly true of organic matter, which is biodegradable. However, radioactive wastes and toxins that may accumulate in living tissue are a different matter. Vigorous attempts to re-

move these materials are often justified because of the materials' potential harm to humans and other organisms.

Sources of pollution are classified as either point sources or nonpoint sources. When a source of pollution can be readily identified because it has a definite source and place where it enters the water, it is said to come from a **point source.** Municipal and industrial discharge pipes are good examples of point sources. Diffuse pollutants, such as from agricultural land and urban paved surfaces, acid rain, and runoff, are said to come from **nonpoint sources** and are much more difficult to identify and control. Initial attempts to

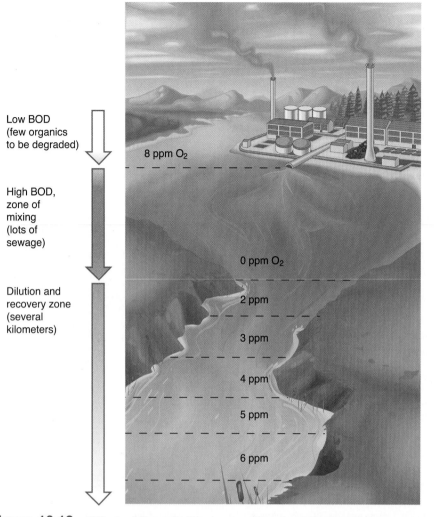

Low BOD
(few organics
to be degraded)

8 ppm O₂

High BOD,
zone of
mixing
(lots of
sewage)

0 ppm O₂

Dilution and
recovery zone
(several
kilometers)

2 ppm

3 ppm

4 ppm

5 ppm

6 ppm

Figure 16.12 **Effect of Organic Wastes on Dissolved Oxygen** Sewage contains a high concentration of organic materials. When these are degraded by organisms, oxygen is removed from the water. This is called the biochemical oxygen demand (BOD). An inverse relationship exists between the amount of organic matter and oxygen in the water. The greater the BOD, the more difficult it is for aquatic animals to survive and the less desirable the water is for human use. The more the organic pollution, the greater the BOD.

control water pollution were focused on point sources of pollution, since these were readily identifiable and economic pressure and adverse publicity could be brought to bear on companies that continued to pollute from point sources. In North America, most point sources of water pollution have been identified and are regulated. In the United States, the EPA is responsible for identifying point sources of pollution, negotiating the permissible levels of pollution allowed from each source, and enforcing the terms of the permits.

Nonpoint sources of water pollution are being addressed, but this is much more difficult to do since regulating many small, individual human acts is necessary.

In 1998, President Bill Clinton announced the formation of the Clean Water Action Plan. This plan requires the cooperation of several levels of government and focuses on watersheds and nonpoint sources of pollution, particularly runoff from urban areas and agricultural land. In addition, plans have been developed to address several long-term water-quality issues, such as acid runoff from mines, contaminated sediments, the presence of toxic chemicals in fish, and restoring degraded wetlands.

Municipal Water Pollution

Municipalities are faced with the double-edged problem of providing suitable drinking water and disposing of wastes. These wastes consist of stormwater runoff, wastes from industry, and wastes from homes and commercial establishments. Wastes from homes consist primarily of organic matter from garbage, food preparation, cleaning of clothes and dishes, and human wastes. Human wastes are mostly undigested food material and a concentrated population of bacteria, such as *Escherichia coli* and *Streptococcus faecalis*. These particular bacteria normally grow in the large intestine (colon) of humans and are present in high numbers in the feces of humans; therefore, they are commonly called **fecal coliform bacteria.** Fecal coliform bacteria are also present in the wastes of other warm-blooded animals, such as birds and mammals. Low numbers of these bacteria in water are not harmful to healthy people. However, because they can be easily identified, their presence in the water is used to indicate the amount of pollution from the fecal wastes of humans and other warm-blooded animals. The numbers of these types of bacteria present in water are directly related to the amount of fecal waste entering the water.

When human wastes are disposed of in water systems, potentially harmful bacteria from humans may be present in amounts too small to detect by sampling. Even in small numbers, these harmful bacteria may cause disease epidemics. It is estimated that some 1.5 million people in the United States become ill each year from infections caused by fecal contamination. In 1993, for example, a protozoan pathogen called *Cryptosporidium* was identified in the Milwaukee, Wisconsin, public water system. This resulted in over 400,000 people becoming ill and at least 100 deaths. The total costs of such diseases amount to billions of dollars per year in the United States alone. The greater the amount of wastes deposited in the water, the more likely it is that there will be populations of disease-causing bacteria. Therefore, the presence of fecal coliform bacteria is used as an indication that other, more harmful organisms may be present as well.

Wastewater from cleaning dishes and clothing contains some organic material along with the soap or detergent, which

The Cleanup of the Holy Ganges

Every day, millions of Hindus use the Ganges River in India. There, they drink and bathe in what they believe to be holy water, as partially cremated corpses float past them and nearby drains emit millions of liters of raw sewage. Clean water is one of India's most scarce resources, but, as in many developing nations, the money and technology needed to properly treat sewage are not available in most cities and villages. With the country's population of slightly more than 1 billion people expected to double in 39 years, officials are concerned that Indians will have no choice but to continue dumping raw waste into local waterways, contributing to epidemics of diarrhea and other diseases that kill thousands of people annually.

Keeping the Ganges clean is made especially difficult because faith in the river's incorruptible purity has generated complacency and ambivalence about its pollution among many of the 300 million people who live in the Ganges Basin. More than 1600 million liters (425 million gallons) of untreated municipal sewage, industrial waste, agricultural runoff, and other pollutants are discharged into the river every day. At the same time, officials estimate that more than 1 million people a day bathe or take a "holy dip" in the Ganges, and thousands drink straight from its banks.

The Hindu belief in cremation has led to several environmental problems and has had an impact on the Ganges. A wood cremation takes more than 50 pounds of wood, costing two week's wages for the typical Indian. Thus, not only are wood cremations helping to devour India's forests, but due to the high cost of wood, many bodies are not completely cremated. These partially burned corpses are disposed of in the Ganges.

One solution to the problem of corpses in the river has been the introduction of 25,000 specially raised snapping turtles to devour the corpses. While this solution may seem extreme, there is another that is more acceptable in the long run. In 1992, as part of the Ganges Action Plan, an electric crematorium was built at the city of Varanasi that charges less than $2 per body and does a thorough job of turning the bodies to ash. There is some concern, however, that many Hindus will not want to abandon the traditional ritual of a wood cremation for the more efficient and less costly electric cremation.

Although this entire scenario may seem somewhat unusual to North Americans, it is important to keep in mind how culture and religion affect our environment. To many Indians, the cultural and religious practices of North Americans are equally puzzling.

The Ganges River flows through the northeastern corner of India.

People drink from and bathe in the Ganges River even though it contains large amounts of pollutants.

helps to separate the contaminant from the dishes or clothes. Soaps and detergents are useful because one end of the molecule dissolves in dirt or grease and the other end dissolves in water. When the soap or detergent molecules are rinsed away by the water, the dirt or grease goes with them.

At one time, many detergents contained phosphates as a part of their chemical makeup, which contributed to eutrophication. However, because of the environmental effects of phosphate on aquatic environments, since 1994 most major detergent manufacturers in North America and other developed countries have eliminated phosphates from most of their formulations. Today, the majority of

Figure 16.13 **Sources of Groundwater Contamination** A wide variety of activities have been identified as sources of groundwater contamination.

every facet of social, agricultural, and industrial activities. (See figure 16.13.) Once groundwater pollution has occurred, it is extremely difficult to remedy. Pumping groundwater and treating it is very slow and costly, and it is difficult to know when all of the contaminated water has been removed. A much better way to deal with the issue of groundwater pollution is to work very hard to prevent the pollution from occurring in the first place.

Major sources of groundwater contamination include:

1. Agricultural products. Pesticides contribute to unsafe levels of organic contaminants in groundwater. Seventy-three different pesticides have been detected in the groundwater in Canada and the United States. Accidental spills or leaks of pesticides pollute groundwater sources with 10 to 20 additional pesticides. Other agricultural practices contributing to groundwater pollution include animal-feeding operations, fertilizer applications, and irrigation practices.

2. Underground storage tanks. For many years in North America, a large number of underground storage tanks containing gasoline and other hazardous substances have leaked. Four liters (1 gallon) of gasoline can contaminate the water supply of a community of 50,000 people. A major program of replacing leaking underground storage tanks recently was completed in the United States. However, the effects of past leaks and abandoned tanks will continue to be a problem for many years.

3. Landfills. Even though recently constructed landfills have special liners and water collection systems, approximately 90 percent of the landfills in North America have no liners to stop leaks to underlying groundwater, and 96 percent have no system to collect the leachate that seeps from the landfill. Sixty percent of landfills place no restrictions on the waste accepted, and many landfills are not inspected even once a year.

4. Septic tanks. Poorly designed and inadequately maintained septic systems have contaminated groundwater with nitrates, bacteria, and toxic cleaning agents. Over 20 million septic tanks are in use in the United States, and up to a third have been found to be operating improperly.

5. Surface impoundments. Over 225,000 pits, ponds, and lagoons are used in North America to store or treat wastes. Seventy-one percent are unlined, and only 1 percent use a plastic or other synthetic, nonsoil liner. Ninety-nine percent of these impoundments have no leak-detection systems. Seventy-three percent have no restriction on the waste placed in the impoundment. Sixty percent are not even inspected annually. Many of these ponds are located near groundwater supplies.

Other sources of groundwater contamination include mining wastes, salting for controlling road ice, land application of treated wastewater, open dumps, cemeteries, radioactive disposal sites, urban runoff, construction excavation, fallout from the atmosphere, and animal feedlots.

Water-Use Planning Issues

In the past, wastes were discharged into waterways with little regard for the costs imposed on other users by the resulting decrease in water quality. Furthermore, as the population has grown and the need for irrigation and domestic water has intensified, in many parts of the world, there has not been enough water to satisfy everyone's needs. With today's increasing demands for high-quality water, unrestrained waste disposal and unlimited withdrawal of water could lead to serious conflicts about water uses, causing social, economic, and environmental losses at both local and international levels. (Table 16.2 summarizes some of the areas of controversy involving water throughout the world.)

Metropolitan areas must deal with a variety of issues and maintain an extensive infrastructure to provide three basic water services:

1. Water supply for human and industrial needs
2. Wastewater collection and treatment
3. Stormwater collection and management

Water sources must be identified and preserved for use. Some cities obtain all their municipal water from groundwater and must have a thorough understanding of the size and characteristics of the aquifer they use. Some cities, such as New York City, obtain potable water by preserving a watershed that supplies the water needed by the population. Other cities have abundant water in the large rivers that flow by them but must deal with pollution problems caused by upstream users. In many places where water is in short supply, municipal and industrial-agricultural needs for water conflict.

Water for human and industrial use must be properly treated and purified. It is then pumped through a series of pipes to consumers. After the water is used, it flows through a network of sewers to a wastewater treatment plant, where it is treated before it is released. Maintaining the infrastructure of pipes, pumps, and treatment plants is expensive.

Metropolitan areas must also deal with great volumes of excess water during storms. This water is known as **stormwater runoff.** Because urban areas are paved and little rainwater can be absorbed into the ground, managing stormwater is a significant problem. Cities often have severe local flooding because the water is channeled along streets to storm sewers. If these sewers are overloaded or blocked with debris, the water cannot escape and flooding occurs.

The Water Quality Act of 1987 requires that municipalities obtain permits for discharges of stormwater runoff so that nonpoint sources of pollution are controlled. In the past, many cities had a single system to handle both sewage and stormwater runoff. During heavy precipitation or spring thaws, the runoff from streets could be so large that the wastewater treatment plant could not handle the volume. The wastewater was then diverted directly into the receiving body of water without being treated. Because of these new requirements, some cities have created areas in which to store this excess

Table 16.2 International Water Disputes

River/Lakes	Countries Involved	Issues
Asia		
Brahmaputra, Ganges, Farakka	Bangladesh, India, Nepal	Alluvial deposits, dams, floods, irrigation, international quotas
Mekong	Cambodia, Laos, Thailand, Vietnam	Floods, international quotas
Salween	Tibet, China (Yunan), Burma	Alluvial deposits, floods
Middle East		
Euphrates, Tigris	Iraq, Syria, Turkey	International quotas, salinity levels
West Bank Aquifer, Jordan, Litani, Yarmuk	Israel, Jordan, Lebanon, Syria	Water diversion, international quotas
Africa		
Nile	Mainly Egypt, Ethiopia, Sudan	Alluvial deposits, water diversion, floods, irrigation, international quotas
Lake Chad	Nigeria, Chad	Dam
Okavango	Namibia, Angola, Botswana	Water diversion
Europe		
Danube	Hungary, Slovak Republic	Industrial pollution
Elbe	Germany, Czech Republic	Industrial pollution, salinity levels
Meuse, Escaut	Belgium, Netherlands	Industrial pollution
Szamos (Somes)	Hungary, Romania	Water allocation
Tagus	Spain, Portugal	Water allocation
Americas		
Colorado, Rio Grande	United States, Mexico	Chemical pollution, international quotas, salinity levels
Great Lakes	Canada, United States	Pollution
Lauca	Bolivia, Chile	Dams, salinity
Paran	Argentina, Brazil	Dams, flooding of land
Cenepa	Ecuador, Peru	Water allocation

Sources: Walter H. Corson (ed.), *The Global Ecology Handbook* (Boston: Beacon Press, 1990), pp. 160–61; and Peter H. Fleick (ed.), *Water in Crisis: A Guide to the World's Freshwater Resources* (New York: Oxford University Press, 1993).

water until it can be treated. This is expensive and, therefore, is done only if federal or state funding is available. Many cities have also gone through the expensive process of separating their storm sewers from their sanitary sewers. A good example of the long and costly process of separation of sewers is Portland, Oregon. By 2001, Portland had completed only half of a 20-year project to separate its storm and sanitary sewer systems. The final costs are estimated to exceed the project's $1 billion budget.

Providing water services is expensive. We must understand that water supplies are limited. We must also understand that water's ability to dilute and degrade pollutants is limited and that proper land-use planning is essential if metropolitan areas are to provide services and limit pollution.

In pursuing these objectives, city planners encounter many obstacles. Large metropolitan areas often have hundreds of local jurisdictions (governmental and bureaucratic areas) that divide responsibility for management of basic water services. The Chicago metropolitan area is a good example. This area is composed of six counties and approximately 2000 local units of government. It has 349 separate water-supply systems and 135 separate wastewater disposal systems. Efforts to implement a water-management plan, when so many layers of government are involved, can be complicated and frustrating.

To meet future needs, urban, agricultural, and national interests will need to deal with a number of issues, such as the following:

- Increased demand for water will generate pressure to divert water to highly populated areas or areas capable of irrigated agriculture.
- Increased demand for water will force increased treatment of wastewater and reuse of existing water supplies.
- In many areas where water is used for irrigation, evaporation of water from the soil over many years results in a buildup of salt in the soil. When the water used to flush the salt from the soil is returned to a stream, the quality of the water is lowered.

- In some areas, wells provide water for all categories of use. If the groundwater is pumped out faster than it is replaced, the water table is lowered.
- In coastal areas, seawater may intrude into the aquifers and ruin the water supply.
- The demand for water-based recreation is increasing dramatically and requires high-quality water, especially for activities involving total body contact, such as swimming.

In 2002, the National Research Council issued a report entitled *Envisioning the Agenda for Water Resources Research in the 21st Century*. The report addresses the future of U.S. water resources and the research needed to support efforts to manage those resources in a sustainable fashion. The purpose of the report is to draw attention to the urgency and complexity of the water-resources issues facing the United States in the twenty-first century and to inform decision makers, researchers, and the public about these issues and the challenges they bring.

The report is organized around the general categories of water availability, water use, water institutions, and water research organization. It makes some 43 recommendations, which can be summarized into four key themes:

1. The challenge of solving the nation's water problems requires a renewed national commitment, which must entail changes in the way research agendas and priorities are established and a significant infusion of new federal funding.
2. Water quality and water quantity need to be viewed as interrelated. Research priorities in these areas, therefore, should be developed in an integrated fashion.
3. More attention must be given to water-related research in the social sciences and research focused on the development of innovative institutions. Such institutions must translate public preferences into water-management goals and objectives, and facilitate the application of science and technology to achieve them.

4. Research on water-related environmental issues needs to become a major and continuing part of the water research agenda.

Water Diversion

Water diversion is the physical process of transferring water from one area to another. The aqueducts of ancient Rome are early examples of water diversion. Thousands of diversion projects have been constructed since then. New York City, for example, receives 90 percent of its water supply from the Catskill Mountains (Schoharie and Ashokan Reservoirs) and from four reservoirs collecting water from tributaries to the Delaware River west of the Catskills. About 10 percent of its supply comes from the Croton watershed east of the Hudson River. The most distant watershed serving New York City is some 200 kilometers (125 miles) away. Los Angeles is another example. It began importing water from the Owens Valley, 400 kilometers (250 miles) to the north, as early as 1913. (See Environmental Close-up: The California Water Plan on page 361.)

While diversion is seen as a necessity in many parts of the world, it often generates controversy. An example of this is the Garrison Diversion Unit in North Dakota, which was originally envisioned as a way to divert water from the Missouri River system for irrigation. (See figure 16.14.) The initial plan to irrigate portions of the Great Plains was developed during the Dust Bowl era of the 1930s. The Federal Flood Control Act of 1944 authorized the construction of the Garrison Diversion Unit. However, one of the original intents of the plan—to divert water from the Missouri River to the Red River—has not been met. Since the Red River flows north into Canada, the project requires international cooperation. The Canadian government has concerns about the effects on water quality of releasing additional water into the Red River. There are also concerns about environmental consequences. Portions of wildlife refuges, native grasslands, and waterfowl breeding marshes would be damaged or destroyed. Some states also have expressed concern about diverting water from the Missouri River,

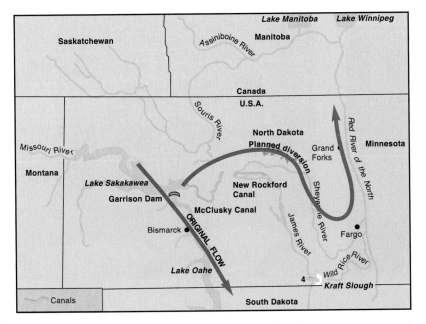

Figure 16.14 **The Garrison Diversion Unit** The original intent of this plan was to divert water from the Missouri River to the McClusky Canal and the Sheyenne River and eventually into Lake Winnepeg by way of the Red River. In the process, additional land could be irrigated, and growing populations in the Red River Valley could be served with adequate water. Opposition has stalled the project, and only portions of the canal system and a pumping station have been completed.

Area	Percent
North America	90
Europe	72
Mediterranean Sea	30
Caribbean Basin	Less than 10
Southeast Pacific	Almost zero
South Asia	Almost zero
South Pacific	Almost zero
West and Central Africa	Almost zero

Table 16.3 Percent of Sewage Treated in Selected Areas

Source: Data from World Resources Institute, 2000–01.

since any water diverted is not available for those downstream on the Missouri River.

While the plan has been modified several times, two sections of canal and a pumping station at Lake Sakakawea have been completed but not used. Proponents continue to push for legislation to complete the connection between the two canals that would allow water to be diverted to the Red River for municipal use. The original intent of diverting water for irrigation has been eliminated from the most recent proposal.

One major consequence of diverting water for irrigation and other purposes is that the water bodies downstream of the diversion are deprived of their source of water. This often has major ecological consequences. Water levels in Lake Chad in Africa are falling due to drought and increased demand for irrigation water. This has affected the populations of fish and other wildlife. In Mexico, 2500 kilometers (1550 miles) of rivers have dried up because the water was diverted for other purposes, resulting in the extinction of 15 species of fish and threatening half of the remaining species.

As people recognize the significance of wildlife habitat, plans are being worked out to balance societal needs with the need for water to maintain wildlife habitat. For example, Mono Lake in the Sierra Mountains in central California began to shrink in 1941 when much of the water that fed the lake was diverted to supply Los Angeles. Since Mono Lake has no outlet, its size is determined by the balance between the water flowing into the lake and evaporation from the lake surface. Consequently, the water level in the lake dropped 13 meters (43 feet), the lake's volume decreased by half, and its salinity doubled. These changes resulted in a loss of habitat important for many ducks and other waterfowl. In 1994, an agreement was reached to increase the amount of water flowing to the lake so that the level of the lake would rise. The plan is to increase the level of the lake by 5 meters (17 feet) over a 20-to-30-year period.

In New South Wales, Australia, an area known as the Macquarie Marshes was affected by water diversion. The original extent of the marshes was reduced by about 50 percent because the Macquarie River, which feeds the marsh, was

dammed to provide irrigation water. In the mid-1990s, because of concerns about the loss of wildlife, an agreement was reached to provide additional water to the area to maintain the marsh and protect the breeding areas of waterfowl.

Wastewater Treatment

Because water must be cleaned before it is released, most companies and municipalities in the developed world maintain wastewater treatment facilities. The percentage of sewage that is treated, however, varies greatly throughout the world. (See table 16.3.) Treatment of sewage is usually classified as primary, secondary, or tertiary. **Primary sewage treatment** is primarily a physical process that removes larger particles by filtering water through large screens and then allowing smaller particles to settle in ponds or lagoons. Water is removed from the top of the settling stage and either released to the environment or to a subsequent stage of treatment. If the water is released to the environment, it does not have any sand or grit; but it still carries a heavy load of organic matter, dissolved salts, bacteria, and other microorganisms. The microorganisms use the organic material for food, and as long as there is sufficient oxygen, they will continue to grow and reproduce. If the receiving body of water is large enough and the organisms have enough time, the organic matter will be degraded.

a.

c.

b.

Figure 16.15 **Primary and Secondary Wastewater Treatment** Primary treatment is physical; it includes filtrating and settling of wastes. Photograph (a) is of a settling tank in which particles settle to the bottom. Secondary treatment is mostly biological and includes the concentration of dissolved organics by microorganisms. Two major types of secondary treatment are trickling filter and activated sludge methods. Photograph (b) shows a trickling filter system, and photograph (c) shows an activated sludge system.

In crowded areas, where several municipalities take water and return it to a lake or stream within a few kilometers of each other, primary water treatment is not adequate and major portions of the receiving body of water are affected.

Secondary sewage treatment is a biological process that usually follows primary treatment. It involves holding the wastewater until the organic material has been degraded by the bacteria and other microorganisms. Secondary treatment facilities are designed to promote the growth of microorganisms. To encourage this action, the wastewater is mixed with large quantities of highly oxygenated water, or the water is aerated directly, as in a trickling filter system or an activated sludge system. In a **trickling filter system,** the wastewater is sprayed over the surface of rock or other substrate to increase the amount of dissolved oxygen. The rock also provides a place for a film of bacteria and other microbes to attach so they are exposed simultaneously to the organic material in the water and to oxygen. These microorganisms feed on the dissolved organic matter and small suspended particles, which then become incorporated into their bodies as part of their cell structure. The bodies of the microorganisms are larger than the dissolved and suspended organic matter, so this process concentrates the organic wastes into particles that are large enough to settle out. This mixture of organisms and other particular matter is called **sewage sludge.** The sludge that settles consists of living and dead microorganisms and their waste products.

In **activated-sludge sewage treatment** plants, the wastewater is held in tanks and has air continuously bubbled through it. The sludge eventually is moved to settling tanks where the water and sludge can be separated. To make sure that the incoming wastewater has appropriate kinds and amounts of decay organisms, some of the sludge is returned to aeration tanks, where it is mixed with incoming wastewater. This kind of process uses less land than a trickling filter. (See figure 16.15.) Both processes produce a sludge that settles out of the water.

The sludge that remains is concentrated and often dewatered (dried) before disposal. Sludge disposal is a major problem in large population centers. In the San Francisco Bay area, 2500 metric tons of sludge are produced each day. Most of this is carried to landfills and lagoons, and some is composted and returned to the land as fertilizer. Some municipalities incinerate their sludge. In other areas, if the sewage sludge is free of heavy metals and other contaminants that might affect plant growth or the quality and safety of food products, it is applied directly to agricultural land as a fertilizer and soil conditioner.

In North America and much of the developed world, wastewater receives both primary and secondary sewage treatment.

Restoring the Everglades

Everglades National Park is a unique, subtropical, freshwater wetland visited by about a million people per year. This unique ecosystem exists because of an unusual set of conditions. South Florida is a nearly flat landscape with a slight decrease in elevation from the north to the south. In addition, the substrate is a porous limestone that allows water to flow through rather easily. Originally, water drained in a broad sheet from Lake Okeechobee southward to Florida Bay. This constant flow of water sustained a vast, grassy wetland interspersed with patches of trees.

After Everglades Park was established in 1947, about 800,000 hectares (2 million acres) of wetlands to the north were converted to farms and urban development. South Florida boomed. Some 4.5 million people currently live in the horseshoe crescent around the Everglades region, and new residents arrive each day. Conversion of land to agriculture and urban development required changes in the natural flow of water. Dams, drainage canals, and water diversion supported and protected the human uses of the area but cut off the essential, natural flow of freshwater to the Everglades. Changes in the normal pattern of water flow to the Everglades resulted in periods of drought and a general reduction in the size of this unique wetland region. Populations of wading birds in the southern Everglades fell drastically as their former breeding and nesting areas dried up. Other wildlife, such as alligators, Florida panthers, snail kites, and wood storks, also were negatively affected because drought reduces suitable habitat during parts of the year and the animals are forced to congregate around the remaining sources of water. The quality of the water is also important. The original wetland ecosystem was a nutrient-poor system. The introduction of nutrients into the water from agricultural activities encouraged the growth of exotic plants that replaced natural vegetation.

As people recognized that the key element necessary to preserve the Everglades was a constant, reliable source of clean freshwater, several steps were taken to modify water use to preserve the Everglades. The Kissimmee River, which flows into Lake Okeechobee, was channelized into an arrow-straight river in the late 1960s. This project destroyed the marshes and allowed nutrients from dairy farms and other agricultural activities to pollute the lake and Everglades Park, to which the water from Lake Okeechobee eventually flowed. To help alleviate this problem, in 1990, the U.S. Army Corps of Engineers began to return the Kissimmee to its natural state, with twisting oxbow curves and extensive wetlands. This allows the plants in the natural wetlands to remove much of the nutrient load before it enters the lake. To reduce the likelihood of further development near the park, in 1989, Congress approved the purchase of 43,000 hectares (100,000 acres) for an addition to the east section of the park. Florida obtained an additional 60,000 hectares (150,000 acres) as an additional buffer zone for the park.

For several years, the U.S. Army Corps of Engineers and the South Florida Water Management District cooperated in the development of a comprehensive restoration plan that was finished in 2000. The development of the plan involved the input of scientists, politicians, various business interests, and environmentalists. Key components of the plan are:

1. Developing facilities to store surface water and pump water into aquifers so that it can be released when needed

2. Developing wetlands to treat municipal and agricultural runoff so that nutrient loads are reduced

3. Using clean wastewater to recharge aquifers and supply water to wetlands in the Miami area

4. Reducing the amount of water lost through levees and redirecting water to the Everglades

5. Removing barriers to the natural flow of water through the Everglades

The plan received strong support from Congress in the fall of 2001, when it allocated $1.4 billion to begin implementing the plan. It will require many more billions of dollars and up to 30 years to accomplish all aspects of the plan, but if the plan continues to be implemented over the next few decades, the Everglades ecosystem will be restored to a more stable condition and will have a more hopeful future.

Everglades National Park is in southern Florida.

Everglades map: Copyright 1990 U.S. News & World Report, L.P. Reprinted with permission.

Everglades wading birds once numbered more than a million, but pollution and drought brought on by water diversion have decimated their number.

Construction threatens to destroy the fragile ecosystem of the Everglades.

While the water has been cleansed of its particles and dissolved organic matter, it still has microorganisms that might be harmful. Therefore, the water discharged from these sewage treatment plants must be disinfected. The least costly method of disinfection is chlorination. However, many people feel that the use of chlorine should be discontinued, since chlorination may be responsible for the creation of chlorinated organic compounds that are harmful. Therefore, other methods of killing microorganisms are being explored. The chemical ozone also kills microorganisms and has been substituted for chlorine by some facilities. Ultraviolet light and ultrasonic energy can also be used. Vancouver, Washington, for example, has been using ultraviolet light treatment since 1998 as a final treatment before returning water to the Columbia River. However, chlorine is inexpensive and very effective, so it continues to be the primary method used.

A growing number of larger sewage treatment plants use additional processes called tertiary sewage treatment. **Tertiary sewage treatment** involves a variety of techniques to remove dissolved pollutants left after primary and secondary treatments. (See table 16.4.) The tertiary treatment of municipal wastewater is often used to remove phosphorus and nitrogen that could increase aquatic plant growth. Some municipalities are using natural or constructed wetlands to serve as tertiary sewage treatment systems. In other cases, the effluent from the treatment facility is used to irrigate golf courses, roadside vegetation, or cropland. The vegetation removes excess nutrients and prevents them from entering streams and lakes where they would present a pollution problem. Tertiary treatment of industrial and other specialized wastewater streams is very costly because it requires specific chemical treatment of the water to eliminate specific problem materials. Many industries maintain their own wastewater facilities and design specific tertiary treatment processes to match the specific nature of their waste products.

As water has become scarce in many parts of the world, people have looked at wastewater as a source of water for other purposes. Water from wastewater

Table 16.4 Tertiary Treatment Methods

Kind of Tertiary Treatment	Problem Chemicals	Methods
Biological	Phosphorus and nitrogen compounds	1. Large ponds are used to allow aquatic plants to assimilate the nitrogen and phosphorus compounds from the water before the water is released. 2. Columns containing denitrifying bacteria are used to convert nitrogen compounds into atmospheric nitrogen.
Chemical	Phosphates and industrial pollutants	1. Water can be filtered through calcium carbonate. The phosphate substitutes for the carbonate ion, and the calcium phosphate can be removed. 2. Specific industrial pollutants, which are nonbiodegradable, may be removed by a variety of specific chemical processes.
Physical	Primarily industrial pollutants	1. Distillation 2. Water can be passed between electrically charged plates to remove ions. 3. High-pressure filtration through small-pored filters 4. Ion-exchange columns

plants can be used for irrigation, industrial purposes, cooling, and many other activities. Ultimately, it is possible to have a closed loop system for domestic water in which the output of the wastewater plant becomes the input for the drinking water supply.

Salinization

Another water-use problem results from **salinization,** an increase in salinity caused by growing salt concentrations in soil. This is primarily a problem in areas where irrigation has been practiced for several decades. As water evaporates from soil or plants extract the water they need, the salts present in all natural waters become concentrated. Since irrigation is most common in hot, dry areas that have high rates of evaporation, there is generally an increase in the concentration of salts in the soil and in the water that runs off the land. (See figure 16.16.) Every river increases in salinity as it flows to the ocean. The salinity of the Colorado River water increases 20 times as it passes through irrigated cropland between Grand Lake in north-central Colorado and the Imperial

Dam in southwest Arizona. The problem of salinity will continue to increase as irrigation increases.

Groundwater Mining

Groundwater mining means that water is removed from an aquifer faster than it is replaced. When this practice continues for a long time, the water table eventually declines. Groundwater mining is common in areas of the western United States and throughout the world. In North America, it is a particular problem due to growing cities and increasing irrigation. In aquifers with little or no recharge, virtually any withdrawal constitutes mining, and sustained withdrawals will eventually exhaust the supply. This problem is particularly serious in communities that depend heavily on groundwater for their domestic needs.

Groundwater mining can also lead to problems of settling or subsidence of the ground surface. Removal of the water allows the ground to compact, and large depressions may result. For example, in the San Joaquin Valley of California, groundwater has been withdrawn for

Table 16.5 Groundwater Depletion in Major Regions of the World

Region/Aquifer	Estimates of Depletion
California	Groundwater overdraft exceeds 1.7 billion cubic meters (60 billion cubic feet) per year. The majority of the depletion occurs in the Central Valley, which is referred to as the vegetable basket of the United States.
Southwestern United States	In parts of Arizona, water tables have dropped more than 120 meters (400 feet). Projections for parts of New Mexico indicate that water tables will drop an additional 22 meters (70 feet) by 2020.
High Plains aquifer system, United States	The Ogallala aquifer underlies nearly 20 percent of all the irrigated land in the United States. To date, the net depletion of the aquifer is in excess of 350 billion cubic meters (12 trillion cubic feet), or roughly 15 times the average annual flow of the Colorado River. Most of the depletion has been in the Texas high plains, which have witnessed a 26 percent decline in irrigated land from 1979 to 1989. Current depletion is estimated to be in excess of 13 billion cubic meters (450 billion cubic feet) a year.
Mexico City and Valley of Mexico	Use exceeds natural recharge by 60 to 85 percent, causing land subsidence and falling water tables.
African Sahara	North Africa has vast nonrecharging aquifers where current depletion exceeds 12 billion cubic meters (425 billion cubic feet) a year.
India	Water tables are declining throughout much of the most productive agricultural land in India. In parts of the country, groundwater levels have declined 90 percent during the past two decades.
North China	The water table underneath portions of Beijing has dropped 40 meters (130 feet) during the past 40 years. A large portion of northern China has significant groundwater overdraft.
Arabian peninsula	Groundwater use is nearly three times greater than recharge. At projected depletion rates, exploitable groundwater reserves could be exhausted within the next 50 years. Saudi Arabia depends on nonrenewable groundwater for roughly 75 percent of its waters. This includes irrigation of 2 million to 4 million metric tons of wheat per year.

Figure 16.16 Salinization As water evaporates from the surface of the soil, the salts it was carrying are left behind. In some areas of the world, this has permanently damaged cropland.

irrigation and cultivation since the 1850s, and groundwater levels have fallen over 100 meters (300 feet). More than 1000 hectares (approximately 2500 acres) of ground have subsided, some as much as 6 meters (20 feet). Currently, the ground surface in that area is sinking 30 centimeters (12 inches) per year. London, Mexico City, Venice, Houston, and Las Vegas are some other cities that have experienced subsidence as a result of groundwater withdrawal. Table 16.5 lists some estimated amounts of groundwater depletion in certain areas of the United States and the world. As people recognize the severity of the problem, public officials are beginning to develop water conservation plans for their cities. Albuquerque, New Mexico, which relies on groundwater for its water supply, has an extensive public education program to encourage people to reduce their water consumption. Since grass demands water, people are encouraged to use desert plants for landscaping or collect rainwater to water their lawns. Finding and correcting leaks, reducing the amount of water used in bathing, and recycling water from swimming pools are other conservation strategies.

Groundwater mining poses a special problem in coastal areas. As the fresh groundwater is pumped from wells along the coast, the saline groundwater moves inland, replacing fresh groundwater with unusable saltwater. This process, called **saltwater intrusion,** is shown in figure 16.17. Saltwater intrusion is a serious problem in heavily populated coastal areas throughout the world.

Preserving Scenic Water Areas and Wildlife Habitats

Some bodies of water have unique scenic value. To protect these resources, the way in which the land adjacent to the water is used must be consistent with preserving these scenic areas.

The U.S. Federal Wild and Scenic Rivers Act of 1968 established a system to protect wild and scenic rivers from development. All federal agencies must consider the wild, scenic, or recreational values of certain rivers in planning for the use and development of the rivers and adjacent land. The process of designating a river or part of a river as wild or scenic is complicated. It often encounters local opposition from businesses dependent on growth. Following reviews by state and federal agencies, rivers may be designated as wild and scenic by action of either Congress or the secretary of the Interior. Sections of over 150 streams comprising about 12,000 kilometers (7700 miles) in the United States have been designated as wild or scenic.

Many unique and scenic shorelands have also been protected from future development. Until recently, estuaries and shorelands have been subjected to significant physical modifications, such as dredging and filling, which may improve conditions for navigation and construction but destroy fish and wildlife habitats. Recent actions throughout North America have attempted to restrict the development of shorelands. Development has been restricted in some particularly scenic areas, such as Cape Cod National Seashore in Massachusetts and the Bay of Fundy in the Atlantic provinces of Canada.

Historically, poorly drained areas were considered worthless. Subsequently, many of these wetlands were filled or drained and used for building sites. At the time of European settlement, the area that is now the conterminous United States contained an estimated 89 million hectares (221 million acres) of wetlands. Over time, wetlands have been drained, dredged, filled, leveled, and flooded to the extent that less than half of the original wetland acreage remains. Figure 16.18 shows the most recent causes of wetlands loss.

Today, 95 percent of the remaining 20 million hectares (50 million acres) of wetlands in the United States are inland freshwater wetlands. The remaining 5 percent are in saltwater estuarine environments. Freshwater forested wetlands make up the single largest category of all wetlands in the conterminous United States. The natural and economic importance of wetlands has been recognized only recently. In addition to providing spawning and breeding habitats for many species of wildlife, wetlands act as natural filtration systems by trapping nutrients and pollutants and preventing them from entering adjoining lakes, streams, or estuaries. Wetlands also slow down floodwaters and permit nutrient-rich particles to settle out. In

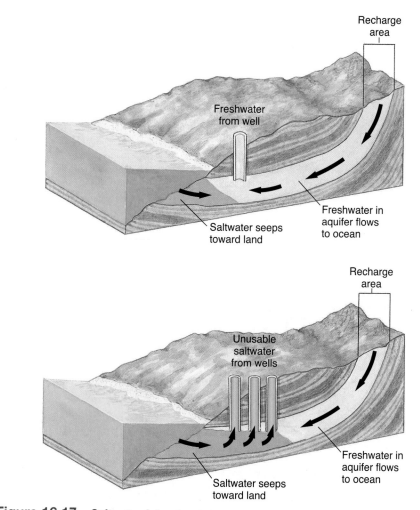

Figure 16.17 **Saltwater Intrusion** When saltwater intrudes on fresh groundwater, the groundwater becomes unusable for human consumption and for many industrial processes.

Causes of Wetlands Loss (1986–2000)

Figure 16.18 **Wetlands Conversion** The loss of wetlands occurs because people convert wetlands to other uses. Conversion to urban and rural housing and other infrastructure accounts for just over 50 percent of the wetlands loss. Draining wetlands for agriculture accounts for an additional 26 percent.

Source: U.S. Geological Survey.

GLOBAL PERSPECTIVE

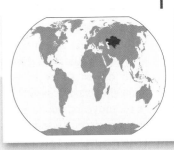

Death of a Sea

"At the beginning you drink water, at the end you drink poison" is an old Uzbek proverb about living on the wrong end of a watershed. The Aral Sea lies on the border between Uzbekistan and Kazakhstan in the former Soviet Union. It was once larger than any of the Great Lakes except Superior. Today, it has actually split into two waterbodies (the North and South Aral Seas), both of which are disappearing. Since the 1920s, agricultural planners have used up the Aral Sea, diverting its waters for irrigation. The two rivers feeding the Aral were drawn off to irrigate millions of hectares of cotton. An irrigation canal, the world's longest, stretches over 1300 kilometers (800 miles) into Turkmenistan, paralleling the boundaries of Afghanistan and Iran. The cotton production plan worked, and by 1937, the Soviet Union was a net exporter of cotton. The success of the cotton program, however, spelled the end for the Aral Sea.

For a long time, the ecological impact on the sea and surrounding area was largely hidden from public view. Since the 1960s, however, the Aral has lost 75 percent of its volume and over 50 percent of its surface area, or over 22,000 square kilometers (8500 square miles) of what are now largely dry, salt-encrusted wastelands. The once-thriving fishing industry that depended on the water is all but gone. Some 20 of the 24 fish species there have disappeared. The fish catch, which totaled 44,000 metric tons a year in the 1950s and supported some 60,000 jobs, has dropped to zero. Abandoned fishing villages dot the sea's former coastline.

Another apparent consequence of the dried-up sea is a host of human illnesses. A high rate of throat cancer is attributed to dust from the drying sea. Each year, winds pick up a million metric tons of a toxic dust-salt mixture from the dry sea bed and deposit it on the surrounding farmland, harming or killing crops. The low river flows have concentrated salts and toxic chemicals, making water supplies hazardous to drink and contributing to disease. In the northwest part of the Republic of Uzbekistan, the infant mortality rate is the highest in the former Soviet Union.

The former fishing center of the sea was a town named Muynak. The town is now landlocked more than 30 kilometers (20 miles) from the water. Less than 25 years ago, Muynak was a seaport. The population of Muynak is down from 40,000 in 1970 to 12,000 today. In 1990, the mayor and last harbormaster of Muynak commented:

> The water continued to go away while the salinity increased. The weather changed for the worse, with the summers getting hotter and the winters colder. The people feel salt on their lips and in their eyes all the time. It's getting harder to open your eyes here.

In 1993, the central Asian republics of Uzbekistan, Kazakhstan, Turkmenistan, Tajikistan, and Kaygyzstan signed an international agreement to save the Aral Sea. By the mid-1990s, the mayor of Aralsk was desperate to save the North Aral Sea, so he had a huge pile of sand stacked into a makeshift dam to prevent any further water loss. Even though the dam had to be rebuilt many times, it succeeded in increasing the water level and restoring some fishing. The World Bank was so impressed by these results that it approved a $64 million loan to build a permanent dam and channel system on the North Aral Sea and the river that flows into it. The government of Kazakhstan raised the additional $21 million needed to complete the project, and construction started in 2002. The hope is that the dam and channel system will refill the North Aral Sea, restore fishing and agriculture, and improve human health.

These ships once sailed on the Aral Sea.

addition, wetlands can act as reservoirs and release water slowly into lakes, streams, or aquifers, thereby preventing floods. (See figure 16.19.) Coastal estuarine zones and adjoining sand dunes also provide significant natural flood control. Sand dunes act as barriers and absorb damaging waves caused by severe storms. In recent years, public appreciation of the ecological, social, and economic values of wetlands has increased substantially. The increased awareness of how much wetland acreage has been lost or damaged since the time of European settlement and the consequences of those losses has led to the development of many federal, state, and local wetland protection programs and laws.

- Filter toxic wastes, excess nutrients, sediments, and other pollutants
- Help prevent erosion
- Reduce flooding by storing stormwater

- Reduce storm damage by absorbing waves
- Provide feeding and resting spots for migratory waterfowl

- Provide food and habitat for other aquatic species

- Provide nursery sites for the young of a number of species, including oysters, clams, crabs, and shrimp

Figure 16.19 **The Value of Wetlands** Wetlands are areas covered with water for most of the year that support aquatic plant and animal life. Wetlands can be either fresh or saltwater and may be isolated potholes or extensive areas along rivers, lakes, and oceans. We once thought of wetlands as only a breeding site for mosquitoes. Today, we are beginning to appreciate their true value.

The Dead Zone of the Gulf of Mexico

Each summer, a major "dead zone" of about 18,000 square kilometers (7000 square miles) develops in the Gulf of Mexico off the mouth of the Mississippi River. This dead zone contains few fish and bottom-dwelling organisms. It is caused by low oxygen levels (hypoxia) brought about by the rapid growth of algae and bacteria in the nutrient-rich waters. The nutrients can be traced to the extensive use of fertilizer in the major farming areas of the central United States, farming areas drained by the Mississippi River and its tributaries. About 1.6 million metric tons of nitrogen—mostly fertilizer runoff from Midwestern farms flows out of the Mississippi River every year.

The hypoxia problem begins when nitrogen and other nutrients wash down the Mississippi River to the Gulf of Mexico, where they trigger a bloom of microscopic plants and animals. The dead cells and fecal matter of the organisms then fall to the seafloor. As growing colonies of bacteria digest this waste, they consume dissolved oxygen faster than it can be replenished. The flow of oxygen-rich water from the Mississippi cannot rectify the problem, because differences in temperature and density cause the warm freshwater to float above the colder, salty ocean water. Crustaceans, worms, and any other animals that cannot swim out of the hypoxic zone die.

According to the EPA, nutrient pollution has degraded more than half of U.S. estuaries. In 2002, the National Research Council named nutrient pollution and the sustainability of fisheries as the most important problems facing the U.S. coastal waters in the next decade.

Gulf of Mexico fisheries, which generate some $2.8 billion a year in revenue, are one potential casualty of hypoxia. Hypoxia can block crucial migration of shrimp, which must move from inland nurseries to feed and spawn offshore. In other places in the world, such as the Black and Baltic Seas, hypoxia has been responsible for the collapse of some commercial fisheries.

Several approaches could reduce the amount of hypoxia-causing nitrogen released into the Mississippi River Basin. These include:

- Reduce use of nitrogen-based fertilizers and improve storage of manure. Reduce runoff from feedlots.
- Plant perennial crops instead of fertilizer-intensive corn and soybeans on 10 percent of the acreage.
- Remove nitrogen and phosphorus from domestic wastewater.
- Restore 2 million to 4 million hectares (5–10 million acres) of wetlands, which absorb nitrogen runoff.

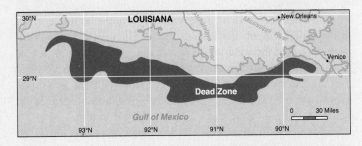

Nitrogen Delivered to Gulf (kilograms per hectare per year)

- Low (less than 3)
- Medium (3 to 8)
- High (more than 8)

Nitrogen delivered to Gulf: Reprinted by permission from *Nature*, 403; 761, copyright © 2000 Macmillan Publishers Ltd.

Source: Dead Zone map modified from R. B. Alexander, R. A. Smith, and G. E. Schwartz, "Effect of Stream Channel Size on the Delivery of Nitrogen to the Gulf of Mexico," *Nature*, 17 (February 2000): 761.

Water is a renewable resource that circulates continually between the atmosphere and the Earth's surface. The energy for the hydrologic cycle is provided by the sun. Water loss from plants is called evapotranspiration. Water that infiltrates the soil and is stored underground in the tiny spaces between rock particles is called groundwater, as opposed to surface water that enters a river system as runoff. There are two basic kinds of aquifers. Unconfined aquifers have an impervious layer at the bottom and receive water that infiltrates from above. The top of the layer of water is called the water table. A confined aquifer is sandwiched between two impervious layers and is often under pressure. The recharge area may be a great distance from where the aquifer is tapped for use. The way in which land is used has a significant impact on rates of evaporation, runoff, and infiltration.

The four human uses of water are domestic, agricultural, instream, and industrial. Water use is measured by either the amount withdrawn or the amount consumed. Domestic water is in short supply in many metropolitan areas. Most domestic water is used for waste disposal and washing, with only a small amount used for drinking. The largest consumptive use of water is for agricultural irrigation. Major in-stream uses of water are for hydroelectric power, recreation, and navigation. Most industrial uses of water are for cooling and for dissipating and transporting waste materials.

Major sources of water pollution are municipal sewage, industrial wastes, and agricultural runoff. Nutrients, such as nitrates and phosphates from wastewater treatment plants and agricultural runoff, enrich water and stimulate algae and aquatic plant growth. Organic matter in water requires oxygen for its decomposition and, therefore, has a large biochemical oxygen demand (BOD). Oxygen depletion can result in fish death and changes in the normal algal community, which leads to visual and odor problems.

Point sources of pollution are easy to identify and resolve. Nonpoint sources of pollution, such as agricultural runoff and mine drainage, are more difficult to detect and control than those from municipalities or industries.

Thermal pollution occurs when an industry returns heated water to its source. Temperature changes in water can alter the kinds and numbers of plants and animals that live in it. The methods of controlling thermal pollution include cooling ponds, cooling towers, and dry cooling towers.

Wastewater treatment consists of primary treatment, a physical settling process; secondary treatment, biological degradation of the wastes; and tertiary treatment, chemical treatment to remove specific components. Two major types of secondary wastewater treatments are the trickling filter and the activated sludge sewage methods.

Groundwater pollution comes from a variety of sources, including agriculture, landfills, and septic tanks. Marine oil pollution results from oil drilling and oil-tanker accidents, runoff from streets, improper disposal of lubricating oil from machines and car crankcases, and intentional discharges from oil tankers during loading or unloading.

Reduced water quality can seriously threaten land use and in-place water use. In the United States and other nations, legislation helps to preserve certain scenic water areas and wildlife habitats. Shorelands and wetlands provide valuable services as buffers, filters, reservoirs, and wildlife areas. Water management concerns of growing importance are groundwater mining, increasing salinity, water diversion, and managing urban water use. Urban areas face several problems, such as providing water suitable for human use, collecting and treating wastewater, and handling stormwater runoff in an environmentally sound manner. Water planning involves many governmental layers, which makes effective planning difficult.

Issues – Analysis

Is There Lead in Our Drinking Water?

A reader of a major metropolitan newspaper wrote a letter to the editor outlining her family's decision to move from the suburbs to the city to experience arts, culture, and greater diversity, and to help prevent urban sprawl. Her family was very happy with their decision until they received a letter from the city water department informing them that everyone in their neighborhood had lead in their drinking water, and it would cost each homeowner thousands of dollars to replace old pipes— a measure that might or might not fix the problem. The lead level was 400 parts per billion (ppb), which is more than 20 times higher than safe levels. The Environmental Protection Agency takes action if the level exceeds 15 ppb for 90 percent of homes tested.

What is the health risk from lead exposure? Children, babies, and pregnant women face the highest risk because children absorb lead at a faster rate than adults and lead has a greater impact on developing bodies. Studies from the Centers for Disease Control and Prevention (CDC) have established that high lead exposure levels are likely to result in learning disorders in babies and children under age 6, and may even lead to behavioral problems. Adults are not exempt from effects either. Exposure to high lead levels is associated with high blood pressure, infertility, decreased kidney and muscular function, and possibly even cancer. Lead mimics calcium in the body, moving through the liver, kidneys, and brain. It can be stored in the bones and teeth over long periods of time. In the brain, lead disrupts the electrical messages sent to cells throughout the body.

How widespread is the problem? Lead paint in older buildings is the leading cause of lead exposure; however, drinking water accounts for 20 percent of exposure. The EPA estimates that 98 percent of homes in the country have plumbing that can slowly leach lead into tap water as the pipes and fixtures deteriorate over time. Heat and acidity can accelerate the process, as can changes in the chemicals used to treat drinking water. Lead exposure levels from water increase to 50 percent for babies who are fed formula. While residential drinking water is routinely tested, there is no federal mandate to test for lead at schools and daycare centers. Concern by parents and school officials has led to voluntary testing for lead in selected areas of the country.

What has been done? The federal government has long recognized the severe health problems for children and pregnant women associated with lead. Leaded gasoline and house paint were banned in the 1970s. The Safe Drinking Water Act amendments of 1986 banned lead pipes and solder in new construction and repairs. However, there are miles of old lead pipes in drinking water systems throughout the country. For example, in Washington, D.C., the city is trying to replace 23,000 lead service lines. The good news is that these actions by the federal government have resulted in a significant decrease in lead exposure levels. In 2001, 2.2 percent of children ages 1 to 5 had blood lead levels that exceeded CDC standards, compared to 90 percent in 1976. The bad news is that urban and suburban areas in Maryland, Pennsylvania, Washington, New York, and the District of Columbia, to name a few, have recently found high lead levels in drinking water in homes, schools, and daycare centers. In 2004, Senator Jim Jeffords of Vermont introduced the Lead-Free Drinking Water Act, which would ban all lead plumbing fixtures and grant $200 million to the District of Columbia to replace lead plumbing throughout the city.

What do you think should be done?

- Should schools and daycare centers be required to monitor drinking water for lead?
- If high lead levels are found, who should pay for short-term solutions such as bottled water and additional testing or for permanent solutions such as replacing all plumbing?
- Who should pay to replace lead pipes in homes and cities throughout the nation?

Key Terms

activated-sludge sewage treatment *374*
aquiclude *355*
aquifer *354*
aquitard *355*
artesian well *355*
biochemical oxygen demand (BOD) *364*
confined aquifer *355*
domestic water *357*
eutrophication *365*
evapotranspiration *354*
fecal coliform bacteria *366*
groundwater *354*

groundwater mining *376*
hydrologic cycle *354*
industrial water use *359*
in-stream water use *360*
irrigation *358*
limiting factor *364*
nonpoint source *365*
point source *365*
porosity *355*
potable water *353*
primary sewage treatment *373*
runoff *354*

salinization *376*
saltwater intrusion *377*
secondary sewage treatment *374*
sewage sludge *374*
stormwater runoff *371*
tertiary sewage treatment *376*
thermal pollution *369*
trickling filter system *374*
unconfined aquifer *354*
vadose zone *355*
water diversion *372*
water table *354*

Review Questions

1. Describe the hydrologic cycle.
2. Distinguish between withdrawal and consumption of water.
3. What are the similarities between domestic and industrial water use? How are they different from in-stream use?
4. How is land use related to water quality and quantity? Can you provide local examples?
5. What is biochemical oxygen demand? How is it related to water quality?
6. How can the addition of nutrients such as nitrates and phosphates result in a reduction of the amount of dissolved oxygen in the water?
7. Differentiate between point and nonpoint sources of water pollution.
8. How are most industrial wastes disposed of? How has this changed over the past 25 years?
9. What is thermal pollution? How can it be controlled?
10. Describe primary, secondary, and tertiary sewage treatment.
11. What are the types of wastes associated with agriculture?
12. Why is stormwater management more of a problem in an urban area than in a rural area?
13. Define groundwater mining.
14. How does irrigation increase salinity?
15. What are the three major water services provided by metropolitan areas?

Critical Thinking Questions

1. Leakage from freshwater distribution systems accounts for significant losses. Is water so valuable that governments should require systems that minimize leakage to preserve the resource? Under what conditions would you change your evaluation?
2. Do nonfarmers have an interest in how water is used for irrigation? Under what conditions should the general public be involved in making these decisions along with the farmers who are directly involved?
3. Should the United States allow Mexico to have water from the Rio Grande and the Colorado River, both of which originate in the United States and flow to Mexico?
4. Do you believe that large-scale hydroelectric power plants should be promoted as a renewable alternative to power plants that burn fossil fuels? What criteria do you use for this decision?
5. What are the costs and the benefits of the proposed Garrison Diversion Unit? What do you think should happen with this project?
6. How might you be able to help save freshwater in your daily life? Would the savings be worth the costs?
7. Look at the hydrologic cycle in figure 16.3. If global warming increases the worldwide temperature, how should increased temperature directly affect the hydrologic cycle?
8. After reading the Issues—Analysis concerning the California Water Plan, do you believe water should be diverted from northern California to southern California?

Welcome to McGraw-Hill's Online Learning Center

Location: www.mhhe.com/enger10e

Interactive Exploration

Check out the website at www.mhhe.com/enger10e and click on the cover of this textbook for quizzing, career information, case studies, and hot links for the following topics:

Water Use and Management

Groundwater

Water Pollution

Impact of Humans on the Sea; Pollution

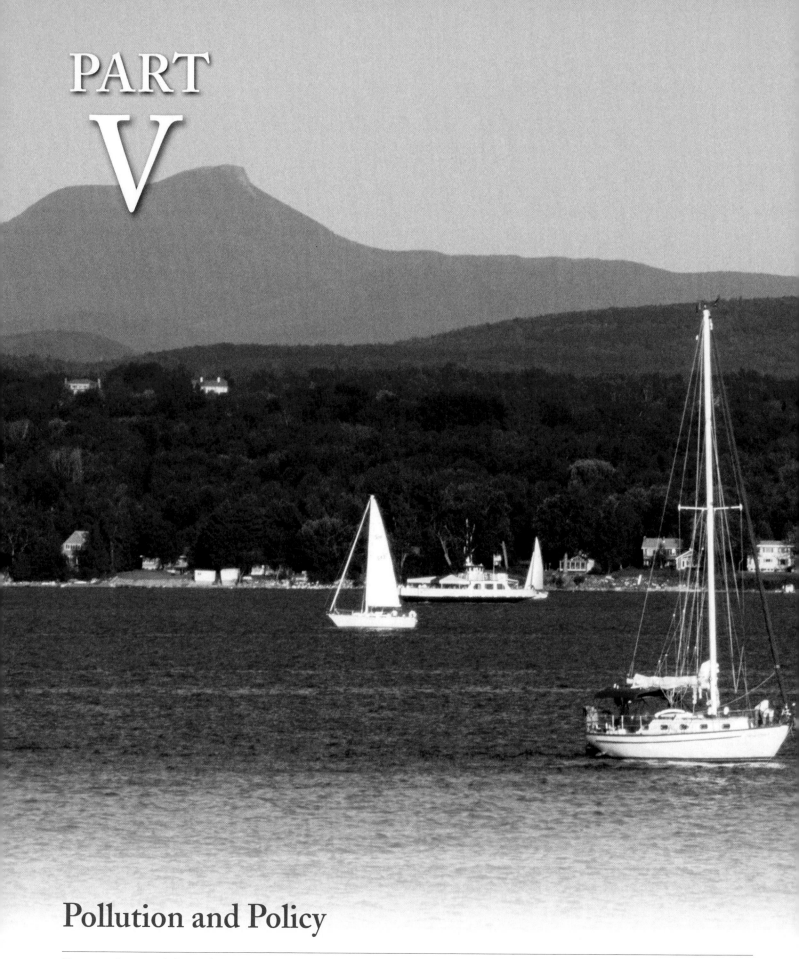

PART V

Pollution and Policy

Human use has caused changes in the water quality of Lake Champlain.

Lake Champlain: It Isn't Easy Stayin' Clean

Alan W. McIntosh

University of Vermont

Lake Champlain is a magnificent, deep glacial lake stretching 177 kilometers (110 miles) from southern Vermont to its outlet, the Richelieu River, in southern Quebec. Its watershed of 21,408 square kilometers (8234 square miles) lies in one of the most sparsely developed parts of the eastern United States. The lake's remarkable beauty has long been celebrated. In 1870, novelist Henry James described Lake Champlain as "...delightfully free, noble and open...its great beauty is the really great style of its landscape."

The reality, however, is that Lake Champlain today is far from pristine. Its waters face a triple threat from phosphorus enrichment, persistent toxic substances such as mercury, and invasive species such as the zebra mussel. Let's look at the challenge posed by each of these issues.

The problems caused by high concentrations of phosphorus are partly a result of human perception: We don't like to swim and boat in waters colored green by algae. Simply put, the more phosphorus that enters Lake Champlain, the greener the water becomes. A more sinister symptom of overenrichment has appeared in recent years. Toxins released by blue-green algae blooms in the summer have killed several dogs that ingested lake water.

The fight to control phosphorus inputs to the lake was fairly easy at first. Sewage treatment plants were improved and phosphate detergents were banned. Then came the hard part: 80 percent of the phosphorus entering Lake Champlain now comes from nonpoint sources, with agriculture accounting for 57 percent of the total.

While best management practices, such as constructing manure pits on farms and sweeping suburban streets, can help reduce phosphorus levels in runoff, much of the problem boils down to changing human behavior. Convincing people to pick up their pet waste or reduce their use of lawn chemicals remains a daunting task. While the large point sources of phosphorus in the Lake Champlain basin have mostly been controlled, the real battle now extends from house to house and farm to farm.

The challenges posed by mercury are tougher yet. Long used in products ranging from barometers to milking machines, mercury, if it escapes into the environment, can take many forms. The most menacing is methylmercury, which may be formed in lake sediments by bacteria. This fat-soluble organic compound biomagnifies up food webs, reaching dangerously high levels in some sport fish. In Lake Champlain, walleye contain such high concentrations that women of childbearing age are advised by the Vermont Health Department to avoid consuming the fish.

While Vermont officials have tried to remove as much mercury from circulation as possible by collecting thermometers and recycling mercury-containing fluorescent lightbulbs, about 50 percent of the mercury in Lake Champlain fish is carried by winds to the watershed from distant fossil-fuel power plants and incinerators. Imagine the challenge for Vermont officials trying to control those sources!

The final threat comes from invasive species. Introduced to the Great Lakes in the 1980s, zebra mussels, D-shaped striped bivalves the size of a thumbnail, spread rapidly, reaching southern Lake Champlain in 1993. With each female mussel able to release 1 million eggs annually, the growth of mussel populations after introduction can be explosive. Underwater photos reveal that entire sections of the lake's bottom are now covered with these pesky mollusks.

While their sharp-edged shells can cut the feet of unsuspecting swimmers, zebra mussels can also clog water intake pipes. Since each mussel can filter about a liter of water daily, there may even be ecological consequences. If mussels filter out the plankton that many small fish rely on for food, entire food webs may be at risk of collapse. While some ecosystems, such as portions of the Hudson River, changed rapidly after the invasion of the zebra mussel, similar changes aren't yet apparent in Lake Champlain—but only time will tell.

The challenge is to keep invasive species out of Lake Champlain in the first place. Once there, most invaders are almost impossible to remove. It is possible that the zebra mussels first entered the lake on the hull of a boat coming from an already-infested area elsewhere. While efforts are being made to educate boaters and anglers about how to prevent the introduction of these invaders, it is very hard to ensure the complete success of these programs.

A final thought about the fight to protect Lake Champlain: Remember that ecosystems are complex. With over 80 species of fish and countless other forms of life in the lake, it is hard enough to figure out who's feeding on whom. What happens when algae at the base of the food web are being stimulated by excessive phosphorus, while fish at the top of the web such as walleye may produce fewer offspring due to the effects of mercury? Add an invasive species or two to the mix, and it becomes almost impossible to predict how the lake's ecosystem will behave. This is perhaps the greatest challenge for those trying to protect the quality of lakes like Champlain.

What Do You Think?

1. When considering what an ecosystem is, would you say it is the lake itself or the entire watershed of the lake? Why?

2. Should the use of toxic substances such as mercury be banned in the Lake Champlain watershed? Why or why not?

3. Is it fair to expect farmers and suburban homeowners living within the Lake Champlain basin but far from the lake to spend money to control phosphorus? Why or why not?

4. Can you think of a way you might try to control the spread of the zebra mussel?

5. Of the three challenges discussed in this essay, which do you think is the easiest to solve? Why?

17

Air Quality Issues

Chapter Outline

Objectives

After reading this chapter, you should be able to:

- Recognize that air can accept and disperse significant amounts of pollutants.
- List the major sources and effects of the six criteria air pollutants.
- Describe how photochemical smog is formed and how it affects humans.
- Explain how acid rain is formed.
- Understand that human activities can alter the atmosphere in such a way that they can change climate.

- Describe the kinds of changes that could occur as a result of global warming.
- Describe the link between chlorofluorocarbon use and ozone depletion.
- Recognize that there are many positive actions that have improved air quality.
- Recognize that enclosed areas can trap air pollutants that are normally diluted in the atmosphere.

PAST PRESENT FUTURE

1987 Montreal Protocol mandated phaseout of ozone-depleting chemicals.

1988 International Panel on Climate Change was established.

1997 Kyoto Protocol on climate change was established.

1985 U.S. emissions of volatile organic compounds equaled 24,300 metric tons.

1982 U.S. emissions of lead equaled 55,000 metric tons.

2004 Antarctic ozone hole measured about 27 million square kilometers (10 million square miles), but did not appear to be increasing.

2004 Scientists reached consensus that human activity is causing the climate to get warmer.

2004 120 countries have ratified the Kyoto Protocol. The United States has withdrawn.

2002 U.S. emissions of volatile organic compounds equaled 14,800 metric tons (a 40 percent decrease).

2002 U.S. emissions of lead equaled 3750 metric tons (a 93 percent decrease).

The Atmosphere

The atmosphere (air) is composed of 78.1 percent nitrogen, 20.9 percent oxygen, and a number of other gases such as argon, carbon dioxide, methane, and water vapor that total about 1 percent. Most of the atmosphere is held close to the Earth by the pull of gravitational force, so it gets less dense with increasing distance from the Earth. Throughout the various layers of the atmosphere, nitrogen and oxygen are the most common gases present, although the molecules are farther apart at higher altitudes.

The atmosphere is composed of four layers. (See figure 17.1.) The *troposphere* extends from the Earth's surface to about 10 kilometers (about 6.2 miles) above the Earth. It actually varies from about 8 to 18 kilometers (5–11 miles), depending on the position of the Earth and the season of the year. The temperature of the troposphere declines by about 6°C (11°F) for every kilometer above the surface. The troposphere contains most of the water vapor of the atmosphere and is the layer in which weather takes place. The *stratosphere* extends from the top of the troposphere to about 50 kilometers (about 31 miles) and contains most of the ozone. The ozone is in a band between 15 and 30 kilometers (9–19 miles) above the Earth's surface. Because the ozone layer absorbs sunlight, the upper layers of the stratosphere are warmer than the lower layers. The *mesosphere* is a layer with decreasing temperature from 50 to 80 kilometers (31–50 miles) above the Earth. The *thermosphere* is a layer with increasing temperature that extends to about 300 kilometers (186 miles) above the Earth's surface.

Even though gravitational force keeps the majority of the air near the Earth, the air is not static. As it absorbs heat from the Earth, it expands and rises. When its heat content is radiated into space, the air cools, becomes more dense, and flows toward the Earth. As the air circulates vertically due to heating and cooling, it also moves horizontally over the surface of the Earth because the Earth rotates on its axis. The combination of all air movements creates the wind and weather patterns characteristic of different regions of the world. (See figure 17.2.)

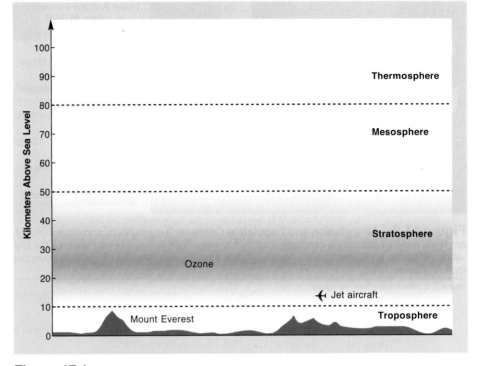

Figure 17.1 **The Atmosphere** The atmosphere is divided into the troposphere, the relatively dense layer of gases close to the surface of the Earth; the stratosphere, more distant with similar gases but less dense; the mesosphere; and the thermosphere. Weather takes place in the troposphere, and the important ozone layer is present in the stratosphere.

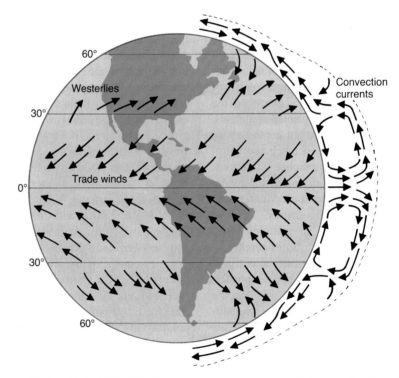

Figure 17.2 **Global Wind Patterns** Wind is the movement of air caused by the rotation of the Earth and atmospheric pressure changes brought about by temperature differences. Both of these contribute to the patterns of world air movement. In North America, most of the winds are westerlies (from the west to the east).

GLOBAL PERSPECTIVE

The Kyoto Protocol on Greenhouse Gases

On December 10, 1997, 160 nations reached agreement in Kyoto, Japan, on limiting emissions of carbon dioxide and other greenhouse gases. The Kyoto Protocol was a significant agreement in focusing the attention of world leaders on the issues of climate change. The Kyoto Protocol called for the industrialized nations to reduce their average national emissions over the period 2008–2012 to about 5 percent below 1990 levels. The United States pledged to achieve a level 7 percent below the 1990 level, slightly less than the European Union's pledge and slightly more than Japan's. None of the developing countries was required to set any limits.

The protocol initially covered only three greenhouse gases: carbon dioxide, methane, and nitrous oxide. Three more compounds—hydrofluorocarbons, perfluorocarbons, and sulfur hexafluoride—were to be added in subsequent years. The protocol also contains the elements of a program for international trading of greenhouse gas emissions. Such trading would employ market incentives to help ensure that the lowest-cost opportunities for reduction are pursued.

While the Kyoto climate agreement was important in bringing a new level of international attention to greenhouse gas emissions, there remain many important issues to be resolved, issues pertaining to the agreement. These include:

1. The rules and institutions that are to govern international trading of greenhouse gas emissions among industrialized countries must be better established.

2. The criteria used to judge compliance, and any penalties for non-compliance, must be clearly articulated.

3. To make longer-term objectives more credible, moderate but specific near-term goals should be set for industrialized countries, and these countries should be able to use early emissions reductions to meet longer-term requirements.

One thing is for sure: If the goals of the protocol are to be met by 2008, the developed nations of the world will be forced to focus attention on their energy policies, and this in turn will affect all of us.

Key dates in the global warming story

1898—Swedish scientist Svante Ahrrenius warns that the Industrial Revolution's carbon dioxide emissions, from coal and oil, could accumulate in atmosphere and lead to global warming.

1961—New observatory atop Hawaii's Mauna Loa volcano detects rise in atmospheric carbon dioxide.

1980s—Computer models of world climate project temperature rises.

1988—UN establishes an authoritative network of climate scientists, the Intergovernmental Panel on Climate Change, to study the problem and make recommendations.

1990— Intergovernmental Panel on Climate Change certifies scientific basis for "greenhouse effect" and global warming predictions.

1992—Climate change treaty is signed, setting voluntary goals for industrial nations to lower greenhouse gas emissions to 1990 levels by 2000. Almost 170 nations eventually ratify. The United States signs the treaty, but the Senate does not ratify it.

1996—The Intergovernmental Panel on Climate Change concludes that climate change is occurring and that it is highly probable that human activity is an important cause of the change.

October 1997—Negotiators end two years of preliminary tasks with major issues, including level of binding targets, unresolved; preparations begin for final conference in Kyoto.

July 2001—Representatives from 180 countries meet in Bonn, Germany, to continue work on the details of implementing the global warming treaty. The European Union supports the treaty, but few countries within the EU are moving toward ratification. The United States withdraws from the treaty but continues to participate in the discussions.

2003—Neither Russia nor the United States ratified the treaty, and its future was in doubt.

2004—Russia agreed to ratify the treaty and it will go into effect although the United States, the largest producer of cabon dioxide, has not approved the treaty.

Table 17.4 Indoor Air Pollution Summary

Pollutant	Description	Sources	Effects*
Asbestos	Light, fibrous mineral; fireproof and insulative	Ceilings and floor tiles, insulation, and spackling compounds	Easily inhaled to cause lung damage Cancer
Carbon monoxide (CO)	Odorless, colorless gas	Combustion sources such as charcoal, kerosene heaters, attached garages with poor ventilation, tobacco smoke	Reduces ability of blood to carry oxygen; impairs vision and alertness Dizziness, headaches, fatigue, death by suffocation
Formaldehyde	Pungent gas; preservative and disinfectant	Foam insulation, resin in particleboards, plywood paneling, fiberboard, some carpets, drapery and upholstery fabrics	Headaches, dizziness, nausea, lethargy, rashes, upper respiratory irritations
Lead (Pb)	Metallic element	House paints manufactured prior to 1976, old plumbing lines and solder, leaded crystal and pottery, old toys	Learning and behavior problems in children, high blood pressure, joint pain, concentration problems, reproductive problems
Biological pollutants and microorganisms	Pollen, dust mites, and pet dander Bacteria, molds, fungi, and viruses	Improperly maintained heating and cooling systems Washrooms, humidifiers, and dehumidifiers Pets and plants Wall-to-wall carpeting	Allergic diseases and skin irritations, influenza and Legionnaires' disease, acute cases of asthma
Nitrogen dioxide (NO$_2$)	Brownish gas	Gas appliances, fireplaces, wood and coal stoves	Eye and respiratory tract irritations Lowers resistance to respiratory infections Chronic bronchitis
Radon (Rn)	Naturally occurring, radioactive gas	Rocks and soils that contain the decaying radioactive elements of uranium and radium Enters through cracks in foundation and basements	Lung cancer
Tobacco smoke	Mixture of several substances including known human carcinogens	Cigarettes, pipes and cigars, secondhand smoke	Eye, nose and throat irritations, headaches and nausea, coughing and chest discomfort, respiratory infections, lung cancer
Volatile organic compounds (VOCs)	Substances produced by synthetic chemical industry and naturally; vaporize at ordinary temperatures	Some furniture, paint, adhesives, solvents, upholstery and drapery fabrics, construction materials, cleaning compounds, deodorizers, felt-tipped markers, correction fluid, dry-cleaned clothes, pesticides, wool preservatives, tobacco smoke, etc.	Eye and respiratory irritations, kidney and liver damage in animals Long-term effects are still being studied.

*Effects from prolonged exposure or high concentrations.

Source: Chart compiled by Earth Force, Inc., (www.earthforce.org) with data from the U.S. Environmental Protection Agency and Wisconsin Department of Natural Resources. All rights reserved. Reprinted by permission.

than outdoor air in even the largest and most industrialized cities. Many indoor air pollutants and pollutant sources are thought to have an adverse effect on human health. These pollutants include asbestos; formaldehyde, which is associated with many consumer products, including certain wood products and aerosols; airborne pesticide residues; chloroform; perchloroethylene (associated particularly with dry cleaning); paradichlorobenzene (from mothballs and air fresheners); and many disease-causing or allergy-producing organisms. (See table 17.4.) Smoking is the most important air pollutant source in the United States in terms of

human health. The surgeon general estimates that 350,000 people in this country die each year from emphysema, heart attacks, strokes, lung cancer, or other diseases caused by tobacco smoking. Banning smoking probably would save more lives than would any other pollution-control measure.

A recent contributing factor to the concern about indoor air pollution is the weatherizing of buildings to reduce heat loss and save on fuel costs. In most older homes, a complete exchange of air occurs every hour. This means that fresh air leaks in around doors and windows and through cracks and holes in the building.

In a weatherized home, a complete air exchange may occur only once every five hours. Such a home is more energy efficient, but it also tends to trap air pollutants.

Even though we spend on average almost 90 percent of our time indoors, the movements to reduce indoor air pollution lag behind regulations governing outdoor air pollution. In the United States, the Environmental Protection Agency is conducting research to identify and rank the human health risks that result from exposure to individual indoor pollutants or mixtures of multiple indoor pollutants.

Environmental Close-Up

Radon

In 1985, the clothing worn by an engineer at the Limerick Nuclear Generating Station in Pottstown, Pennsylvania, registered a high radiation level. Initially, the generating station was believed to be the source of the radiation. However, subsequent studies indicated that radon 222 from the engineer's home was the source. Following this incident, interest in radon and its effects has increased.

The source of radon is uranium-238, a naturally occurring element that makes up about 3 parts per million of the Earth's crust. Uranium-238 goes through 14 steps of decay before it becomes stable, nonradioactive lead 206. **Radon,** an inert radioactive gas having a half-life of 3.8 days, is one of the products formed during this process.

Since radon is an inert gas, it does not enter into any chemical reactions within the body, but it can be inhaled. Once in the lungs, it may undergo radioactive decay, producing other kinds of atoms called "daughters" of radon. These decay products (daughters) of radon—plutonium 218, which has a 3-minute half-life; lead 214, which has a 27-minute half-life; bismuth 214, which has a 20-minute half-life; and polonium 214, which has a millisecond half-life—are solid materials that remain in the lungs and are chemically active.

Increased incidence of lung cancer is the only known health effect associated with radon decay products. It is estimated that the decay products of radon are responsible for about 15,000 lung cancer deaths annually in the United States. This is about 10 percent of lung cancer deaths. The radon risk evaluation chart indicates that individuals exposed to greater than 4 picocuries per liter are at increased risk of developing lung cancer.

As the radon gas is formed in the rocks, it usually diffuses up through the rocks and soil and escapes harmlessly into the atmosphere. It can also diffuse into groundwater. Radon usually enters a home through an open space in the foundation. A crack in the basement floor or the foundation, the gap around a water or sewer pipe, or a crawl space allows the radon to enter the home. It may also enter in the water supply from wells.

About 10 percent of the homes in the United States have a potential radon problem. The map shows those regions of the United States that are likely to have elevated levels of radon. However, the increased publicity about radon has many people worried. In addition, the Environmental Protection Agency and the U.S. surgeon general recommend that all Americans (other than those living in apartment buildings above the second floor) test their homes for radon. If the tests indicate radon levels at or above 4 picocuries per liter, the EPA recommends that the homeowner take action to lower the level. This is usually not expensive and consists of blocking the places where radon is entering or venting radon sources to the outside. People who are concerned about radon should contact their state's public health department or environmental protection agency.

Radon Risk Evaluation Chart

pCi/L	WL	Estimated Lung-Cancer Deaths Due to Radon Exposure (out of 1000)	Comparable Exposure Levels	Comparable Risk
200	1.0	440-770	One thousand times average outdoor level	More than 60 times nonsmoker risk
100	0.5	270-630	One hundred times average indoor level	Four-pack-a-day smoker
				Two thousand chest X rays per year
40	0.2	120-380		
				Two-pack-a-day smoker
20	0.1	60-120	One hundred times average outdoor level	
				One-pack-a-day smoker
10	0.05	30-120	Ten times average indoor level	
				Five times nonsmoker risk
4	0.02	13-50	Level at which EPA suggests remedial action	
				Two hundred chest X rays per year
2	0.01	7-30	Ten times average outdoor level	
				Nonsmoker risk of dying from lung cancer
1	0.005	3-13	Average indoor level	
				Twenty chest X rays per year
0.2	0.001	1-3	Average outdoor level	

Note: Measurement results are reported in one of two ways: (1) pCi/L (picocuries per liter) — measurement of *radon gas*, or (2) WL (working levels) — measurement of *radon decay products*.

Source: Data from U.S. Environmental Protection Agency, Office of Air and Radiation Programs.

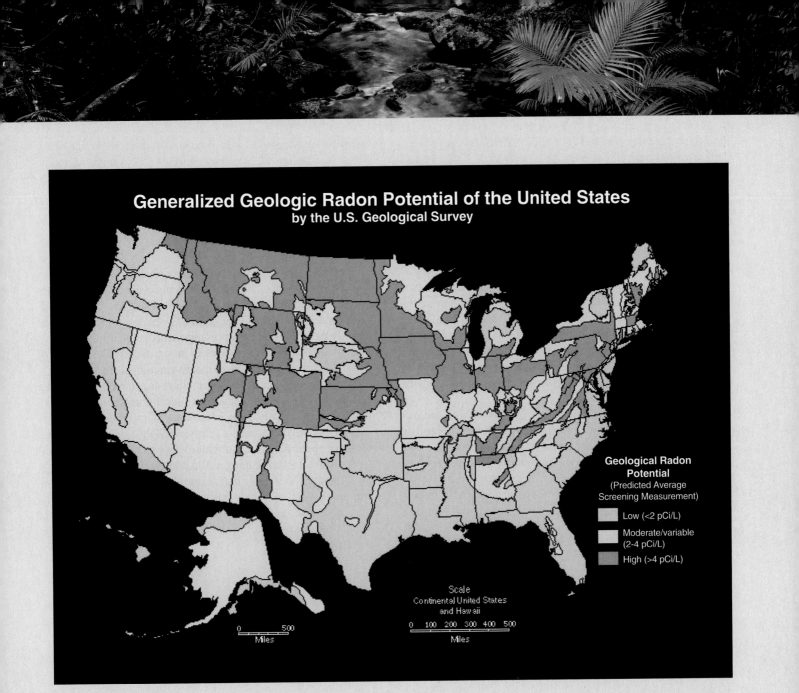

Generalized Geologic Radon Potential of the United States
by the U.S. Geological Survey

Geological Radon Potential
(Predicted Average
Screening Measurement)

Low (<2 pCi/L)

Moderate/variable
(2-4 pCi/L)

High (>4 pCi/L)

Scale
Continental United States
and Hawaii

0   500
Miles

0   100   200   300   400   500
Miles

Source: U.S. Geological Survey.

Summary

The atmosphere has a tremendous ability to disperse pollutants. Carbon monoxide, hydrocarbons, particulate matter, sulfur dioxide, and oxides of nitrogen compounds are the primary air pollutants. They can cause a variety of health problems. The U.S. Environmental Protection Agency establishes standards for six pollutants known as criteria air pollutants. They are carbon monoxide, nitrogen dioxide, sulfur dioxide, volatile organic compounds (hydrocarbons), ozone, and lead. The EPA also regulates hazardous air pollutants.

Photochemical smog is a secondary pollutant, formed when hydrocarbons and oxides of nitrogen are trapped by thermal inversions and react with each other in the presence of sunlight to form peroxyacetyl nitrates and ozone. Elimination of photochemical smog requires changes in technology, such as more fuel-efficient automobiles, special devices to prevent the loss of hydrocarbons, and catalytic converters to more completely burn hydrocarbons in exhaust gases.

Acid rain is caused by emissions of sulfur dioxide and oxides of nitrogen in the upper atmosphere, which form acids that are washed from the air when it rains or snows or settle as particles on surfaces. Direct effects of acid rain on terrestrial ecosystems are difficult to prove, but changes in many forested areas are suspected of being partly the result of additional stresses caused by acid rain. Recent evidence suggests that loss of calcium from the soil may be a major problem associated with acid rain. The effect of acid rain on aquatic ecosystems is easy to quantify. As waters become more acidic, the complexity of the ecosystem decreases, and many species fail to reproduce. The control of acid rain requires the use of scrubbers, precipitators, and filters—or the removal of sulfur from fuels. However, oxides of nitrogen are still a problem.

Currently, many are concerned about the damaging effects of greenhouse gases: carbon dioxide, methane, and chlorofluorocarbons. These gases are likely to be causing an increase in the average temperature of the Earth and, consequently, are leading to major changes in the climate. Human and ecological systems are already vulnerable to a range of environmental pressures, including climate extremes and variability. Global warming is likely to amplify the effects of other pressures and disrupt our lives in numerous ways. Significant impacts on our health, the vitality of forests and other natural areas, the distribution of freshwater supplies, and the productivity of agriculture are among the probable consequences of climate change. Chlorofluorocarbons also lead to the destruction of ozone in the upper atmosphere, which results in increased amounts of ultraviolet light reaching the Earth. Concern about the effects of chlorofluorocarbons has led to international efforts that have resulted in significant reductions in the amount of these substances reaching the atmosphere. Many commonly used materials release gases into closed spaces (indoor air pollution), where they cause health problems. The most important of these health problems are associated with tobacco smoking.

Issues – Analysis

Pollution, Policy, and Personal Choice

The air pollution problem that has shown the least improvement in North America is the control of ground-level ozone. Some geographic regions are more prone to developing conditions that generate ground-level ozone, and we know that the presence of nitrogen oxides and volatile organic compounds triggers the events leading to the development of unhealthy levels of ground-level ozone. Furthermore, it is clear that automobiles are the major source of the volatile organic compounds and nitrogen oxides.

It is also clear that a large proportion of the population is regularly exposed to unhealthy levels of ozone and that many millions of people become ill or have their activities restricted as a direct result of poor air quality.

There are several ways to attack this problem, but all of them involve restricting the freedom of people to use automobiles or restricting the kinds of automobiles they can drive. The following list describes some of the approaches used and their effects on the driving public.

1. *Restrict the kinds of cars that people can buy to less-polluting types.* Requirements that automobiles obtain better mileage with reduced emissions have successfully produced less pollution per mile driven. However, people are driving more and a loophole in the Clean Air Act allows sport utility vehicles (SUVs) to be classified as trucks and avoid the more stringent mileage requirements. The majority of new automobiles purchased are SUVs and light trucks.

2. *Provide incentives to encourage fewer people to drive.* Many metropolitan areas have carpool lanes restricted to cars with two or more passengers. These lanes are largely unused. People choose to drive separately.

3. *Provide economic penalties for the use of automobiles.* In many cities of the world, people pay a fee to drive in parts of the city. In addition, toll roads and bridges are an economic incentive not to drive. However, this only reduces traffic if the fees are high and there are no "free" roads that people can use.

4. *Provide affordable, safe, and convenient alternative means of transport.* For every dollar spent on public transport in the United States, about $6 are spent on roads for cars.

These approaches to dealing with ground-level ozone raise several issues that affect personal choice.

- Should the use of automobiles be restricted to protect the health of people who are highly sensitive to ground-level ozone?
- Should governments determine what kinds of automobiles are allowed?
- Should fees, taxes, or assessments be used to increase the cost of using a car?
- A typical driver spends $500 to $1000 per year for gasoline. Would you pay an annual fee of $1000 to use public transport if it were available?

Key Terms

acid deposition *400*
acid rain *400*
air toxics *391*
carbon dioxide (CO$_2$) *405*
carbon monoxide (CO) *391*
carcinogenic *393*
chlorofluorocarbons (CFCs) *405*
criteria air pollutants *391*
decibel *399*
greenhouse effect *405*

greenhouse gas *403*
hazardous air pollutants (HAP) *391*
hydrocarbons (HC) *394*
methane *406*
nitrogen dioxide (NO$_2$) *393*
nitrogen monoxide (NO) *393*
nitrous oxide (N$_2$O) *407*
oxides of nitrogen (NO$_X$) *393*
ozone(O$_3$) *394*
particulate matter *392*

photochemical smog *394*
PM$_{2.5}$ *393*
PM$_{10}$ *392*
primary air pollutants *390*
radon *414*
secondary air pollutants *390*
sulfur dioxide (SO$_2$) *393*
thermal inversion *396*
volatile organic compounds (VOCs) *394*

Review Questions

1. List the five primary air pollutants commonly released into the atmosphere and their sources.
2. List the six criteria air pollutants and their sources.
3. Define *secondary air pollutants* and give an example.
4. List three health effects of air pollution.
5. Why is air pollution such a large problem in urban areas?
6. What is photochemical smog? What causes it?
7. Describe three actions that can be taken to control air pollution.
8. What causes acid rain? List three probable detrimental consequences of acid rain.
9. Why is carbon dioxide (a nontoxic normal component of the atmosphere) called a "greenhouse gas"?
10. What would the consequences be if the ozone layer surrounding the Earth were destroyed?
11. How does energy conservation influence air quality?

Critical Thinking Questions

1. What could you do to limit the air pollution you create?
2. Do you agree with a ban on smoking, as in California, that includes all indoor public places, even privately owned restaurants and bars? Why or why not?
3. Some developing countries argue that they should be exempt from limits on the production of greenhouse gases and that developed countries should bear the brunt of the changes that appear to be necessary to curb global climate change. What values, beliefs, and perspectives underlie this argument? What do you think about this argument?
4. As a nation, the United States provides many subsidies to make energy cheap because policy makers feel that economic development depends on cheap energy. If these subsidies were withdrawn or taxes on energy were added, what effect would this have on your own energy consumption? Would you be willing to support high gasoline prices, in the $3- to $4-a-gallon range as in many European countries, if it would cut greenhouse gas emissions?
5. Why do you think air pollution is so much worse in developing countries than in developed countries? What should developed countries do about this, if anything?
6. What common indoor air pollutants are you exposed to? What can you do to limit this exposure?
7. What kinds of noise pollution do you encounter? How important is noise pollution to you? What can you do to reduce noise pollution? How can we make cities quieter?
8. Is it possible to have zero emissions of pollutants? What level of risk are you willing to live with?

Welcome to McGraw-Hill's Online Learning Center

Location: www.mhhe.com/enger10e

Mc Graw Hill

Interactive Exploration

Check out the website at www.mhhe.com/enger10e and click on the cover of this textbook for quizzing, career information, case studies, and hot links for the following topics:

Air Pollution

Acid Deposition

Atmosphere, Climate, and Weather

Global Warming

Human Respiratory Topics

Solid Waste Management and Disposal

18

Objectives

After reading this chapter, you should be able to:

- Explain why solid waste is a problem throughout the world.
- Understand that the management of municipal solid waste is directly affected by economics, changes in technology, and citizen awareness and involvement.
- Describe the various methods of waste disposal and the problems associated with each method.
- Understand the difficulties in developing new municipal landfills.
- Define the problems associated with incineration as a method of waste disposal.
- Describe some methods of source reduction.
- Describe composting and how it fits into solid waste disposal.
- List some benefits and drawbacks of recycling.

Chapter Outline

PAST PRESENT FUTURE

1979 U.S. Environmental Protection Agency prohibited open dumps and set landfill standards.

1979 There were about 20,000 landfills and open dumps in the United States.

1980 Of all municipal solid waste, 9.6 percent was recycled in the United States.

1980 The waste produced per person per day was 1.7 kilograms (3.75 pounds) in the United States.

2004 Open dumps and poorly designed landfills have been eliminated in the United States.

2004 Fewer than 2000 landfills remain (about an 80 percent reduction).

2004 Of all municipal solid waste, 30 percent was recycled.

2004 The waste produced per person per day was 2 kilograms (4.4 pounds).

Kinds of Solid Waste

The several kinds of waste produced by a technological society can be categorized in many ways. Some kinds of wastes are released into the air and water. Some are purposely released, while others are released accidentally. Many wastes that are purposely released are treated before their release. Wastes that are released into the water and air are discussed in chapters 16 and 17, respectively. There are wastes with particularly dangerous characteristics, such as nuclear wastes, medical wastes, industrial hazardous wastes, and household hazardous wastes. Chapter 11 deals with nuclear wastes, and chapter 19 considers hazardous waste issues. The focus of this chapter is solid waste.

Solid waste is generally made up of objects or particles that accumulate on the site where they are produced, as opposed to water- and airborne wastes that are carried away from the site of production. Solid wastes are typically categorized by the sector of the economy responsible for producing them, such as mining, agriculture, manufacturing, and municipalities.

We have very good information about those waste streams that are tightly regulated (hazardous wastes, municipal solid waste, medical wastes, nuclear wastes) but only general estimates for many of the other kinds of wastes, such as mining and agricultural waste.

Mining waste is generated in three primary ways. First, in most mining operations, large amounts of rock and soil need to be removed to get to the valuable ore. This waste material is generally left on the surface at the mine site. Second, milling operations use various technologies to extract the valuable material from the ore. These techniques vary from relatively simple grinding and sorting to sophisticated chemical separation processes. Regardless of the technique involved, once the valuable material is recovered, the remaining waste material, commonly known as tailings, must be disposed of. Solid materials are typically dumped on the land near the milling site, and liquid wastes are typically stored in ponds. It is difficult to get vegetation to grow on these piles of waste rock and tailings, so they are unsightly and remain exposed to rain and wind. Finally, the water that drains or is pumped from mines or that flows from piles of waste rock or tailings often contains hazardous materials (such as asbestos, arsenic, lead, and radioactive materials) or high amounts of acid that must be contained or treated—but often are not. In addition, failures of the earthen dams used to form waste ponds result in the release of contaminated water into local streams. Between 1993 and 2003, about 30 such failures occurred worldwide, seven of them in the United States.

It is difficult to get more than a rough estimate of the amount of mining waste produced, but the U.S. Environmental Protection Agency estimates that between 1000 million and 2000 million metric tons of mining waste are produced each year in the United States. Of the total waste produced, about 700 million to 800 million metric tons are considered hazardous.

Many types of mining operations require vast quantities of water for the extraction process. The quality of this water is degraded, so it is unsuitable for drinking, irrigation, or recreation. Since mining disturbs the natural vegetation in an area, water may carry soil particles into streams and cause erosion and siltation. Some mining operations, such as strip mining, rearrange the top layers of the soil, which lessens or eliminates its productivity for a long time. Strip mining has disturbed approximately 75,000 square kilometers (30,000 square miles) of U.S. land, an area equivalent to the state of Maine.

Agricultural waste is the second most common form of waste and includes waste from the raising of animals and the harvesting and processing of crops and trees. The amount of animal manure produced annually is estimated at about 1240 million metric tons. Other wastes associated with agriculture, such as waste from processing operations (peelings, seeds, straw, stems, sludge, and similar materials), might bring the total agricultural waste to about 1500 million metric tons per year. Since most agricultural waste is organic, approximately 90 percent is used as fertilizer or for other soil-enhancement activities. Other materials are burned as a source of energy, so little of this waste needs to be placed in landfills. However, when too much waste is produced in one place, there may not be enough farmland available to accept the agricultural waste without causing water pollution problems associated with runoff or groundwater contamination due to infiltration. (See Environmental Close-Up: Industrial Production of Livestock in chapter 15.)

Industrial solid waste from sources other than mining is variously estimated to be between 200 million and 600 million metric tons of solid waste per year. It includes a wide variety of materials such as demolition waste, foundry sand, scraps from manufacturing processes, sludge, ash from combustion, and other similar materials. These materials are tested to determine if they are hazardous. If they are classified as hazardous waste, their disposal requires that they be placed in special hazardous waste landfills. Hazardous wastes are discussed in chapter 19. In addition to solid wastes, industries produce several billion metric tons of aquatic waste. See chapter 16 for a discussion of industrial use of water.

Municipal solid waste (MSW) consists of all the materials that people in a region no longer want because they are broken, spoiled, or have no further use. It includes waste from households, commercial establishments, institutions, and some industrial sources, and amounts to about 210 million metric tons per year. Table 18.1 summarizes estimates of the quantity of various kinds of solid waste produced in the United States. The remainder of this chapter will focus on the generation and disposal of municipal solid waste.

Municipal Solid Waste

Wherever people exist, waste disposal is a problem. Archaeologists are eager to find the middens, or waste heaps, of ancient civilizations. The kinds of articles discarded can tell us a great deal about the nature of the society that produced them. In modern society, many products are discarded when they are broken or worn out, while others have only a temporary use. Those that have only temporary uses make up the majority of solid waste.

The United States produces about 210 million metric tons of municipal solid waste each year. This equates to about 2 kilograms (4.4 pounds) of trash per person per day, or 0.73 metric tons per person per year. The amount of municipal solid waste has more than doubled since 1960, and the per capita rate has increased by nearly 70 percent in that same time, although per capita rates began to stabilize about 1990. When recycling is included, the net waste produced has actually fallen since 1990. (See figure 18.1.)

Nations with high standards of living and productivity tend to have more municipal solid waste per person than less-developed countries. (See figure 18.2.) The United States and Canada, therefore, are world leaders in waste production. Large metropolitan areas have the greatest difficulty dealing with their solid waste because of the large volume and the challenge of finding suitable landfill sites near the city. For example, Toronto, Canada, has no local landfill for its waste. As a result, Toronto developed an ambitious plan to divert 100 percent of its solid waste from landfills by 2010. By 2004, it had met its intermediate target and was diverting about 40 percent. Toronto has a long-term contract with a landfill site in Michigan to dispose of its remaining solid waste (about 4000 metric tons per day) at a cost of about $US40 per metric ton. In response to public concern, the Michigan

legislature has enacted several laws that restrict the nature of the waste that can be imported into the state.

In March 2001, New York City closed its Fresh Kills Landfill on Staten Island. (It was reopened for a time following the September 11, 2001, terrorist attack to serve as a place to take the debris from the World Trade Center so that it could be processed.) Before its closure, it was the largest landfill in the world and received about 12,600 metric tons of trash each day. Today, New York City is exporting all of its solid waste to landfills in other parts of New York state, Pennsylvania, Virginia, and other states. In contrast to Toronto which diverts about 40 percent of its waste, New York City diverts less than 20 percent of its waste.

Archaeologists rely on the waste of past societies to tell them about the nature of the culture and lifestyle of ancient civilizations. In the same way today, our municipal solid waste is a reflection of our society. Figure 18.3 shows how the composition of our trash has changed since 1960. Notice particularly the increase in the amount of paper and plastic and the effect that recycling has had on the amount of glass in the trash. In the United States, the two most common items in the waste stream are paper products and yard waste, and other significant segments are wood, metal, glass, plastics, and food waste. (See figure 18.4.) An analysis of the composition of our waste will present us with possible approaches to reducing the amount of waste we generate.

Table 18.1 Estimates of Solid Waste Produced per Year in the United States

Category of Waste	Amount of Waste (million metric tons)
Mining waste	1000–2000
Agricultural waste	1500
Industrial waste	200–600
Municipal solid waste	210

Source: Estimates from U.S. Environmental Protection Agency, U.S. Geological Survey, and U.S. Department of Agriculture.

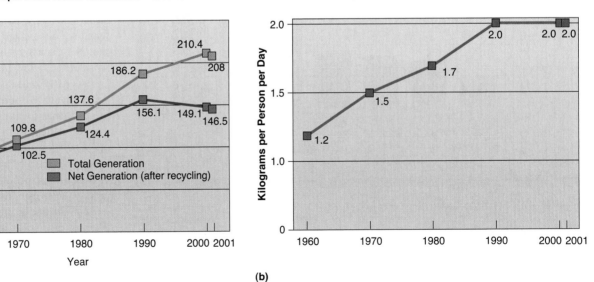

Total and Net Municipal Solid Waste Generation—1960 to 2001

(a)

Per Capita Municipal Solid Waste Generation

(b)

Figure 18.1 Municipal Solid Waste Generation Rates The generation of municipal solid waste in the United States has increased steadily. However, because of increased recycling rates, the net production rates (after recyclables have been removed) has actually fallen since 1990 (a), and the per capita rate has stabilized (b).

Source: U.S. Environmental Protection Agency, Washington, D.C.

CHAPTER 18 Solid Waste Management and Disposal

421

Mountain Top Removal

Most seams of coal are found in horizontal layers under various depths of overlying rock and soil known as overburden. To get to the coal, several methods have been used. In mountainous regions, this coal may be exposed in the valleys, and for many generations, miners have penetrated into the mountain along these seams to extract the coal. The invention of large machines made it possible to move huge amounts of overburden to get to the coal seams. The overburden was removed in strips, followed by the removal of the coal, and the overburden from a new strip was then deposited in the hole where the coal had been; thus, the method became known as strip mining, or surface mining. About 60 percent of the coal mined in the United States is from surface mines.

Mountain top removal is a relatively new variation of surface mining being used in Appalachia that has generated a considerable amount of controversy. The method uses explosives and heavy machinery to remove the large amount of overburden to get at the coal. However, the disposal of the large amount of overburden is a problem and involves depositing the waste material in the valleys adjacent to the mountain. In addition to changing the topography from mountainous to flat, the method buries valley vegetation and streams.

This new variation of surface mining has generated intense lobbying in Congress by both opponents and proponents. Coal companies argue that mountain top removal is the only cost-effective way to mine the coal and that they are providing cheap energy at minimal environmental cost. Furthermore, they argue that they must reclaim the land and that the flatter land left behind is suitable for industrial development, which would bring large numbers of jobs to areas that need them.

Others argue that the destruction of the landscape costs many tourism-related jobs because the scenery is degraded. Furthermore, the exposed mountain surfaces and piles of waste rock in the valleys are subject to severe erosion during periods of heavy rain. The streams are degraded by runoff from the waste materials pushed into the valleys, and flooding can cause the unstable material to wash down the valley, destroying homes and the landscape. They argue that these environmental damages are too much to pay for cheap coal. Both sides will continue to lobby Congress and introduce legislation. Meanwhile, mountain top removal continues in several Appalachian states.

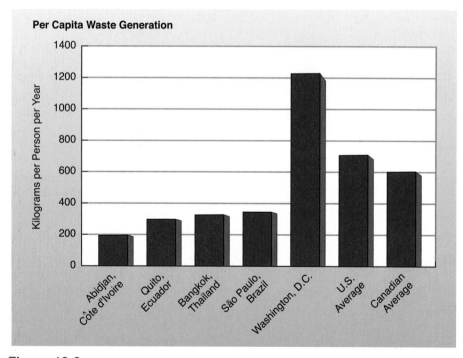

Figure 18.2 **Waste Generation and Lifestyle** The waste generation rates of people are directly related to their economic condition. People in richer countries produce more garbage than those in poorer countries.

Source: Data from *World Resources* 1996–1997, and U.S. Environmental Protection Agency.

Methods of Waste Disposal

From prehistory through the present day, the favored means of disposal was simply to dump solid wastes outside of the city or village limits. Frequently, these dumps were in wetlands adjacent to a river or lake. To minimize the volume of the waste, the dump was often burned. Unfortunately, this method is still being used in remote or sparsely populated areas in the world. (See figure 18.5.)

As better waste-disposal technologies were developed and as values changed, more emphasis was placed on the environment and quality of life. Dumping and open burning of wastes is no longer an acceptable practice from an environmental or health perspective. While the technology of waste disposal has evolved during the past several decades, our options are still limited. Realistically, there are no ways of dealing with waste that have not been known for many thousands of years. Essentially, five techniques are used: (1) landfills, (2) incineration, (3) source reduction, (4) composting, and (5) recycling.

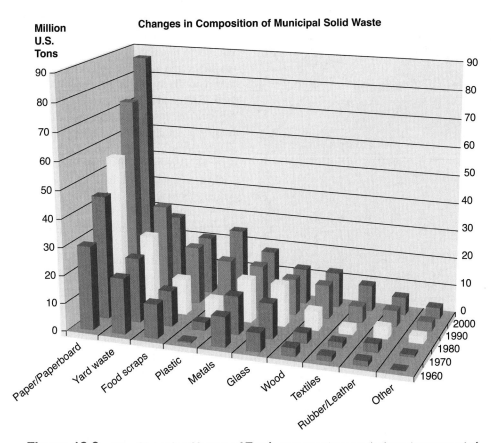

Changes in Composition of Municipal Solid Waste

Figure 18.3 **The Changing Nature of Trash** Paper products are the largest component of the waste stream. Changes in lifestyle and packaging have led to a change in the nature of trash. Note the increase in the amount of plastics in the waste stream. Most of what is currently disposed of could be recycled.

Source: Data from the U.S. Environmental Protection Agency.

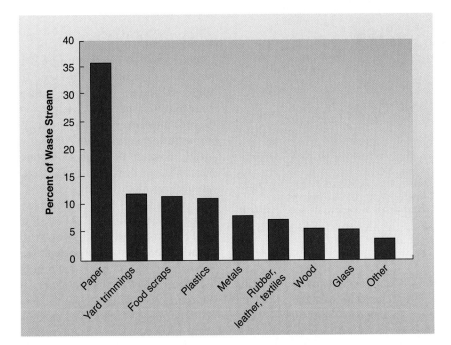

Figure 18.4 **Composition of Trash in the United States (2001)** Paper and yard waste are the most common materials disposed of, accounting for about 50 percent of the waste stream.

Source: Data from the U.S. Environmental Protection Agency.

Landfills

Landfills have historically been the primary method of waste disposal because this method is the cheapest and most convenient, and because the threat of groundwater contamination was not initially recognized. As we have recognized some of the problems associated with poorly designed landfills, efforts to reduce the amount of material placed in landfills have been substantial. Although the amount of waste has increased, composting and recycling have removed significant amounts of materials from the waste stream, and the amount of material entering landfills has declined. (See figure 18.6.) However, the landfill of today is far different from a simple hole in the ground into which garbage is dumped.

A modern **municipal solid waste landfill** is typically constructed above an impermeable clay layer that is lined with an impermeable membrane and includes mechanisms for dealing with liquid and gas materials generated by the contents of the landfill. Each day's deposit of fresh garbage is covered with a layer of soil to prevent it from blowing around and to discourage animals from scavenging for food. Selection of landfill sites is based on an understanding of local geologic conditions such as the presence of a suitable clay base, groundwater geology, and soil type. In addition, it is important to address local citizens' concerns. Once the site is selected, extensive construction activities are necessary to prepare it for use. New landfills have complex bottom layers to trap contaminant-laden water, called **leachate,** leaking through the buried trash. The water that leaches through the site must be collected and treated. In addition, monitoring systems are necessary to detect methane gas production and groundwater contamination. In some cases, methane produced by decomposing waste is collected and used to produce heat or generate electricity. As a result of the technology involved, new landfills are becoming increasingly more complex and expensive. They currently cost up to $1 million per hectare ($400,000 per acre) to prepare. (See figure 18.7.)

Today, about 55 percent of the municipal solid waste from the United States

Figure 18.5 **Burning Landfills** In the past, it was common practice to burn the waste in landfills to reduce the volume. Waste is still being burned in sparsely populated areas in North America and other parts of the world.

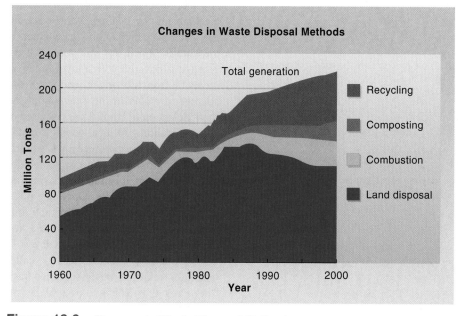

Figure 18.6 **Changes in Waste Disposal Methods** The landfill is still the primary method of waste disposal. Historically, landfills have been the cheapest means of disposal, but this may not be the case in the future. Notice that recycling and composting have grown over the past decade, while the amount of waste going to landfills has declined somewhat.

Source: Data from the U.S. Environmental Protection Agency.

and about 80 percent of Canadian municipal solid waste goes into landfills. The number of landfills is declining. In 1988, there were about 8000 landfills, and in 2001, there were about 1800 active sanitary landfills. (See figure 18.8.) The number of landfills has decreased for two reasons. Many small, poorly run landfills have been closed because they were not meeting regulations. Others have closed because they reached their capacity. The overall capacity, however, has remained relatively constant because new landfills are much larger than the old ones.

A prolonged public debate over how to replace lost landfill capacity is developing where population density is high and available land is scarce. Selecting sites for new landfills in locations such as Toronto, New York, and Los Angeles is extremely difficult because of (1) the difficulty in finding a geologically suitable site and (2) local opposition, which is commonly referred to as the NIMBY, or "not-in-my-backyard," syndrome. Resistance by the public comes from concern over groundwater contamination, rodents and other vectors of disease, odors, and truck traffic. Public officials look for alternatives to landfills to avoid controversy over landfill site selection. Although sites must be chosen for new landfills, politicians are often unwilling to take strong positions that might alienate their constituents.

Japan and many Western European countries have already moved away from landfills as the primary method of waste disposal because of land scarcities and related environmental concerns. Switzerland and Japan dispose of less than 15 percent of their waste in landfills, compared to 55 percent in the United States. Instead, recycling and incineration are the primary methods. (See figure 18.9.) In addition, the energy produced by incineration can be used for electric generation or heating.

Incineration

Incineration is the process of burning refuse in a controlled manner. Today, about 15 percent of the municipal solid waste in the United States is incinerated; Canada incinerates about 8 percent. While some incinerators are used just to burn trash, most are designed to capture the heat, which is then used to make steam to produce electricity. (See figure 18.10.) The production of electricity partially offsets the cost of disposal. In 2001, about 100 combustors with energy recovery existed in the United States, with the capacity to burn up to 86,000 metric tons of MSW per day. Most incineration facilities burn unprocessed municipal solid waste. This is often referred to as **mass burn** technology. About one-fourth of the incinerators use refuse-derived fuel—collected refuse that has been processed into pellets prior to combustion. This is particularly

PART FIVE   Pollution and Policy

How a Modern Landfill Works

Figure 18.7 **A Well-Designed Modern Landfill** A modern sanitary landfill is far different from a simple hole in the ground filled with trash. A modern landfill is a self-contained unit that is separated from the soil by impermeable membranes and sealed when filled. Methane gas and groundwater are continuously monitored to ensure that wastes are not escaping to the air or the groundwater.

Source: National Solid Waste Management Association.

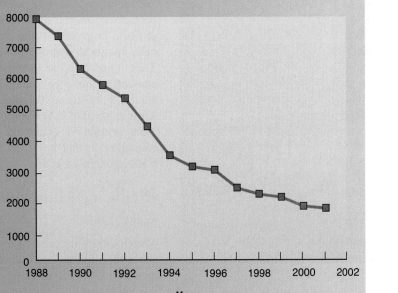

Figure 18.8 **Reducing the Number of Landfills** The number of landfills in the United States is declining because they are filling up or because their design and operation do not meet environmental standards.

Source: Data from the U.S. Environmental Protection Agency.

useful with certain kinds of materials, such as tires.

Incinerators drastically reduce the amount of municipal solid waste—up to 90 percent by volume and 75 percent by weight. Primary risks of incineration, however, involve air-quality problems and the toxicity and disposal of the ash.

Modern incinerators have many pollution control devices that trap nearly all of the pollutants produced. However, tiny amounts of pollutants are released into the atmosphere, including certain metals, acid gases, and classes of chemicals known as dioxins and furans, which have been implicated in birth defects and several kinds of cancer. The long-term risks from the emissions are still a subject of debate.

Ash from incineration is also an important issue. Small concentrations of heavy metals are present in both the fly ash captured from exhaust stacks and the bottom ash collected from these facilities. Because the ash contains lead, cadmium, mercury, and arsenic in varying concentrations from such items as batteries, lighting fixtures, and pigments, the ash is tested to determine if it should be designated as a

Figure 18.9 **Disposal Methods Used in Various Countries** Many countries have diffi-culty finding adequate space for landfills. Therefore, they rely on other technologies, such as incineration and recycling, to reduce the amount of waste that must be placed in a landfill.

Source: Data from the U.S. Environmental Protection Agency.

Figure 8.10 **Waste-to-Energy** Incineration of municipal solid waste reduces its weight and volume significantly. In many cases, the heat that results is used to produce electricity. However, there are concerns about air quality and the toxicity and disposal of the ash.

hazardous waste. This is a concern be-cause the toxic substances are more con-centrated in the ash than in the original garbage and can seep into groundwater from poorly sealed landfills. In nearly all cases, the ash is not designated as haz-ardous and can be placed in a landfill or used as aggregate for roads and other purposes.

The cost of the land and construction for new incinerators are also major con-cerns facing many communities. Incinera-tor construction is often a municipality's single largest bond issue. Incinerator con-struction costs in North America in 2000 ranged from $45 million to $350 million, and the costs are not likely to decline.

Incineration is also more costly than landfills in most situations. As long as landfills are available, they will have a cost advantage. When cities are unable to dispose of their trash locally in a landfill and must begin to transport the trash to distant sites, incinerators become more cost effective. The U.S. Environmental Protection Agency has not looked favor-ably on the construction of new waste-to-energy facilities and has encouraged recycling and source reduction as more ef-fective ways to reduce the solid waste problem. Critics have argued that cities and towns have impeded waste reduction and recycling efforts by putting a priority on incinerators and committing resources to them. Proponents of incineration have been known to oppose source reduction. They argue that incinerators need large amounts of municipal solid waste to oper-ate and that reducing the amount of waste generated makes incineration impractical. Many communities that have opposed in-cineration say that they support a vigorous waste-reduction and recycling effort.

Producing Mulch and Compost

Mulch is organic material that is used to cover the soil. It is often used to protect areas where the soil is disturbed or to con-trol the growth of unwanted vegetation in certain kinds of plantings. Typically, large branches, bark, and other organic materi-als are chopped or shredded into smaller pieces. Mulches can be sorted by the size of the pieces and can be colored for

426

Resins Used in Consumer Packaging

Thermoplastics are plastics that can be remelted and reprocessed, usually with only minor changes in their properties. About 30 percent of thermoplastic resins produced are used in consumer packaging applications. The most commonly used resins are the following:

1. Polyethylene terephthalate (PET) is used extensively in rigid containers, particularly beverage bottles for carbonated beverages and medicine containers.

2. High-density polyethylene (HDPE) is used for rigid containers, such as milk and water jugs, household-product containers, and motor oil bottles.

3. Polyvinyl chloride (PVC) is a tough plastic often used in construction and plumbing. It is also used in some food, shampoo, oil, and household-product containers.

4. Low-density polyethylene (LDPE) is often used in films and bags.

5. Polypropylene (PP) is used in a variety of areas, from yogurt containers to battery cases to disposable diaper linings. It is frequently interchanged for polyethylene or polystyrene.

6. Polystyrene (PS) is used in form cups, trays, and food containers. In its rigid form, it is used in plastic cutlery.

7. Other. These usually contain layers of different kinds of resins and are most commonly used for squeezable bottles (for example, for ketchup).

Currently, HDPE and PET are the two most commonly recycled resins because containers made of these resins are typically recovered by municipal recycling programs. Other plastics are less frequently accepted. Most of the recycled LDPE is from commercial establishments that receive large numbers of shipments wrapped in LDPE.

Recycling Rates for Plastic Resins Used in Packaging

Type of Plastic Resin	Percent Recycled (2002)
High-density polyethylene (HDPE)	24.2%
Polyethylene terephthalate (PET)	19.8%
Polypropylene (PP)	3.9%
Low-density polyethylene (LDPE) films	3–4%
Polyvinyl chloride (PVC)	Less than 1%
Polystyrene (PS)	Less than 1%

Source: Data from American Plastics Council.

PET

PVC

PP

Other

HDPE

LDPE

PS

decorative purposes. Since mulch is organic matter, it eventually breaks down and becomes part of the soil.

Composting is the process of allowing the natural process of decomposition to transform organic materials—anything from manure and corncobs to grass and soiled paper—into compost, a humuslike material with many environmental benefits. In nature, leaves and branches that fall to the forest floor form a rich, moist layer that protects the roots of plants and provides a habitat for worms, insects, and a host of bacteria, fungi, and other microorganisms.

In composting operations, proper management of air and moisture provides ideal conditions for these organisms to transform large quantities of organic material into compost in a few weeks. A good small-scale example is a backyard compost pile. Green materials (grass, kitchen vegetable scraps, and flower clippings) mixed with brown materials (twigs, dry leaves, and soiled paper towels) at a ratio of 1:3 provide a balance of nitrogen and carbon that helps microbes efficiently decompose these materials.

Large-scale municipal composting uses the same principles of organic decomposition to process large volumes of organic materials. Composting facilities of various sizes and technological sophistication accept materials such as yard trimmings, food scraps, biosolids from sewage treatment plants, wood shavings, unrecyclable paper, and other organic materials. These materials undergo processing—shredding, turning, and mixing—and, depending on the materials, can be turned into compost in a period ranging from 8 to 24 weeks. About 3800 composting facilities are in use in the United States. In 2001, 57 percent of yard trimmings were converted into mulch or composted in the United States through municipal programs. (See figure 18.11.) Most municipal programs involve one of three composting methods: windrows, aerated piles, or enclosed vessels.

- *Windrow* systems involve placing the compostable materials into long piles or rows called windrows. Tractors with front-end loaders or other kinds of specialized machinery are used to

States with Yard Waste Bans (1999)

Figure 18.11   Diverting Yard Waste Through Composting Since yard waste is such an important segment of the solid waste stream, many states have passed laws that prohibit the deposition of yard waste in landfills. The states shown in green have laws that ban yard waste from going into a landfill. This will extend the useful life of the landfill. To accommodate their citizens, many communities still collect yard waste but have instituted composting programs to deal with that waste. Even states that do not ban yard waste from landfills can extend the life of landfills by diverting yard waste to composting programs. The numbers on the map indicate the number of composting programs in each state.

Source: Data from the U.S. Environmental Protection Agency.

turn the piles periodically. Turning mixes the different kinds of materials and aerates the mixture.

- *Aerated piles* are large piles of material that have air pumped through them (aeration) so that no mechanical turning or agitation is necessary. They also typically are covered with a layer of mature compost or other material to insulate the pile to keep it at an optimal temperature.

- *Enclosed vessels* can also be used to compost materials very rapidly (within days). However, these systems are much more technologically complex. In such systems, compostable material is fed into a drum, silo, or other structure where the environmental conditions are closely controlled, and the material is mechanically aerated and mixed.

In addition to keeping wastes from entering a landfill, composting has other significant benefits. The addition of compost to soil will improve it by making clay soils more porous or increasing the

water-holding capacity of sandy soils. Nitrogen, potassium, iron, phosphorus, sulfur, and calcium are all common in compost and are beneficial to plant growth.

Source Reduction

Throughout the life cycle of a product—from extraction of raw materials, to transport to processing and manufacturing facilities, to manufacture and use—waste is generated. The simplest way to reduce waste is to prevent it from ever becoming waste in the first place. **Source reduction** is the practice of designing, manufacturing, purchasing, using, and reusing materials so that the amount of waste or its toxicity is reduced.

Design changes to soft drink bottles and milk jugs are good examples of source reduction. Since 1977, the weight of a 2-liter plastic soft drink bottle has been reduced from 68 grams (2.4 ounces) to 51 grams (1.8 ounces). That translates to 114 million kilograms (250 million

Environmental Close-Up

What You Can Do to Reduce Waste and Save Money

You can make a difference. While this statement is sometimes overused, it does speak the truth when it comes to your ability to lessen the stream of solid waste being generated every day. Here are a few ideas that are easy to follow, will save you money, and will help reduce waste.

- Buy things that last, keep them as long as possible, and have them repaired, if possible.
- Buy things that are reusable or recyclable, and be sure to reuse and recycle them.
- Buy beverages in refillable glass containers instead of cans or throwaway bottles.
- Use plastic or metal lunch boxes and metal or plastic garbage containers without throwaway plastic liners.
- Use rechargeable batteries.
- Skip the bag when you buy only a quart of milk, a loaf of bread, or anything you can carry with your hands.

- Buy recycled goods, and then recycle them.
- Recycle all newspaper, glass, and aluminum, and any other items accepted for recycling in your community.
- Reduce the amount of junk mail you get. This can be accomplished by writing to Mail Preference Service, Attn: Dept: 13885751, Direct Marketing Association, P. O. Box 282, Carmel, NY 10512. Ask that your name not be sold to large mailing-list companies. You may also register online by going to the website. Type Mail Preference Service into your browser. Of the junk mail you do receive, recycle as much of the paper as possible.
- Push for mandatory trash separation and recycling programs in your community and schools.
- Choose items that have the least packaging or, better yet, no packaging ("nude products").
- Compost your yard and food wastes, and pressure local officials to set up a community composting program.

---

pounds) of plastic per year that has been kept out of the waste stream. The weight of a plastic milk jug has been reduced 30 percent.

Manufacturing processes have been changed in many industries to reduce the amount of waste produced. One of the simplest ways to reduce waste is to pay careful attention to leaks, spills, and accidents during the manufacturing process. All of these incidents generate waste, and their prevention reduces the amount of raw material needed and the amount of waste generated.

Purchasing decisions can significantly reduce the amount of waste produced. In many cases, consumers and businesses can choose to purchase things that have reduced packaging waste. You can choose to purchase products in larger sizes so that the amount of waste produced is reduced. In addition, careful planning of the quantities purchased can prevent unused surplus materials from becoming part of the waste stream.

Using materials in such a way that waste is not generated is an important means of curbing waste. Using less-hazardous alternatives for certain items (e.g., cleaning products and pesticides), sharing products that contain hazardous chemicals instead of throwing out leftovers, following label directions carefully, and using the smallest amount necessary are ways to reduce waste or its toxicity.

Reusing items is a way to reduce waste at the source because it delays or prevents the entry of reused items into the waste collection and disposal system. For example, many industries participate in waste exchanges that allow a waste product from one industry to be used as a raw material in another industry. In such cases, both industries benefit because the waste producer does not need to pay to dispose of the waste and the industry using the waste has an inexpensive source of a raw material.

Most businesses and manufacturers have a strong economic incentive to make

sure that they get the most from all the materials used in their operations. As regulations on the production of waste become more stringent, industries will need to become even more efficient. Most businesses also see the value in reducing waste disposal costs. Any activities that reduce the amount of waste produced reduce the cost of waste disposal, the amount of raw materials needed, and the amount of pollution generated. This economic incentive also works at the consumer level. In the United States, over 4000 communities have instituted "pay-as-you-throw" programs in which citizens pay for each can or bag of trash they set out for disposal rather than through the tax base or a flat fee. When these households reduce waste at the source, they dispose of less trash and pay lower trash bills.

Recycling

Recycling is one of the best environmental success stories of the late twentieth century. (See figure 18.12.) In the United States, recycling, including composting, diverted about 30 percent of the solid waste stream from landfills and incinerators in 2001, up from about 16 percent in

1990. Several kinds of programs have contributed to the increase in recycling rate. Some benefits of recycling are resource conservation, pollutant reduction, energy savings, job creation, and reduced need for landfills and incinerators. However, incentives are needed to encourage people to participate in recycling programs. Several kinds of programs have been successful.

Container laws provide an economic incentive to recycle. In October 1972, Oregon became the first state to enact a "bottle bill." The law required a deposit of two to five cents on all beverage containers that could be reused. It banned the sale of one-time-use beverage bottles and cans. One of the primary goals of the law was to reduce the amount of litter, and it worked. Within two years of when it went into effect, beverage-container litter decreased by about 49 percent.

Today, 11 states—Oregon, Vermont, Maine, Connecticut, New York, Iowa, Rhode Island, Michigan, Delaware, California, and Hawaii—have enacted legislation requiring deposits on bottles, specifically beverage containers. In 2002, Hawaii became the most recent state to pass a bottle deposit law. It will go into effect in January 2005 and will require a five-cent deposit on beer and soft drink containers.

Many argue that a national bottle bill is long overdue. A national bottle bill would reduce litter, save energy and money, and create jobs. It would also help to conserve natural resources. But the lobbying efforts of the soft drink and brewing industries are very strong, and Congress currently has failed to pass a national container law.

Mandatory recycling laws provide a statutory incentive to recycle. Many states and cities have passed mandatory recycling laws. Some of these laws simply require that residents separate their recyclables from other trash. Other laws are aimed at particular products such as

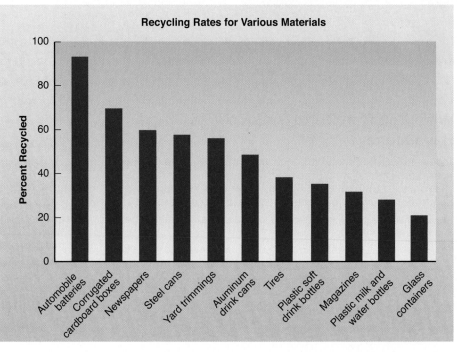

Figure 18.12   Recycling Percentage for Selected Materials (2001) Recycling rates for materials that have high value such as automobile batteries are extremely high. Other materials are more difficult to market. But recycling rates today are much higher than in the past as technology and markets have found uses for materials that once were considered valueless.

Source: Data from the U.S. Environmental Protection Agency, *Characterization of Municipal Solid Waste in the United States,* 2001.

beverage containers and require that they be recycled. Some are aimed at businesses and require them to recycle certain kinds of materials such as cardboard or batteries. Finally, some laws forbid the disposal of certain kinds of materials in landfills. Therefore, the materials must be recycled or dealt with in some other way. For example, banning yard waste from landfills has resulted in extensive composting programs.

Those states and cities with mandatory recycling laws understandably have high recycling rates. For example, California mandated 50 percent diversion of waste through recycling and other means by 2001. Although many cities did not meet the goal by 2002, the statewide recycling rate was 48 percent.

Curbside recycling provides a convenient way for people to recycle. In 1990, a thousand U.S. cities had curbside recycling programs. By

2000, the number had grown to about 9000 cities. By 1999, mandatory recycling laws for all materials had been passed in 15 states. In Canada, Toronto, Mississauga, and the province of Ontario have comprehensive recycling programs.

Some large cities—such as Portland, Oregon; San Jose, California; Los Angeles, California; and Minneapolis, Minnesota—have achieved recycling rates of 50 percent or more. In general, these cities have curbside recycling and accept a wide variety of materials, including junk mail and cereal boxes. By contrast, cities that do not provide curbside recycling, such as El Paso, Texas, and Detroit, Michigan, have recycling rates of less than 10 percent.

Recycling Concerns

Although recycling programs have successfully reduced the amount of material

Recycling Is Big Business

In 1994, Weyerhaeuser Company, one of the world's largest forest products producers, began building a new paper mill in Cedar Rapids, Iowa, an area of North America not known for its abundant forests. The Cedar River Paper Company is a joint venture between Weyerhaeuser and Midwest Recycling Company and is the largest paper recycling plant in the United States. Even with a lack of local trees, the mill is not hurting for raw material to make paper and corrugated cardboard.

The mill is supplied with old paper, including scrap boxes from K-Mart. K-Mart signed an agreement with Weyerhaeuser in 1992 to have Weyerhaeuser purchase its cardboard. The agreement provides Weyerhaeuser with a supply of paper for its recycled paper plants and solves a waste disposal problem for K-Mart. The fact that a supply of recycled paper can determine where new paper-making factories will be built indicates just how important the recycling business is becoming. Weyerhaeuser is one of the largest companies involved in paper recycling in the United States. It recycles about 6 million metric tons of waste paper each year through about 20 reprocessing plants. Currently, the average recycled content of Weyerhaeuser paper products is 38 percent. Clearly, recycling paper is a big business.

The recycling operation at the Cedar River Paper Company.

that needs to be trucked to a landfill or incinerated, there are many technical and economic problems associated with recycling. Technical questions are of particular concern when recycling plastics. (See figure 18.13.) While the plastics used in packaging are recyclable, the technology to do so differs from plastic to plastic. There are many different types of plastic polymers. Since each type has its own chemical makeup, different plastics cannot be recycled together. In other words, a milk container is likely to be high-density polyethylene (HDPE), while an egg container is polystyrene (PS), and a soft-drink bottle is polyethylene terephthalate (PET).

Plastic recycling is still a relatively new field. Industry is researching new technologies that promise to increase the quality of plastics produced from recycled materials and that will allow mixing of different plastics. Until such technology is developed, separation of different plastics before recycling will be necessary.

The economics of recycling are also a primary area of concern. The stepped-up commitment to recycling in many developed nations has produced a glut of certain materials on the market. Markets for collected materials fill up just like landfills. Unless the demand for recycled products keeps pace with the growing supply, recycling programs will face an uncertain future. The prices for selected recycled materials are listed in figure 18.14. Prices for materials can vary widely from year to year, depending on demand.

Markets for materials collected in recycling programs grew dramatically during the 1990s. The establishment of a recyclables exchange on the Chicago Board of Trade allows for consumers and producers of recyclable materials to participate in an efficient market so that it is less likely that recyclable materials will be left unclaimed.

The long-term success of recycling programs is also tied to other economic incentives, such as taxing issues and the development of and demand for products manufactured from recycled material. Government policies need to be readjusted to encourage recycling efforts. Such policies as subsidizing the building of roads for extracting timber, taxing forest land at low rates, or selling trees from federal lands at low prices artificially lower the cost of virgin forest materials compared to recycled materials. Such policies severely inhibit the cost-effectiveness of paper recycling. In addition, on an individual level, we can have an impact by purchasing products made from recycled materials. The demand for recycled products must grow if recycling is to succeed on a large scale.

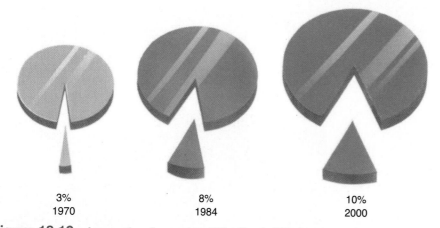

| 3% | 8% | 10% |
| 1970 | 1984 | 2000 |

Figure 18.13 **Increasing Amounts of Plastics in Trash** Plastics are a growing component of municipal solid waste in North America. Increased recycling of plastics could reverse this trend.

Source: Data from Franklin Associates, Ltd.

Recycling Composite Prices		1999	2004
Metal			
Ferrous ($/ton)	Used steel cans	71	161
Nonferrous (¢/lb.)	Aluminum cans	50	57
	Auto batteries	6	5
Plastic (¢/lb.)			
	Green PET	7	16
	Clear PET	6	17
	Mixed HDPE	9	11
Paper ($/ton)			
	Corrugated	92	83
	Newspaper	31	58
	High-grade office	93	128
	Computer laser	163	181
Glass ($/ton)			
	Clear	39	29
	Green	14	9
	Brown	24	17

Figure 18.14 **Recycling Composite Prices** Prices for materials can vary widely from year to year, depending on demand.

Source: Data from American Metal Market LLC, a division of Metal Bulletin Plc.

Municipal solid waste is managed by landfills, incineration, composting, waste reduction, and recycling. Landfills are the primary means of disposal; however, a contemporary landfill is significantly more complex and expensive than the simple holes in the ground of the past. The availability of suitable landfill land is also a problem in large metropolitan areas.

About 15 percent of the municipal solid waste in the United States is incinerated. While incineration does reduce the volume of municipal solid waste, the problems of ash disposal and air quality continue to be major concerns. There are several forms of composting that can keep organic wastes from entering a landfill.

The most fundamental way to reduce waste is to prevent it from ever becoming waste in the first place. Using less material in packaging, reusing items, and composting yard waste are all examples of source reduction. On an individual level, we can all attempt to reduce the amount of waste we generate.

About 30 percent of the waste generated in North America is handled through recycling. Recycling initiatives have grown rapidly in North America during the past several years. As a result, the markets for some recycled materials have become very volatile. Recycling of municipal solid waste will be successful only if markets exist for the recycled materials. Another problem in recycling is the current inability to mix various plastics.

Future management of municipal solid waste will be an integrated approach involving landfills, incineration, composting, source reduction, and recycling. The degree to which any option will be used will depend on economics, changes in technology, and citizen awareness and involvement.

Issues – Analysis

Paper or Plastic?

The plastic shopping bag has become a common feature of the landscape. It is marketed as an inexpensive, lightweight, strong, and recyclable alternative to the paper bag. Plastic bags require much less energy to make than paper bags, and because they are so thin, their disposal results in about 80 percent less solid waste than an equivalent number of paper bags. Their cost is about one-fourth that of paper bags. On the other hand, paper bags can hold 25 to 50 percent more items than a plastic bag and will decompose in a few months, while plastic bags last for years. Although plastic bags are recyclable, less than 1 percent actually are recycled.

The presence of plastic shopping bags in the environment has many negative effects. Because they are so lightweight, they are easily picked up and carried off by the wind and become unsightly trash on the ground and "flags" on vegetation. Furthermore, since they do not decompose easily, they remain in the environment, where they plug sewers and endanger wildlife such as marine turtles that eat them. Bangladesh banned plastic bags because they were blocking drainage systems and causing flooding.

In Ireland, where tourism is a major source of income, the presence of plastic bags in the landscape is an important issue. In 2002, Ireland introduced a tax on plastic shopping bags of 0.15 euros per bag (about 20 cents). Retailers must charge the fee, and it must show on the sales slip. As a consequence, the use of plastic bags has dropped by 90 percent. Australia has mandated that retailers reduce the use of plastic bags by 25 percent or a tax on the bags will be imposed. Other countries are considering prohibiting the use of plastic bags or taxes to reduce their negative environmental impacts.

- Do you use paper, plastic, or a carrier bag? How did you make your choice?
- Would you favor a tax on plastic bags?
- How would a ban of plastic bags alter your lifestyle?

Key Terms

agricultural waste *420*
composting *428*
incineration *424*
industrial solid waste *420*
leachate *423*

mass burn *424*
mining waste *420*
mulch *426*
municipal solid waste landfill *423*
municipal solid waste (MSW) *420*

recycling *429*
solid waste *420*
source reduction *428*

Review Questions

1. How is lifestyle related to the quantity of municipal solid waste generated?
2. What conditions favor incineration over landfills?
3. Describe some of the problems associated with modern landfills.
4. What are four concerns associated with incineration?
5. Describe examples of source reduction.

6. Describe the importance of recycling household solid wastes.
7. Name several strategies that would help to encourage the growth of recycling.
8. Describe the various types of composting and the role of composting in solid waste management.

Critical Thinking Questions

1. How can you help solve the solid waste problem?
2. Given that you have only so much time, should you spend your time acting locally, as a recycling coordinator, for example, or advocating for larger political and economic changes at the national level, changes that would solve the waste problems? Why? Or should you do nothing? Why?
3. How does your school or city deal with solid waste? Can solid waste production be limited at your institution or city? How? What

barriers exist that might make it difficult to limit solid waste production?
4. It is possible to have a high standard of living, as in North America and Western Europe, and not produce large amounts of solid waste. How?
5. Incineration of solid waste is controversial. Do you support solid waste incineration in general? Would you support an incineration facility in your neighborhood?

Welcome to McGraw-Hill's Online Learning Center

Location: www.mhhe.com/enger10e

McGraw Hill

Interactive Exploration

Check out the website at www.mhhe.com/enger10e and click on the cover of this textbook for quizzing, career information, case studies, and hot links for the following topics:

Solid, Toxic, and Hazardous Waste

Recycling

Solid Waste

Regulating Hazardous Materials

19

Objectives

After reading this chapter, you should be able to:

- Distinguish between hazardous substances and hazardous wastes.
- Distinguish between hazardous and toxic substances.
- Describe the four characteristics used to identify hazardous materials.
- Describe the kinds of environmental problems caused by hazardous and toxic materials.
- Understand the difference between persistent and nonpersistent pollutants.
- Describe the difference between chronic and acute exposures to hazardous wastes.
- Describe why hazardous-waste dump sites developed.
- Describe how hazardous wastes are managed, and list five technologies used in their disposal.
- Describe the importance of source reduction with regard to hazardous wastes.

Chapter Outline

PAST PRESENT FUTURE

1980 Superfund legislation was passed by the U.S. Congress.

1984 A chemical release in Bhopal, India, killed about 4000 people.

1986 The Toxic Release Inventory was initiated in the United States. In 1988, it documented the release of 1434 million kilograms (3161 million pounds) of 273 chemicals.

2004 Of 1499 Superfund sites, 846 (56 percent) have been cleaned up.

2004 Lawsuits related to the chemical release were still in progress.

2001 Six hundred and fifty-three million kilograms (1440 million pounds) of 273 chemicals were released (about a 55 percent reduction).

Hazardous and Toxic Materials in Our Environment

Our modern technological society makes use of a large number of substances that are hazardous or toxic. The benefits gained from using these materials must be weighed against the risks associated with their use. When these materials are released into the environment in an inappropriate way through accident or neglect, significant environmental and human health consequences result. The following items indicate the magnitude of the problem.

- Pesticides thought to degrade in soils are found in rural drinking-water wells.
- There are elevated levels of mercury downwind of coal-fired power plants.
- Chemicals leaching from abandoned waste sites contaminate city water supplies.
- Pesticides spilled into the Rhine River from a warehouse near Basel, Switzerland, in 1988 destroyed a half million fish, disrupted water supplies, and caused considerable ecological damage.
- The collapse of an oil storage tank in Pennsylvania in 1988 spilled over 3 million liters (750,000 gallons) of oil into the Monongahela River and threatened the water supply of millions of residents.
- Release of polychlorinated biphenyls (PCBs) from electrical parts manufacturing plants before the 1970s contaminated hundreds of hectares of marine sediments in New Bedford Harbor, Massachusetts, resulting in marine life being reduced and the area being closed to fishing.

Toxic and hazardous products and by-products are becoming a major issue of our time. At sites around the world, accidental or purposeful releases of hazardous and toxic chemicals are contaminating the land, air, and water. The potential health effects of these chemicals range from minor, short-term discomforts, such as

Figure 19.1　**The Life Cycle of Toxic Substances** Controlling the problems of hazardous substances is complicated because of the many steps involved in a substance's life cycle.

headaches and nausea, to serious health problems, such as cancers and birth defects (that may not manifest themselves for years), to major accidents that cause immediate injury or death. Today, such names as Love Canal, New York, and Times Beach, Missouri, in the United States; Lekkerkerk in the Netherlands; Vác, Hungary; and Minamata Bay, Japan, are synonymous with the problems associated with the release of hazardous and toxic wastes into the environment.

Increasingly, governments and international agencies are attempting to control the growing problem of hazardous substances in our environment. Controlling the release of these substances is difficult since there are so many places in their cycles of use at which they may be released. (See figure 19.1.)

Hazardous and Toxic Substances—Some Definitions

To begin, it is important to clarify various uses of the words *hazardous* and *toxic*, as well as to distinguish between things that are wastes and those that are not. **Hazardous substances** or **hazardous materials** are those that can cause harm to humans or the environment. Different government agencies have slightly different definitions for what constitutes a hazardous substance. The U.S. Environmental Protection Agency defines hazardous substances as having one or more of the following characteristics:

1. **Ignitability**—Describes materials that pose a fire hazard during

routine management. Fires not only present immediate dangers of heat and smoke but also can spread harmful particles over wide areas. Common examples are gasoline, paint thinner, and alcohol.

2. **Corrosiveness**—Describes materials requiring special containers because of their ability to corrode standard materials or requiring segregation from other materials because of their ability to dissolve toxic contaminants. Common examples are strong acids and bases.

3. **Reactivity** (or explosiveness)—Describes materials that, during routine management, tend to react spontaneously, to react vigorously with air or water, to be unstable to shock or heat, to generate toxic gases, or to explode. Common examples are gunpowder, which will burn or explode; the metal sodium, which reacts violently with water; and nitroglycerine, which explodes under a variety of conditions.

4. **Toxicity**—Describes materials that, when improperly managed, may release toxicants (poisons) in sufficient quantities to pose a substantial hazard to human health or the environment. Almost everything that is hazardous is toxic in high enough quantities. For example, tiny amounts of carbon dioxide in the air are not toxic, but high levels are.

Some hazardous materials fall into several of these categories. Gasoline, for example, is ignitable, can explode, and is toxic. It is even corrosive to certain kinds of materials. Other hazardous materials meet only one of the criteria. PCBs are toxic but will not burn, explode, or corrode other materials. While the terms *toxic* and *hazardous* are often used interchangeably, there is a difference. **Toxic** commonly refers to a narrow group of substances that are poisonous and cause death or serious injury to humans and other organisms by interfering with normal body physiology. **Hazardous,** the broader term, refers to all dangerous materials, including toxic ones, that present an immediate or long-term human health risk or environmental risk.

Another important distinction is the difference between hazardous substances and hazardous wastes. Although the health and safety considerations regarding hazardous substances and hazardous wastes are similar, the legal and regulatory implications are quite different. Hazardous substances are materials that are used in business and industry for the production of goods and services. Typically, hazardous substances are consumed or modified in industrial processes. **Hazardous wastes** are by-products of industrial, business, or household activities for which there is no immediate use. These wastes must be disposed of in an appropriate manner, and there are stringent regulations pertaining to their production, storage, and disposal.

Defining Hazardous Waste

Although the definition of hazardous waste varies from one country to another, one of the most widely used definitions is contained in the U.S. **Resource Conservation and Recovery Act** of 1976 (RCRA). The RCRA considers wastes toxic and/or hazardous if they:

> cause or significantly contribute to an increase in mortality or an increase in serious irreversible, or incapacitating reversible, illness; or pose a substantial present or potential hazard to human health or the environment when improperly treated, stored, transported, disposed of, or otherwise managed.

This very complex definition gives one an appreciation for the complexity of hazardous-waste regulation.

Because of the difficulty in developing simple definitions of hazardous materials, listing is the most common method for defining hazardous waste in many countries. The U.S. Environmental Protection Agency has compiled such a list of hazardous wastes. The EPA also requires that a waste be tested to determine if it possesses any one of the four characteristics discussed earlier: ignitability, corrosiveness, reactivity, and toxicity. If it does, it is subject to regulation under RCRA.

There are numerous types of hazardous waste, ranging from materials contaminated with dioxins and heavy metals (such as mercury, cadmium, and lead) to organic wastes. These wastes can also take many forms, such as barrels of liquid waste or sludge, old computer parts, used batteries, and incinerator ash. In industrialized countries, industry and mining are the main sources of hazardous wastes, though small-scale industry, hospitals, military establishments, transport services, and small workshops also contribute to the generation of significant quantities of such wastes in both the industrialized and developing worlds.

Improper handling and disposal of hazardous wastes can affect human health and the environment by contaminating groundwater, soil, waterways, and the atmosphere. The environmental and health effects can be immediate (acute), as when exposure to toxins at a particular site causes sudden illness, or long-term (chronic), as when contaminated waste leaches into groundwater or soil and then works its way into the food chain. Although chronic exposures to low levels of toxic materials do not cause immediate health effects, they may cause serious health problems later in life. The damage caused by the release of hazardous wastes into the environment also takes an economic toll, since cleaning up contaminated sites can be costly for local authorities. The handling of hazardous waste requires special training to prevent exposure of workers.

Exact figures regarding the amounts of hazardous waste generated worldwide are difficult to obtain. However, the United Nations Environment Programme estimates between 300 million and 500 million metric tons are generated annually. A large proportion comes from developed countries. Data from the EPA indicate that the United States generates about 36 million metric tons of hazardous waste each year. The European Union generates about 40 million metric tons. The rapidly industrializing countries of Asia are generating increasing amounts of hazardous waste, but it is difficult to obtain estimates of the amounts.

Figure 19.2 Common Materials Can Produce Hazardous Wastes Many commonly used materials generate hazardous wastes during their manufacture. Some consumer products such as pesticides, paints, and solvents are also hazardous and should be disposed of in an approved manner.

The table within the figure reads:

The products we use...	The potentially hazardous waste they generate...
Plastics	Organic chlorine compounds
Pesticides	Organic chlorine compounds, organic phosphate compounds
Medicines	Organic solvents and residues, heavy metals (mercury and zinc, for example)
Paints	Heavy metals, pigments, solvents, organic residues
Oil, gasoline, and other petroleum products	Oils, phenols, and other organic compounds, heavy metals, ammonia salts, acids, caustics
Metals	Heavy metals, fluorides, cyanides, acid and alkaline cleaners, solvents, pigments, abrasives, plating salts, oils, phenols
Leather	Heavy metals, organic solvents
Textiles	Heavy metals, dyes, organic chlorine compounds, solvents

Issues Involved in Setting Regulations

Whether a hazardous substance is a raw material, an ingredient in a product, or a waste, problems are associated with determining regulations that pertain to it. Mining, electric generation, and chemical and petrochemical industries are the major producers of hazardous wastes. These in-dustries produce many useful materials that are converted into the products we use every day, such as metals, energy sources, and many synthetic materials. (See figure 19.2.)

Identification of Hazardous and Toxic Materials

In attempting to regulate the use of toxic and hazardous substances and the genera-tion of toxic and hazardous wastes, most countries simply draw up a list of specific substances that have been scientifically linked to adverse human health or envi-ronmental effects. However, since many potentially harmful chemical compounds have yet to be tested adequately, most lists include only the known offenders. Histor-ically, we have often identified toxic mate-rials only after their effects have shown up in humans or other animals. Asbestos was

Determining Toxicity

We are all exposed to potentially harmful toxins. The question is, at what levels is such exposure harmful? One measure of toxicity of a substance is its **LD$_{50}$**, the dosage of a substance that will kill (lethal dose) 50 percent of a test population (LD$_{50}$ = lethal dose 50 percent). Toxicity is measured in units of poisonous substance per kilogram of body weight. For example, the deadly chemical that causes botulism, a form of food poisoning, has an LD$_{50}$ in adult human males of 0.0014 milligrams per kilogram. This means that if each of 100 human adult males weighing 100 kilograms consumed a dose of only 0.14 milligrams—about the equivalent of a few grains of table salt—approximately 50 of them will die.

Lethal doses are not the only danger from toxic substances. During the past decade, concern has been growing over minimum harmful dosages, or threshold dosages, of poisons, as well as their sublethal effects.

The length of exposure further complicates the determination of toxicity values. Acute exposure refers to a single exposure lasting from a few seconds to a few days. Chronic exposure refers to continuous or repeated exposure for several days, months, or even years. Acute exposure usually is the result of a sudden accident, such as the tragedy at Bhopal, India, mentioned at the beginning of the chapter. Acute expo-

sures often make disaster headlines in the press, but chronic exposure to sublethal quantities of toxic materials presents a much greater hazard to public health. For example, millions of urban residents are continually exposed to low levels of a wide variety of pollutants. Many deaths attributed to heart failure or such diseases as emphysema may actually be brought on by a lifetime of exposure to sublethal amounts of pollutants in the air.

---

identified as a cause of lung cancer in humans who were exposed on the job, and DDT was identified as toxic to birds when robins began to die and eagles and other fish-eating birds failed to reproduce. Once these substances were identified as toxic, their use was regulated. Countries contemplating regulation of hazardous and toxic materials and wastes must consider not only how toxic each one is but also how flammable, corrosive, and explosive it is, and whether it will produce mutations or cause cancer.

Setting Exposure Limits

Even after a material is identified as hazardous or toxic, there are problems in determining appropriate exposure limits. Nearly all substances are toxic in sufficiently high doses. The question is, When does a chemical cross over from safe to toxic? There is no easy way to establish acceptable levels. Several government agencies set limits for different purposes. The Occupational Safety and Health Administration, the Food and Drug Administration, the U.S. Public Health Service, the

Environmental Protection Agency, and others publish guidelines or set exposure limits for hazardous substances in the air, water, and soil. Furthermore, it is important to recognize that people can be exposed in three primary ways: by breathing the material, by consuming the material through the mouth, or by absorbing the material through the skin. Each of these routes may require different exposure classifications. Regardless, for any new compounds that are to be brought on the market, extensive toxicology studies must be done to establish their ability to do harm. Usually these are tests on animals. (See Environmental Close-Up: Determining Toxicity.) Typically, the regulatory agency will determine the level of exposure at which none of the test animals is affected (**threshold level**) and then set the human exposure level lower to allow for a safety margin. This safety margin is important because it is known that threshold levels vary significantly among species, as well as among members of the same species. Even when concentrations are set, they may vary considerably from country to country. For example, in the Nether-

lands, 50 milligrams of cyanide per kilogram of waste is considered hazardous; in neighboring Belgium, the toxicity standard is fixed at 250 milligrams per kilogram.

Acute and Chronic Toxicity

Regulatory agencies must look at both the effects of one massive dose of a substance (**acute toxicity**) and the effects of exposure to small doses over long periods (**chronic toxicity**). Acute toxicity is readily apparent because organisms respond to the toxin shortly after being exposed. Chronic toxicity is much more difficult to determine because the effects may not be seen for years. Furthermore, an acute exposure may make an organism ill but not kill it, while chronic exposure to a toxic material may cause death. A good example of this effect is alcohol toxicity. Consuming extremely high amounts of alcohol can result in death (acute toxicity and death). Consuming moderate amounts may result in illness (acute toxicity and full recovery). Consuming moderate amounts over a number of years may result in liver damage and death (chronic toxicity and death).

Another example of chronic toxicity involves lead. Lead was used in paints, gasoline, and pottery glazes for many years, but researchers discovered that it has harmful effects. The chronic effects on the nervous system are most noticeable in children, particularly when children eat paint chips.

Synergism

Another problem in regulating hazardous materials is assessing the effects of mixtures of chemicals. Most toxicological studies focus on a single compound, even though industry workers may be exposed to a variety of chemicals; and hazardous wastes often consist of mixtures of compounds. Although the materials may be relatively harmless as separate compounds, once mixed, they may become highly toxic and cause more serious problems than do individual pollutants. This is referred to as **synergism.** For example, all uranium miners are exposed to radioactive gases, but those who smoke tobacco and thus are exposed to the toxins in tobacco smoke have unusually high incidences of lung cancer. Apparently, the radioactive gases found in uranium mines interact synergistically with the carcinogens found in tobacco smoke.

Persistent and Nonpersistent Pollutants

The regulation of hazardous and toxic materials is also influenced by the degree of persistence of the pollutant. **Persistent pollutants** are those that remain in the environment for many years in an unchanged condition. Most of the persistent pollutants are human-made materials. An estimated 30,000 synthetic chemicals are used in the United States. They are mixed in an endless variety of combinations to produce all types of products used in every aspect of daily life. They are part of our food, transportation, clothing, building materials, home appliances, medicine, recreational equipment, and many other items. Our way of life depends heavily on synthetic materials.

An example of a persistent pollutant is DDT. It was used as an effective pesticide worldwide and is still used in some countries because it is so inexpensive and very effective in killing pests. However, once released into the environment, it accumulates in the food chain and causes death when its concentration is high enough. (See chapter 15 for a discussion of DDT as a pesticide.)

Another widely used group of synthetic compounds of environmental concern are polychlorinated biphenyls. PCBs are highly stable compounds that resist changes from heat, acids, bases, and oxidation. These characteristics make PCBs desirable for industrial use but also make them persistent pollutants when released into the environment. At one time, these materials were commonly used in transformers and electrical capacitors. Other uses included inks, plastics, tapes, paints, glues, waxes, and polishes. Although the manufacture of PCBs in the United States stopped in 1977, these persistent chemicals are still present in the soil and sediments and continue to do harm. PCBs are harmful to fish and other aquatic forms of life because they interfere with reproduction. In humans, PCBs produce liver ailments and skin lesions. In high concentration, they can damage the nervous system, and they are suspected carcinogens.

In addition to synthetic compounds, our society uses heavy metals for many purposes. Mercury, beryllium, arsenic, lead, and cadmium are examples of heavy metals that are toxic. These metals are used as alloys with other metals, in batteries, and have many other special applications. In addition, these metals may be released as a by-product of the extraction and use of other metals. When released into the environment, they enter the food chain and become concentrated. In humans, these metals can produce kidney and liver disorders, weaken the bone structure, damage the central nervous system, cause blindness, and lead to death. Because these materials are persistent, they can accumulate in the environment even though only small amounts might be released each year. When industries use these materials in a concentrated form, it presents a hazard not found naturally.

A **nonpersistent pollutant** does not remain in the environment for very long. Most nonpersistent pollutants are biodegradable. Others decompose as a result of inorganic chemical reactions. Still others quickly disperse to concentrations that are too low to cause harm. A biodegradable material is chemically changed by living organisms and often serves as a source of food and energy for decomposer organisms, such as bacteria and fungi. Phenol and many other kinds of toxic organic materials can be destroyed by decomposer organisms.

Other toxic materials, such as many insecticides, are destroyed by sunlight or reaction with oxygen or water in the atmosphere. These include the "soft biocides." For example, organophosphates usually decompose within several weeks. As a result, organophosphates do not accumulate in food chains because they are pollutants for only a short period of time.

Other toxic and hazardous materials such as carbon monoxide, ammonia, or hydrocarbons can be dispersed harmlessly into the atmosphere (as long as their concentration is not too great), where they eventually react with oxygen.

Because persistent materials can continue to do harm for a long time (chronic toxicity), their regulation is particularly important. Nonpersistent materials need to be kept below threshold levels to protect the public from acute toxicity. They are not likely to present a danger of chronic toxicity since they either disperse or decompose.

Environmental Problems Caused by Hazardous Wastes

Hazardous wastes enter the environment in several ways. Many molecules that evaporate readily are vented directly to the atmosphere. Many kinds of solvents used in paints and other industrial processes fall into this category. Other materials escape from faulty piping and valves. These materials are often not even thought of as hazardous waste but are

GLOBAL PERSPECTIVE

Lead and Mercury Poisoning

Lead and mercury are naturally present in rock, and as the rock weathers, they are released into the soil and water. They have been present in the environment and probably have been a source of pollution for centuries. For example, the lead drinking and eating vessels used by the wealthy Romans may have caused the death of many of them. Another example, believe it or not, comes from *Alice in Wonderland,* published in 1865. One of the characters was the Mad Hatter. At that period in history, mercury was widely used in the treatment of beaver skins for making hats. As a result of exposure to mercury, hat makers often suffered from a variety of mental problems; hence, the phrase "mad as a hatter."

In 1953, a number of physical and mental disorders in the Minamata Bay region of Japan were diagnosed as being caused by mercury: 52 people developed symptoms of mercury poisoning, 17 died, and 23 became permanently disabled. In 1970, an outbreak of mercury poisoning in North America was traced to mercury in the meat of swordfish and tuna. In both incidents, the toxic material was not metallic mercury but a mercurous compound, methylmercury. Metallic mercury is converted to methylmercury by bacteria in the water. Because methylmercury is lost from organisms relatively slowly, it bioaccumulates in aquatic organisms and becomes part of the food chain. Once methylmercury enters the food chain, it can become further concentrated as the result of biological amplification. This explains why larger predator fish such as tuna and swordfish are likely to have higher quantities of methylmercury than organisms that are lower on the food chain. Sufficient amounts in humans can cause brain damage, kidney damage, or birth defects. Today, regulations reduce the release of mercury into the environment and set allowable levels in foods. The problem persists, however, because it is impossible to eliminate the large amounts of mercury already present in the environment, and it is difficult to prevent the release of mercury in all cases. For example, burning coal releases 3200 metric tons of mercury into the Earth's atmosphere each year, and mercury is still "lost" when it is used for various industrial purposes.

Like mercury, lead is a heavy metal and has been a pollutant for centuries. Studies of the Greenland ice cap indicate a 1500 percent increase in the lead content today as compared to 800 B.C. These studies reveal that the first large increase occurred during the Industrial Revolution, and the second great increase occurred after the invention of the automobile. Oil companies added lead to gasoline to improve performance, and burning gasoline is a major source of lead pollution. There has been a reduction in airborne lead as a result of North America and Europe and many other countries eliminating lead from gasoline. Another source of lead pollution is older paints. Before 1940, indoor and outdoor paints often contained lead.

Disagreement still exists over what levels of lead and mercury can cause human health problems. Although ingested lead from paint can cause death or disability, such a strong correlation cannot be made for atmospheric lead.

There are concerns about fish contamination and public health as a result of methylmercury.

called fugitive emissions. Uncontrolled or improper incineration of hazardous wastes, whether on land or at sea, can contaminate the atmosphere and the surrounding environment.

Other wastes are in solid or liquid form and are more easily contained. Some of these wastes can be treated to reduce or eliminate their hazardous nature, and these treated wastes can be released to the environment. Many kinds of liquid wastes, such as sewage and acid discharges, are handled in this manner. Other wastes cannot be treated and must be stored.

In the past, many kinds of hazardous wastes were improperly disposed of on land, in containers, storage lagoons, or landfills. Groundwater contamination has resulted from leaking land disposal facilities. Once groundwater is polluted with hazardous wastes, the cost of reversing the damage is prohibitive. In fact, if an aquifer is contaminated with organic chemicals, restoring the water to its original state is seldom physically or economically feasible. Chapter 16 deals with many other aspects of groundwater contamination.

Even improper labeling and recordkeeping procedures can result in releases. Because workers are unable to distinguish hazardous waste from other kinds of waste materials, hazardous wastes are not properly disposed of.

Table 19.1 Top 20 Hazardous Substances, 2003

Substance	Source	Toxic Effects
Arsenic	From elevated levels in soil or water	Multiple organ systems affected. Heart and blood vessel abnormalities, liver and kidney damage, impaired nervous system function
Lead	Lead-based paint Lead additives in gasoline	Neurological damage. Affects brain development in children Large doses affect brain and kidneys in adults and children.
Mercury	Air or water at contaminated sites Methylmercury in contaminated fish and shellfish	Permanent damage to brain, kidneys, developing fetus
Vinyl chloride	Plastics manufacturing Air or water at contaminated sites	Acute effects: dizziness, headache, unconsciousness, death Chronic effects: liver, lung, and circulatory damage
Polychlorinated biphenyls (PCBs)	Eating contaminated fish Industrial exposure	Probable carcinogens; acne and skin lesions
Benzene	Industrial exposure Glues, cleaning products, gasoline	Acute effects: drowsiness, headache, death at high levels Chronic effects: damage to blood-forming tissues and immune system; also carcinogenic
Cadmium	Released during combustion Living near a smelter or power plant Picked up in food	Probable carcinogen; kidney damage, lung damage, high blood pressure
Polycyclic aromatic hydrocarbons	Exposure to smoke from a variety of sources	Probable carcinogen; possible birth defects
Benzo[a]pyrene	Product of combustion of gasoline or other fuels In smoke and soot	Probable carcinogen; possible birth defects
Benzo[b]fluoranthene	Product of combustion of gasoline and other fuels Inhaled in smoke	Probable carcinogen
Chloroform	Contaminated air and water Many kinds of industrial settings	Affects central nervous system, liver, and kidneys; probable carcinogen
DDT	From food with low levels of contamination Still used as pesticide in parts of world	Probable carcinogen; possible long-term effect on liver; possible reproductive problems
Aroclor 1254 (a mixture of PCBs)	From food and air	Probable carcinogens; acne and skin lesions
Aroclor 1260 (a mixture of PCBs)	From food and air	Probable carcinogens; acne and skin lesions
Dibenzo[a,h]anthracene	Product of combustion in smoke	Probable carcinogen
Trichloroethylene	Used as a degreaser, evaporates into air	Dizziness, numbness, unconsciousness, death
Chromium (hexavalent)	Air emissions from industry	Breathing chromium dust causes nosebleeds; ingestion causes ulcers and liver and kidney damage.
Dieldrin	Past use of the persistent pesticides Aldrin and Dieldrin	Dieldrin bioaccumulates and affects the liver and immune system.
White phosphorus	Industrial sources and explosives	Contact and ingestion retard wound healing.
Chlordane	Past use of this persistent pesticide	Affects nervous, digestive, and liver tissue

Source: Data from Agency for Toxic Substances and Disease Registry.

Health Risks Associated with Hazardous Wastes

Because most hazardous wastes are chemical wastes, controlling chemicals and their waste products is a major issue in most developed countries. Every year, roughly 1000 new chemicals join the nearly 70,000 in daily use. Many of these chemicals are toxic, but they pose little threat to human health unless they are used or disposed of improperly. For example, many insecticides are extremely toxic to humans. However, if they are stored, used, and disposed of properly, they do not constitute a human health hazard. Unfortunately, at the center of the hazardous-waste problem is the fact that the products and by-products of industry are often handled and disposed of improperly. Table 19.1 is a list of 20 top toxic materials as identified by the federal Agency for Toxic Substances and Disease Registry.

Establishing the medical consequences of exposure to toxic chemicals is extremely complicated. The problem of linking a particular chemical to specific injuries or diseases is further compounded by the lack of toxicity data on most hazardous substances.

Although assessing environmental contamination from toxic substances and determining health effects is extremely difficult, what little is known is cause for concern. Most older hazardous-waste dump sites, for example, contain dangerous and toxic chemicals along with heavy-metal residues and other hazardous substances.

Hazardous-Waste Dumps—A Legacy of Abuse

In the United States prior to the passage of the Resource Conservation and Recovery Act (RCRA) in 1976, hazardous waste was essentially unregulated. Similar conditions existed throughout most of the industrial nations of the world. Hazardous wastes were simply buried or dumped without any concern for potential environmental or health risks. Such uncontrolled sites included open dumps, landfills, bulk storage containers, and surface impoundments. These sites were typically located convenient to the industry and were often in environmentally sensitive areas, such as floodplains or wetlands. Rain and melting snow soaked through the sites, carrying chemicals that contaminated underground waters. When these groundwaters reached streams and lakes, they were contaminated as well. When the sites became full or were abandoned, they were frequently left uncovered, thus increasing the likelihood of water pollution from leaching or flooding and of people having direct contact with the wastes. At some sites, specifically the uncovered ones, the air was also contaminated by toxic vapors from evaporating liquid wastes or from uncontrolled chemical reactions. (See figure 19.3.) In North America alone, the number of abandoned or uncontrolled sites was over 25,000.

Europeans are also paying a heavy price for their negligence. The Netherlands is a good example. Authorities estimate that up to 8 million metric tons of hazardous chemical wastes may be buried in the Netherlands. Estimates for cleaning up those wastes run as high as $7 billion. In the republics of the former Soviet Union and Eastern Europe, many hazardous waste sites have been identified recently, and there is no money to pay for cleanup. It is assumed that there are many hazardous waste sites in countries of the developing world, but these are poorly known.

In the United States, the federal government has become the principal participant in the cleanup of hazardous-waste

Figure 19.3 **Improper Toxic Chemical Storage** This is a site where toxic wastes were improperly stored. Often such sites are abandoned, and it is difficult to assign responsibility for cleanup because the company that produced the waste is unknown or no longer exists.

sites. The program that deals with the cleanup has popularly become known as **Superfund.** Superfund was established when Congress responded to public pressure to clean up hazardous-waste dumps and protect the public against the dangers of such wastes. The **Comprehensive Environmental Response, Compensation, and Liability Act (CERCLA)** (Superfund) was enacted in 1980. CERCLA had several key objectives:

1. To develop a comprehensive program to set priorities for cleaning up the worst existing hazardous-waste sites.
2. To make responsible parties pay for those cleanups whenever possible.
3. To set up a $1.6 billion Hazardous Waste Trust Fund—popularly known as Superfund—to support the identification and cleanup of abandoned hazardous-waste sites.

4. To advance scientific and technological capabilities in all aspects of hazardous-waste management, treatment, and disposal.

A **National Priorities List** of hazardous-waste sites requiring urgent attention was drawn up for Superfund action. Under CERCLA, over 44,000 sites were evaluated, and about 11,000 were considered serious enough to warrant further investigation. The number of sites listed on the National Priorities List fluctuates as new sites are added and sites are removed because they are cleaned up or deleted from the list. Currently about 1200 sites are on the list.

Although the purpose of the Superfund program was clear, it became controversial during its early history. Millions of dollars were spent by both the federal government and industry, but most of the

Environmental Close-Up

Computers—A Hazardous Waste

They once cost thousands of dollars and were considered a major investment meant to last. Now computers sell for hundreds of dollars and are obsolete 18 months after they are out of the bubble wrap. Rapid innovation in computer hardware is dramatically cutting the cost and the useful life of modern computers, creating a solid waste problem in the process.

Computers are more than just household waste. They contain large amounts of substances such as lead, cadmium, mercury, and chromium that can leach into soil and contaminate groundwater or, if incinerated, can be released into the air. The average personal computer contains nearly 2 kilograms (about 4 pounds) of lead. Much of the lead is associated with the cathode ray tube screen. In addition, electronic circuit boards typically contain significant amounts of cadmium, mercury, and chromium. Each year, approximately 50 million computers become obsolete in the United States, but only 11 percent are recycled. It will no doubt get worse as computers become increasingly cheaper, faster, and more disposable. The good news is that most computers are highly reusable, and up to 97 percent of the parts can be recycled, either as up-

graded components for use in other computers or melted down as scrap.

In 2000, Massachusetts became the first state to initiate a ban on the disposal of household computer screens, TV sets, and other glass picture tubes in landfills and incinerators. The state set up six collection centers to handle the items, and cities and towns must now transport the items to those centers. From there, they will either be refurbished or sent on for recycling. Many corporations that manufacture and sell electronic products now participate in "e-cycling programs" in which they take back obsolete electronic products for recycling.

Obsolete computer hardware

money went for litigation and technical studies to support or disprove the claims of the parties. Little of the money paid for cleanup. One of the primary reasons for this problem was the way CERCLA was written. It provided that anyone who contributed to a specific hazardous-waste site could be required to pay for the cleanup of the entire site, regardless of the degree to which they contributed to the problem. Since many industries that contributed to the problem had gone out of business or could not be identified, those that could be identified were asked to pay for the cleanup. Most businesses found it cost-effective to hire lawyers to fight their inclusion in a cleanup effort rather than to pay for the cleanup. Consequently, little cleanup occurred.

After a slow start, however, the Superfund program is showing significant results. By 2004, there were about 1200 active sites on the National Priorities List and about 65 were being considered for inclusion. About 900 sites have been cleaned up. Most of the remaining sites were in the process of being cleaned up or were under study to determine the best way to proceed. This has been an ex-

pensive undertaking. Total expenditures for the Superfund program have been about $27 billion. In addition, the EPA estimates that the settlements it has reached with responsible parties amount to over $20 billion.

Toxic Chemical Releases

In 1987, as the result of EPA requirements, certain industries in the United States had to begin reporting toxic chemicals released into the environment. Any industrial plant that released 23,000 kilograms (50,000 pounds) or more of toxic pollutants was required to file a report. These were primarily manufacturing industries. The information collected allowed the EPA to target specific industries for enforcement action. Since the information is public, many industries were encouraged to take action to reduce their emissions. These activities have been successful, and industrial emissions of toxic chemicals have been reduced by about 54 percent from 1988 through 2001. In the intervening years, changes have been made in who must report. The quantity that required reporting was reduced to

11,400 kilograms (25,000 pounds). In addition, industries that were originally exempt from reporting now must report toxic releases. About 2.8 billion kilograms (6.2 billion pounds) of toxic chemicals were reported released into the environment by industry in 2001.

Today, the primary industries involved in toxic releases to the environment are the mining, power generation, chemical, and metal manufacturing industries. (See figure 19.4.) The large amount of toxic waste produced by the mining industry is also reflected in figure 19.5. Mining waste typically contains metals and is deposited on the surface of the land. In addition, you would expect those states with large mining operations to have high amounts of waste.

Hazardous-Waste Management Choices

In discussing the kinds of hazardous wastes released to the environment, it is

Figure 19.4 **Sources of Toxic Releases (2002)** Mining industries are responsible for about 26 percent of all toxic waste material released to the environment. These are primarily deposited on the surface of the land. Electric utilities are responsible for about 23 percent of releases. These are primarily releases to the atmosphere. The chemical industry and metal processing industries are also significant sources of toxic releases.

Source: Data from the U.S. Environmental Protection Agency.

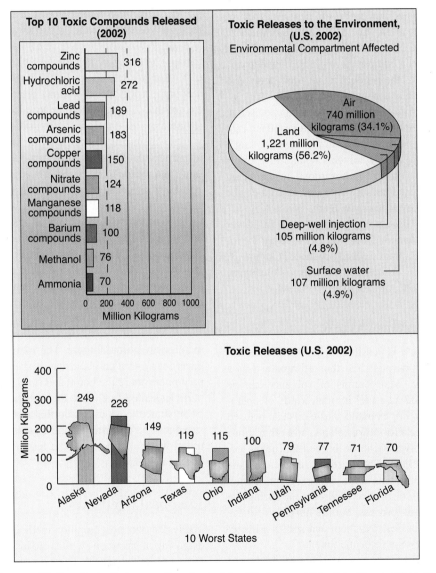

Figure 19.5 **Toxic Releases Dominated by Mining and Electric Generation** The large amount of waste generated by the mining industry is reflected in the kinds of waste produced (toxic metals), the environmental compartment affected (land disposal), and the states with the most toxic releases. Electric power generation is responsible for large amounts of toxic materials released to the air.

Source: Data from the U.S. Environmental Protection Agency.

important to point out that the amount released is less than one-quarter (about 23 percent) of the total amount of hazardous waste produced. In 2001, a total of about 12 billion kilograms (about 27 billion pounds) of hazardous waste was produced by industries in the United States. About 77 percent of this material was recycled, burned for energy, or treated so that it was not released to the environment. What forces contribute to this trend toward reduced amounts of waste being released?

The environmental costs related to the uncontrolled release of toxic materials to the environment are astronomical. Recognition of these costs has resulted in the passage of laws and institution of regulations that regulate release of hazardous wastes, require industries to report the hazardous waste they release, and force industries to pay for the measures necessary to eliminate or reduce the waste they produce. Strong, enforceable laws have forced industries to absorb the cost of waste disposal and look for better and cheaper ways to treat and dispose of hazardous wastes. These legal and economic forces are driving industrial behavior toward pollution prevention and waste minimization.

In the past, the management of hazardous waste was always added on to the end of the industrial process. The effluents from pipes or smokestacks were treated to reduce their toxicity or concentration.

In recent years, it has become obvious that a better way to deal with the problem of hazardous waste is to not produce it in the first place. To this end, the EPA and regulatory agencies in other countries have emphasized pollution prevention and waste minimization. Strong regulatory control requires that industries report the hazardous wastes they produce and that the wastes be stored, transported, and disposed of properly.

The EPA now promotes a **pollution-prevention hierarchy** that emphasizes reducing the amount of hazardous waste produced. This involves the following strategy:

1. Reduce the amount of pollution at the source.
2. Recycle wastes wherever possible.
3. Treat wastes to reduce their hazard or volume.
4. Dispose of wastes on land or incinerate them as a last resort. (See figure 19.6.)

Reducing the Amount of Waste at the Source

Pollution prevention encourages changes in the operations of business and industry that prevent hazardous wastes from being produced in the first place. Many of these actions are simple to perform and cost little. Primary among them are activities that result in fewer accidental spills, leaks from pipes and valves, loss from broken containers, and similar mishaps. These reductions often can be achieved at little cost through better housekeeping and awareness training for employees. Many industries actually save money because they need to buy less raw material because less is being lost.

Pollution prevention can be applied in unusual ways. In 2000, the U.S. Army announced that it would begin issuing an environmentally friendly "green bullet" that contains no lead. The bullet contains a nonpolluting tungsten core and does not contain lead, which contaminates the soil and air around firing ranges. The U.S. military uses about 700 million rounds of small-caliber ammunition each year at about 3000 firing ranges. Lead contamination has closed hundreds of outdoor firing ranges on military bases across the United States. In 1998, when lead concentrated in firing berms was found to be leaching into Cape Cod's water supply, the Environmental Protection Agency ordered the Massachusetts Military Reservation to stop live-fire training.

Waste minimization involves changes that industries could make in the way they manufacture products, changes that would reduce the amount of waste produced. For example, it may be possible to change a process so that a solvent that is a hazardous material is replaced with water, which is not a hazardous material. This is an example of source reduction: any change or strategy that reduces the amount of waste produced.

Another strategy is to use the waste produced in a process in another aspect of the process, thus reducing the amount of waste produced. For example, water used to clean equipment might be included as a part of the product rather than being discarded as a contaminated waste.

Another technique that can be used to reduce the amount of waste produced is to clean solvents used in processes. Using a still to purify solvents results in a lower total volume of hazardous waste being produced because the same solvent can be used over and over again.

The simple process of allowing water to evaporate from waste can reduce the total amount of waste produced. Obviously, the hazardous components of the waste are concentrated by this process.

Recycling Wastes

Often it is possible to use a waste for another purpose and thus eliminate it as a waste. Many kinds of solvents can be burned as a fuel in other kinds of operations. For example, waste oils can be used as fuels for power plants, and other kinds of solvents can be burned as fuel in cement kilns. Care needs to be taken that the contaminants in the oils or solvents are not released into the environment during the burning process, but the burning of these wastes destroys them and serves a useful purpose at the same time.

Similarly, many kinds of acids and bases are produced as a result of industrial activity. Often these can be used by other industries. Ash or other solid wastes can often be incorporated into concrete or other building materials and therefore do not require disposal. Thus, the total amount of waste is reduced.

Treating Wastes

Wastes can often be treated in such a way that their amount is reduced or their hazardous nature is modified.

Neutralization of dangerous acids and bases by reacting them with one another can convert hazardous substances to materials that are not hazardous.

Biodegradation of organic materials by the actions of microorganisms can convert hazardous chemicals to innocuous substances. Sewage treatment plants perform their function by biodegradation.

Air stripping is sometimes used to remove volatile chemicals from water. Volatile chemicals, which have a tendency to vaporize easily, can be forced out of liquid when air passes through it. Steam stripping works on the same principle, except that it uses heated air to raise the temperature of the liquid and force out volatile chemicals that ordinary air would not. The volatile compounds can be captured and reused or disposed of.

Carbon absorption tanks contain specifically activated particles of carbon to treat hazardous chemicals in gaseous and liquid waste. The carbon chemically combines with the waste or catches hazardous particles just as a fine wire mesh catches grains of sand. Contaminated carbon must then be disposed of or cleaned and reused.

Precipitation involves adding special materials to a liquid waste. These bind to hazardous chemicals and cause them to precipitate out of the liquid and form large particles called floc. Floc that settles can be separated as sludge; floc that remains suspended can be filtered, and the concentrated waste can be sent to a hazardous-waste landfill.

Disposal Methods

Today, the two most common methods for disposing of hazardous wastes are incineration and land disposal. The primary factors that determine which method is used are economic concerns and acceptance by the public.

Incineration (thermal treatment) involves burning waste at high temperatures and can be used to destroy a variety of kinds of wastes. A hazardous-waste

Figure 19.6 Pollution-Prevention Hierarchy The simplest way to deal with hazardous wastes is to not produce them in the first place. The pollution-prevention hierarchy stresses reductions in the amount of hazardous waste produced by employing several different strategies.

incinerator can be used to burn organic wastes but is unable to destroy inorganic wastes. A well-designed and well-run incinerator can destroy 99.9999 percent of the hazardous materials that go through it. The relatively high costs of incineration (compared with landfills) and concerns about the emissions affecting surrounding areas have kept incineration from becoming a major method of treatment or disposal in North America. In North America, abundant land is available for land disposal, making it the most economical and most widely used method. Incineration accounts for the disposal of only about 4 percent of the hazardous wastes. In Europe and Japan, where land is in short supply and expensive, incineration is more cost-effective and is used to dispose of large amounts of hazardous wastes but still accounts for the disposal of less than 50 percent of the hazardous wastes produced. However, because of concerns about the emissions from incinerators, significant amounts of hazardous waste are incinerated at sea on specially designed ships.

Land disposal is still the primary method for the disposal of hazardous wastes when all other options have been exhausted. Land disposal can take several forms:

1. Deep-well injection into porous geological formations or salt caverns
2. Discharge of treated and untreated liquids into municipal sewers, rivers, and streams
3. Placement of liquid wastes or sludges in surface pits, ponds, or lagoons
4. Storage of solid wastes in specially designed hazardous-waste landfills

Of these methodologies, deep-well injection is the most important. Nearly 50 percent of all hazardous waste disposed of on land is injected into deep wells. About 20 percent is released to aquatic environments, and about 5 percent is stored in landfills and other surface sites.

There are techniques that reduce the chance that hazardous materials will escape from these locations and become a problem for the public. Immobilizing a waste puts it into a solid form that is easier to handle and less likely to enter the surrounding environment. Waste immobilization is useful for dealing with wastes, such as certain metals, that cannot be destroyed. Two popular methods of immobilizing waste are fixation and solidification. Engineers and scientists mix materials such as fly ash or cement with hazardous wastes. This either "fixes" hazardous particles, in the sense of immobilizing them or making them chemically inert, or "solidifies" them into a solid mass. Solidified waste is sometimes made into solid blocks that can be stored more easily than can a liquid.

International Trade in Hazardous Wastes

The growth in uncontrolled movement of hazardous wastes between countries has been one of the most contentious environmental issues on the international political

GLOBAL PERSPECTIVE

Hazardous Wastes and Toxic Materials in China

Large areas of land are currently used in China for the uncontrolled disposal of industrial wastes. Approximately 600 million metric tons of industrialized waste, of which 50 to 70 percent is hazardous, are generated in China annually. Estimates indicate that in recent years, a total of 5.9 billion metric tons of industrial waste, occupying 540 million cubic meters (19,000 million cubic feet), have been improperly stored or discarded. The majority of the waste is simply piled on unprotected areas, which causes leaching to surface and groundwater bodies. As a result, environmental accidents are prevalent. For instance, a chromium-residue disposal site in Jinzhou caused groundwater pollution in a 12.5-square-kilometer (4.8-square-mile) area: as a result, water from 1800 wells in nine villages is no longer safe to drink.

The production and use of chemicals is developing rapidly in China. More than 30,000 classes of chemicals are now produced, of which many are toxic. During production, transportation, storage, and use, many releases and spills occur. For example, in 1995, the toxic chemical storage in Shenzhen exploded, causing significant damage to life, property, and the environment.

Management of hazardous and toxic materials in China is still in its early stages; therefore, treatment and disposal technologies are primitive and equipment is poor. None of China's current hazardous-waste disposal sites meets environmental standards. In part, this is due to a lack of funds and management capabilities. A demonstration project, however, is under way in China. It is hoped that this will assist China in developing a management system for hazardous and toxic materials.

The "Law of Pollution Prevention and Control of Solid Wastes" has been made a priority item of legislation by the National People's Congress. Research on hazardous-waste management and disposal has been classified as key in national scientific and technological development plans. Experiments with solid waste declaration and registration, and waste exchanges are being performed, and standard protocols for chemical testing, toxicity evaluation of synthetic chemicals, laboratory analysis, and risk assessment have begun to be formulated. The long- and short-term objectives of the new project include the following:

- Formulate and strengthen China's hazardous- and toxic-materials control laws and regulations, and establish criteria for sound environmental management of wastes.

- Establish technical support system for hazardous-waste management.

- Design demonstration projects for hazardous-waste treatment and disposal.

- Formulate laws and regulations, criteria, and policies for toxic-chemicals management.

- Construct in Beijing a hazardous-waste incineration plant with an annual capacity of 3000 metric tons.

This project will introduce hazardous- and toxic-materials control laws and regulations, antipollution criteria, and mitigation measures. The capabilities for the management and control of hazardous materials will be developed, and regulatory enforcement will be initiated. It is hoped that the establishment of declaration, registration, licensing, and wastes exchange for hazardous-waste procedures will result in a reduction in the volume of waste generated in China while providing an incentive for recovery and reuse.

The hazardous-waste disposal demonstration project will provide a model, and increased capabilities and experience can be used by China in establishing regional central-disposal facilities for hazardous wastes. Dissemination of the experience gained in the demonstration project should produce a significant improvement in waste-disposal management strategies, resulting in greater awareness, regulation, and environmental protection.

---

agenda. There is particular concern about rich, industrialized countries exporting such wastes to poorer, developing countries lacking the administrative and technological resources to safely dispose of or recycle the waste. For example, in 1999, between 3000 and 4000 metric tons of mercury-contaminated concrete waste packed in plastic bags was found in an open dump in a small town in Cambodia. The waste, labeled as "construction waste" on import documents, came from a Taiwanese petrochemical company. In this case, the waste was tracked down and returned to its point of origin. Unfortunately, most such cases are not reported or detected. International awareness of the problems associated with the trade in hazardous wastes has increased noticeably owing to several factors: the growing amounts of such wastes being generated; closure of old waste disposal facilities and political opposition to the development of new ones; and the dramatically higher costs associated with the disposal of hazardous wastes in industrialized countries (and thus the potential to earn profits by exporting such wastes to developing countries with low disposal costs). The debate over controlling hazardous-waste movements between countries culminated in 1989 with the creation of the Basel Convention.

The Basel Convention was negotiated under the auspices of the United Nations Environmental Programme between 1987 and 1989. The objectives of the convention are to minimize the generation of hazardous wastes and to control and reduce their transboundary movements to protect human health and the environment. To achieve these objectives, the convention prohibits exports of hazardous waste to Antarctica, to countries that have banned such imports as a national policy, and to nonparties to the convention (unless those transactions are subject to an agreement that is as stringent as the Basel Convention). Though not part of the original agreement, there is now a broad ban on the export of hazardous wastes from the Northern to the Southern Hemisphere. The waste transfers that are permitted under the Basel regime are subject to the mechanism of prior notification and consent, which requires parties to not export hazardous wastes unless a "competent authority" in the importing country has been properly informed and has consented to the trade.

While the Basel regime may not be perceived as being as successful or significant as some other multilateral environmental agreements, it remains an important part of the international community's attempt to protect the global environment and human health from hazardous materials. Now that the convention has been in place for more than a decade, the focus is on assisting parties with the environmentally sound management of hazardous wastes and with reducing the amount of wastes generated.

The exportation of hazardous wastes for recycling from industrialized nations to developing nations raises other questions. Those who support the export of hazardous substances for recycling argue that this practice offers two major benefits: reducing the quantity of such substances that get into the environment through final disposal and slowing down the depletion of natural resources. This argument is undoubtedly correct, provided the receiving country has the proper recycling facilities and adequate environmental standards. Environmentally beneficial trade in hazardous wastes ordinarily requires that there be an established market for these wastes and that the trade be economically viable.

Hazardous-Waste Management Program Evolution

Fundamentally, the goal of a hazardous-waste management program is to change the behavior of those who generate hazardous wastes so that they routinely store, transport, treat, and dispose of them in an environmentally safe manner. The focus on hazardous-waste management typically comes in the second phase of countries' environmental programs, after efforts to address more immediate threats to public health, such as safe drinking water. In the early years of many countries' hazardous-waste management programs, uncontrolled disposal of hazardous waste is the norm. Few, if any, proper treatment and disposal facilities exist. Information on who is generating waste, what types are being generated, and where it is being disposed of is meager or nonexistent.

The transition from an unregulated environment to a regulated one is complex, but hazardous-waste management programs typically evolve through the following major stages: identifying the problem and enacting legislation, designating a lead agency, establishing rules and regulations, developing treatment and disposal capacity, and creating a compliance and enforcement program. Each of these stages takes a number of years, and at each stage, there are many difficult issues to be resolved. Denmark, Germany, and the United States began this process by passing their first major hazardous-waste laws between 1972 and 1976; Canada did so in 1980. In the decade that followed, all four of these countries developed hazardous-waste regulations and requirements, so that by the end of the 1980s, their regulatory systems were largely operational. Laws and policies developed during the 1990s have focused mainly on waste minimization and recycling, as well as on harmonization with international standards and the cleanup of contaminated sites.

The more recent evolution of regulatory programs in Hong Kong, Indonesia, Malaysia, and Thailand has followed a similar pattern. By the early 1980s, all of these countries had enacted some form of environmental legislation providing at least limited authority to regulate hazardous waste. However, hazardous-waste management received little attention until the late 1980s, after periods of rapid economic growth and the expansion of the countries' manufacturing sectors. From 1989 to 1998, all of these countries passed major legislation that addressed hazardous waste or developed regulations outlining comprehensive programs: Malaysia did so in 1989, Hong Kong in 1991, Thailand in 1992, and Indonesia in 1995–98. All four countries now have at least one modern hazardous-waste treatment, storage, and disposal facility.

While all countries pass through the same stages of program development, no two countries follow precisely the same path. Differences in geography, demographics, industrial profile, politics, and culture lead countries to make different choices at each stage.

Public awareness of the problems of hazardous substances and hazardous wastes is relatively recent. The industrialized countries of Europe and North America began major regulation of hazardous materials only during the past 30 years, and most developing countries exercise little or no control over such substances. As a result, many countries are living with serious problems from prior uncontrolled dumping practices, while current systems for management of hazardous and toxic waste remain incomplete and incapable of even identifying all hazardous waste.

A number of fundamental problems are involved in hazardous-waste management. First, there is no agreement as to what constitutes a hazardous waste. Moreover, little is known about the amounts of hazardous wastes generated throughout the world. The issue is further complicated by our limited understanding of the health effects of most hazardous wastes and the fact that large numbers of potentially hazardous chemicals are being developed faster than their health risks can be evaluated.

Hazardous-waste management must move beyond burying and burning. Industries need to be encouraged to generate less hazardous waste in their manufacturing processes. Although toxic wastes cannot be entirely eliminated, technologies are available for minimizing, recycling, and treating wastes. It is possible to enjoy the benefits of modern technology while avoiding the consequences of a poisoned environment. The final outcome rests with governmental and agency policy makers, as well as with an educated public.

Issues – Analysis

Household Hazardous Waste

Since the passage of the Resource Conservation and Recovery Act in 1978 and other environmental legislation, industries have been required to document the amounts of hazardous wastes they produce and to account for their appropriate disposal. As a result, the amount of industrial hazardous waste released to the environment has been significantly reduced. At the same time that citizens insist on business and industry compliance with hazardous-waste regulations, our personal behavior does not always reflect the same level of concern.

Americans generate about 1.45 million metric tons of household hazardous waste per year. Frequently, these materials become waste when they become contaminated or are unwanted. Common forms of household hazardous waste are unwanted paints, varnishes, and other finishes; contaminated solvents such as alcohol and paint thinner; automobile fluids such as waste oil, brake fluid, and antifreeze; appliance and automobile batteries containing heavy metals; caustic cleaning products; unused pesticides; and fluorescent lightbulbs.

We have several options to help us properly handle household hazardous waste. Most containers of such materials carry instructions for the proper disposal of the unwanted material or the empty container. In addition, thousands of communities have regular household hazardous waste collection days staffed by people who are specially trained to assist the public in disposing of such wastes. However, these collection days are often infrequent, which requires that we store and label our waste, find out when and where the waste will be accepted, and deliver it to a particular place on a particular day. Often a reservation is required. This is obviously inconvenient.

We can reduce the amount of hazardous waste we produce by purchasing products that are not hazardous such as water-based paints instead of oil-based paints, carefully choosing the quantities we purchase so that we don't have leftover materials, and choosing not to do our own automobile maintenance.

Therefore, we need to examine our behavior toward the use and disposal of hazardous materials. Some pertinent questions are:

- Do we think about the potential a product has to become hazardous waste when we purchase it?

- Do we read label directions pertaining to proper use and disposal of products and follow them?

- Do we dispose of these materials in our household trash or by dumping them down the drain, or do we look for a more proper method of disposal?

Key Terms

acute toxicity *439*
chronic toxicity *439*
Comprehensive Environmental Response, Compensation, and Liability Act (CERCLA) *443*
corrosiveness *437*
hazardous *437*
hazardous materials *436*
hazardous substances *436*
hazardous wastes *437*

ignitability *436*
incineration *446*
land disposal *447*
LD$_{50}$ *439*
National Priorities List *443*
nonpersistent pollutant *440*
persistent pollutant *440*
pollution prevention *446*
pollution-prevention hierarchy *446*
reactivity *437*

Resource Conservation and Recovery Act (RCRA) *437*
Superfund *443*
synergism *440*
threshold level *439*
toxic *437*
toxicity *437*
waste minimization *446*

Review Questions

1. Explain the problems associated with hazardous-waste dump sites and how such sites developed.
2. Distinguish between acute and chronic toxicity.
3. Give two reasons why regulating hazardous wastes is difficult.
4. In what ways do hazardous wastes contaminate the environment?
5. Describe how hazardous wastes contaminate groundwater.
6. Why is it often a problem to link a particular chemical or hazardous waste to a particular human health problem?
7. Describe what is meant by the U.S. National Priorities List.
8. Describe five technologies for managing hazardous wastes.
9. What is meant by pollution prevention and waste minimization?
10. Describe the pollution-prevention hierarchy.
11. What are RCRA and CERCLA? Why is each important for managing hazardous wastes?

Critical Thinking Questions

1. Scientists at the EPA have to make decisions about thresholds in order to identify which materials are toxic materials. What thresholds would you establish for various toxic materials? What is your reasoning for establishing the limits you do?
2. Go to the EPA's website (www.epa.gov/enviro/html/ef_overview.html) and identify the major releasers of toxic materials in your area. Were there any surprises? Are there other releasers of toxic materials that might not be required to list their releases?
3. In North America alone, there are over 25,000 abandoned and uncontrolled hazardous-waste dumps. Many were abandoned before the RCRA of 1976 was passed. Who should be responsible for cleaning up these dumps?
4. Look at this chapter's section called "Hazardous-Waste Dumps—A Legacy of Abuse." Do the authors present the information from a particular point of view? What other points of view might there be on this issue? What information do you think these other viewpoints would provide?
5. Many economically deprived areas, Native American reservations, and developing countries that need an influx of cash have agreed, over significant local opposition, to site hazardous-waste facilities in their areas. What do you think about this practice? Should "outsiders" have a say in what happens within these sovereign territories?
6. After reading about the problem with hazardous wastes and toxic materials in China, do you think that the United States or any other country should have the right to intervene if another country is creating significant environmental damage? Why or why not?

Welcome to McGraw-Hill's Online Learning Center

Location: www.mhhe.com/enger10e

Interactive Exploration

Check out the website at **www.mhhe.com/enger10e** and click on the cover of this textbook for quizzing, career information, case studies, and hot links for the following topics:

Solid, Toxic, and Hazardous Waste   Hazardous and Toxic Wastes

20

Environmental Policy and Decision Making

Chapter Outline

Objectives

After reading this chapter, you should be able to:

- Explain how the executive, judicial, and legislative branches of the U.S. government interact in forming policy.
- Understand how environmental laws are enforced in the United States.
- Describe the forces that led to changes in environmental policy in the United States during the past three decades.
- Understand the history of the major U.S. environmental legislation.
- Understand why some individuals in the United States are concerned about environmental regulations.
- Understand what is meant by "green" politics.
- Describe the reasons environmentalism is a growing factor in international relations.
- Understand the factors that could result in "ecoconflicts."
- Understand why it is not possible to separate politics and the environment.
- Explain how citizen pressure can influence governmental environmental policies.

PAST PRESENT FUTURE

1983	There were 7000 international nongovernmental organizations (NGOs).		**2003**	There were 26,000 international nongovernmental organizations.
1982	There were 61,447 nuclear warheads in the world.		**2003**	There were 18,000 nuclear warheads in the world.
1982	The gross world product was $US 25 trillion (in 2000 dollars).		**2003**	The gross world product was $US 50 trillion.

New Challenges for a New Century

We live in remarkable times. This is an era of rapid and often bewildering alterations in the forces and conditions that shape human life. This is evident both in the altered nature of geopolitics in the post–Cold War era and in the growing understanding of the relationship between human beings and the natural world. This relationship varies, however, between the developing and developed countries of the world. One of the major challenges of the foreseeable future will be centered on worldwide environmental impact as developing nations evolve economically.

The end of the Cold War has been accompanied by the swift advance of democracy in places where it was previously unknown and an even more rapid spread of market-based economies. The authority of central governments is eroding, and power has begun to shift to local governments and private institutions. In some countries, freedom and opportunity are flourishing, while in others, these changes have unleashed the violence of old conflicts and new ambitions.

Internationally, trade, investment, information, and even people flow across borders largely outside of governmental control. Domestically, deregulation and the shift of responsibilities from federal to state and local governments are changing the relationships among levels of government and between government and the private sector.

Communications technology has enhanced people's ability to receive information and influence events that affect them. This has sparked explosive growth in the number of organizations, associations, and networks formed by citizens, businesses, and communities seeking a greater voice for their interests. As a result, society outside of government—civil society—is demanding a greater role in governmental decisions, while at the same time impatiently seeking solutions outside government's power to decide.

But technological innovation is changing much more than communication. It is changing the ways in which we live, work, produce, and consume. Knowl-

Figure 20.1 **Population Increase Aggravates Scarcity** Struggling to survive in places that can no longer sustain them, growing populations overfish, overharvest, and overgraze.

edge has become the economy's most important and dynamic resource. It has rapidly improved efficiency as those who create and sell goods and services substitute information and innovation for raw materials. During the past 30 years, the amount of energy and natural resources the U.S. economy uses to produce each constant dollar of output has steadily declined, as have many forms of pollution. When U.S. laws first required industry to control pollution, the response was to install cleanup equipment. The shift to a knowledge-driven economy has emphasized the positive connection among efficiency, profits, and environmental protection and helped launch a trend in profitable pollution prevention. More and more people now understand that pollution is waste, waste is inefficient, and inefficiency is expensive.

Tomorrow's world will be shaped by the aspirations of a much larger global population. The number of people living on Earth has doubled in the last 50 years. Growing populations demand more food, goods, services, and space. Where there is scarcity, population increase aggravates it. Where there is conflict, rising demand for land and natural resources exacerbates it. Struggling to survive in places that can no longer sustain them, growing populations overfish, overharvest, and overgraze. (See figure 20.1.)

As we begin the new century, it is important that we recognize that economic, environmental, and social goals are integrally linked and that we develop policies that reflect that interrelationship. Thinking narrowly about jobs, energy, transportation, housing, or ecosystems—as if they

were not connected—creates new problems even as it attempts to solve old ones. Asking the wrong questions is a sure way to get misleading answers that result in short-term remedies for symptoms, instead of cures for long-term problems.

All of this will require new modes of decision making, ranging from the local to the international level. While trend is not always destiny, the trend that has been evolving over the past several years has been toward more collaborative forms of decision making. Perhaps such collaborative structures will involve more people and a broader range of interests in shaping and making public policy. It is hoped that this will improve decisions, mitigate conflict, and begin to counteract the corrosive trends of cynicism and civic disengagement that seem to be growing.

More collaborative approaches to making decisions can be arduous and time-consuming, and all of the players must change their customary roles. For government, this means using its power to convene and facilitate, shifting gradually from prescribing behavior to supporting responsibility by setting goals, creating incentives, monitoring performance, and providing information.

For their part, businesses need to build the practice and skills of dialogue with communities and citizens, participating in community decision making and opening their own values, strategies, and performance to their community and the society.

Advocates, too, must accept the burdens and constraints of rational dialogue built on trust, and communities must create open and inclusive debate about their future.

CHAPTER 20 Environmental Policy and Decision Making

Figure 20.2 **Trend Is Not Destiny** Scenes such as these were very common in North America only a short time ago. Fortunately, for the most part, such photos are today only historic in nature. Positive change is possible.

Does all of this sound too idealistic? Perhaps it is; however, without a vision for the future, where would we be? As was stated previously, trend is not destiny. In other words, we are capable of change regardless of the status quo. This is perhaps nowhere more important than in the world of environmental decision making. (See figure 20.2.)

Governance and Government

"Who let this happen? Who's responsible for this problem?" These are typical questions people ask in reaction to a local environmental problem or to the steady deterioration of global environmental conditions. For most people, it is not obvious who is "in charge" of the environment or how decisions are made about developing, using, or managing ecosystems.

Governance is about decisions and how we make them. It is about the exercise of authority, about being in charge. It relates to decision makers at all levels—government managers and ministers, businesspeople, property owners, farmers,

and consumers. In short, governance deals with who is responsible, how they wield their power, and how they are held accountable.

Environmental governance is inevitably associated with institutions—the organizations in which official authority often resides. These commonly include government departments of environment, agriculture, and mining, and environmental regulatory agencies. But governance also encompasses oversight or advisory groups, corporate councils and trade groups, and even private think tanks and advocacy groups that help to formulate policy. Overall, then, environmental governance takes in the whole range of institutions and decision-making practices that communities use to manage their environment and control natural resources.

Sometimes we use the term *governance* very broadly to describe not just the process of decision making but the actual management actions—where and when to log or how to limit fishing or distribute grazing permits—that result. In other words, in our day-to-day experience, we intertwine environmental governance and ecosystem management, which is where

the real impact of decisions becomes visible.

A common mistake is to confuse governance with *government*—the set of institutions we normally associate with political authority. Clearly, governments are important players in how ecosystems are managed and how natural resources are exploited or conserved. National laws and regulatory frameworks set the formal rules for managing natural resources by recognizing discrete property, mineral, or water rights. They also establish the legal mandates of government agencies with responsibility for environmental protection and resource management. It is these government institutions that we frequently associate with big environmental decisions and the responsibility to govern natural resources.

Governments also act internationally (often through the United Nations) to set ground rules about pollution, water use, fishing rights, and other activities that affect resources across political boundaries. One of the most visible aspects of this global environmental governance is a large set of international environmental treaties, such as the Convention on

Biological Diversity, the Convention on the Law of the Sea, and the Montreal Protocol to protect the stratospheric ozone layer. Multinational bodies such as the World Bank and the World Trade Organization are also assuming greater environmental significance in an increasingly globalized and interdependent world economy.

But environmental governance goes beyond the official actions of governments. Sometimes, corporations or individuals act in the state's place to harvest or manage resources. For instance, states may grant forest or mining concessions to companies for a fee, allowing them broad discretion to cut trees, build roads, or make other important land-use decisions. Or the state may privatize once-public functions such as the delivery of water, electricity, or wastewater treatment, again putting a host of environmental choices—from water pricing to power plant construction—into private hands.

Nongovernmental organizations (NGOs), such as environmental organizations, civic groups, land trusts, labor unions, and neighborhood groups, have become strong advocates for better and fairer environmental decisions in the past few decades. The actions of industry groups, trade associations, and shareholder groups also influence the way companies do business by promoting or obstructing cleaner processes, better environmental accounting practices, or pointing out the financial liabilities of business practices that harm the environment.

Governance includes our individual choices and actions when these influence larger public policies or affect corporate behavior. Voting, lobbying, participating in public hearings, or joining an environmental watchdog or monitoring group are typical ways that individuals can influence environmental decisions. Our actions as consumers are powerful governance forces. For example, the choice to purchase environmentally friendly products such as organic produce, certified lumber, or a fuel-efficient car influences the environmental behavior of businesses through the marketplace. Consumer choices can sometimes be as powerful as government regulations in tempering business decisions that affect the environment.

Learning from the Past

For the past quarter century, the basic pattern of environmental protection in economically developed nations has been to react to specific crises. Institutions have been established, laws passed, and regulations written in response to problems that already were posing substantial ecological and public health risks and costs or that already were causing deep-seated public concern.

The United States is no exception. The U.S. Environmental Protection Agency (EPA) has focused its attention almost exclusively on present and past problems. The political will to establish the agency grew out of a series of highly publicized, serious environmental problems, such as the fire on the Cuyahoga River in Ohio, smog in Los Angeles, and the near extinction of the bald eagle. During the 1970s and 1980s, Congress enacted a series of laws intended to solve these problems, and the EPA, which was created in 1970, was given the responsibility for enforcing most environmental laws.

Despite success in correcting a number of existing environmental problems, there has been a continuing pattern of not responding to environmental problems until they pose immediate and unambiguous risks. Such policies, however, will not adequately protect the environment in the future. People are recognizing that the agencies and organizations whose activities affect the environment must begin to anticipate future environmental problems and take steps to avoid them. One of the most important lessons learned during the past quarter century of environmental history is that the failure to think about the future environmental consequences of prospective social, economic, and technological changes may impose substantial and avoidable economic and environmental costs on future generations.

Thinking About the Future

Thinking about the future is more important today than ever before because the accelerating rate of change is shrinking the distance between the present and the future. Technological capabilities that seemed beyond the horizon just a few years ago are now outdated. Scientific developments and the flow of information are accelerating. For example, who would have envisioned cellular phones, voice-activated computers, or a widely used Internet only 20 years ago? Similarly, the environmental effects of changes in global economic activity are being felt more rapidly by both nations and individuals. Examples include the collapse of the cod-fishing industry in the Maritime provinces of Canada and the northeastern United States; the threatened salmon-fishing industry in the Pacific Northwest of the United States and British Columbia, Canada; the severe air pollution problems in Mexico; and the shortage of safe drinking water supplies in Russia and China.

Thinking about the future is valuable also because the cost of avoiding a problem is often far less than the cost of solving it later. The U.S. experience with hazardous-waste disposal provides a compelling example. Some private companies and federal facilities undoubtedly saved money in the short term by disposing of hazardous wastes inadequately, but those savings were dwarfed by the cost of cleaning up hazardous-waste sites years later. In that case, foresight could have saved private industry, insurance companies, and the federal government (i.e., taxpayers) billions of dollars, while reducing exposure to pollutants and public anxieties in the affected communities.

Environmental foresight can preserve the environment for future generations. When one generation's behavior necessitates environmental remediation in the future, an environmental debt is bequeathed to future generations just as surely as unbalanced government budgets bequeath a burden of financial debt. (See figure 20.3.)

By anticipating environmental problems and taking steps now to prevent them, the present generation can minimize the environmental and financial debts that its children will incur.

Today, we face new classes of environmental problems that are more diffuse than those of the past and thus demand different approaches. Since the first Earth Day in 1970, the vast majority of the significant "point" sources of air and water pollution—large industrial facilities and municipal sewage systems—that once

Figure 20.3 Environmental Debt The plight of these fishermen, resulting from the collapse of the cod fishery in the northeast United States and Canada and the salmon fishery in the northwest United States and Canada, is an environmental debt inherited from past generations of abuse and misuse.

spewed untreated wastes into the air, rivers, and lakes have been controlled. The most important remaining sources of pollution are diffuse and widespread: sediment, pesticides, and fertilizers that run off farmland; oil and toxic heavy metals that wash off city streets and highways; and air pollutants from automobiles, outdoor grills, and woodstoves. Pollution from these sources cannot always be controlled with sewage treatment plants or the same regulatory techniques used to check emissions from large industries. Furthermore, we now have the global environmental problems of biodiversity loss, ozone depletion, and climate change. These problems will require cooperative international responses. We are also recognizing that controlling pollutants alone, no matter how successful, will not achieve an environmentally sustainable economy, since many global concerns are related to the size of the human population and the unequal distribution of resources.

For the most fortunate, the past 50 years have produced a quality of life unprecedented in human history. For another 3 billion people, it has brought about marked improvements in living standards, including significant increases in life expectancy, infant survival, literacy, and access to safe drinking water. This progress notwithstanding, however, more than 20 percent of today's human population still lives in poverty, 15 percent experience persistent hunger, and at least 10 percent is homeless. Moreover, the gap between the very rich and very poor is widening, with whole regions of the world clearly losing ground.

Through both its successes and failures, modern human development has transformed the planet. Human activities have doubled the planet's rate of nitrogen fixation, tripled the rate of invasion by exotic organisms, increased sediment loads in rivers fivefold, and vastly increased natural rates of species extinctions. Clearly, in the future, we will need a different vision from that which shaped our past.

Defining the Future

We are progressing from an environmental paradigm based on cleanup and control to one including assessment, anticipation, and avoidance. Expenditures to develop technologies that prevent environmental harm are beginning to pay off. Agricultural practices are becoming less wasteful

and more sustainable, manufacturing processes are becoming more efficient in the use of resources, and consumer products are being designed with the environment in mind. The infrastructures that supply energy, transportation services, and water supplies are becoming more resource efficient and environmentally benign. Remediation efforts are cleaning up a large portion of existing hazardous-waste sites. Our ability to respond to emerging problems is being aided by more advanced monitoring systems and data analysis tools that continually assess the state of the local, regional, and global environment. Finally, we are developing effective ways of restoring or recreating severely damaged ecosystems to preserve the long-term health and productivity of our natural resource base.

This trend will continue if we are willing to develop strong environmental policies, develop new strategies, and closely coordinate the actions of the public and private sectors toward a set of shared, long-term goals. Such a strategy should include three aims. The first is to articulate a vision of the future and the role environmental technology will play in shaping that future. This vision must be built on an understanding of the strengths and weaknesses of past policies and actions. The second is to define the roles of the many individuals and organizations that are needed to implement the goals. The third is to chart a course by offering suggestions for strategic goals for all the partners in this endeavor.

In the long run, environmental quality is not determined solely by the actions of government, regulated industries, or non-government organizations. It is largely a function of the decisions and behavior of individuals, families, businesses, and communities everywhere. Consequently, the extent of environmental awareness and the strength of environmental institutions will be two critical factors driving changes in environmental quality in the future.

A concerned, educated public, acting through responsive local, national, and international institutions, will serve as effective agents for avoiding future environmental problems, no matter what they are. Environmental institutions,

Figure 20.4 **Major Agencies of the Executive Branch** Major agencies of the executive branch are shown with their environmental responsibility.

strengthened by informed public support, will play a critical role in devising and implementing effective national and international responses to emerging issues.

Looking ahead, consensus is growing that the next 50 years will see a world in which people are more crowded, more connected, and more consuming than at any time in human history. The natural environment in which those people live will be stressed as never before. It will be almost certainly warmer, more polluted, and less species-rich. Many of these trends have generated a good deal of discussion recently, under headings ranging from "globalization" to "climate change." Less remarked upon is the fundamental transition under way in the growth of human populations: Rates of increase are now falling almost everywhere in the world, with the result that

the number of people on the planet is expected to level off at 10 billion or 11 billion by the end of the twenty-first century, reaching around 9 billion—still half again as many as today's count—by 2050.

This transition toward a stable human population, if brought to fruition, could fundamentally transform the challenges of environmentally sustainable human development and how people think about them. It could allow people to focus, for the first time in modern history, on questions of sustaining increases in the *quality,* not just the *quantity,* of human life. Can the transition to a stabilizing human population also be a transition to sustainability in which the people living on Earth over the next half century meet their needs while nurturing and restoring the planet's life support systems? These are the questions yet to be answered.

The Development of Environmental Policy in the United States

Public **policy** is the general principle by which government **branches**—the **legislative, executive,** and **judicial**—are guided in their management of public affairs. The legislature (Congress) is directed to declare and shape national policy by passing legislation, which is the same as enacting law. The executive (president) is directed to enforce the law, while the judiciary (the court system) interprets the law when a dispute arises. (See figure 20.4.)

When Congress considers certain conduct to be against public policy and against

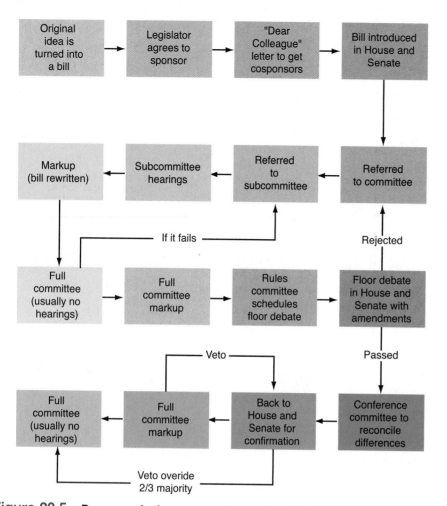

Figure 20.5 **Passage of a Law** This figure illustrates the path of a bill in the U.S. Congress from organization to becoming a law. As we can see, the process is not a quick one.

the public good, it passes legislation in the form of acts or statutes. Congress specifically regulates, controls, or prohibits activity in conflict with public policy and attempts to encourage desirable behavior.

Through legislation, Congress regulates behavior, selects agencies to implement new programs, and sets general procedural guidelines. When Congress passes environmental legislation, it also declares and shapes the national environmental policy, thus fulfilling its policy-making function. (See figure 20.5.)

Over 90 years ago, President Teddy Roosevelt declared that nothing short of defending your country in wartime "compares in leaving the land even better land for our descendants than it is for us." The environmental issues that Roosevelt

strongly believed in, however, did not become major political issues until the early 1970s.

While the publication of Rachel Carson's *Silent Spring* in 1962 is considered to be the beginning of the modern environmental movement, the first Earth Day on April 22, 1970, was perhaps the single event that put the movement into high gear. In 1970, as a result of mounting public concern over environmental deterioration—cities clouded by smog, rivers on fire, waterways choked by raw sewage—many nations, including the United States, began to address the most obvious, most acute environmental problems.

Public opinion polls indicate that a permanent change in national priorities

followed Earth Day 1970. When polled in May 1971, 25 percent of the U.S. public declared protecting the environment to be an important goal—a 2500 percent increase over the proportion in 1969.

During the 1970s, many important pieces of environmental legislation were enacted in the United States. In addition, the percent of discretionary U.S. budget spent on natural resources and the environment grew significantly (See figure 20.6.) Many of the identified environmental problems were so immediate and so obvious that it was relatively easy to see what had to be done and to summon the political will to do it. (See table 20.1.)

The United States is commonly recognized as having been an environmental innovator in the 1970s. U.S. environmental laws and institutions became models for consideration by many other countries. Important to U.S. prominence as an environmental leader was the growth of environmental nongovernmental organizations. Environmental groups were aided by amendments to tax laws that made it relatively easy for groups to incorporate as not-for-profit organizations even while engaged in some lobbying activities. This provided them with important benefits, including the possibility of receiving contributions that are tax-deductible for the donor and favorable postage rates for reaching out to the public. Environmental groups could turn to both members and large, private financial institutions for donations; they were also eligible to receive governmental grants and contracts. U.S. environmental groups also won the important right to stand in court to sue on behalf of environmental interests. Using these new rights, environmental groups became increasingly professionalized and diversified.

The development of a strong environmental NGO community in the 1970s and capacity building in environmental research helped the United States stay at the forefront of new environmental policy developments and environmental research into the 1980s. Yet there was an industry-led backlash in the United States in the 1980s against the regulatory successes of the environmental community. Thus, although the United States has perhaps the strongest environmental NGO community in the world, continues to lead the world in environmental research, and is the largest exporter of pollution-control equipment, it entered the 1990s with an uneven environmental record.

Just as it was beginning to gain momentum, however, the environmental movement began to decline. When the energy crisis threatened to stall the North American economy in the early 1970s, environmental concerns quickly faded. By 1974, President Gerald Ford had proposed accelerating his administration's leasing program for offshore gas and oil drilling. A turnaround in environmental policy was even more pronounced in the 1980s during the Reagan administration. Former Vice President Walter Mondale was fond of noting that President Ronald Reagan "would rather take a polluter to lunch than

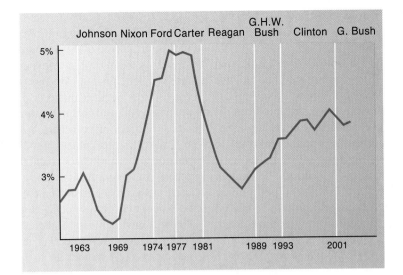

Figure 20.6 **Proportions of Environmental Spending** The share of the discretionary budget spent on natural resources and the environment has increased. The amount of federal dollars available for discretionary spending, however, is shrinking.

Source: *Conservation in Practice*, vol. 4, no. 2, Spring 2003.

Table 20.1 Major U.S. Environmental and Resource Conservation Legislation

Wildlife conservation
Anadromous Fish Conservation Act of 1965
Fur Seal Act of 1966
National Wildlife Refuge System Act of 1966, 1976, 1978
Species Conservation Act of 1966, 1969
Marine Mammal Protection Act of 1972
Marine Protection, Research, and Sanctuaries Act of 1972
Endangered Species Act of 1973, 1982, 1985, 1988, 1995
Fishery Conservation and Management Act of 1976, 1978, 1982, 1996
Whale Conservation and Protection Study Act of 1976
Fish and Wildlife Improvement Act of 1978
Fish and Wildlife Conservation Act of 1980 (Nongame Act)
Fur Seal Act Amendments of 1983

Land use and conservation
Taylor Grazing Act of 1934
Wilderness Act of 1964
Multiple Use Sustained Yield Act of 1968
Wild and Scenic Rivers Act of 1968
National Trails System Act of 1968
National Coastal Zone Management Act of 1972, 1980
Forest Reserves Management Act of 1974, 1976
Forest and Rangeland Renewable Resources Act of 1974, 1978
Federal Land Policy and Management Act of 1976
National Forest Management Act of 1976
Soil and Water Conservation Act of 1977
Surface Mining Control and Reclamation Act of 1977
Antarctic Conservation Act of 1978
Endangered American Wilderness Act of 1978
Alaskan National Interests Lands Conservation Act of 1980
Coastal Barrier Resources Act of 1982
Food Security Act of 1985
Emergency Wetlands Resources Act of 1986
North American Wetlands Conservation Act of 1989
Coastal Development Act of 1990
California Desert Protection Act of 1994
Federal Agriculture Improvement and Reform Act of 1996

General
National Environmental Policy Act (NEPA) of 1969
International Environmental Protection Act of 1983

Energy
Energy Policy and Conservation Act of 1975
National Energy Act of 1978, 1980
Northwest Power Act of 1980
National Appliance Energy Conservation Act of 1987
Energy Policy Act of 1992

Water quality
Refuse Act of 1899
Water Quality Act of 1965
Water Resources Planning Act of 1965
Federal Water Pollution Control Acts of 1965, 1972
Ocean Dumping Act of 1972
Safe Drinking Water Act of 1974, 1984, 1996
Clean Water Act of 1977, 1987
Great Lakes Toxic Substance Control Agreement of 1986
Great Lakes Critical Programs Act of 1990
Oil Spill Prevention and Liability Act of 1990

Air quality
Clean Air Act of 1963, 1965, 1970, 1977, 1990

Noise control
Noise Control Act of 1965
Quiet Communities Act of 1978

Resources and solid waste management
Solid Waste Disposal Act of 1965
Resources Recovery Act of 1970
Resource Conservation and Recovery Act of 1976
Waste Reduction Act of 1990

Toxic substances
Toxic Substances Control Act of 1976
Resource Conservation and Recovery Act of 1976
Comprehensive Environmental Response, Compensation, and Liability (Superfund) Act of 1980, 1986, 1990
Nuclear Waste Policy Act of 1982

Pesticides
Food, Drug, and Cosmetics Act of 1938
Federal Insecticide, Fungicide, and Rodenticide Control Act of 1972, 1988
Food Quality Protection Act of 1996

to court." During the mid-1980s, the environment was not a priority in the Reagan administration.

The period from 1970 to 1990, however, did bring forth some very tangible accomplishments in environmental policy. Among the most visible and quantifiable is the expansion of protected areas. During this period, federal parklands in the United States—excluding Alaska—increased 800,000 hectares (2 million acres), to 10.5 million (25.9 million acres). In Alaska, 18.3 million additional hectares (45 million acres) were protected, bringing the state's total to over 232 million hectares (573 million acres). Also, the extent of the waterways included in the National Wild and Scenic Rivers System increased by more than 12 times, to some 15,000 kilometers (9300 miles).

By the late 1980s, however, a new environmental awareness and concern began to surface as a major political issue. This was in part due to a number of highly visible environmental problems that appeared nightly on the evening news. Images of toxic waste (including hospital waste, such as used syringes) washing up on the nation's beaches and of the pristine waters of Alaska covered in oil from the *Exxon Valdez* spill made an impact on the public. Once again, the public reacted by organizing and putting pressure on the political system, and, as in 1970, the politicians began to respond. For the first time in the history of the United States, the environment became a key issue in a presidential campaign. In 1988, the environmental records of the two major candidates were hotly debated. Environmentalism was evolving as a major public issue. By the 1992 U.S. presidential election, the environment was established as a major campaign issue, a trend that continues today.

In many respects, the environmental movement of the 1970s and 1980s came of age in the 1990s. It has been said that politics have always forged science. Prioritization of issues and political will determine where money will be spent. By the mid-1990s, it appeared that the political will in the United States to address environmental concerns was on the rise.

During the late 1980s, however, a backlash or anti-environmental attitude began to develop among sectors of the U.S. populace. A loose-knit organization of several hundred of these groups became known as the Wise Use Movement. The majority of funding for this coalition comes from various interests, including the timber, oil, and coal industries, real estate developers, and ranchers.

Wise Use pamphlets argue that extinction is a natural process and that some species were not meant to survive. It has been argued that the movement's signature public relations tactic is to frame complex environmental and economic issues in simple, scapegoating terms that benefit its corporate backers. In the Pacific Northwest, for example, the movement has continually dwelled on a supposed "battle" for survival between spotted owls and the families of the workers involved in harvesting the old-growth timber. "Jobs versus owls" is a good sound bite, but it is not the issue. The real issue is much more complicated.

The Changing Nature of Environmental Policy

Over the past nearly four decades, public opinion polls conducted in the United States have consistently revealed lopsided majorities—over 80 percent—in favor of maintaining or strengthening clean air and water laws. Whether people vote the environment or not, and most probably do not, they still want the environment protected and improved. It should be noted, however, that these polls, like most polls, often reflect attitudes and not necessarily actions. For example, saying that you support stronger environmental laws does not always translate into individual actions such as purchasing environmentally "friendly" products, active recycling, or support of higher taxes for more governmental regulation of the environment.

While the public was supportive of environmental protection, the 1990s also witnessed a decline in membership of the major environmental organizations. By 1998, membership declines in many of the large environmental organizations were forcing the layoff of staff and the closure of offices. A major part of the problem with the larger environmental groups may be that they have grown into large bureaucracies and, in the process, have lost the trust of many grassroots environmentally concerned individuals.

While many of the larger, established environmental organizations are still declining somewhat, many newer, smaller, local, and grassroots organizations are being created or are expanding, often in response to environmental threats in their own communities. Estimates are that some 7000 local environmental organizations are active in the United States. On college campuses, environmental studies programs are expanding at both the undergraduate and graduate levels. Students are not just concerned with local recycling programs or nuclear power plants but are focusing on the broader issues. The primary issue is the achievement of long-term sustainability and the fundamental changes in society that this will entail.

Environmental Policy and Regulation

Environmental laws are not a recent phenomenon. As early as 1306, London adopted an ordinance limiting the burning of coal because of the degradation of local air quality. Such laws became more common as industrialization created many sources of air and water pollution throughout the world. In the United States, environmental laws often evolved from ordinances passed by local governments. Interested in protecting public health, officials of towns and cities enacted local laws to limit activities of private citizens for the common good. For example, to have "healthy air," many communities enacted ordinances in the 1880s to regulate rubbish burning within city limits. Public health issues were the foundations on which the environmental laws of today were built.

Environmental law in the United States is governed by administrative law. Administrative law is a relatively new concept, having been developed only during the twentieth century. In 1946, Congress passed the Federal Administrative Procedure Act. This act designated general procedures to be used by federal agencies when they exercised their rule-making, adjudicatory, and enforcement powers. This is a rapidly expanding area of law and defines how governmental organizations such as agencies, boards, and commissions develop and implement the regulatory programs they are legislatively authorized to create. Some of the many U.S. federal agencies that impact environmental issues include the Environmental Protection Agency, the Council on Environmental Quality, the National Forest Service, and the Bureau of Land Management.

Administrative law applies to government agencies and to those that are affected by agency actions. In the United States, many federal environmental programs are administered by the states under the authority of federal and related state laws. States often differ from both the federal government and each other in the way they interpret, implement, and enforce federal laws. In addition, each state has its own administrative guidelines that govern and define how state agencies act. All actions of federal agencies must comply with the 1946 Administrative Procedure Act.

The National Environmental Policy Act (NEPA) was enacted in 1969 and signed into law by President Richard Nixon on New Year's Day, 1970. It is a short, general statute designed to institutionalize within the federal government a concern for the "quality of the environment." NEPA helps encourage environmental awareness among all federal agencies, not just those that prior to NEPA had to consider environmental factors in their planning and decision making. Until 1970, most federal agencies acted within their delegated authority without considering the environmental impacts of their actions. However, in the 1960s, Congress seriously began to study pollution problems. Because Congress has found that the federal government is both a major cause of environmental degradation and a major source of regulatory activity, all actions of the federal government now fall under NEPA.

NEPA forces federal agencies to consider the environmental consequences of their actions before implementing a proposal or recommendation. NEPA has two purposes: first, to advise the president on the state of the nation's environment; and second, to create an advisory council called the Council on Environmental Quality (CEQ). The CEQ outlines NEPA compliance guidelines. The CEQ also provides the president with consistent expert advice on national environmental policies and problems.

Between 1970 and 1977, the CEQ served only as an advisory council to the president. However, in 1977, through an executive order, President Jimmy Carter granted the CEQ authority to issue binding regulations. These regulations set out details for matters that are broadly addressed by NEPA. The CEQ applies to all federal agencies except Congress, the judiciary, and the president.

NEPA has been interpreted narrowly by the federal courts. As a result, many states have passed much stronger state environmental protection acts (SEPAs) as well. Today, NEPA analysis is undertaken as part of almost every recommendation or proposal for federal action. This includes not only actions by agencies of the federal government but also actions of states, local municipalities, and private corporations. NEPA is Congress's mission statement that mandates the means by which the federal government, through the guidance of the CEQ, will achieve its national environmental policy.

In addition to the passage of NEPA, the 1970s saw a series of new environmental laws passed, including the Resource Conservation and Recovery Act (RCRA), the Comprehensive Environmental Response, Compensation and Liability Act (CERCLA), and the Clean Air Act (CAA). All of these acts are broadly worded to identify existing problems that Congress believes can be corrected to protect human health, welfare, and the environment.

Protecting human health, welfare, and the environment is the national policy that Congress has chosen to encourage. For example, under NEPA, the national policy is to "promote efforts which will prevent or eliminate damage to the environment and biosphere and stimulate the health and welfare of [humans]." Similarly, under the Clean Air Act, the policy is to "protect and enhance the quality of the Nation's air resources so as to promote the public health and welfare and the productive capacity of its population."

Each of the above statutes declares national policy on environmental issues and addresses distinct problems. These statutes also authorize the use of some or all of the administrative functions discussed above, such as rule making, adjudication, administrative, civil and criminal enforcement, citizen suits, and judicial review. (See figure 20.7.)

Congress established the Environmental Protection Agency in 1970 as the primary agency to implement the statutes. Administrative functions empower EPA, the states, and private citizens to take responsibility for enforcing the various authorized programs. These administrative functions not only shape environmental law but also control the daily operations of both the regulated industry and agencies authorized to protect the environment.

To date, much environmental law has reflected the perception that environmental problems are localized in time, space, and media (i.e., air, water, soil). For example, many hazardous-waste sites in the United States have been "cleaned" by simply shipping the contaminated dirt someplace else, which not only does not solve the problem but creates the danger of incidents during removal and transportation. Environmental regulation has focused on specific phenomena and adopted the so-called "command-and-control" approach, in which restrictive and highly specific legislation and regulation are implemented by centralized authorities and used to achieve narrowly defined ends.

Such regulations generally have very rigid standards, often mandate the use of specific emission-control technologies, and generally define compliance in terms of "end-of-pipe" requirements. (See figure 20.8.) Examples in the United States include the Clean Water Act (which applied only to surface waters), the Clean Air Act (urban air quality), and the Comprehensive Environmental Response, Compensation, and Liability Act (Superfund), which applied to specific landfill sites.

If properly implemented, command-and-control methods can be effective in

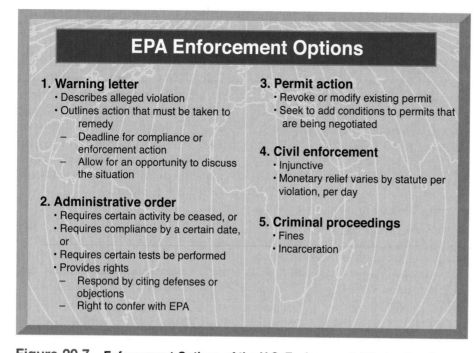

EPA Enforcement Options

1. Warning letter
- Describes alleged violation
- Outlines action that must be taken to remedy
  - Deadline for compliance or enforcement action
  - Allow for an opportunity to discuss the situation

2. Administrative order
- Requires certain activity be ceased, or
- Requires compliance by a certain date, or
- Requires certain tests be performed
- Provides rights
  - Respond by citing defenses or objections
  - Right to confer with EPA

3. Permit action
- Revoke or modify existing permit
- Seek to add conditions to permits that are being negotiated

4. Civil enforcement
- Injunctive
- Monetary relief varies by statute per violation, per day

5. Criminal proceedings
- Fines
- Incarceration

Figure 20.7 **Enforcement Options of the U.S. Environmental Protection Agency** The enforcement options of the U.S. Environmental Protection Agency range from a warning letter to a jail sentence.

Source: U.S. Environmental Protection Agency, Office of Enforcement, Washington, D.C.

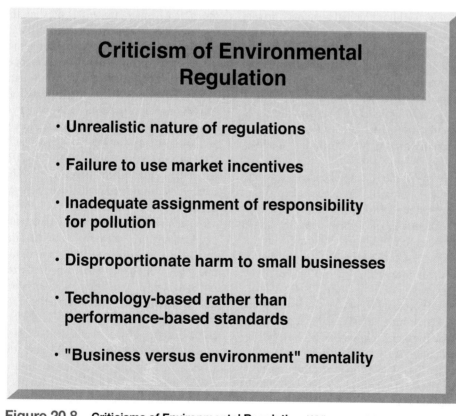

Criticism of Environmental Regulation

- **Unrealistic nature of regulations**

- **Failure to use market incentives**

- **Inadequate assignment of responsibility for pollution**

- **Disproportionate harm to small businesses**

- **Technology-based rather than performance-based standards**

- **"Business versus environment" mentality**

Figure 20.8 **Criticisms of Environmental Regulation** While some of the above criticisms are being addressed, others still need to be studied and discussed.

Source: U.S. Environmental Protection Agency, Office of Enforcement, Washington, D.C.

addressing specific environmental problems. For example, rivers such as the Potomac and the Hudson in the United States are much cleaner as a result of the Clean Water Act. (See figure 20.9.) Moreover, where applied against particular substances, such as the ban on tetraethyl lead in gasoline in the United States, the command-and-control approach has clearly worked well.

In 2003, the White House Office of Management and Budget (OMB) reported that the societal benefits of Environmental Protection Agency regulations significantly outweighed the costs of environmental compliance between 1992 and 2002. After analyzing the costs and benefits of 107 regulations promulgated by agencies across the federal government, OMB found that total benefits of federal rules were three to five times greater than total costs. Benefits were estimated at $146 billion to $230 billion annually, while costs were estimated at $36 billion to $42 billion annually. Among the government regulations evaluated, the majority of the quantified benefits were attributable to the clean air rules issued by the EPA pursuant to the 1990 amendments to the Clean Air Act.

In particular, the OMB noted four EPA regulations with high benefit-to-cost ratios: two that limit particulate matter and NO_X emissions from heavy-duty engines, one that limits emissions from light-duty vehicles, and one that limits sulfur dioxide release as part of the acid rain provisions of the Clean Air Act. These four rules provided combined benefits of $101 billion to $119 billion per year, with costs of only $8 billion to $8.8 billion per year.

In addition to analyzing regulations in place, the OMB also issued guidelines on how federal agencies should analyze their own regulations in the future. It recommended more emphasis on both cost-effectiveness and cost-benefit analyses, more systematic evaluation of benefit and cost factors that cannot be quantified or expressed in dollars (e.g., impacts on privacy and the natural environment), more use of formal probability analysis for important rule making where the underlying science is uncertain, and more public access to regulatory analysis documents.

a.   b.

Figure 20.9 **Effect of the Command-and-Control Approach** The environmental quality of rivers has dramatically improved over the past 30 years as a direct result of the Clean Water Act. (*a*) A visibly impaired river in New England in the 1970s. (*b*) An environmentally healthy Hudson River in the late 1990s.

The Greening of Geopolitics

Environmental or "green" politics have emerged from minority status and become a political movement in many nations. Issues such as transboundary water supply and pollution, acid precipitation, and global climate change have served to bolster the emergence of green politics.

Concern about the environment is not limited to developed nations. A 1989 treaty signed in Switzerland limits what poorer nations call toxic terrorism—use of their lands by richer countries as dumping grounds for industrial waste. In 1990, more than 100 developing nations called for a "productive dialogue with the developed world" on "protection of the environment." As was covered in chapter 2, the Earth Summit in Rio brought together nearly 180 governments to address world environmental concerns, and the 1997 conference on global warming in Kyoto brought together some 120 nations.

Although the Kyoto Protocol, the global treaty that would limit greenhouse gas emissions, has yet to be ratified, some form of regulation seems inevitable. The Kyoto Protocol must be ratified by at least 55 countries, including those responsible for 55 percent of the world's emissions,

before it can go into effect. The first part of that requirement has already been met—120 nations have ratified the treaty. The second awaits a decision from Russia, which produces 17.4 percent of global greenhouse emissions. At present, only two developed countries, Australia and the United States, have opted not to participate. But regardless of the fate of the Kyoto Protocol, participating countries and regions, including the European Union, Britain, Canada, Denmark, and China, have already started developing their own regulatory programs to control emissions of these gases.

Even without congressional consensus, many U.S. states have pressed on with initiatives of their own. California created the nation's first bill to regulate carbon dioxide emissions from automobiles by the 2008 model year. Ten northeastern states announced their intention to come to an agreement by April 2005 on a plan for a regional system that would limit carbon dioxide emissions from power plants. The New England states and eastern Canadian provinces have adopted a plan to reduce their greenhouse gas emissions to 1990 levels by 2010 and even lower levels later, and Maine has put this goal into law. New Jersey has pledged to reduce statewide emissions by 3.5 percent from 1990 levels by 2005 and has signed agreements with its biggest utility and 56

colleges and universities to reduce emissions to below 1990 levels. And the state of Washington has recently announced its intention to join Oregon and California in pursuing greenhouse gas reductions.

Environmental concern is also a growing factor in international relations. Many world leaders see the concern for environment, health, and natural resources as entering the policy mainstream. A sense of urgency and common cause about the environment is leading to cooperation in some areas. Ecological degradation in any nation is now understood almost inevitably to impinge on the quality of life in others. Drought in Africa and deforestation in Haiti have resulted in large numbers of refugees whose migrations generate tensions both within and between nations. From the Nile to the Rio Grande, conflicts flare over water rights. The growing megacities of the developing world are areas of potential civil unrest. Sheer numbers of people overwhelm social services and natural resources. The government of the Maldives has pleaded with the industrialized nations to reduce their production of greenhouse gases, fearing that the polar ice caps may melt and inundate the island nation.

Economic progress in developing nations could also bring the possibility for environmental peril and international tension. China, which accounts for 21 percent

Changing the Nature of Environmental Regulation—The Safe Drinking Water Act

According to the American Water Works Association, the 1996 amendments to the Safe Drinking Water Act (SDWA) amount to a radical rewrite of the 1986 amendments. The amendments were written with little substantive input from the regulated community. The 1996 amendments, however, were developed with significant contributions from water suppliers and state and local officials. The new amendments embody a partnership approach that includes major new infusions of federal funds to help water utilities, especially the thousands of smaller systems, comply with the law.

What all this means is that the amended SDWA is more focused on what it was initially intended to achieve. For example, rather than require the EPA to set standards for 25 new contaminants every three years, regardless of need or actual threat to public health (in some cases, it has been argued, this requirement impeded public health protection by diverting resources from the most critical needs), the amended law focuses the EPA's efforts on regulating contaminants known to pose health risks and requires cost-benefit analyses and risk assessments to take place before new standards are set. The act also allows the EPA to adopt interim regulations for contaminants on an emergency basis if they pose urgent health threats. The act provides millions of dollars in new funding for crucial research; links source water to drinking water (and sets out a state-administered system designed to protect drinking water from becoming contaminated in the first place); authorizes a new system of federal grants to states that will enable water utilities to borrow funds to upgrade their systems; and requires water systems to provide customers with annual updates delineating the sources and quality of the water they provide. It has been stated that the 1996 amendments are good for the public and drinking water industry alike. The law has more flexibility, more state responsibility, more cooperative approaches, and is achieved through partnerships.

All too often we take clean, safe drinking water for granted.

of the world's population, has the world's third-largest recoverable coal reserves. If China's current "modernization" campaign succeeds, the boom will be fueled by coal, to the possible detriment of the planet as a whole. (See figure 20.10.)

Some experts estimate that the developing world, which today produces one-fourth of all greenhouse gas emissions, could be responsible for nearly two-thirds by the middle of this next century. Developing nations have repeatedly indicated that they are not prepared to slow down their own already weak economic growth to help compensate for decades of environmental problems caused largely by the industrialized world.

Some developing countries may resist environmental action because they see a chance to improve their bargaining leverage with foreign aid donors and international bankers. Where before the poor nations never had a strategic advantage, they now may have an ecological edge. Ecologically, there could be more parity than there ever was economically or militarily.

National security may no longer be about fighting forces and weaponry alone. It also relates increasingly to watersheds, croplands, forests, climate, and other factors rarely considered by military experts and political leaders but that, when taken together, deserve to be viewed as equally crucial to a nation's security as are military factors. It is interesting to note that the North Atlantic Treaty Organization (NATO) has developed an office for Scientific and Environmental Affairs, and the U.S. Department of Defense has created an Office of Environmental Security.

In 2000, a dike holding millions of liters of cyanide-laced wastewater gave way at a gold-extraction operation in northwestern Romania, sending a deadly waterborne plume across the Hungarian border and down the nation's second-largest river. Cyanide separates gold from ore, and mining operations often store cyanide-laced sludge in diked-off lagoons. After devastating the upper Tisza River, the 50-kilometer (30-mile) long pulse of cyanide and heavy metals spilled into the Danube River in northern Yugoslavia. The diluted plume finally filtered into the Danube delta at the Black Sea, more than 1000 kilometers (620 miles) from and three weeks after the spill. Scientists across Europe warned that the accident could leave the upper Tisza and the nearby Somes Rivers a poisonous legacy for several years if heavy metals are left to linger in the river sediments.

The Hungarian Academy of Sciences has pointed out that 90 percent of the water entering the tributaries of Hungary's rivers flows through Romania, Ukraine, Slovakia, and Austria. Numerous mines, chemical plants, oil refineries, and other

Figure 20.10   Developing Concerns
If China's current "modernization" continues, the boom will be fueled by coal to the possible detriment of the planet as a whole.

sources of pollutants line those tributaries. The Hungarian government stated that the management of environmental security cannot be stopped at the borders. The government has filed lawsuits seeking monetary damage against the operators of the Baia-Mare gold extraction lagoon, and it has threatened to sue the Romanian government to help recover the cleanup costs. The accident stoked bilateral tensions: Romanian officials accused the Hungarian side of exaggerating the extent of the damage, while Hungarians asserted that the Romanians are downplaying the spill. The issue of environmental security continues to expand in the dialogue between the two nations.

The increased attention to the environment as a foreign policy and national security issue is only the beginning of what will be necessary to avert problems in the future. The most formidable obstacle may be the entrenched economic and political interests of the world's most advanced nations. If North Americans wish to stem the supply of hardwood from a fragile jungle or furs from endangered species, then they will have to stem demand for fancy furniture and fur coats. If they wish to preserve wilderness from the intrusions of the oil industry, then they will have to find alternative sources of energy and use all fuels more efficiently.

What may be needed is self-discipline on the part of the world's haves and increased assistance to the have-nots. In the world today, a billion people live in a degree of poverty that forces them to deplete the environment without regard to its future. Their governments often are too crippled by international debt to afford the short-term costs of environmental safeguards.

William Ruckelshaus, a former administrator of the U.S. Environmental Protection Agency, believes a historical watershed may be at hand. If the industrialized and developing countries did everything they should, he says, the resulting change would represent a "modification of society comparable in scale to the agricultural revolution of the Neolithic age and to the Industrial Revolution of the past two centuries."

Let's put all of this into another perspective that may ring closer to home. By even conservative estimates, the population of the United States is expected to double in the next 50 years. It could be argued, then, that with twice as many people projected for 2050, it will be necessary to double the total U.S. infrastructure in the next 50 years. Potentially, this translates to the following: twice as many cars, trucks, planes, airports, parking lots, streets, and freeways; twice as many houses and apartment buildings; twice as many landfills, wastewater treatment plants, hazardous-waste treatment facilities, and chemicals (pesticides and herbicides) for agriculture; in short, twice as much of everything. Under this scenario, it will be necessary to take over and develop in the next 50 years an amount of farmland and scenic countryside equal to the total already developed in the past 200 years. (See figure 20.11.) Megacities such as New York and Los Angeles will double in size, and many new megacities will develop. In short, this vision is not very appealing. What happens to wilderness areas, remote and quiet places, and habitat for songbirds, waterfowl, and other wild creatures? What happens to our quality of life?

Another way of looking at the future would be along the lines of what was previously mentioned about the analogy to the agricultural and industrial revolutions. This translates into a future of profound change—a future in which virtually everything we do will change. This future of profound change will also be one of phenomenal opportunity and excitement. It is a future in which we will farm, build, and transport in entirely new ways. This vision of the future could well become known as the environmental revolution.

Terrorism and the Environment

A study of environmental terrorism requires understanding motivations, identifying vulnerabilities and risks, and working on effective solutions. At a time when world populations are increasing, the existing resource base (water, energy, soils) is being stretched to provide for more people and consumed at a faster rate. As the value and vulnerability of these resources increase, so does their attractiveness as terrorism targets. Terrorists often choose targets because of what they represent, for example, skyscrapers and government buildings. Rivaling both of those, however, for the amount of long-term damage that can be inflicted on a country, environmental resources should be considered as being at risk. **Environmental terrorism** is defined as the unlawful use of force against environmental resources so as to deprive populations of their benefits or destroy other property. Over the past several decades, several types of environmental terrorism have been waged, from acts of war to acts by individuals or small groups.

The effects of war on the natural environment are catastrophic. By its nature, war is destructive, and the environment is an innocent victim. While the effects of war on the environment have been with us over the centuries, the 1990–91 conflict in the Persian Gulf witnessed a new role for the environment in war: its degradation as a weapon.

The term *ecoterrorism* was applied to former Iraqi leader Saddam Hussein's use of oil as a weapon in the war. Iraq dumped an estimated 1.1 billion liters (290 million gallons) of crude oil into the Persian Gulf

Figure 20.11 **Options and Tradeoffs for the New Century** Unless we become more creative in areas such as transportation and land use, it will be necessary to develop in the United States alone, in the next 50 years, an amount of farmland and scenic countryside equal to the amount developed over the past 200 years.

from Kuwait's Sea Island terminal. The oil spill that resulted from the dumping was the world's largest—over 30 times the size of the 1989 *Exxon Valdez* disaster in Alaska. (See figure 20.12a.)

Saddam Hussein may have engineered the spill to block allied plans for an amphibious invasion of Kuwait, but he was also probably trying to shut down seaside desalination plants that provide much of the freshwater for Saudi Arabia's Eastern Province. Whatever the military or political effect, the environmental effect was much greater.

The Persian Gulf waters, shores, and islands are dotted with coral reefs, mangrove swamps, and beds of sea grass alive with many unique species of birds, fish, and marine mammals. The gulf is also rather isolated, with only one narrow outlet—the Strait of Hormuz, just 55 kilometers (34 miles) across. The gulf takes up to five years to flush out. Many of the region's species, such as the bottlenose dolphin, dugong, green turtle, and Caspian tern, were already classified as threatened. The 1991 Gulf War oil spill provided opportunities to monitor the effects of the

oil slick on marine organisms and biotic communities. Despite the devastating effects on coastal habitats such as salt marshes in Saudi Arabia, postwar coastal surveys showed that the recovery of these habitats had been relatively rapid. By 1997, several coastal habitats had attained normal to 60 percent diversity. Despite considerable population fluctuations, postwar coastal bird surveys also suggest no overall decline of coastal bird populations when compared to prewar population levels. The long-term fate and possible ecological effects of sunken oil and oil buried under sediments still are uncertain.

Another act of ecoterrorism that took place during the Iraq-Kuwait war was the intentional burning of hundreds of oil wells in Kuwait. (See figure 20.12b.) The air pollution resulting from the burning oil covered a large area. It took several months to extinguish the fires following the war. Using environmental damage as a weapon of war is a frightening concept. The environmental consequences of such acts of ecoterrorism will not be fully understood for years.

Not all acts of environmental terrorism are as dramatic as those of the 1991 Gulf War. Individual acts affecting a more localized region are also documented, such as in July 2000, when workers at the Cellatex chemical plant in northern France dumped 3000 liters (793 gallons) of sulfuric acid into the Meuse River when they were denied health benefits.

Concerns over the use of biological and chemical weapons as a means of terrorism became, unfortunately, all too real following the events of September 11, 2001. The use of such weapons, however, is not new. During the Vietnam War, a chemical defoliant called Agent Orange was used extensively. During the First World War, several chemicals were used as poison gas. History is full of examples of both chemical and biological weapons being used. There is a difference, however, between biological and chemical weapons.

Biological weapons are naturally occurring organisms that cause disease. The two most common examples are the bacterium *Bacillus anthracis* (anthrax), which produces a toxin, and the virus that causes

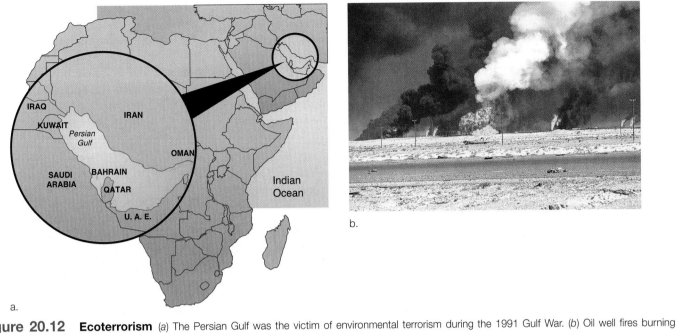

a.

b.

Figure 20.12 **Ecoterrorism** (a) The Persian Gulf was the victim of environmental terrorism during the 1991 Gulf War. (b) Oil well fires burning in Kuwait.

Source: United Nations Development Program.

smallpox, a highly infectious disease. Anthrax bacteria produce spores that allow them to live in a dormant state in soil. When used as a weapon, the spores enter the lungs, where they are carried into the blood and the immune system. The spores become active, reproduce in large numbers, and release a devastating toxin that is lethal to cells. If enough spores are inhaled, anthrax can be fatal.

The history of biological weapons is a long one. Almost as soon as humans figured out how to make arrows, they were dipping them in animal feces to poison them. The Roman Empire used animal carcasses to contaminate their enemies' water wells. This practice was used again in Europe's many wars, in the American Civil War, and even into the twentieth century. In the fifteenth century, Spanish conquistador Francisco Pizarro gave clothing contaminated with the smallpox virus to natives in South America. Britain's Lord Amherst continued the practice into the late eighteenth century, spreading smallpox among Native Americans during the French-Indian War by giving them blankets that had been used at a hospital treating smallpox victims.

Chemical weapons are poisons such as mustard gas and nerve gases such as sarin. In the First World War, poison gas was used on both the Eastern and Western fronts from 1915 to 1918. Chlorine gas was the most widely used of the poison gases. Chlorine gas burns and destroys lung tissue. Chlorine is not an exotic chemical. Most municipal water systems use it to kill bacteria.

Modern chemical weapons tend to be made with agents having much greater power, meaning that it takes a lot less of the chemical to kill the same number of people. Many of them use the types of chemicals found in insecticides. When you spray your lawn or garden with a chemical to control aphids, you are, in essence, waging a chemical war on aphids.

In 1972, 103 countries signed the Biological Weapons Convention (BWC), which prohibited the development and use of biological and chemical weapons. Even so, several countries are known to have developed biological weapons since the convention. The BWC still allows research for defense, such as vaccines, against biological weapons. In 2001, the United States announced it was developing a new form of anthrax for defensive research.

Controlling a biological agent is the most troublesome part of using biological weapons and one of the most important reasons they have not been widely used. Bacteria and viruses do not discriminate between ally and foe, and the so-called boomerang effect—the biological agent affecting those who released it—is a potential problem. Still, the use of biological and chemical agents for terrorist activities in the recent past has been alarming.

Fortunately, turning chemical, biological, or nuclear materials into usable weapons is not that easy. First, you have to acquire or manufacture sufficient quantities of the lethal agent. Second, you have to deliver it to the target. And third, you have to either detonate it or spread it around in a way that will actually harm a lot of people.

Of the three classes of weapons of mass destruction, those based on chemicals should be the easiest to make. Their ingredients are often commercially available, and their manufacturing techniques are well known. As mentioned, these weapons have been used from time to time in real warfare, so their deployment is also understood. Biological weapons are trickier; and nuclear weapons trickier still. Germs need to be coddled and are hard to spread. The manufacture and detonation of a nuclear device are extremely complicated.

Numerous attempts have been made to smuggle nuclear materials out of the former Soviet Union. To date, police and customs officers have seized mostly low-grade nuclear waste. This could not be turned into a proper atomic bomb, but with enough of it, a terrorist group might hope to build a device to spread radioactive contamination in an area.

While war is obviously harmful to the environment, the reverse could also be argued in certain circumstances. Armed conflict—or the threat of it—can sometimes be good for the environment. The demilitarized zone between North and South Korea is a 250-kilometer (155-mile) long strip of mountains, jungle, and wetlands untouched by humans since 1953. It is now home to wildlife that is extinct elsewhere on the peninsula. Landmines laid in civil wars in Africa have discouraged hunters and allowed game to flourish in areas from which it had previously disappeared. In Congo, anarchy has prevented mining companies and timber firms from spreading into the country's remaining wild areas. Although many large animals have been killed by gun-toting soldiers, recent aerial surveys suggest that Congo's rhinos have survived the conflict well; some 6000 elephants remain as well. The people of Congo would doubtless prefer less anarchy, even if it meant fewer elephants. But the fact remains that when men are busy killing each other, nature sometimes gains.

International Environmental Policy

If there must be a war, let it be against environment contamination, nuclear contamination, chemical contamination; against the bankruptcy of soil and water systems; against the driving of people away from the lands as environmental refugees. If there must be war, let it be against those who assault people and other forms of life by profiteering at the expense of nature's capacity to support life. If there must be war, let the weapons be your healing hands, the hands of the world's youth in defense of the environment.

Mustafa Tolba
Former Secretary General
United Nations Environment
Programme

There are many institutions that address the global environment. In the United Nations system, 21 separate agencies deal with environmental issues. In addition to the United Nations, the World Bank and other institutions charged with economic development play an important role in the implementation of policies and projects that affect the global environment. Institutions that deal with trade such as the World Trade Organization (WTO) and the North American Free Trade Agreement (NAFTA) also affect the global environment.

While there have been examples of successful initiatives, global organizations have not been able to achieve significant progress in reversing global environmental degradation. There are several reasons for this. Some fail because they are controlled by a disparate membership with competing interests who are unable to reach consensus on complex and difficult issues. An example of this is the UN Food and Agriculture Organization (FAO), under whose policies only limited progress has been made in dealing with global fishing and agriculture issues. Although the FAO has pursued the collection of statistics on fish stocks and it is clear that most fisheries are being fished at or above capacity, it has not acted to develop systems that would conserve fish stocks, and individual governments continue to set catch quotas above sustainable levels.

Other institutions do not succeed because they are responsible only for specific functions or activities and are unable to address whole issues on their own. The World Bank, for example, can provide guidelines for dealing with air pollution or biodiversity preservation only for development projects that use World Bank funds. Still other institutions have failed to take even initial steps to weave the environment into their policies and programs as a permanent concern. For example, the WTO focuses on trade issues but has yet to confront the nexus of trade—sustainable economic growth and the environment.

International coordination and political resolve are necessary if the goals of preserving and protecting the global environment are to be realized. A step in that direction was the 1972 United Nations Conference in Stockholm, Sweden. This was the first international conference specifically dealing with global environmental concerns. The UN Environment Programme, a separate department of the United Nations that deals with environmental issues, developed out of that conference. The United Nations Conference on the Law of the Sea produced a comprehensive convention that addressed many of the issues concerning jurisdiction over ocean waters and use of ocean resources. This treaty is viewed by many as a model for international environmental protection. Positive results are already evident from application of the agreements on pollution control, marine mammal protection, navigation safety, and other aspects of the marine environment. The issue of deep ocean mining, primarily of manganese nodules, has to date kept many industrial nations, such as Germany and the United States, from ratifying the treaty.

Over 150 global environmental treaties have been negotiated since the start of the twentieth century. In addition, at least 500 bilateral agreements are now in effect dealing with cross-border environmental issues. These agreements generally address single issues, such as the Framework Convention on Climate Change or the Convention on International Trade in Endangered Species of Wild Fauna and Flora. History has shown that most of the existing agreements are lacking in ambition or are framed in such a way that they result in little concrete action.

However, there have been several successful international conventions and treaties. The Antarctic Treaty of 1961

GLOBAL PERSPECTIVE

ISO Standards for Environmental Management Systems

The International Organization for Standardization (ISO) was established in 1947 in Geneva, Switzerland, to promote the development of voluntary standards that would facilitate international trade. ISO is a nongovernmental organization made up of national standards bodies from more than 100 countries. Since its formation, ISO has developed more than 10,000 standards that govern products, such as the size of paper, the format of credit cards, and the symbols on automobile dashboards. ISO standards ensure that products share uniform characteristics no matter where they are produced.

In the late 1970s, the focus of ISO's standard setting began to shift. While product standards helped to ensure that specific products performed as expected, they did not address the operations of the facilities that manufactured them. ISO committees set to work on a series of quality-management standards known as ISO 9000. The ISO 9000 standards were drafted to help ensure that facilities had in place competent management processes that were capable of manufacturing high-quality products, regardless of how business conditions might change. Thousands of organizations in Europe, Asia, and the United States quickly adopted these standards.

In the early 1990s, the ISO began work on a series of environmental management standards. However, leaders in business and government were concerned that the national environmental management standards under development in England, Ireland, France, and Spain, as well as European Union standards, could result in unintended barriers to international trade as each country developed it own potentially inconsistent requirements for manufacturers. As part of the preparations for the United Nations Conference on Environment and Development in Brazil in 1992, business leaders called on the ISO to provide guidance for sustainable business development. This task, combined with the success of ISO 9000, led the ISO in 1992 to establish subcommittees to work on aspects of environmental management, including environmental management systems (EMSs), auditing, labeling, performance evaluation, and life-cycle assessment. The work of these committees resulted in a series of standards known as the 14000 series, published beginning in 1996. The first of these standards, ISO 14001, established criteria for EMSs. It is the only standard in the 14000 series to which a facility may be certified as having an EMS in compliance with ISO criteria.

The American National Standards Institute represents the United States in the ISO. Work on ISO 14000 takes place under the auspices of the U.S. Technical Advisory Group, which includes business, government, environmental advocacy, and academic organizations. The advisory group currently is working on proposed changes to ISO 14001—the first such revision since its publication. Additional areas under debate include ISO 14001's requirements for regulatory compliance and public disclosure of information.

Summary of the International Organization for Standardization's ISO 14001

Policy Objectives	Requirements
Improve environmental performance	Managers must establish a procedure to identify the environmental impacts of "activities, products, and services" over which they have control. They must establish specific and measurable goals to reduce impacts they deem "significant" and programs to accomplish these goals. Workers must be aware of their personal contribution to these impacts.
Comply with regulations	Managers must establish an environmental policy that includes a commitment to regulatory compliance, establish procedures that will inform them of their legal obligations, and set targets and objectives with compliance in mind. However, actual regulatory compliance is not required.
Prevent pollution	ISO 14001 requires "commitment" to "prevention of pollution," defined to include pollution control.
Audit performance	Audits must be conducted routinely to identify and correct violations of the system and must include an evaluation of the facility's regulatory compliance program.
Disclose information to the public	The only information that must be disclosed is a facility's environmental policy.

Source: International Organization for Standardization, Technical Committee 207, Subcommittee 1, "Environmental Management Systems—Specification with Guidance for Use" (Geneva, 1996).

GLOBAL PERSPECTIVE

Overview of an International Organization—The International Whaling Commission

The first records of whaling can be traced back hundreds of years to the twelfth century in the Bay of Biscay of Spain. During whaling's heydays of the 1930s, 85 percent of the world's catch of whales was coming from the Antarctic, with 95 percent of these catches dominated by Norway and the United Kingdom.

With the possible exhaustion of whale stock, the problem of whales as a common property resource came to the forefront. When a resource is common property, single users have no right to the resource, nor can they prevent others from sharing in its exploitation. These doctrines of freedom of the seas and common property resources date back to Roman times and were constantly used by states as an argument to justify the overfishing of whale stocks in previous centuries. Seeing a need to keep statistical data on the populations of different whale species, Norway, at the request of the International Council for the Exploration of the Seas, set up the Bureau of International Whaling Statistics in 1930 to deal with whaling problems. This eventually led to the drafting and signing of the Convention for the Regulation of Whaling in 1931. The convention "applied to all waters . . . but was only applicable to baleen whale" and provided exemptions for aboriginal subsistence whaling. This convention had little effect on the amount of whaling because important nations such as Germany and Japan failed to adhere to the rules. A new draft of the convention was brought into effect in 1937 and included such countries as Norway, the United Kingdom, and Germany. In 1946, the International Whaling Conference (IWC) was convened by the United States to address the post–World War II whaling issues.

The measures to be taken by this conference were to completely protect certain whale species, specify certain whale sanctuaries, set limits on the number of whales taken, decide when hunting seasons would open and close, and specify the size and age of those whales that could be taken. It is important to point out that the majority of the nations that were original members of the IWC were whaling states and had no real interest in protecting the whale to preserve biologic diversity. The formation and actions of the IWC in its early years were purely economically motivated.

The ineffectiveness of the IWC to control the slaughter of whale populations continued into the 1970s until environmentally oriented nongovernmental organizations began to champion the struggle of the whale. Soon millions of people in the world were chanting "save the whale" and began demanding that the IWC pass a moratorium to end whaling. But a powerful veto coalition made up of Japan, Norway, the Soviet Union, Iceland, Chile, and Peru was able to control the vote in the IWC and defeat any measures that were harmful to the whaling industry as a whole.

Seeing that it could not influence the voting of the IWC alone, the United States began to form a new strategy. To ensure a moratorium on whaling in the IWC, nonwhaling states from the developing nations were brought in to outnumber the whaling states. The state leading this group of new members was the small island nation of Seychelles. Despite economic threats from Japan, Seychelles held strong and was able to push for a moratorium that was eventually passed by the IWC in 1982. Although many believed that this was the final victory in the struggle to end whaling, the war continued. Nations such as Norway and Iceland threatened to leave the IWC and eventually did so. Other nations found loopholes in the moratorium and continued to kill large numbers of whales. Today, the future of both the moratorium and the IWC is unclear. Japan rallies developing nations onto its side with promises of economic aid. Other nations such as Norway are refusing to abide by the ruling handed down by the IWC and state that they will resume whaling. With a decline in the United States' ability to influence the IWC, perhaps the only hope for the whale in the future is the changing attitude of the Japanese public toward the killing of whales.

Despite the efforts of the International Whaling Commission, whales are not safe from being slaughtered.

reserves the Antarctic continent for peaceful scientific research and bans all military activities in the region. The 1979 Convention on Long-Range Transboundary Air Pollution was the first multilateral agreement on air pollution and the first environmental accord involving all the nations of Eastern and Western Europe and North America. The 1987 Montreal Protocol on Substances That Deplete the Stratospheric Ozone Layer addresses the ozone protective shield problem.

Agreed to in 1987, the Montreal Protocol's objective was to phase out the manufacture and use of chemicals depleting the Earth's protective ozone layer (see chapter 17). Since its adoption over a decade ago, 160 countries are now parties, representing over 95 percent of the Earth's population. Production and consumption of chlorofluorocarbons (CFCs), carbon tetrachloride, halons, and methyl chloroform have been phased out in developed countries, with reduction schedules set for their use in developing countries. In 1999, developing countries ended the production and consumption of CFCs, with phaseout scheduled for 2010.

By the late 1990s, the concentration of some CFCs in the atmosphere had started to decline, and predictions are that the ozone layer could recover by the middle of this century. Among the many reasons for these achievements, three merit attention:

- *Global agreement on the nature and seriousness of the threat.* Even the strongest skeptics could not deny the Antarctic ozone hole, which was first brought to international attention by British scientists in 1985. It was understood for the first time that emissions of ozone-depleting substances were in reality putting our lives and the lives of future generations at risk. Decisive action was required.

- *A cooperative approach, especially between developed and developing countries.* In the developed countries, it was recognized that their industries had contributed significantly to this global problem and that they had to take the lead in stopping emissions and finding alternatives. It was also recognized that solving the ozone problem required a global solution, with all countries committed to eliminating ozone-depleting substances. Innovative partnerships, including early controls for developed countries and a grace period, funding, and technology transfer for developing countries, were set in place.

- *Policy based on expert and impartial advice.* The parties to the Montreal Protocol were able to receive impartial advice from their science, technical, and economic committees. These drew together experts from around the world to evaluate the need for further action and propose options that are technically and economically feasible.

Put these three factors together—acknowledgment of the threat, agreement to cooperate, and commitment to take effective action based on expert advice—and you have a potentially strong recipe to solve global issues, such as climate change and biodiversity loss.

Even in cases where countries have ratified treaties that have entered into force, parties to the treaties do not always comply with their provisions. An example is Russia's noncompliance with the Montreal Protocol. There are few penalties for noncompliance other than public anger, in part because countries are unwilling to give up their sovereignty.

Why is the implementation of global environmental agreements so difficult? One answer is that for most of these accords to function effectively, participation must be truly global. Yet because of the unique nature of individual nations and their differing economies, it is difficult to reach meaningful agreement among the necessary participants. Another obstacle to implementing global environmental agreements is poorer nations' lack of capacity to comply with international treaty requirements. Some treaties have addressed this problem by offering financial assistance to countries that are unable to comply; for example, the Multilateral Fund of the Montreal Protocol. Other agreements have established more relaxed timetables for compliance for these countries. In many cases, however, the financing has not been forthcoming, and differentiating among countries based on their ability to act has become politically controversial. The Kyoto Protocol, for example, deals with emissions reductions from developed countries only, which has opened the treaty to criticism in the United States. The challenge of developing accords that are both effective and fair has proved to be a major obstacle to progress.

Despite these problems, treaties are the best traditional tools of global environmental governance. Frameworks for action that are negotiated among different nations are key to leveling the playing field and establishing the rules under which governments, businesses, nongovernmental organizations, and citizens can work together toward a common goal.

In 1997, the General Assembly of the United Nations held a special session and adopted a comprehensive document entitled Programme for the Further Implementation of Agenda 21 prepared by the Commission on Sustainable Development. It also adopted the program of work of the commission for 1999–2003. The Commission on Sustainable Development (CSD) was created in 1992 to ensure effective follow-up of the United Nations Conference on Environment and Development held in Rio de Janeiro, Brazil.

The Commission on Sustainable Development consistently generates a high level of public interest. Over 50 national leaders attend the CSD each year, and more than 1000 nongovernmental organizations are accredited to participate in the commission's work. The commission ensures the high visibility of sustainable development issues within the UN system and helps to improve the UN's coordination of environment and development activities. The CSD also encourages governments and international organizations to host workshops and conferences on different environmental and cross-sectoral issues. The results of these expert-level meetings enhance the work of CSD and help it work better with national governments and various nongovernmental partners in promoting sustainable development worldwide.

There is no international legislature with authority to pass laws; nor are there international agencies with power to regulate resources on a global scale. An international court at The Hague in the Netherlands has no power to enforce its decisions. Nations can simply ignore the court if they wish. However, a network is growing of multilateral environmental organizations that have developed a greater sense of their roles and a greater incentive to work together. These include not only the United Nations Environment Programme but also the Environment Committee of the Organization for Economic Cooperation and Development and the Senior Advisors on Environmental Problems of the Economic Commission for Europe. Such institutions perform unique functions that cannot be carried out by governments acting alone or bilaterally.

Is the goal of an environmentally healthy world realistic? There is growing optimism that the community of nations is slowly maturing with regard to our common environment. We have all suffered a loss of innocence about "Earth management." Laissez-faire may be good economics, but it can be a prescription for disaster in ecology.

This environmental "coming of age" is reflected in the broadening of intellectual perspective. Governments used to be preoccupied with domestic environmental affairs. Now, they are beginning to broaden their scope to confront problems that cross international borders, such as transboundary air and water pollution, and threats of a planetary nature, such as stratospheric ozone depletion and climatic warming. It is becoming increasingly evident that only decisive mutual action can secure the kind of world we seek.

Earth Summit on Environment and Development

In June 1992, representatives from 178 countries, including 115 heads of state, met in Rio de Janeiro, Brazil, at the Earth Summit. Officially, the meeting was titled the United Nations Conference on Environment and Development (UNCED), and it was the largest gathering of world leaders ever held. The first Earth Summit had been held 20 years earlier in Stockholm, Sweden. At that time, the planet was divided into rival East and West blocs and was preoccupied with the perils of the nuclear arms race. With the collapse of the eastern bloc and the thawing of the Cold War, a fundamental shift in the global base of power had occurred.

Today, the more important division, especially on environmental issues, is not between East and West but between North (Europe, North America, and Japan) and South (most of Asia, Africa, and Latin America). And, although the immediate threat of nuclear destruction has lifted, the planet is still at risk.

The idea behind the Earth Summit was that the relaxation of Cold War tensions, combined with the growing awareness of ecological crises, offered a rare opportunity to persuade countries to look beyond their national interests and agree to some basic changes in the way they treat the environment. The major issues

are clear: The developed countries of the North have grown accustomed to lifestyles that are consuming a disproportionate share of natural resources and generating the bulk of global pollution. Many of the developing countries of the South are consuming irreplaceable global resources to provide for their growing populations.

The Earth Summit was intended to promote better integration of nations' environmental goals with their economic aspirations. Although the hopes of some developing nations for large commitments of new foreign assistance did not fully materialize, much was accomplished during the summit.

- The *Rio Declaration on Environment and Development* sets out 27 principles to guide the behavior of nations toward more environmentally sustainable patterns of development. The declaration, a compromise between developing and industrialized countries that was crafted at preparatory meetings, was adopted in Rio without negotiation due to fears that further debate would jeopardize any agreement.

- States at UNCED also adopted a voluntary action plan called *Agenda 21,* named because it is intended to provide an agenda for local, national, regional, and global action into the twenty-first century. UNCED Secretary General Maurice Strong called Agenda 21 "the most comprehensive, the most far-reaching and, if implemented, the most effective program of international action ever sanctioned by the international community." Agenda 21 includes hundreds of pages of recommended actions to address environmental problems and promote sustainable development. It also represents a process of building consensus on a "global work-plan" for the economic, social, and environmental tasks of the United Nations as they evolve over time.

- The third official product of UNCED was a *"non-legally binding authoritative statement of principles for a global consensus on the management, conservation, and sustainable devel-*

opment of all types of forests." Negotiations on the forest statement, begun as negotiations for a legally binding convention on forests, were among the most difficult of the UNCED process. Many states and experts, dissatisfied with the end result, came away from UNCED seeking further negotiations toward agreement on a framework convention on forests.

Environmental Policy and the European Union

"The environment knows no frontiers" was the slogan of the 1970s, when the European Community—known today as the European Union—began to develop its first environmental legislation.

The early laws focused on the testing and labeling of dangerous chemicals, testing drinking water, and controlling air pollutants such as sulfur dioxide, oxides of nitrogen, and particulates from power plants and automobiles. Many of the directives from the 1970s and 1980s were linked to Europe's desire to improve the living and working conditions of its citizens.

In 1987, the Single European Act gave this growing body of environmental legislation a formal legal basis and set three objectives: protection of the environment, protection of human health, and prudent and rational use of natural resources.

The treaty reflected what many governments had already understood: that countries are part of an interconnected and interdependent world of people who are bound together by the air they breathe, the food they eat, the products they use, the wastes they throw away, and the energy they consume.

Similarly, a factory in one European nation may import supplies and raw material from several other neighboring nations; it may consume energy produced from imported gas, produce wastes that affect the air and water quality across the border or downstream, and export products whose wastes become the risk and responsibility of governments and peoples several hundred or many thousand kilometers away.

The 1992 Maastricht Treaty formally established the concept of sustainable development in European Union law. Then,

Table 20.2 Major Worldwide Environmental Treaties Since 1975

Treaty	Status
Convention on International Trade in Endangered Species, in force since 1975	The U.S. is a party.
Geneva Convention on Long-Range Transboundary Air Pollution, in force since 1983	The U.S. is a party.
Bonn Convention on Conservation of Migratory Species, in force since 1983	The U.S. is not a party.
Vienna Convention for the Protection of the Ozone Layer, in force since 1988	The U.S. is a party.
Montreal Protocol to the Vienna Convention, in force since 1989. This protocol set the first explicit limits to emissions of gases that erode the ozone layer.	The U.S. is a party.
Basel Convention on Transboundary Movements of Hazardous Wastes and Their Disposal, in force since 1992. This convention was a response to the growing practice of dumping wastes in developing countries and Eastern Europe.	The U.S. is a party.
Convention on Biological Diversity, in force since 1993	The U.S. is not a party.
Framework Convention on Climate Change, in force since 1994	The U.S. is a party.
Convention on the Law of the Sea, in force since 1994	The U.S. is not a party.
Convention to Combat Desertification, in force since 1996	The U.S. is a party.
Kyoto Protocol to the Framework Convention. The negotiators adopted the text in 1997, but it has not yet collected enough ratifications to go into force.	The U.S. has said that it will not ratify.
Rotterdam Convention on the Prior Informed Consent Procedure for Certain Hazardous Chemicals and Pesticides in International Trade. The text was adopted in 1998 but is not yet in force.	The U.S. has not ratified.
Stockholm Convention on Persistent Organic Pollutants. The text was adopted in 2001 but is not yet in force.	The U.S. has not ratified.

Source: *Resources for the Future*, Summer 2002, Issue 148.

in 1997, the Amsterdam Treaty made sustainable development one of the overriding objectives of the European Union. The treaty considerably strengthened the commitment to the principle that the European Union's future development must be based on the principle of sustainable development and a high level of protection of the environment. The environment must be integrated into the definition and implementation of all of the Union's other economic and social policies, including trade, industry, energy, agriculture, transport, and tourism.

Under certain conditions, the member states may maintain or introduce environmental standards and requirements that are stricter than those of the Union. The European Commission verifies the compatibility of these stricter national programs and the potential effects on other member states.

New International Instruments

Over the past few decades, the global community has responded to emerging environmental problems with unprecedented international agreements—notably the Montreal Protocol, the Framework Convention on Climate Change, the Convention on Biological Diversity, and the Convention to Combat Desertification, among others. (See table 20.2.)

In the course of crafting these global agreements, many important lessons have been learned. As the experience with the Montreal Protocol shows, the scientific community can play a crucial role in two ways: first, confirming the links between human activities and global environmental problems; and second, showing what could happen to human health and the global environment if nothing is done.

When the evidence is in hand, an international consensus to act can emerge quickly. The same process is underway in the current international debate on climate change, but the process has been more difficult because the linkages between human activities and global environmental impacts are more complex and still not completely understood. Nevertheless, a global consensus for action is emerging. Another important lesson learned has to do with the structure of international

agreements and the elements that can contribute to an effective structure. In the case of the Montreal Protocol, the agreement was not punitive and favored incentives and results-oriented approaches. All nations participated in the agreement but at different levels of responsibility in recognition of their differing conditions. The Kyoto Protocol to the Climate Change Convention has benefited from the experience with the Montreal Protocol and included many of the same elements.

A third vital lesson is that, to the extent possible, every interested party must have an opportunity to participate as full partners in the process and to voice their concerns. It is particularly helpful for environmental advocacy groups and the business community to be part of this process. International agreements need to provide incentives to foster public-private partnerships and to provide a role for business leaders to seek innovative technical solutions.

A fourth lesson concerns the role of governments in implementing these conventions. Government actions need to be consistent and predictable; provide sufficient lead times; favor government-led incentives over direct industry subsidies; and use flexible, market-based solutions where they are appropriate.

Finally, a fifth lesson is that agreements mark the beginning of a process, not the end. Scientists and nongovernmental organizations must continue to further global understanding of environmental problems and communicate what they have learned to the public and policy makers. Policy makers, in turn, must be flexible and respond to changing circumstances with new or modified policy solutions.

One example of new policy solutions is grounded in consumerism. *Green consumerism* is the concept of rational consumption of our scarce resources for the benefit of the environment and future generations. The old saying "the world is enough for everyone's need but not for everyone's greed" calls for a change in our behavior and lifestyle in favor of a sustainable future. Ecolabels have been introduced in a number of countries to help consumers choose products with a proven environmental edge, determined by the product's choice of raw materials, production process, product life cycle, and associated disposal problems. Ecolabels provide evidence that products have met the safety, quality, and environmental protection requirements of the authority that issues the label.

The first ecolabel was introduced in Germany in 1978. This distinctive Blue Angel label was followed by other environmental certification, including the White Swan of Northern European countries, Environmental Choice of Canada, Green Label of Singapore, and Environmental Label of China. By 1995, over a dozen countries had adopted this system of providing consumers with information enabling them to become responsible green consumers. The ecolabel of the European Union is significant because it is the world's first regional scheme to apply the same minimum standards across national markets. (See figure 20.13.)

1992
European Union - 'Eco-label'

1992
Singapore - 'Green Label'

1993
China - 'Environmental Label'

1978
Germany - 'Blue Angel'

1988
Canada - 'Environmental Choice'

1989
Nordic Council - 'White Swan'

Figure 20.13   **Ecolabels**   A sampling of ecolabels from around the world.

Source: ECCO, *Bulletin of the Environmental Campaign Committee.*

It All Comes Back to *You*

No set of policies, no system of incentives, no amount of information can substitute for individual responsibility when it comes to ensuring that our grandchildren will enjoy a quality of life that comes from a quality environment. Information can provide a basis for action. Vision and ideas can influence perceptions and inspire change. New ways to make decisions can empower those who seek a role in shaping the future. However, all of this will be meaningless unless individuals acting as citizens, consumers, investors, managers, workers, and professionals decide that it is important to them to make choices on the basis of a broader, longer view of their self-interest; to get involved in turning those choices into action; and, most important, to be held accountable for their actions.

The combination of political will, technological innovation, and a very large investment of resources and human ingenuity in pursuit of environmental goals has produced enormous benefits over the past two decades. This is an achievement to celebrate, but in a world that steadily uses more materials to make more goods for more people, we must recognize that we will have to achieve more in the future for the sake of the future. We must move toward a world in which zero waste will become the ideal for society that zero defects has become for manufacturing. Finally, we must recognize that the pursuit of one set of goals affects others and that we must pursue policies that integrate economic, environmental, and social goals.

Summary

Politics and the environment cannot be separated. In the United States, the government is structured into three separate branches, each of which impacts environmental policy.

The increase in environmental regulations in the United States over the past 30 years has caused concerns in some sectors of society.

The late 1980s and early 1990s witnessed a new international concern about the environment, in both the developed and developing nations of the world. Environmentalism is also seen as a growing factor in international relations. This concern is leading to international cooperation where only tension had existed before.

While there exists no world political body that can enforce international environmental protection, the list of multilateral environmental organizations is growing.

It remains too early to tell what the ultimate outcome will be, but progress is being made in protecting our common resources for future generations. Several international conventions and treaties have been successful. In the final analysis, however, each of us has to adjust our lifestyle to clean up our own small part of the world.

Issues – Analysis

Gasoline, Taxes, and the Environment

In 2004, the average price of a gallon of gasoline in the United States crossed the $2 line. Given the increase in the price of gasoline, the thought of higher taxes on each gallon is not a welcome one. However, the question remains: Should there be higher taxes on gasoline? Would higher taxes encourage conservation, stimulate mass transit use, and, hence, reduce local air pollution?

Gasoline taxes vary dramatically in different countries. The United Kingdom has a gasoline tax equivalent to $2.80 per gallon—the highest among industrial countries. The United States, on the other hand, has the lowest tax—40 cents per gallon (18 cents federal tax and, on average, 22 cents state tax). It is commonly thought that Europeans have a greater tolerance for high fuel taxes than Americans because they have shorter distances to travel and better access to public transportation.

A number of arguments are made for implementing high gasoline taxes. By discouraging driving and fuel combustion, gasoline taxes help reduce local air pollution, emissions of carbon dioxide (a greenhouse gas), traffic congestion, traffic accidents, and oil dependency. Taxing gasoline is one way of forcing people to take into account the social costs of these problems when deciding how much—and what type of vehicle—to drive.

Gasoline taxes also provide a source of government revenues. In Britain, gasoline tax revenues are several times the amount of highway spending, and the government has argued that if gasoline taxes are reduced, some schools and hospitals will have to close. This argument is somewhat misleading because the revenues could always be made up through other sources, such as income taxes. The real issue is what level of gasoline taxation might be justified when taking into account the full social costs of driving and the appropriate balance between gasoline taxes and other taxes in raising revenues for the government.

- Is an increase in gasoline taxes feasible in our current political climate?
- Would you vote for a candidate who advocated higher gasoline taxes?
- Would you support increasing the taxes on gasoline?
- Would higher taxes on gasoline encourage you to travel by mass transit more often? To purchase vehicles that are more energy-efficient?

Key Terms

environmental terrorism *465*
executive branch *457*

judicial branch *457*
legislative branch *457*

policy *457*

Review Questions

1. What are the major responsibilities of each of the three branches of the U.S. government?
2. What are some of the enforcement options in U.S. environmental policy?
3. What role does administrative law play in U.S. environmental policy?
4. What are some of the criticisms of U.S. environmental policy?
5. In the past 10 years, how has public opinion in the United States changed concerning the protection of the environment?
6. Why is environmentalism a growing factor in international relations?
7. Give some examples of international environmental conventions and treaties.

Critical Thinking Questions

1. Does chapter 20 have an overall point of view? If you were going to present the problems of environmental policy making and enforcement to others, what framework would you use?

2. The authors of this text say that "we are progressing from an environmental paradigm based on cleanup and control to one including assessment, anticipation, and avoidance." Do you agree with this assessment? Are there environmental problems that are harder to be proactive about than others?

3. Does a command-and-control approach to environmental problems, an approach that emphasizes regulation and remediation, make sense with global environmental problems such as global climate change, habitat destruction, and ozone depletion?

4. What values, perspectives, and beliefs does the Wise Use Movement exhibit in its response to environmental legislation? How are they similar to or different from your own?

5. How is it best, as a global society with many political demarcations, to preserve the resources that are held in common? What special problems does this kind of preservation entail?

6. Do you agree with William Ruckelshaus that current environmental problems require a change on the part of industrialized and developing countries that would be "a modification in society comparable in scale to the agricultural revolution . . . and Industrial Revolution"? What kinds of changes might that mean in your life? Would these be positive or negative changes?

7. New treaties regarding free trade might enable some nations to argue that other nations' environmental legislation is too restrictive, thereby imposing a barrier to trade that is subject to sanction. What special problems and possibilities might the new global economy present for environmental preservation? What do you think about that?

Welcome to McGraw-Hill's Online Learning Center

Location: www.mhhe.com/enger10e

McGraw Hill

Interactive Exploration

Check out the website at www.mhhe.com/enger10e and click on the cover of this textbook for quizzing, career information, case studies, and hot links for the following topics:

Environmental Policy and Decision Making

The Ultimate Ecological Answers

Individual Contributions to Environmental Issues

Environmental Organizations

Miscellaneous Environmental Resources

Environmental and Ecological Organization Sites

Environmental Philosophy

Glossary

A

abiotic factors   Nonliving factors that influence the life and activities of an organism.

abyssal ecosystem   The collection of organisms and the conditions that exist in the deep portions of the ocean.

acid   Any substance that, when dissolved in water, releases hydrogen ions.

acid deposition   The accumulation of potential acid-forming particles on a surface.

acid mine drainage   A kind of pollution, associated with coal mines, in which bacteria convert the sulfur in coal into compounds that form sulfuric acid.

acid rain (acid precipitation)   The deposition of wet acidic solutions or dry acidic particles from air.

activated sludge sewage treatment
Method of treating sewage in which some of the sludge is returned to aeration tanks, where it is mixed with incoming wastewater to encourage degradation of the wastes in the sewage.

activation energy   The initial energy input required to start a reaction.

active solar system   A system that traps sunlight energy as heat energy and uses mechanical means to move it to another location.

acute toxicity   A serious effect, such as a burn, illness, or death, that occurs shortly after exposure to a hazardous substance.

age distribution   The comparative percentages of different age groups within a population.

agricultural products   Any output from farming: milk, grain, meat, etc.

agricultural runoff   Surface water that carries soil particles, nutrients, such as phosphate, nitrates, and other agricultural chemicals, as it runs off agricultural land to lakes and streams.

agricultural waste   Waste from the raising of animals and harvesting and processing of crops and trees.

air stripping   The process of pumping air through water to remove volatile materials dissolved in the water.

air toxics   See hazardous air pollutants.

alpha radiation   A type of radiation consisting of a particle with two neutrons and two protons.

alpine tundra   The biome that exists above the tree line in mountainous regions.

alternative agriculture   All nontraditional agricultural practices.

anthropocentric   Human-centered; a theory of moral responsibility that views the environment as a resource for humankind.

aquiclude   An impervious confining layer of an aquifer.

aquifer   A porous layer of earth material that becomes saturated with water.

aquitard   A partially permeable layer in an aquifer.

artesian well   The result of a pressurized aquifer being penetrated by a pipe or conduit, within which water rises without being pumped.

atom   The basic subunit of elements, composed of protons, neutrons, and electrons.

auxin   A plant hormone that stimulates growth.

B

base   Any substance that, when dissolved in water, removes hydrogen ions from solution; forms a salt when combined with an acid.

benthic   Describes organisms that live on the bottom of marine and freshwater ecosystems.

benthic ecosystem   A type of marine or freshwater ecosystem consisting of organisms that live on the bottom.

beta radiation   A type of radiation consisting of electrons released from the nuclei of many fissionable atoms.

bioaccumulation   The buildup of a material in the body of an organism.

biocentric   Life-centered; a theory of moral responsibility that states that all forms of life have an inherent right to exist.

biogeochemical cycles   Movement of matter within or between ecosystems; caused by living organisms, geological forces, or chemical reactions. The cycling of nitrogen, carbon, sulfur, oxygen, phosphorus, and water are examples.

biochemical oxygen demand (BOD)   The amount of oxygen required by microbes to degrade organic molecules in aquatic ecosystems.

biocide   A kind of chemical that kills many different types of living things.

biodegradable   Able to be broken down by natural biological processes.

biodiversity   A measure of the variety of kinds of organisms present in an ecosystem.

biomagnification   The increases in the amount of a material in the bodies of organisms at successively higher trophic levels.

biomass   Any accumulation of organic material produced by living things.

biome   A kind of plant and animal community that covers large geographic areas. Climate is a major determiner of the biome found in a particular area.

biotechnology   Inserting specific pieces of DNA into the genetic makeup of organisms.

biotic factors   Living portions of the environment.

biotic potential   The inherent reproductive capacity.

birthrate   The number of individuals born per thousand individuals in the population per year.

black lung disease   A respiratory condition resulting from the accumulation of large amounts of fine coal dust particles in miners' lungs.

boiling-water reactor (BWR)   A type of light water reactor in which steam is formed directly in the reactor, and is used to generate electricity.

boreal forest   A broad band of mixed coniferous and deciduous trees that stretches across northern North America (and also Europe and Asia); its northernmost edge is integrated with the Arctic tundra.

brownfields   Buildings and land that have been abandoned because they are contaminated and the cost of cleaning up the site is high.

brownfields cleanup   Cleaning a contaminated industrial site to the point that it is safe to use for specific purposes.

brownfields development   The concept that abandoned contaminated sites can be cleaned up sufficiently to allow some specified uses without totally removing all of the contaminants.

bush meat   Meat from wild animals.

C

carbamate   A class of soft pesticides that work by interfering with normal nerve impulses.

carbon absorption   The use of carbon particles to treat chemicals by having the chemicals attach to the carbon particles.

carbon cycle   The cyclic flow of carbon from the atmosphere to living organisms and back to the atmospheric reservoir.

carbon dioxide (CO_2)   A normal component of the Earth's atmosphere that in elevated concentrations may interfere with the Earth's heat budget.

carbon monoxide (CO)   A primary air pollutant produced when organic materials, such as gasoline, coal, wood, and trash, are incompletely burned.

carcinogen   A substance that causes cancer.

carcinogenic   The ability of a substance to cause cancer.

carnivores   Animals that eat other animals.

carrying capacity   The optimum number of individuals of a species that can be supported in an area over an extended period of time.

catalyst   A substance that alters the rate of a reaction but is not itself changed.

chemical bond   The physical attraction between atoms that results from the interaction of their electrons.

chemical weathering   Processes that involve the chemical alteration of rock in such a manner that it is more likely to fragment or to be dissolved.

chlorinated hydrocarbon   A class of pesticide consisting of carbon, hydrogen, and chlorine; these pesticides are very stable.

chlorofluorocarbons (CFC)   Stable compounds containing carbon, hydrogen, chlorine, and fluorine. They were formerly used as refrigerants, propellants in aerosol containers, and expanders in foam products. They are linked to the depletion of the ozone layer.

chronic toxicity   A serious effect, such as an illness or death, that occurs after prolonged exposure to small doses of a toxic substance.

clear-cutting   A forest harvesting method in which all the trees in a large area are cut and removed.

climax community   Last stage of succession; a relatively stable, long-lasting, complex, and interrelated community of plants, animals, fungi, and bacteria.

coevolution   Two or more species of organisms reciprocally influencing the evolutionary direction of the other.

combustion   The process of releasing chemical bond energy from fuel.

commensalism   The relationship between organisms in which one organism benefits while the other is not affected.

community   Interacting groups of different species.

competition   An interaction between two organisms in which both require the same limited resource, which results in harm to both.

competitive exclusion principle   A theory that no two populations of different species will occupy the same niche and compete for exactly the same resources in the same habitat for very long.

composting   The process of harnessing the natural process of decomposition to transform organic materials into compost, a humuslike material with many environmental benefits.

compound   A kind of matter composed of two or more different kinds of atoms bonded together.

Comprehensive Environmental Response, Compensation, and Liability Act (CERCLA)   The 1980 U.S. law that addressed the cleanup of hazardous-waste sites.

confined aquifer   An aquifer that is bounded on the top and bottom by impermeable confining layers.

conservation   To use in the best possible way so that the greatest long-term benefit is realized by society.

conservation ethic   An environmental ethic that stresses a balance between total development and absolute preservation.

consumers   Organisms that use other organisms as food.

contour farming   A method of tilling and planting at right angles to the slope, which reduces soil erosion by runoff.

controlled experiment   An experiment in which two groups are compared. One, the control, is used as a basis of comparison and the other, the experimental, has one factor different from the control.

coral reef ecosystem   A tropical, shallow-water, marine ecosystem dominated by coral organisms that produce external skeletons.

corporation   A business structure that has a particular legal status.

corrosiveness   Ability of a chemical to degrade standard materials.

cost-benefit analysis   A method used to determine the feasibility of pursuing a particular project by balancing estimated costs against expected benefits.

cover   A term used to refer to any set of physical features that conceals or protects animals from the elements or their enemies.

criteria air pollutants   Those air pollutants for which specific air quality standards have been set by the U.S. Environmental Protection Agency.

crust   The thin, outer, solid surface of the Earth.

D

death phase   The portion of the population growth curve of some organisms that shows the population declining.

death rate   The number of deaths per thousand individuals in the population per year.

debt-for-nature exchange   The purchase of a nation's debt by a third party that requires conservation on the part of the debtor nation in exchange for relief from the debt.

decibel   A unit used to measure the loudness of sound.

decommissioning   Decontaminating and disassembling a nuclear power plant and safely disposing of the radioactive materials.

decomposers   Small organisms, such as bacteria and fungi, that cause the decay of dead organic matter and recycle nutrients.

deforestation   Activities that destroy forests.

deferred cost   A cost that is not paid immediately when an economic decision is made but must be paid at a later date.

demand   Amount of a product that consumers are willing and able to buy at various prices.

demographic transition   The hypothesis that economies proceed through a series of stages, beginning with growing populations with high birth and death rates and low economic development and ending with stable populations with low birth and death rates and high economic development.

demography   The study of human populations, their characteristics, and their changes.

denitrifying bacteria   Bacteria that convert nitrogen compounds into nitrogen gas.

density-dependent limiting factors   Those limiting factors that become more severe as the size of the population increases.

density-independent limiting factors   Those limiting factors that are not affected by population size.

desert A biome that receives less than 25 centimeters (10 inches) of precipitation per year.

desertification The conversion of arid and semiarid lands into deserts by inappropriate farming practices or overgrazing.

detritus Tiny particles of organic material that result from fecal waste material or the decomposition of plants and animals.

development ethic Philosophy that states that the human race should be the master of nature and that the Earth and its resources exist for human benefit and pleasure.

dispersal Migration of organisms from a concentrated population into areas with lower population densities.

domestic water Water used for domestic activities, such as drinking, air conditioning, bathing, washing clothes, washing dishes, flushing toilets, and watering lawns and gardens.

E

ecocentrism An approach to environmental responsibility that maintains that the environment deserves direct moral consideration rather than consideration derived merely from human interests.

ecology A branch of science that deals with the interrelationship between organisms and their environment.

economic growth The perceived increase in monetary growth within a society.

economics The study of how people choose to use resources to produce goods and services and how these goods and services are distributed to the public.

ecosystem A group of interacting species along with their physical environment.

ecosystem diversity A measure of the number of kinds of ecosystems present in an area.

ectoparasite A parasite that is adapted to live on the outside of its host.

electron The lightweight, negatively charged particle that moves around at some distance from the nucleus of an atom.

element A form of matter consisting of a specific kind of atom.

emergent plants Aquatic vegetation that is rooted on the bottom but has leaves that float on the surface or protrude above the water.

emigration Movement out of an area that was once one's place of residence.

endangered species Those species that are present in such small numbers that they are in immediate jeopardy of becoming extinct.

endoparasite A parasite that is adapted to live within a host.

endothermic reaction Chemical reaction in which the newly formed chemical bonds contain more energy than was present in the compounds from which they were formed.

energy The ability to do work.

entropy The degree of disorder in a system. All systems tend toward a high degree of disorder or entropy.

environment Everything that affects an organism during its lifetime.

environmental cost Damage done to the environment as a resource is exploited.

environmental justice Fair application of laws designed to protect the health of human beings and ecosystems; that no groups suffer unequal environmental harm.

Environmental Protection Agency (EPA) U.S. government organization responsible for the establishment and enforcement of regulations concerning the environment.

environmental resistance The combination of all environmental influences that tend to keep populations stable.

environmental science An interdisciplinary area of study that includes both applied and theoretical aspects of human impact on the world.

environmental terrorism The unlawful use of force against environmental resources so as to deprive populations of their benefits or destroy other property.

enzyme Protein molecules that speed up the rate of specific chemical reactions.

erosion The processes that loosen and move particles from one place to another.

estuaries Marine ecosystems that consist of shallow, partially enclosed areas where freshwater enters the ocean.

ethics A discipline that seeks to define what is fundamentally right and wrong.

euphotic zone The upper layer in the ocean where the sun's rays penetrate.

eutrophication The enrichment of water (either natural or cultural) with nutrients.

eutrophic lake A usually shallow, warm-water lake that is nutrient rich.

evapotranspiration The process of plants transporting water from the roots to the leaves where it evaporates.

evolution A change in the structure, behavior, or physiology of a population of organisms as a result of some organisms with favorable characteristics having greater reproductive success than those organisms with less favorable characteristics.

executive branch The office of the president of the United States.

exothermic reaction Chemical reaction in which the newly formed compounds have less chemical energy than the compounds from which they were formed.

experiment An artificial situation designed to test the validity of a hypothesis.

exponential growth phase The period during population growth when the population increases at an ever-increasing rate.

extended product responsibility The concept that the producer of a product is responsible for all the negative effects involved in its production, including the ultimate disposal of the product when its useful life is over.

external costs Expenses, monetary or otherwise, borne by someone other than the individuals or groups who use a resource.

extinction The death of a species; the elimination of all the individuals of a particular kind.

F

fecal coliform bacteria Bacteria found in the intestines of humans and other animals, often used as an indicator of water pollution.

first law of thermodynamics A statement about energy that says that under normal physical conditions, energy is neither created nor destroyed.

fissionable The property of the nucleus of some atoms that allows them to split into smaller particles.

fixation A form of waste immobilization in which materials, such as fly ash or cement, are mixed with hazardous waste to prevent the waste from dispersing.

floodplain Lowland area on either side of a river that is periodically covered by water.

floodplain zoning ordinances Municipal laws that restrict future building in floodplains.

food chain The series of organisms involved in the passage of energy from one trophic level to the next.

food web Intersecting and overlapping food chains.

fossil fuels The organic remains of plants, animals, and microorganisms that lived millions of years ago that are preserved as natural gas, oil, and coal.

free-living nitrogen-fixing bacteria Bacteria that live in the soil and can convert nitrogen gas (N_2) in the atmosphere into forms that plants can use.

freshwater ecosystem Aquatic ecosystems that have low amounts of dissolved salts.

friable A soil characteristic that describes how well a soil crumbles.

fungicide A pesticide designed to kill or control fungi.

G

gamma radiation A type of electromagnetic radiation that comes from disintegrating atomic nuclei.

gas-cooled reactor (GCR) A type of nuclear reactor that uses graphite as a moderator and carbon dioxide or helium as a coolant.

gene A unit of heredity; a segment of DNA that contains information for the synthesis of a specific protein, such as an enzyme.

genetic diversity A term used to describe the number of different kinds of genes present in a population or a species.

genetically modified organisms Organisms that have had their genetic makeup modified by biotechnology.

genetic engineering Inserting specific pieces of DNA into the genetic makeup of organisms.

geothermal energy The heat energy from the Earth's molten core.

Global Reporting Initiative Guidelines for reporting on the economic, environmental, and social performance of corporations.

grasslands Areas receiving between 25 and 75 centimeters (10-30 inches) of precipitation per year. Grasses are the dominant vegetation, and trees are rare.

greenhouse effect The property of carbon dioxide (CO_2) that allows light energy to pass through the atmosphere but prevents heat from leaving; similar to the action of glass in a greenhouse.

greenhouse gas Gas in the atmosphere that allows sunlight to enter but retards the outward flow of heat from the Earth.

Green Revolution The introduction of new plant varieties and farming practices that increased agricultural production worldwide during the 1950s, 1960s, and 1970s.

gross national income (GNI) An index that measures the total goods and services generated within a country as well as income earned by citizens of the country who are living in other countries.

gross national product (GNP) An index that measures the total goods and services generated annually within a country.

groundwater Water that infiltrates the soil and is stored in the spaces between particles in the earth.

groundwater mining Removal of water from an aquifer faster than it is replaced.

H

habitat The specific kind of place where a particular kind of organism lives.

habitat management The process of changing the natural community to encourage the increase in populations of certain desirable species.

hard pesticide A pesticide that persists for long periods of time; a persistent pesticide.

hazardous All dangerous materials, including toxic ones, that present an immediate or long-term human health risk or environmental risk.

hazardous air pollutants Certain airborne compounds with high toxicity.

hazardous materials or substances Substances that can cause harm to humans or the environment.

hazardous-waste dump A site in which hazardous waste is disposed in a dump, landfill, or surface impoundment without any concern for potential environmental or health risks.

hazardous wastes Substances that could endanger life if released into the environment.

heavy-water reactor (HWR) A type of nuclear reactor that uses the hydrogen isotope deuterium in the molecular structure of the coolant water.

herbicide A pesticide designed to kill or control plants.

herbivores Primary consumers; animals that eat plants.

horizon A horizontal layer in the soil. The top layer (*A* horizon) has organic matter. The lower layer (*B* horizon) receives nutrients by leaching. The *C* horizon is partially weathered parent material.

host The organism a parasite uses for its source of food.

humus Partially decomposed organic matter typically found in the top layer of the soil.

hydrocarbons (HC) Group of organic compounds consisting of carbon and hydrogen atoms that are evaporated from fuel supplies or are remnants of the fuel that did not burn completely and that act as a primary air pollutant.

hydrologic cycle Constant movement of water from surface water to air and back to surface water as a result of evaporation and condensation.

hydroxide ion A negatively charged particle consisting of a hydrogen and an oxygen atom, commonly released from materials that are bases.

hypothesis A logical statement that explains an event or answers a question that can be tested.

I

ignitability Characteristic of materials that results in their ability to combust.

immigration Movement into an area where one has not previously resided.

incineration Method of disposing of solid waste by burning.

industrial ecology A concept that stresses cycling resources rather than extracting and eventually discarding them.

Industrial Revolution A period of history during which machinery replaced human labor.

industrial solid waste A wide variety of materials such as demolition waste, foundry sand, scraps from manufacturing processes, sludge, ash from combustion, and other similar materials produced by industry.

industrial water uses Uses of water for cooling and for dissipating and transporting waste materials.

infrastructure Permanent structural foundations of a society such as highways and bridges.

insecticide A pesticide designed to kill or control insects.

in-stream water uses Use of a stream's water flow for such purposes as hydroelectric power, recreation, and navigation.

integrated pest management A method of pest management in which many aspects of the pest's biology are exploited to control its numbers.

interspecific competition Competition between members of different species for a limited resource.

intraspecific competition Competition among members of the same species for a limited resource.

ion An atom or group of atoms that has an electric charge because it has either gained or lost electrons.

irrigation Adding water to an agricultural field to allow certain crops to grow where the lack of water would normally prevent their cultivation.

isotope Atoms of the same element that have different numbers of neutrons.

J

judicial branch That portion of the U.S. government that includes the court system.

K

keystone species One that has a critical role to play in the maintenance of specific ecosystems.

kinetic energy Energy of moving objects.

kinetic molecular theory The widely accepted theory that all matter is made of small particles that are in constant movement.

K-strategists Large organisms that have relatively long lives, produce few offspring, provide care for their offspring, and typically have populations that stabilize at the carrying capacity.

L

lag phase The initial stage of population growth during which growth occurs very slowly.

land The surface of the Earth not covered by water.

land disposal The placement of unwanted materials on the surface in landfills or impoundments or by injecting them below the surface of the land.

landfill A method of disposing of solid wastes that involves burying the wastes in specially constructed sites.

land-use planning The process of evaluating the needs and wants of the population, the characteristics and values of the land, and various alternative solutions before changes in land use are made.

latent heat Heat transfer that occurs when a substance is changed from one state to another—solid to liquid, gas to liquid—in which heat is transferred but the temperature does not change.

law of conservation of mass States that matter is not gained or lost during a chemical reaction.

LD$_{50}$ A measure of toxicity; the dosage of a substance that will kill (lethal dose) 50 percent of a test population.

leachate Contaminant-laden water that flows from landfills or other contaminated sites.

leaching The movement of minerals from the top layers of the soil to the B horizon by the downward movement of soil water.

legislative branch That portion of the U.S. government that is responsible for developing laws.

less-developed countries Countries of the world that typically have a per capita income of less than $US 5000.

life-cycle analysis The process of assessing the environmental effects associated with the production, use, reuse, and disposal of a product over its entire useful life.

light-water reactor A nuclear reactor that uses ordinary water as a coolant.

limiting factor The one primary condition of the environment that determines population size of an organism.

limnetic zone Region that does not have rooted vegetation in a freshwater ecosystem.

liquefied natural gas Natural gas that has been converted to a liquid by cooling to $-162°C (-260°F)$.

liquid metal fast-breeder reactor (LMFBR) Nuclear fission reactor using liquid sodium as the moderator and heat transfer medium; produces radioactive plutonium-235, which can be used as a nuclear fuel.

lithosphere A combination of the crust and outer layer of the mantle that forms the plates that move over the Earth's surface.

litter A layer of undecomposed or partially decomposed organic matter on the soil surface.

littoral zone Region with rooted vegetation in a freshwater ecosystem.

loam A soil type with good drainage and good texture that is ideal for growing crops.

M

macronutrient A nutrient, such as nitrogen, phosphorus, and potassium, that is required in relatively large amounts by plants.

management ethic An environmental ethic that stresses a balance between total development and absolute preservation.

mangrove swamp ecosystems Marine shoreline ecosystems dominated by trees that can tolerate high salt concentrations.

mantle The layer of the Earth between the crust and the core.

marine ecosystems Aquatic ecosystems that have high salt content.

marsh Area of grasses and reeds that is flooded either permanently or for a major part of the year.

mass burn A method of incineration of solid waste in which material is fed into a furnace on movable metal grates.

matter Substance with measurable mass and volume.

mechanical weathering Physical forces that reduce the size of rock particles without changing the chemical nature of the rock.

Mediterranean shrublands Coastal ecosystems characterized by winter rains and summer droughts that are dominated by low woody vegetation with small leaves.

megalopolis A large, regional urban center.

methane (CH$_4$) An organic compound produced by living organisms that is a greenhouse gas.

micronutrient A nutrient needed in extremely small amounts for proper plant growth; examples are boron, zinc, and magnesium.

migratory birds Birds that fly considerable distances between their summer breeding areas and their wintering areas.

mining waste Waste from the processing of rock from mining operations. It includes solid materials that are typically dumped on the land near the milling site and liquid wastes typically stored in ponds.

mixture A kind of matter consisting of two or more kinds of matter intermingled with no specific ratio of the kinds of matter.

moderator Material that absorbs the energy from neutrons released by fission.

molecule Two or more atoms chemically bonded to form a stable unit.

monoculture A system of agriculture in which large tracts of land are planted with the same crop.

morals Predominant feeling of a culture about ethical issues.

more-developed countries Countries of the world that typically have a per capita income that exceeds $US 10,000; Europe, Canada, United States, Australia, New Zealand, and Japan.

mortality The number of deaths per year.

mulch An organic material that is used to cover the soil.

multiple land use Land uses that do not have to be exclusionary, so that two or more uses of land may occur at the same time.

municipal solid waste landfill A waste storage site constructed above an impermeable clay layer that is lined with an impermeable membrane and includes mechanisms for dealing with liquid and gas materials generated by the contents of the landfill.

municipal solid waste (MSW) All the waste produced by the residents of a community.

mutualism The association between organisms in which both benefit.

mutations Changes in the genetic information of an organism.

mycorrhizae Symbiotic soil fungi, present in most soils, that attach themselves directly onto the roots of most plants. They help the host plants absorb more water and nutrients while the host plants provide food for the fungi.

N

natality The number of individuals added to the population through reproduction.

Index

trickling filter system, 374
trillium, 271*f*
tritium isotopes, 227, 228*f*
trombe wall, 206
trophic levels, 92, 157–159
tropical deforestation, 117, 258–259
tropical dry forests, 116
tropical rainforests, 116–118
    soil quality, 313
    use and resources, 117
troposphere, 389
true *vs.* perceived risks, 42–43
tundra, 120–122
Turkey
    female literacy, fertility, and GNI, 153*f*
    population characteristics, 151*t*

U

Ukraine
    Chernobyl nuclear power plant accident, 230–233
    nuclear power reactors, 226*t*
UNCED, 6–7, 472
unconfined aquifers, 354, 356*f*
underground mining, 192
underground storage tanks, 370
United Church of Christ Commission on Racial Justice, 28
United Kingdom
    electricity consumption and price, 179*t*
    gasoline taxes, 178*f*, 475
    nuclear power reactors, 226*t*
    oil exports, 179
United Nations
    Basel Convention, 448–449
    biodiversity preservation and, 269
    environmental programs, 7, 468
    international security issues, 318
United Nations Conference on Environment and Development (UNCED), 6–7, 472
United Nations Convention to Combat Desertification (UNCCD), 318
United Nations Framework Convention on Climate Change, 32
United States
    agricultural land use, 317*f*
    ambient air quality standards, 391*t*
    auto industry and oil consumption, 173–174
    crude oil imports, 179*t*
    ecosystem losses, 129
    electricity consumption and price, 179*t*

as endangered species market, 30
energy conservation, 213–214
energy consumption, 175*t*, 177*t*, 181–182
environmental movement, 458–460
federal land use issues, 299–301
female literacy, fertility, and GNI, 153*f*
gasoline taxes, 178, 475
geologic radon potential, 415
government. *See* United States government
hydroelectric capacity, 198
International Whaling Commission and, 470
municipal waste production, 420–421, 423*f*
nuclear power facilities, 226*t*, 234–235, 239, 240
nuclear waste disposal, 236–237, 238
population characteristics, 151*t*
population comparisons with Canada and Mexico, 163
population trends, 160–162, 465
standard of living and population growth, 155*f*
United States government
    biodiversity legislation, 270
    Boundary Waters Treaty, 6
    branches of, 457
    environmental legislation, 459*t*
    environmental policy, 457–460
    executive branch agencies, 457*f*
    legislative process, 457–458
uranium
    fission, 222, 223*f*, 226–227
    global requirements, 226*t*
    Hanford facility production and waste, 240
    mining, 228, 233, 234*f*
    radon and, 414
uranium-235, 223*f*
urban centers
    demise of central city, 288
    in developing nations, 283
    energy consumption, 174
    habitat loss and, 260–261
    human populations and, 160
    poverty and, 160
    rural migration to, 281, 282*f*, 283
    solid waste, 212*t*, 420–421, 422*f*, 423*f*. *See also* solid waste disposal
    unplanned growth outcomes, 287–290
    urban-to-suburban migration, 283–285
    water pollution, 366–368
    water-use planning, 371–372

waterways and, 281, 282*f*
urban growth limit, 292
urban planning
    recreation issues, 297, 298*f*
    redevelopment issues, 297–299
    smart growth approach, 299
    transportation issues, 295–297
urban sprawl, 284
    contributing factors, 285–287
    problems associated with, 287–290
    smart growth and, 299
    types of, 284–285
U.S. Geological Survey (USGS)
    wastewater contaminants studies, 368
    World Petroleum Assessment 2003, 189
USDA (United States Department of Agriculture) "organic" standards, 349

V

vadose zone, 355, 356*f*
value(s)
    assigning to natural resources, 45
    of biological and ecosystem services, 250, 254
    environmental, 20
    risk values, 41
variables, 65
vectors, 88
vegetation. *See* plants
vehicles. *See* automobiles
Venezuela
    agriculturally suited land, 318*t*
    oil exports, 179
vertebrates, biodiversity hot spots, 251–253
Vietnam
    female literacy, fertility, and GNI, 153*f*
    Mekong River Basin and, 15
    population characteristics, 151*t*
viral diseases, animal-to-human transmission, 90
Virginia, blue crab seafood fishery, 3
visual pollution, 296
volatile organic compounds (VOCs), 394, 413*t*

W

war, impact of, 465–468
warblers, Kirkland's, 267
Warren County (NC), hazardous waste protest, 28
waste. *See also* hazardous waste; radioactive waste; solid waste
    corporate ethics and, 24–26
    defining, 27
waste accumulation, population growth and, 137

waste disposal. *See also* solid waste disposal
    population growth and, 142
    radioactive material, 233, 236–238, 240
    U.S. legislation, 459*t*
waste heat, 233–234, 235*f*, 360, 369
Waste Isolation Pilot Plant, 236
waste minimization, 446
water, 353–354
    active solar systems and, 207
    bodies of. *See* aquatic ecosystems; waterways; *specific entity*
    carbon cycle in, 97
    conservation methods, 358, 359*f*
    erosive actions, 315
    global shortages, 353–354
    hydrologic cycle, 354–356
    lead content, 383
    supply and demand, climate change and, 408–409
    use. *See* water use; water-use planning
    as weathering agent, 307–308
water diversion, 372–373
water pollution, 364–366
    agricultural, 343, 368
    fertilizer effects, 99–100, 368, 381
    Great Lakes and East Coast, 13
    groundwater, 369–371
    industrial, 368–369
    in industrial *vs.* developing nations, 363
    marine oil, 369
    municipal, 366–368
    thermal, 233–234, 235*f*, 360, 369
    urban growth and, 288, 289*f*
water quality, U.S. legislation, 459*t*
water table, 354
water use
    agricultural, 344–345, 358–359, 360*f*
    categories, 356–357
    consumption as ethical issue, 31
    domestic use, 357–358, 359*f*
    in dry west region of North America, 10–11
    global use, 358*f*
    industrial use, 359–360
    in industrial *vs.* developing nations, 363
    in-stream use, 360, 362, 364
    land-use planning and, 290
    in nuclear fission, 222–223, 224*f*
    U.S. legislation, 459*t*
    wastewater treatment, 368, 373–374, 376

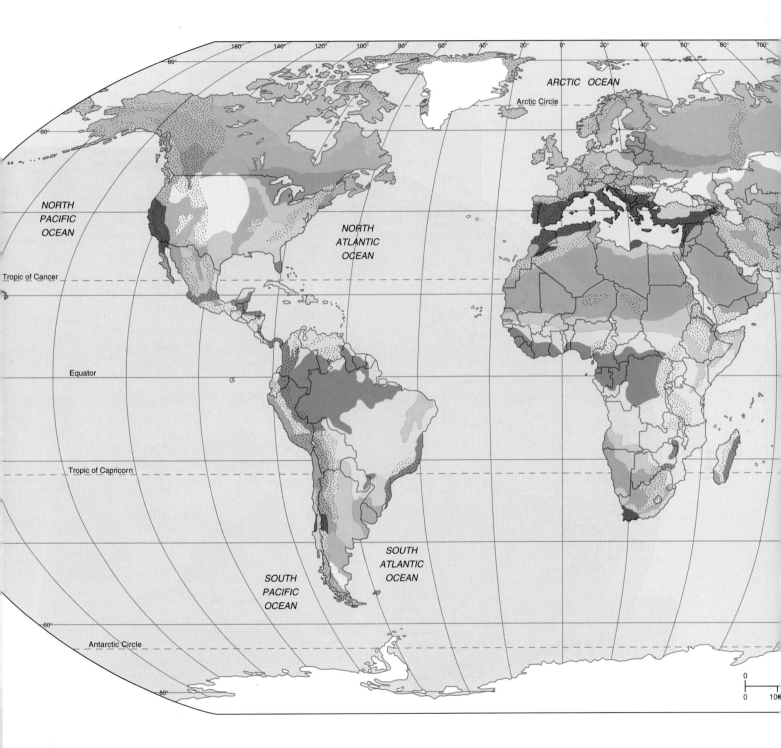

ARCTIC OCEAN

Arctic Circle

NORTH
PACIFIC
OCEAN

NORTH
ATLANTIC
OCEAN

Tropic of Cancer

Equator

Tropic of Capricorn

SOUTH
ATLANTIC
OCEAN

SOUTH
PACIFIC
OCEAN

Antarctic Circle

NORTH
PACIFIC
OCEAN

Tropic of Cancer

Equator

INDIAN
OCEAN

Tropic of Capricorn

Antarctic Circle

120° 140° 160° 80° 60° 40° 20° 0° 20° 40° 60° 80°

1000   2000 Miles
2000   3000 Kilometers